# Statistical Theory and Modeling for Turbulent Flows

# Statistical Theory and Modeling for Turbulent Flows

Second Edition

**P. A. Durbin**
*Iowa State University, USA*

**B. A. Pettersson Reif**
*Norwegian Defence Research Establishment (FFI), Norway*

WILEY
A John Wiley and Sons, Ltd., Publication

This edition first published 2011


First Edition published in 2001

*Registered office*
John Wiley & Sons Ltd, The Atrium, Southern Gate, Chichester, West Sussex, PO19 8SQ, United Kingdom

For details of our global editorial offices, for customer services and for information about how to apply for permission to reuse the copyright material in this book please see our website at www.wiley.com.



*Library of Congress Cataloguing-in-Publication Data*

Durbin, Paul A.
Statistical theory and modeling for turbulent flows / Paul A. Durbin, Bjørn Anders Pettersson Reif. – 2nd ed.
p. cm.
Includes index.
ISBN 978-0-470-68931-8 (cloth)
1. Turbulence–Mathematical models. I. Reif, B. A. Pettersson. II. Title.
QA913.D94 2010
532′.0527015118–dc22

2010020346

A catalogue record for this book is available from the British Library.

Print ISBN: 978-0-470-68931-8
ePDF ISBN 978-0-470-97206-9
oBook ISBN: 978-0-470-97207-6

Set in 10/12pt Times by Laserwords Private Limited, Chennai, India

*to Cinian and Seth*

# Contents

# Preface

## Preface to second edition

The original first edition intentionally avoided the topic of eddy-resolving, computer simulation. The crucial role of numerics in turbulence simulation is why we shied away from the topic. One cannot properly introduce direct numerical simulation or large eddy simulation without discussing discretization schemes.

However, large eddy simulation and detached eddy simulation are now increasingly seen as partners to Reynolds averaged modeling. This revised second edition contains a new Part IV on direct numerical simulation, large eddy simulation, and detached eddy simulation. In keeping with our original perspective, it is not encyclopedic. We address some of the key issues, with sufficient technical content for the reader to acquire concrete understanding. For example, dissipative and dispersive errors are defined in order to understand why central schemes are preferred. The notion of energy-conserving schemes is reviewed. Our discussion of filtering is brief compared to the development of large eddy simulation in other books. Given our concise treatment, we chose instead to focus on the nature of subgrid models. Although the material on simulation was appended at the end of the text as Part IV, it fits just as well before Part II.

Transition modeling is currently seen as a critical complement to turbulence modeling. The original text described the manner in which turbulence models switch from laminar to turbulent solutions, but that is not transition modeling. The research community has moved in the direction of adding either an intermittency equation or an equation for fluctuations in laminar regions. This revised edition discusses these approaches.

Other, smaller, revisions have been made elsewhere in the text.

Ames, Iowa, 2010

## Preface to first edition

This book evolved out of lecture notes for a course taught in the Mechanical Engineering department at Stanford University. The students were at M.S. and Ph.D. level. The course served as an introduction to turbulence and to turbulence modeling. Its scope was single-point statistical theory, phenomenology, and Reynolds averaged closure. In preparing the present book the purview was extended to include two-point, homogeneous turbulence theory. This has been done to provide sufficient breadth for a complete introductory course on turbulence.

Further topics in modeling also have been added to the scope of the original notes; these include both practical aspects, and more advanced mathematical analyses of models. The advanced material was placed into a separate chapter so that it can be circumvented if desired. Similarly, two-point, homogeneous turbulence theory is contained in Part III and could be avoided in an M.S. level engineering course, for instance.

No attempt has been made at an encyclopedic survey of turbulence closure models. The particular models discussed are those that today seem to have proved effective in computational fluid dynamics applications. Certainly, there are others that could be cited, and many more in the making. By reviewing the motives and methods of those selected, we hope to have laid a groundwork for the reader to understand these others. A number of examples of Reynolds averaged computation are included.

It is inevitable in a book of the present nature that authors will put their own slant on the contents. The large number of papers on closure schemes and their applications demands that we exercise judgement. To boil them down to a text requires that boundaries on the scope be set and adhered to. Our ambition has been to expound the subject, not to survey the literature. Many researchers will be disappointed that their work has not been included. We hope they will understand our desire to make the subject accessible to students, and to make it attractive to new researchers.

An attempt has been made to allow a lecturer to use this book as a guideline, while putting his or her personal slant on the material. While single-point modeling is decidedly the main theme, it occupies less than half of the pages. Considerable scope exists to choose where emphasis is placed.

## Motivation

It is unquestionably the case that closure models for turbulence transport are finding an increasing number of applications, in increasingly complex flows. Computerized fluid dynamical analysis is becoming an integral part of the design process in a growing number of industries: increasing computer speeds are fueling that growth. For instance, computer analysis has reduced the development costs in the aerospace industry by decreasing the number of wind tunnel tests needed in the conceptual and design phases.

As the utility of turbulence models for computational fluid dynamics (CFD) has increased, more sophisticated models have been needed to simulate the range of phenomena that arise. Increasingly complex closure schemes raise a need for computationalists to understand the origins of the models. Their mathematical properties and predictive accuracy must be assessed to determine whether a particular model is suited to computing given flow phenomena. Experimenters are being called on increasingly to provide data for testing turbulence models and CFD codes. A text that provides a solid background for those working in the field seems timely.

The problems that arise in turbulence closure modeling are as fundamental as those in any area of fluid dynamics. A grounding is needed in physical concepts and mathematical techniques. A student, first confronted with the literature on turbulence modeling, is bound to be baffled by equations seemingly pulled from thin air; to wonder whether constants are derived from principles, or obtained from data; to question what is fundamental and what is peculiar to a given model. We learned this subject by ferreting around the literature, pondering just such questions. Some of that experience motivated this book.

## Epitome

The prerequisite for this text is a basic knowledge of fluid mechanics, including viscous flow. The book is divided into three major parts.

Part I provides background on turbulence phenomenology, Reynolds averaged equations, and mathematical methods. The focus is on material pertinent to single-point, statistical analysis, but a chapter on eddy structures is also included.

Part II is on turbulence modeling. It starts with the basics of engineering closure modeling, then proceeds to increasingly advanced topics. The scope ranges from integrated equations to second-moment transport. The nature of this subject is such that even the most advanced topics are not rarefied; they should pique the interest of the applied mathematician, but should also make the R&D engineer ponder the potential impact of this material on her or his work.

Part III introduces Fourier spectral representations for homogeneous turbulence theory. It covers energy transfer in spectral space and the formalities of the energy cascade. Finally rapid distortion theory is described in the last section. Part III is intended to round out the scope of a basic turbulence course. It does not address the intricacies of two-point closure, or include advanced topics.

A first course on turbulence for engineering students might cover Part I, excluding the section on tensor representations, most of Part II, excluding Chapter 8, and a brief mention of selected material from Part III. A first course for more mathematical students might place greater emphasis on the latter part of Chapter 2 in Part I, cover a limited portion of Part II – emphasizing Chapter 7 and some of Chapter 8 – and include most of Part III. Advanced material is intended for prospective researchers.

## Acknowledgements

Finally, we would like to thank those who have provided encouragement for us to write this book. Doubts over whether to write it at all were dispelled by Cinian Zheng-Durbin; she was a source of support throughout the endeavor.

We gratefully acknowledge the conducive environment created by the Stanford/NASA Center for Turbulence Research, and its director Prof. P. Moin. This book has benefited from our interactions with visitors to the CTR and with its post-doctoral fellows. Our thanks to Dr. L. P. Purtell of the Office of Naval Research for his support of turbulence modeling research over the years. We have benefited immeasurably from many discussions and from collaboration with the late Prof. C. G. Speziale. Interactions with Prof. D. Laurence and his students have been a continual stimulus. Prof. J. C. R. Hunt's unique insights into turbulence have greatly influenced portions of the book.

Stanford, California, 2000

# Part I

# FUNDAMENTALS OF TURBULENCE

# 1

# Introduction

Where under this beautiful chaos can there lie a simple numerical structure?
– Jacob Bronowski

Turbulence is a ubiquitous phenomenon in the dynamics of fluid flow. For decades, comprehending and modeling turbulent fluid motion has stimulated the creativity of scientists, engineers, and applied mathematicians. Often the aim is to develop methods to predict the flow fields of practical devices. To that end, analytical models are devised that can be solved in computational fluid dynamics codes. At the heart of this endeavor is a broad body of research, spanning a range from experimental measurement to mathematical analysis. The intent of this text is to introduce some of the basic concepts and theories that have proved productive in research on turbulent flow.

Advances in computer speed are leading to an increase in the number of applications of turbulent flow prediction. Computerized fluid flow analysis is becoming an integral part of the design process in many industries. As the use of turbulence models in computational fluid dynamics increases, more sophisticated models will be needed to simulate the range of phenomena that arise. The increasing complexity of the applications will require creative research in engineering turbulence modeling. We have endeavored in writing this book both to provide an introduction to the subject of turbulence closure modeling, and to bring the reader up to the state of the art in this field. The scope of this book is certainly not restricted to closure modeling, but the bias is decidedly in that direction.

To flesh out the subject, the spectral theory of homogeneous turbulence is reviewed in Part III and eddy simulation is the topic of Part IV. In this way an endeavor has been made to provide a complete course on turbulent flow. We start with a perspective on the problem of turbulence that is pertinent to this text. Readers not very familiar

*Statistical Theory and Modeling for Turbulent Flows, Second Edition* P. A. Durbin and B. A. Pettersson Reif

with the subject might find some of the terminology unfamiliar; it will be explicated in due course.

## 1.1 The turbulence problem

The turbulence problem is an age-old topic of discussion among fluid dynamicists. It is not a problem of physical law; it is a problem of description. Turbulence is a state of fluid motion, governed by known dynamical laws – the Navier–Stokes equations in cases of interest here. In principle, turbulence is simply a solution to those equations. The turbulent state of motion is defined by the complexity of such hypothetical solutions. The challenge of description lies in the complexity: How can this intriguing behavior of fluid motion be represented in a manner suited to the needs of science and engineering?

Turbulent motion is fascinating to watch: it is made visible by smoke billows in the atmosphere, by surface deformations in the wakes of boats, and by many laboratory techniques involving smoke, bubbles, dyes, etc. Computer simulation and digital image processing show intricate details of the flow. But engineers need numbers as well as pictures, and scientists need equations as well as impressions. How can the complexity be fathomed? That is the turbulence problem.

Two characteristic features of turbulent motion are its ability to stir a fluid and its ability to dissipate kinetic energy. The former mixes heat or material introduced into the flow. Without turbulence, these substances would be carried along streamlines of the flow and slowly diffuse by molecular transport; with turbulence they rapidly disperse across the flow. Energy dissipation by turbulent eddies increases resistance to flow through pipes and it increases the drag on objects in the flow. Turbulent motion is highly dissipative because it contains small eddies that have large velocity gradients, upon which viscosity acts. In fact, another characteristic of turbulence is its continuous range of scales. The largest size eddies carry the greatest kinetic energy. They spawn smaller eddies via nonlinear processes. The smaller eddies spawn smaller eddies, and so on in a cascade of energy to smaller and smaller scales. The smallest eddies are dissipated by viscosity. The grinding down to smaller and smaller scales is referred to as the *energy cascade*. It is a central concept in our understanding of stirring and dissipation in turbulent flow.

The energy that cascades is first produced from orderly, mean motion. Small perturbations extract energy from the mean flow and produce irregular, turbulent fluctuations. These are able to maintain themselves, and to propagate by further extraction of energy. This is referred to as production and transport of turbulence. A detailed understanding of such phenomena does not exist. Certainly these phenomena are highly complex and serve to emphasize that the true problem of turbulence is one of analyzing an intricate phenomenon.

The term "eddy" may have invoked an image of swirling motion round a vortex. In some cases that may be a suitable mental picture. However, the term is usually meant to be more ambiguous. Velocity contours in a plane mixing layer display both large- and small-scale irregularities. Figure 1.1 illustrates an organization into large-scale features with smaller-scale random motion superimposed. The picture consists of contours of a passive scalar introduced into a mixing layer. Very often the image behind the term "eddy" is this sort of perspective on scales of motion. Instead of vortical whorls, eddies are an impression of features seen in a contour plot. Large eddies are the large lumps

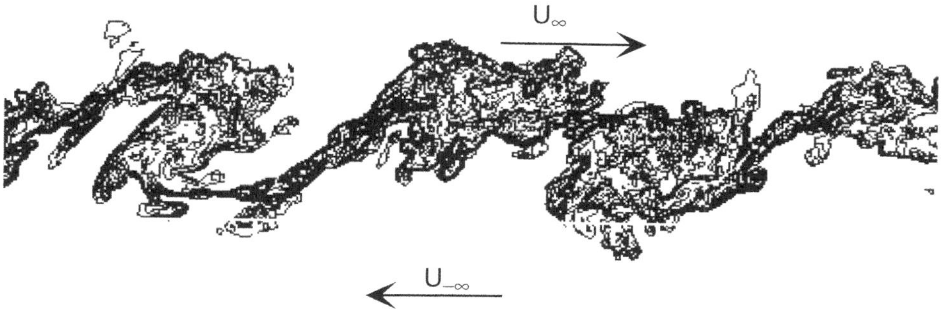


**Figure 1.1** Turbulent eddies in a plane mixing layer subjected to periodic forcing. From Rogers and Moser (1994), reproduced with permission.

seen in the figure, and small eddies are the grainy background. Further examples of large eddies are discussed in Chapter 5 of this book on coherent and vortical structures.

A simple method to produce turbulence is by placing a grid normal to the flow in a wind tunnel. Figure 1.2 contains a smoke visualization of the turbulence downstream of the bars of a grid. The upper portion of the figure contains velocity contours from a numerical simulation of grid turbulence. In both cases the impression is made that, on average, the scale of the irregular velocity fluctuations increases with distance downstream. In this sense the average size of eddies grows larger with distance from the grid.

Analyses of turbulent flow inevitably invoke a statistical description. Individual eddies occur randomly in space and time and consist of irregular regions of velocity or vorticity. At the statistical level, turbulent phenomena become reproducible and subject to systematic study. Statistics, like the averaged velocity, or its variance, are orderly and develop regularly in space and time. They provide a basis for theoretical descriptions and for a diversity of prediction methods. However, exact equations for the statistics do not exist. The objective of research in this field has been to develop mathematical models and physical concepts to stand in place of exact laws of motion. Statistical theory is a way to fathom the complexity. Mathematical modeling is a way to predict flows. Hence the title of this book: "Statistical theory and modeling for turbulent flows."

The alternative to modeling would be to solve the three-dimensional, time-dependent Navier–Stokes equations to obtain the chaotic flow field, and then to average the solutions in order to obtain statistics. Such an approach is referred to as direct numerical simulation (DNS). Direct numerical simulation is not practicable in most flows of engineering interest. Engineering models are meant to bypass the chaotic details and to predict statistics of turbulent flows directly. A great demand is placed on these engineering closure models: they must predict the averaged properties of the flow without requiring access to the random field; they must do so in complex geometries for which detailed experimental data do not exist; they must be tractable numerically; and they must not require excessive computing time. These challenges make statistical turbulence modeling an exciting field.

The goal of turbulence theories and models is to describe turbulent motion by analytical methods. The particular methods that have been adopted depend on the objectives: whether it is to understand how chaotic motion follows from the governing equations, to

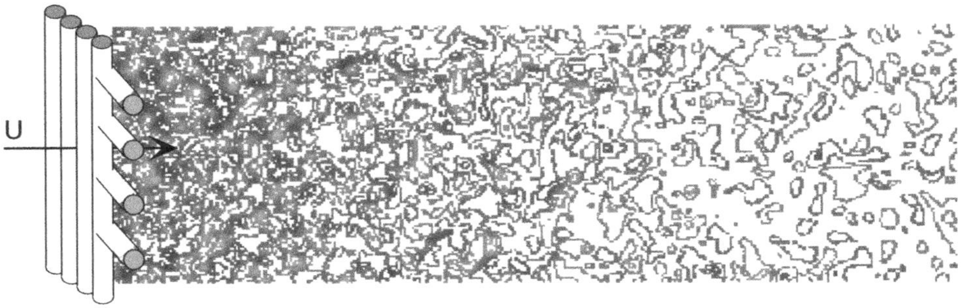


(a)

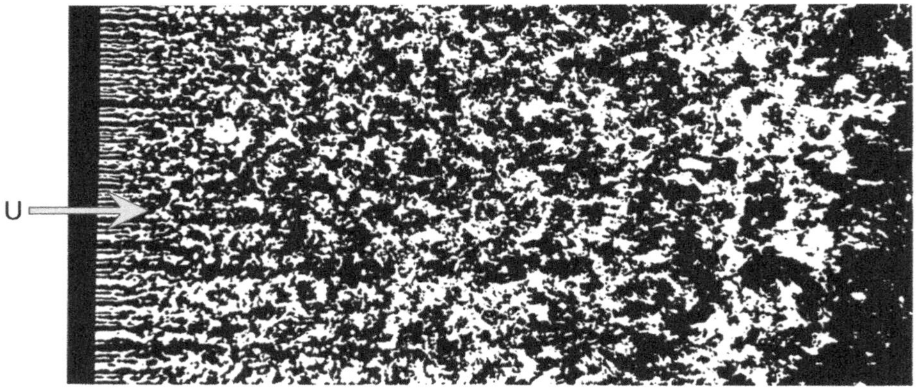


(b)

**Figure 1.2** (a) Grid turbulence schematic, showing contours of streamwise velocity from a numerical simulation. (b) Turbulence produced by flow through a grid. The bars of the grid would be to the left of the picture, and flow is from left to right. Visualization by smoke wire of laboratory flow. Courtesy of T. Corke and H. Nagib.

construct phenomenological analogs of turbulent motion, to deduce statistical properties of the random motion, or to develop semi-empirical calculational tools. The latter two are the subject of this book.

The first step in statistical theory is to greatly compress the information content from that of a random field of eddies to that of a field of statistics. In particular, the turbulent velocity consists of a three-component field $(u_1, u_2, u_3)$ as a function of four independent variables $(x_1, x_2, x_3, t)$. This is a rapidly varying, irregular flow field, such as might be seen embedded in the billows of a smoke stack, the eddying motion of the jet in Figure 1.3, or the more explosive example of Figure 1.4. In virtually all cases of engineering interest, this is more information than could be used, even if complete data were available. It must be reduced to a few useful numbers, or functions, by averaging. The picture to the right of Figure 1.4 has been blurred to suggest the reduced information in an averaged representation. The small-scale structure is smoothed by averaging. A true average in this case would require repeating the explosion many times and summing the images; even the largest eddies would be lost to smoothing. A stationary flow can be

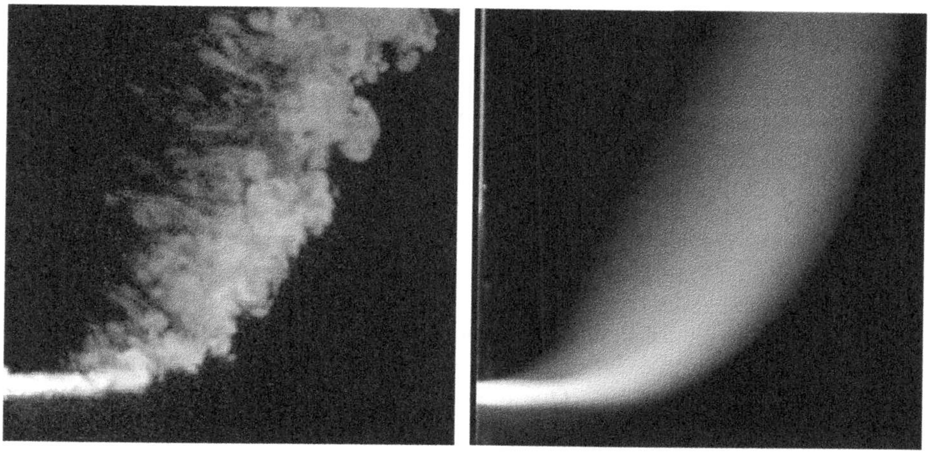

**Figure 1.3** Instantaneous and time-averaged views of a jet in cross flow. The jet exits from the wall at left into a stream flowing from bottom to top (Su and Mungal 1999).

**Figure 1.4** Large- and small-scale structure in a plume. The picture at the right is blurred to suggest the effect of ensemble averaging.

averaged in time, as illustrated by the time-lapse photograph on the right of Figure 1.3. Again, all semblance of eddying motion is lost in the averaged view.

An example of the greatly simplified representation invoked by statistical theory is provided by grid turbulence. When air flows through a grid of bars, the fluid velocity produced is a complex, essentially random, three-component, three-dimensional, time-dependent field that defies analytical description (Figure 1.2). This velocity field might be described statistically by its variance, $q^2$, as a function of distance downwind of the grid; $q^2$ is the average value of $u_1^2 + u_2^2 + u_3^2$ over planes perpendicular to the flow. This statistic provides a smooth function that characterizes the complex field. In fact, the dependence of $q^2$ on distance downstream of the grid is usually represented to good approximation by a power law: $q^2 \propto x^{-n}$ where $n$ is about 1. The average length scale of the eddies grows like $L \propto x^{1-n/2}$. This provides a simple formula that agrees with the impression created by Figure 1.2 of eddy size increasing with $x$.

The catch to the simplification that a statistical description seems to offer is that it is only a simplification if the statistics somehow can be obtained without having first to solve for the whole, complex velocity field and then compute averages. The task is to predict the smooth jet at the right of Figure 1.3 without access to the eddying motion at the left. Unfortunately, there are no exact governing equations for the averaged flow, and empirical modeling becomes necessary. One might imagine that an equation for the average velocity could be obtained by averaging the equation for the instantaneous velocity. That would only be the case if the equations were linear, which the Navier–Stokes equations are not.

The role of nonlinearity can be explained quite simply. Consider a random process generated by flipping a coin, assigning the value 1 to heads and 0 to tails. Denote this value by $\xi$. The average value of $\xi$ is 1/2. Let a velocity, $u$, be related to $\xi$ by the linear equation

$$u = \xi - 1. \tag{1.1.1}$$

The average of $u$ is the average of $\xi - 1$. Since $\xi - 1$ has probability $1/2$ of being 0 and probability $1/2$ of being $-1$, the average of $u$ is $-1/2$. Denote this average by $\overline{u}$. The equation for $\overline{u}$ *can* be obtained by averaging the exact equation: $\overline{u} = \overline{\xi} - 1 = 1/2 - 1 = -1/2$. But if $u$ satisfies a nonlinear equation

$$u^2 + 2u = \xi - 1, \tag{1.1.2}$$

then the averaged equation is

$$\overline{u^2} + 2\overline{u} = \overline{\xi} - 1 = -1/2. \tag{1.1.3}$$

This is not a *closed** equation for $\overline{u}$ because it contains $\overline{u^2}$: squaring, then averaging, is not equal to averaging, then squaring, that is, $\overline{u^2} \neq \overline{u}^2$. In this example, averaging produces a single equation with two dependent variables, $\overline{u}$ and $\overline{u^2}$. The example is contrived so that it first can be solved, then averaged: its solution is $u = \sqrt{\xi} - 1$; the average is then $\overline{u} = \frac{1}{2}(\sqrt{1} - 1) + \frac{1}{2}(\sqrt{0} - 1) = -1/2$. Similarly $\overline{u^2} = 1/2$, but this could not be known without first solving the random equation, then computing the average. In the case of the Navier–Stokes equations, one cannot resort to solving, then averaging. As in this simple illustration, the average of the Navier–Stokes equations are equations for $\overline{u}$ that contain $\overline{u^2}$. Unclosed equations are inescapable.

---

* The terms "closure problem" and "closure model" are ubiquitous in the literature. Mathematically, this means that there are more unknowns than equations. A closure model simply provides extra equations to complete the unclosed set.

## 1.2 Closure modeling

Statistical theories of turbulence attempt to obtain statistical information either by systematic approximations to the averaged, unclosed governing equations, or by intuition and analogy. Usually, the latter has been the more successful: the Kolmogoroff theory of the inertial subrange and the log law for boundary layers are famous examples of intuition.

Engineering closure models are in this same vein of invoking systematic analysis in combination with intuition and analogy to close the equations. For example, Prandtl drew an analogy between the turbulent transport of averaged momentum by turbulent eddies and the kinetic theory of gases when he proposed his "mixing length" model. Thereby he obtained a useful model for predicting turbulent boundary layers.

The allusion to "engineering flows" implies that the flow arises in a configuration that has technological application. Interest might be in the pressure drop in flow through a bundle of heat-exchanger tubes or across a channel lined with ribs. The turbulence dissipates energy and increases the pressure drop. Alternatively, the concern might be with heat transfer to a cooling jet. The turbulence in the jet scours an impingement surface, enhancing the cooling. Much of the physics in these flows is retained in the averaged Navier–Stokes equations. The general features of the flow against the surface, or the separated flow behind the tubes, will be produced by these equations *if* the dissipative and transport effects of the turbulence are represented by a model. The model must also close the set of equations – the number of unknowns must equal the number of equations.

In order to obtain closed equations, the extra dependent variables that are introduced by averaging, such as $\overline{u^2}$ in the above example, must be related to the primary variables, such as $\overline{u}$. For instance, if $\overline{u^2}$ in Eq. (1.1.3) were modeled by $\overline{u^2} = a\overline{u}^2$, the equation would be $a\overline{u}^2 + 2\overline{u} = -1/2$, where $a$ is an "empirical" constant. In this case $a = 2$ gives the correct answer $\overline{u} = -1/2$.

Predicting an averaged flow field, such as that suggested by the time-averaged view in Figure 1.3, is not so easy. Conceptually, the averaged field is strongly affected by the irregular motion, which is no longer present in the blurred view. The influence of this irregular, turbulent motion must be represented if the mean flow is to be accurately predicted. The representation must be constructed in a manner that permits a wide range of applications. In unsteady flows, like Figure 1.4, it is unreasonable to repeat the experiment over and over to obtain statistics; nevertheless, there is no conceptual difficulty in developing a statistical prediction method. The subject of turbulence modeling is certainly ambitious in its goals.

Models for such general purposes are usually phrased in terms of differential equations. For instance, a widely used model for computing engineering flows, the $k$–$\varepsilon$ model, consists of differential transport equations for the turbulent energy, $k$, and its rate of dissipation, $\varepsilon$. From their solution, an eddy viscosity is created for the purpose of predicting the mean flow. Other models represent turbulent influences by a stress tensor, the Reynolds stress. Transport models, or algebraic formulas, are developed for these stresses. The perspective here is analogous to constitutive modeling of material stresses, although there is a difference. Macroscopic material stresses are caused by molecular motion and by molecular interactions. Reynolds stresses are not a material property: they are a property of fluid motion; they are an averaged representation of random convection. When modeling Reynolds stresses, the concern is to represent properties of the flow field, not properties of a material. For that reason, the analogy to constitutive modeling should be tempered by some understanding of the aspects of turbulent motion that models are meant to represent. The various topics covered in this book are intended to provide a tempered introduction to turbulence modeling.

In practical situations, closure relations are not exact or derivable. They invoke empiricism. Consequently, any closure model has a limited range of use, implicitly circumscribed by its empirical content. In the course of time, a number of very useful semi-empirical models has been developed to calculate engineering flows. However, this continues to be an active and productive area of research. As computing power increases, more elaborate and more flexible models become feasible. A variety of models, their motivations, range of applicability, and some of their properties, are discussed in this book; but this is not meant to be a comprehensive survey of models. Many variations on a few basic types have been explored in the literature. Often the variation is simply to add parametric dependences to empirical coefficients. Such variants affect the predictions of the models, but they do not alter their basic analytical form. The theme in this book is the essence of the models and their mathematical properties.

## 1.3 Categories of turbulent flow

Broad categories can be delineated for the purpose of organizing an exposition on turbulent flow. The categorization presented in this section is suited to the aims of this book on theory and modeling. An experimenter, for instance, might survey the range of possibilities differently.

The broadest distinction is between *homogeneous* and *non-homogeneous* flows. The definition of spatial homogeneity is that statistics are not functions of position. Homogeneity in time is called stationarity. The statistics of homogeneous turbulence are unaffected by an arbitrary positioning of the origin of the coordinate system; ideal homogeneity implies an unbounded flow. In a laboratory, only approximate homogeneity can be established. For instance, the smoke puffs in Figure 1.2 are statistically homogeneous in the $y$ direction: their average size is independent of $y$. Their size increases with $x$, so $x$ is not a direction of homogeneity.

Idealized flows are used to formulate theories and models. The archetypal idealization is *homogeneous*, *isotropic* turbulence. Its high degree of statistical symmetry facilitates analysis. *Isotropy* means there is no directional preference. If one were to imagine running a grid every which way through a big tank of water, the resulting turbulence would have no directional preference, much as illustrated by Figure 1.5. This figure shows the instantaneous vorticity field in a box of homogeneous isotropic turbulence, simulated on a computer. At any point and at any time, a velocity fluctuation in the $x_1$ direction would be as likely as a fluctuation in the $x_2$, or any other, direction. Great mathematical simplifications follow. The basic concepts of homogeneous, isotropic turbulence are covered in this book. A vast amount of theoretical research has focused on this idealized state; we will only scratch the surface. A number of relevant monographs exist (McComb 1990) as well as the comprehensive survey by Monin and Yaglom (1975).

The next level of complexity is *homogeneous*, *anisotropic* turbulence. In this case, the intensity of the velocity fluctuations is not the same in all directions. Strictly, it could be either the velocity, the length scale, or both that have directional dependence – usually it

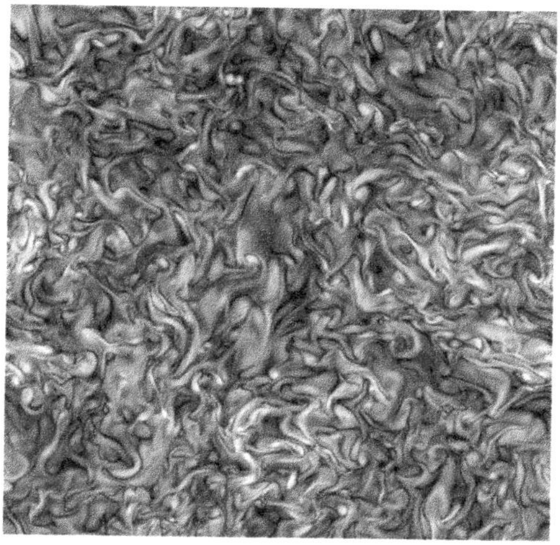

**Figure 1.5** Vorticity magnitude in a box of isotropic turbulence. The light regions are high vorticity. Courtesy of J. Jiménez (Jimenez 1999).

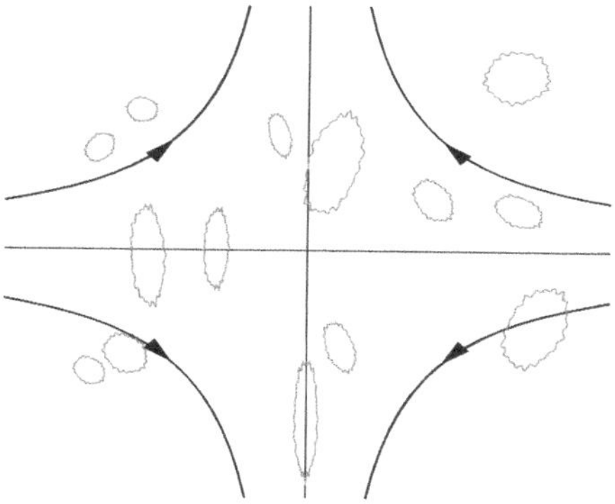

**Figure 1.6** Schematic suggesting eddies distorted by a uniform straining flow.

is both. Anisotropy can be produced by a mean rate of strain, as suggested by Figure 1.6. Figure 1.6 shows schematically how a homogeneous rate of strain will distort turbulent eddies. Eddies are stretched in the direction of positive rate of strain and compressed in the direction of negative strain. In this illustration, it is best to think of the eddies as vortices that are distorted by the mean flow. Their elongated shapes are symptomatic of both velocity and length scale anisotropy.

To preserve homogeneity, the rate of strain must be uniform in space. In general, homogeneity requires that the mean flow have a constant gradient; that is, the velocity should be of the form $U_i = A_{ij}x_j + B_i$, where $A_{ij}$ and $B_i$ are independent of position (but are allowed to be functions of time).† The mean flow gradients impose rates of

† The convention of summation over repeated indices is used herein: $A_{ij}x_j \equiv \sum_{j=1}^{3} A_{ij}x_j$, or, in vector notation, $\boldsymbol{U} = \boldsymbol{A} \cdot \boldsymbol{x} + \boldsymbol{B}$.

rotation and strain on the turbulence, but these distortions are independent of position, so the turbulence remains homogeneous.

Throughout this book, the mean, or average, of a quantity will be denoted by a capital letter and the fluctuation from this mean by a lower-case letter; the process of averaging is signified by an overbar. The total random quantity is represented as the sum of its average plus a fluctuation. This prescription is to write $U + u$ for the velocity, with $U$ being the average and $u$ the fluctuation. In a previous illustration, the velocity was $\xi - 1$, with $\xi$ given by coin toss; thus, $U + u = \xi - 1$. Averaging the right-hand side shows that $U = -1/2$. Then $u = \xi - 1 - U = \xi - 1/2$. By definition, the fluctuation has zero average: in the present notation, $\overline{u} = 0$.

A way to categorize *non-homogeneous* turbulent flows is by their mean velocity. A turbulent shear flow, such as a boundary layer or a jet, is so named because it has a *mean* shear. In a separated flow, the *mean* streamlines separate from the surface. The turbulence always has shear and the flow around eddies near to walls will commonly include separation; so these names would be ambiguous unless they referred only to the mean flow.

The simplest non-homogeneous flows are parallel or self-similar shear flows. The term "parallel" means that the velocity is not a function of the coordinate parallel to its direction. The flow in a pipe, well downstream of its entrance, is a parallel flow, $U(r)$. The mean flow is in the $x$ direction and it is a function of the perpendicular direction, $r$. All statistics are functions of $r$ only. Self-similar flows are analogous to parallel flow, but they are not strictly parallel. Self-similar flows include jets, wakes, and mixing layers. For instance, the width, $\delta(x)$, of a mixing layer spreads with downstream distance. But if the cross-stream coordinate, $y$, is normalized by $\delta$, the velocity becomes parallel in the new variable: $U$ as a function of $y/\delta \equiv \eta$ is independent of $x$. Again, there is dependence on only one coordinate, $\eta$; the dependence on downstream distance is parameterized by $\delta(x)$. Parallel and self-similar shear flows are also categorized as "fully developed." Figure 1.7 shows the transition of the flow in a jet from a laminar state, at the left, to the turbulent state. Whether it is a laminar jet undergoing transition, or a turbulent flow evolving into a jet, the upstream region contains a central core into which turbulence will penetrate as the flow evolves downstream. A fully developed state is reached only after the turbulence has permeated the jet.

Shear flows away from walls, or free-shear flows, often contain some suggestion of large-scale eddying motion with more erratic small-scale motions superimposed; an example is the turbulent wake illustrated by Figure 1.8. All of these scales of irregular motion constitute the turbulence. The distribution of fluctuating velocity over the range

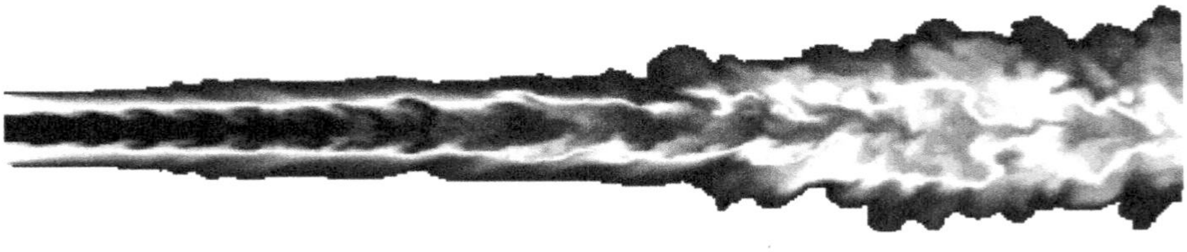

**Figure 1.7** Transition from a laminar to a turbulent jet via computer simulation. The regular pattern of disturbances at the left evolves into the disorderly pattern at the right. Courtesy of B. J. Boersma.

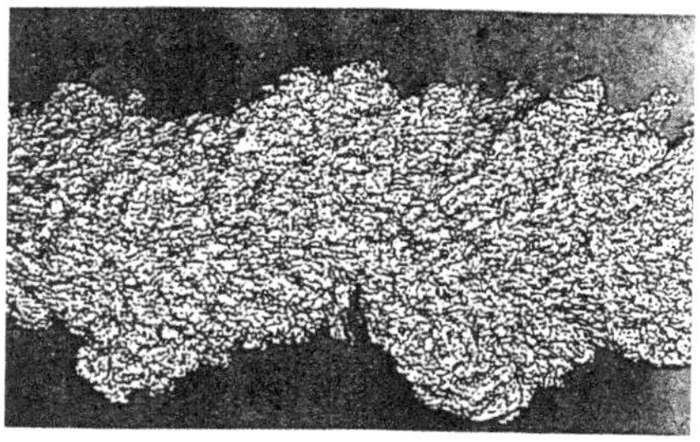

Turbulent wake behind a bullet, visualized by Schlireren photography (Corrsin and Kisler, 1954). Flow is from left to right.

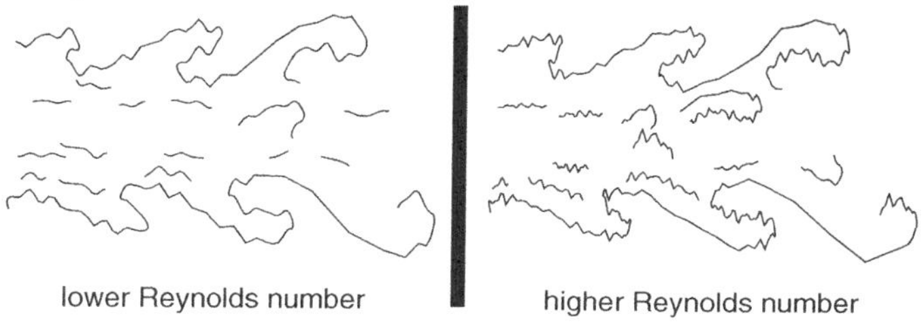


**Figure 1.8** Schematic suggesting large- and small-scale structure of a free-shear layer versus Reynolds number. The large scales are insensitive to Reynolds number; the smallest scales become smaller as Re increases.

of scales is called the spectrum of the turbulence. Fully turbulent flow has a continuous spectrum, ranging from the largest, most energetic scales, that cause the main indentations in Figure 1.8, to the smallest eddies, nibbling at the edges. An extreme case is provided by the dust cloud of an explosion in Figure 1.4. A wide range of scales can be seen in the plume rising from the surface. The more recognizable large eddies have acquired the name "coherent structures."

Boundary layers, like free-shear flows, also contain a spectrum of eddying motion. However, the large scales appear less coherent than in the free-shear layers. The larger eddies in boundary layers are described as "horseshoe" or "hairpin" vortices (Figure 5.9, page 99). In free-shear layers, large eddies might be "rolls" lying across the flow and "rib" vortices, sloping in the streamwise direction (Figure 5.3, page 94). In all cases a background of irregular motion is present, as in Figure 1.8. Despite endeavors to identify recognizable eddies, the dominant feature of turbulent flow is its highly irregular, chaotic nature.

A category of "complex flows" is invariably included in a discussion of turbulence. This might mean *relatively* complex, including pressure-gradient effects on thin shear layers, boundary layers subject to curvature or transverse strain, three-dimensional thin shear layers, and the like. Alternatively, it might mean *quite* complex, and run the whole gamut. From a theoretician's standpoint, complex flows are those in which statistics depend on more than one coordinate, and possibly on time. These include perturbations to basic shear layers, and constitute the case of *relatively* complex turbulence. The category *quite* complex flows includes real engineering flows: impinging jets, separated boundary layers, flow around obstacles, and so on. For present purposes, it will suffice to lump quite and relatively complex flows into one category of *complex turbulent flows*. The models discussed in Chapters 6 and 7 are intended for computing such flows. However, the emphasis in this book is on describing the underlying principles and the processes of model development, rather than on surveying applications. The basic forms of practical models have been developed by reference to simple, canonical flows; fundamental data are integrated into the model to create a robust prediction method for more complex applications. A wealth of computational studies can be found in the literature: many archival journals contain examples of the use of turbulence models in engineering problems.

## Exercises

**Exercise 1.1.** *Origin of the closure problem.* The closure problem arises in any nonlinear system for which one attempts to derive an equation for the average value. Let $\xi$ correspond to the result of coin tossing, as in the text, and let

$$u^3 + 3u^2 + 3u = (u+1)^3 - 1 = 7\xi.$$

Show that, if $\overline{u^3} = \overline{u}^3$ and $\overline{u^2} = \overline{u}^2$ *were* correct, then the mean value of $u$ would be $\overline{u} = (9/2)^{1/3} - 1$. By contrast, show that the correct value is $\overline{u} = 1/2$. Explain why these differ, and how this illustrates the "closure problem."

**Exercise 1.2.** *Eddies.* Identify what you would consider to be large- and small-scale eddies in the photographic portions of Figures 1.4 and 1.8.

**Exercise 1.3.** *Turbulence in practice.* Discuss practical situations where turbulent flows might be unwanted or even an advantage. Why do you think golf balls have dimples?

# 2

# Mathematical and statistical background

> To understand God's thoughts we must study statistics, for these are the measure of his purpose
>
> – Florence Nightingale

While the primary purpose of this chapter is to introduce the mathematical tools that are used in single-point statistical analysis and modeling of turbulence, it also serves to introduce some important concepts in turbulence theory. Examples from turbulence theory are used to illustrate the particular mathematical and statistical material.

## 2.1 Dimensional analysis

One of the most important mathematical tools in turbulence theory and modeling is dimensional analysis. The primary principles of dimensional analysis are simply that all terms in an equation must have the same dimensions and that the arguments of functions can only be non-dimensional parameters: the Reynolds number $UL/\nu$ is an example of a non-dimensional parameter. This might seem trivial, but dimensional analysis, combined with fluid dynamical and statistical insight, has produced one of the most useful results in turbulence theory: the Kolmogoroff $-5/3$ law. The reasoning behind the $-5/3$ law is an archetype for turbulence scale analysis.

The insight comes in choosing the relevant dimensional quantities. Kolmogoroff's insight originates in the idea of a turbulent energy cascade. This is a central conception in the current understanding of turbulent flow. The notion of the turbulent energy cascade pre-dates Kolmogoroff's work (Kolmogoroff, 1941); the origin of the cascade as an analytical theory is usually attributed to Richardson (1922).

*Statistical Theory and Modeling for Turbulent Flows, Second Edition* P. A. Durbin and B. A. Pettersson Reif

Consider a fully developed turbulent shear layer, such as illustrated by Figure 1.8. The largest-scale eddies are on the order of the thickness, $\delta$, of the layer; $\delta$ can be used as a unit of length. The size of the smallest eddies is determined by viscosity, $\nu$. If the eddies are very small, they are quickly diffused by viscosity, so viscous action sets a lower bound on eddy size. Another view is that the Reynolds number of the small eddies, $u\eta/\nu$, is small compared to that of the large eddies, $u\delta/\nu$, so small scales are the most affected by viscous dissipation. For the time being, it will simply be supposed that there is a length scale $\eta$ associated with the small eddies and that $\eta \ll \delta$.

The largest eddies are produced by the mean shear – which is why their length scale is comparable to the thickness of the shear layer. Thus we have the situation that the large scales are being generated by shear and the small scales are being dissipated by viscosity. There must be a mechanism by which the energy produced at large scales is transferred to small scales and then dissipated. Kolmogoroff reasoned that this requires an intermediate range of scales across which the energy is transferred, without being produced or dissipated. In equilibrium, the energy flux through this range must equal the rate at which energy is dissipated at small scales. This intermediate range is called the *inertial subrange*. The transfer of energy across this range is called the *energy cascade*. Energy cascades from large scale to small scale, across the inertial range. The physical mechanism of the energy cascade is somewhat nebulous. It may be a sort of instability process, whereby larger-scale regions of shear develop smaller-scale irregularities; or it might be nonlinear distortion and stretching of large-scale vorticity.

As already alluded to, the rate of transfer across the inertial range, from the large scales to the small, must equal the rate of energy dissipation at small scale. Denote the rate of dissipation per unit volume by $\rho\varepsilon$, where $\rho$ is the density, and $\varepsilon$ is the rate of energy dissipation per unit mass. The latter has dimensions of $\ell^2/t^3$, which follows because the kinetic energy per unit mass, $k \equiv \frac{1}{2}(u_1^2 + u_2^2 + u_3^2)$ has dimensions of $\ell^2/t^2$, and its rate of change has another factor of $t$ in the denominator. The rate $\varepsilon$ plays a dual role: it is the rate of energy dissipation, and it is the rate at which energy cascades across the inertial range. These two are strictly equivalent only in equilibrium. In practice, an assumption of local equilibrium in the inertial and dissipation ranges is usually invoked. Even though the large scales of turbulence might depart from equilibrium, the small scales are assumed to adapt almost instantaneously to them. The validity of this assumption is sometimes challenged, but it has provided powerful guidance to theories and models of turbulence.

Now the application of dimensional reasoning: we want to infer how energy is distributed within the inertial range as a function of eddy size, $E(r)$. The inertial range is an overlap between the large-scale, energetic range and the small-scale dissipative range. It is shared by both. Large scales are not directly affected by molecular dissipation. Because the inertial range is common to the large scales, it cannot depend on molecular viscosity $\nu$. The small scales are assumed to be of universal form, not depending on the particulars of the large-scale flow geometry. Because the inertial range is common to the small scales, it cannot depend on the flow width $\delta$. All that remains is the rate of energy cascade, $\varepsilon$.

Consider an eddy of characteristic size $r$ lying in this intermediate range. Based on the reasoning of the previous paragraph, on dimensional grounds its energy is of order $(\varepsilon r)^{2/3}$. This is the essence of Kolmogoroff's law: in the inertial subrange the energy of the eddies increases with their size to the $2/3$ power.

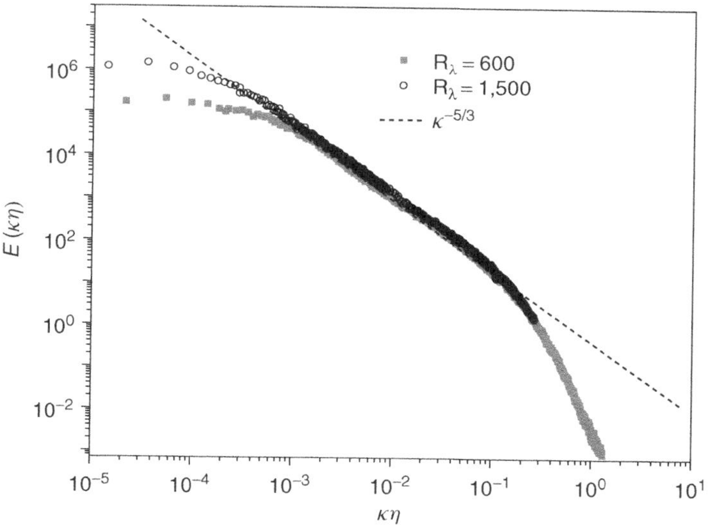


**Figure 2.1** Experimental spectra measured by Saddoughi and Veeravalli (1994) in the boundary layer of the NASA Ames 80 × 100 foot wind tunnel. This enormous wind tunnel gives a very high Reynolds number, so that the −5/3 law can be verified over several decades. In this figure, $\kappa\eta \lesssim 10^{-3}$ is the energetic range and $\kappa\eta \gtrsim 0.1$ is the dissipation range.

This 2/3 law becomes a −5/3 law in Fourier space; that is how it is more commonly known. One motive for the transformation is to obtain the most obvious form in which to verify Kolmogoroff's theory by experimental measurements. The distribution of energy across the scales of eddies in physical space is the inverse Fourier transform of the spectral energy density in Fourier space. This is a loose definition of the energy spectral density, $E(\kappa)$. The energy spectral density is readily measured. It is illustrated by the log–log plot in Figure 2.1.

Equating the inertial-range energy to the inverse transform of the inertial-range energy spectrum (cf. Section 2.2.2.2):

$$(\varepsilon r)^{2/3} \propto \int E(\kappa)[1 - \cos(\kappa r)]\, d\kappa. \tag{2.1.1}$$

Assume that $E(\kappa) \propto \varepsilon^{2/3}\kappa^n$ in (2.1.1). Then

$$\begin{aligned} r^{2/3} \propto \int \kappa^n[1 - \cos(\kappa r)]\, d\kappa &= r^{-n-1} \int (r\kappa)^n[1 - \cos(\kappa r)]\, d(r\kappa) \\ &= r^{-n-1} \int x^n[1 - \cos(x)]\, dx. \end{aligned} \tag{2.1.2}$$

Here, $\kappa$ has dimensions of $1/\ell$, so the integrand of the second expression is non-dimensional. The final integral is just some number, independent of $r$. Equating the exponents on both sides of (2.1.2) gives $n + 1 = -2/3$ or $n = -5/3$. The more famous statement of Kolmogoroff's result is the "$-5/3$ law"

$$E(\kappa) \propto \varepsilon^{2/3}\kappa^{-5/3}. \tag{2.1.3}$$

Note that $E(\kappa)$ has dimensions $(\ell^2/t^3)^{2/3}\ell^{5/3} = \ell^3/t^2$. Here, $E(\kappa)$ is the energy density per unit wavenumber (per unit mass), so $E(\kappa)\,\mathrm{d}\kappa$ has dimensions of velocity squared. Because $\kappa \sim 1/\ell$, *large* scales correspond to *small* $\kappa$ and vice versa. Hence, the spectrum in Figure 2.1 shows how the energy declines as the eddies grow smaller.

These ideas about scaling turbulent spectra expand to a general approach to constructing length and time-scales for turbulent motion. Such scaling is essential both to turbulence modeling for engineering computation, and to more basic theories of fluid dynamical turbulence.

### 2.1.1 Scales of turbulence

The notion of large and small scales, with an intervening inertial range, begs the question: How are "large" and "small" defined? If the turbulent energy $k$ is being dissipated at a rate $\varepsilon$, then a time-scale for energy dissipation is $T = k/\varepsilon$. In order of magnitude this is the time it would take to dissipate the existing energy. This time-scale is sometimes referred to as the eddy lifetime, or *integral time-scale*. Since it is formed from the overall energy and its rate of dissipation, $T$ is a scale of the larger, more energetic eddies.

Formula (2.1.3) and Figure 2.1 show that the large scales make the biggest contribution to $k$. The very small scales of motion have little energy, so it would not be appropriate to use $k$ to infer a time-scale for these eddies. They are diffused by viscosity, so it is more appropriate to form a time-scale from $\nu$ and $\varepsilon$: $T_\eta = \sqrt{\nu/\varepsilon}$ has the right dimensions. The dual role that $\varepsilon$ plays has already been commented on; here it is functioning as a property of the small-scale eddies. This scaling applies to the *dissipation subrange*. The time-scale $T_\eta$ is often referred to as the "Kolmogoroff" time-scale and the small eddies to which it applies are Kolmogoroff scale eddies. However, this is a bit confusing because Kolmogoroff's $-5/3$ law applies at scales large compared to the dissipation scales, so here $T_\eta$ will simply be called the dissipative time-scale.

The time-scales of the large and small eddies are in the ratio

$$T/T_\eta = \sqrt{k^2/\varepsilon\nu}. \tag{2.1.4}$$

A turbulent Reynolds number can be defined as

$$R_\mathrm{T} \equiv k^2/\varepsilon\nu, \tag{2.1.5}$$

so (2.1.4) becomes $T/T_\eta = R_\mathrm{T}^{1/2}$. This can be written in a more familiar form, $R_\mathrm{T} = uL/\nu$, by noting that $u = \sqrt{k}$ is a velocity scale, and by defining the large eddy length scale to be $L = T\sqrt{k} = k^{3/2}/\varepsilon$ (this length scale is sometimes denoted $L_\varepsilon$).

The ratio $T_\eta/T$ varies as $R_\mathrm{T}^{-1/2}$: the time-scale of small eddies, relative to that of the dissipative eddies, decreases with Reynolds number. At high Reynolds number the small eddies die much more quickly than the large ones. This is the conclusion of dimensional analysis, tempered by physical intuition. Figure 1.8 illustrates that the small scales become smaller as $R_\mathrm{T}$ increases, so the dimensional analysis is physically sensible.

The same dimensional reasoning can be applied to length scales. The large scale is $L = k^{3/2}/\varepsilon$. The small scale that can be formed from $\varepsilon$ and $\nu$ is

$$\eta = (\nu^3/\varepsilon)^{1/4}. \tag{2.1.6}$$

This is the dissipative length scale. In defining these scales it has been assumed that they are disparate: that is, $L/\eta \gg 1$, or $R_T^{3/4} \gg 1$ is required.

The inertial subrange lies between the large, energetic scales and the small, dissipative scales. As $R_T$ increases so does the separation between these scales and, hence, so does the length of the inertial subrange. To see the $-5/3$ region of the energy spectrum clearly, the Reynolds number must be quite high. Measurements at high Reynolds number were made by Saddoughi and Veeravalli (1994) in a very large wind tunnel at NASA Ames; these are the data shown in Figure 2.1. These data confirm Kolmogoroff's law very nicely. In the figure, the wave number $\kappa$ is normalized by $\eta$, so that the dissipation range is where $\kappa\eta = O(1)$. The energetic range is where $\kappa\eta = O(\eta/L)$. The present dimensional analysis predicts that the lower end of the $-5/3$ range will decrease as $R_T^{-3/4}$ (see Exercise 2.1).

In the experiments, the spatial spectrum $E(\kappa)$ was not measured; this would require measurements with two probes that have variable separation. In practice, a single, stationary probe was used and the frequency spectrum $E(\omega)$ of eddies convected past the probe was measured. If an eddy of size $r$ is convected by a velocity $U_c$, then it will pass by in time $t = O(r/U_c)$. Hence the $r^{2/3}$ law becomes $(U_c t)^{2/3}$. When this is Fourier-transformed in $t$, the inertial-range spectrum (2.1.3) becomes $E(\omega) \sim (\omega/U_c)^{-5/3}$. The spectra in Figure 2.1 are frequency spectra that were plotted by equating $\kappa$ to $\omega/U_c$, where $U_c$ is the mean velocity at the position of the probe. This use of a convection velocity to convert from temporal to spatial spectra is referred to as *Taylor's hypothesis*. It is an accurate approximation if the time required to pass the probe is short compared to the eddy time-scale. For the large scales, this requires the turbulent intensity to be low, by the following reasoning:

$$L/U_c \ll T \qquad \Rightarrow \qquad L/TU_c \sim \sqrt{k}/U_c \ll 1.$$

Turbulent intensity is usually defined as $\sqrt{2k/3}/U_c$.

For small scales, the probe resolution is usually a greater limit on accuracy than is Taylor's hypothesis. The latter only requires

$$\eta/U_c \ll T_\eta \qquad \Rightarrow \qquad \frac{\eta}{T_\eta U_c} \sim (\nu\varepsilon)^{1/4}/U_c \sim R_T^{-1/4}\sqrt{k}/U_c \ll 1.$$

The former requires the probe size to be of order $\eta$. These inequalities illustrate the use of scale analysis to assess the instrumentation required to measure turbulent flow.

## 2.2 Statistical tools

### 2.2.1 Averages and probability density functions

Averages of a random variable have already been used, but it is instructive to look at the process of averaging more formally. Consider a set of independent samples of a random variable $\{x_1, x_2, \ldots, x_N\}$. To be concrete, these could be the results of a set of coin tosses with $x = 1$ for heads and $x = 0$ for tails. Their average is

$$x_{\rm av} = \frac{1}{N}\sum_{i=1}^{N} x_i. \tag{2.2.1}$$

The mean $\overline{x}$ is defined formally as the limit as $N \to \infty$ of the average. Unfortunately, the average converges to the mean rather slowly, as $1/\sqrt{N}$ (see Section 2.2.2), so experimental estimates of the mean could be inaccurate if $N$ is not very large.

If the random process is *statistically stationary*, the above *ensemble* of samples can be obtained by measuring $x$ at various times, $\{x(t_1), x(t_2), \ldots, x(t_N)\}$. The ensemble average can then be obtained by time averaging. Rather than adding up measurements, the time average can be computed by integration

$$\overline{x} = \lim_{t\to\infty} \frac{1}{t}\int_0^t x(t')\,\mathrm{d}t'. \tag{2.2.2}$$

The caveat of statistical stationarity means that the statistics are independent of the time origin; in other words, $x(t)$ and $x(t+t_0)$ have the same statistical properties for any $t_0$. For instance, suppose one is measuring the turbulent velocity in a wind tunnel. If the tunnel has been brought up to speed and that speed is maintained constant, then it does not matter whether the mean is measured starting now and averaging for a minute, or if the measurements start ten minutes from now and average for a minute: expectations are that the averages will be the same to within experimental uncertainty. In other words, $\overline{x(t)} = \overline{x(t+t_0)}$. Stationarity can be described as translational invariance in time of the statistics. Time averaging is only equivalent to ensemble averaging if the random process is statistically stationary. The failure to recognize this caveat is surprisingly common. One instance where this mistake has been made is in the turbulent flow behind a bluff body. A deterministic oscillation at the Strouhal shedding frequency is usually detectable in the wake. This deterministic time dependence means that the flow is not stationary – the statistics vary periodically with time at the oscillation frequency. Hence the statistical averaging invoked by the Reynolds averaged Navier–Stokes equations is not synonymous with time averaging. The statistical average can be measured by taking samples at a fixed phase of the oscillation.

In the case of (2.2.2) the difference between a finite-time average and the mean decreases like $\sqrt{T/t}$ as $t \to \infty$, where $T$ is the integral time-scale of the turbulence and $t$ is the averaging time. The reasoning is analogous to (2.2.9) in Section 2.2.2: the integral can be written as

$$\overline{x} = \frac{1}{N}\sum_i x_i \qquad \text{where} \qquad x_i = \int_{(i-1)T}^{iT} x\,\mathrm{d}t/T, \quad i = 1, 2, 3, \ldots, N$$

with $N = t/T$. The $x_i$ can be thought of as independent samples to which the estimate (2.2.9) applies.

From here on, mean values simply will be denoted by an overbar, it being understood that this implies an operation like (2.2.2) or (2.2.1) in the limit $N \to \infty$. Ensemble averaging is a linear operation, and hence

$$\begin{aligned}
&\text{(a)} \quad \overline{x+y} = \overline{x} + \overline{y},\\
&\text{(b)} \quad \overline{ax} = a\overline{x},\\
&\text{(c)} \quad \overline{\overline{x}} = \overline{x},\\
&\text{(d)} \quad \overline{x - \overline{x}} = 0,
\end{aligned} \tag{2.2.3}$$

where $a$ is a non-random constant. These follow from (2.2.1). For example

$$\lim_{N\to\infty}\left[\frac{1}{N}\sum_{i=1}^{N} x_i + y_i = \frac{1}{N}\sum_{i=1}^{N} x_i + \frac{1}{N}\sum_{i=1}^{N} y_i\right]$$

proves the first of (2.2.3). These properties motivate the decomposition of the velocity into its mean plus a fluctuation. This decomposition is $U = \overline{U} + u$. Averaging both sides of this using (2.2.3a,c) gives

$$\overline{U} = \overline{\overline{U} + u} = \overline{\overline{U}} + \overline{u} = \overline{U} + \overline{u},$$

so $\overline{u} = 0$. The decomposition of $U$ is into its mean plus a part that has zero average.

It is essential to recognize that in general

$$\overline{xy} \neq \overline{x}\,\overline{y}.$$

Again, this follows from (2.2.1):

$$\frac{1}{N}\sum_{i=1}^{N} x_i y_i \neq \left(\frac{1}{N}\sum_{i=1}^{N} x_i\right) \times \left(\frac{1}{N}\sum_{i=1}^{N} y_i\right).$$

For instance, try $N = 2$.

The above is often a sufficient understanding of averaging for turbulence modeling. However, another approach is to relate averaging to an underlying probability density. That approach is widely used in models designed for turbulent combustion.

When one computes an average like (2.2.1) for a variable with a finite number of possible values (like heads or tails), the mean can be computed as the sum of each value times its probability of occurrence. That probability is estimated by the fraction of the $N$ samples that have the particular value. If the possible values of $x$ are $\{a_1, a_2, \ldots, a_J\}$, then, as $N \to \infty$,

$$\overline{x} = \frac{1}{N}\sum_{i=1}^{N} x_i = \sum_{j=1}^{J} \frac{N_j}{N} a_j = \sum_{j=1}^{J} p_j a_j, \tag{2.2.4}$$

where $N_j$ is the number of times the value $a_j$ appears in the sample $\{x_j\}$. Here $p_j$ is the probability of the value $a_j$ occurring. This probability is defined to be $\lim_{N\to\infty} N_j/N$. The $a_j$ are simply a deterministic set of numbers; for example, for the roll of a die, they would be the numbers 1 to 6. The set $\{x_1, x_2, \ldots, x_N\}$ are random samples that are being averaged: for instance, one might roll a die $10^4$ times and record the outcomes as the $x_i$, with $N_1$ being the number of these for which $x_i = 1$, and so on.

If $x$ has a continuous range of possible values, then the summation in (2.2.4) must be replaced by an integral over this range. The probability $p_j$ that $x = a_j$ must be replaced by the probability $P(a)\,\mathrm{d}a$ that $x$ lies between $a - \frac{1}{2}\mathrm{d}a$ and $a + \frac{1}{2}\mathrm{d}a$. The average is then $\overline{x} = \int aP(a)\,\mathrm{d}a$, where the range of integration includes all possible values of $a$. Common practice is to use the same variable name in the integral and write

$$\overline{x} = \int x' P(x')\,\mathrm{d}x', \tag{2.2.5}$$

but one should be careful to distinguish conceptually between the random variable $x$ and the integration variable $x'$. The former can be sampled and can vary erratically from sample to sample; the latter is a dummy variable that ranges over the possible values of $x$. In Eq. (2.2.5), $P(x)$ is simply a suitable function, called the *probability density function* (PDF). An example is the Gaussian, $P(x) = \mathrm{e}^{-x^2/\sigma^2}/\sigma\sqrt{\pi}$. The PDF is largest at the most likely values of $x$ and is small for unlikely values.

The interpretation of the probability density as a frequency of occurrence of an event is illustrated by Figure 2.2. For a time-dependent random process, $P(x')\,\mathrm{d}x'$ can be described as the average fraction of time that $x(t)$ lies in the interval $(x' - \frac{1}{2}\mathrm{d}x', x' + \frac{1}{2}\mathrm{d}x')$. In the figure, the fraction of time that this event occurs is $\sum_i \mathrm{d}t_i/T$; and $P(x')\,\mathrm{d}x'$ is the limit of this ratio as $T \to \infty$. According to this definition, there is no need for the curve in Figure 2.2 to be random; any smooth function of time has a PDF, $P(x') = \sum_i (\mathrm{d}t_i/\mathrm{d}x)/T$, but the interest here is in random fluid motion.

Instead of averaging $x$, any function of $x$ can be averaged as in (2.2.5):

$$\overline{f(x)} = \int f(x')P(x')\,\mathrm{d}x'. \tag{2.2.6}$$

As a special case, if $f(x) = 1$ for all $x$, then $\int P(x')\,\mathrm{d}x' = 1$, so the PDF must be a function that integrates to unity. The PDF also is non-negative, $P(x) \geq 0$. The PDF is

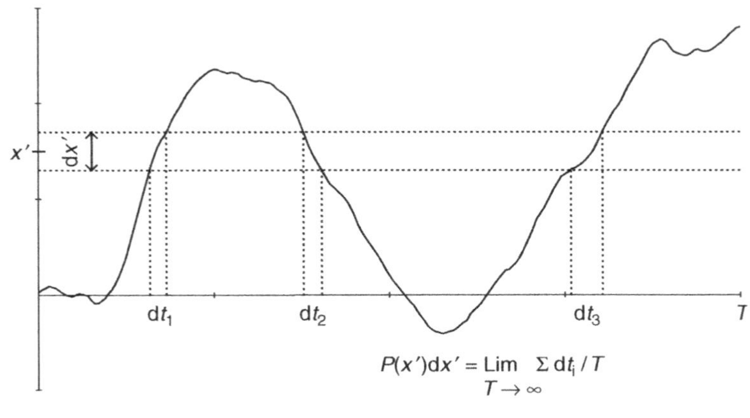


**Figure 2.2** Illustration of the definition of the PDF: it is the fraction of time that the random function lies in the interval $(x' - \frac{1}{2}\mathrm{d}x', x' + \frac{1}{2}\mathrm{d}x')$.

used in statistical sampling theory, but here the interest is only in averaging. Its utility is illustrated by an intriguing application to non-premixed turbulent combustion.

#### 2.2.1.1 Application to reacting turbulent flow

Suppose two chemical species, A and B, react very rapidly. Initially they are unmixed, but as the turbulence stirs them together, they react instantly to form product. The schematic Figure 2.3 suggests the contorted interface between the reactants. The reaction is so fast that A and B can never exist simultaneously in a fluid element; they immediately react to consume whichever has the lower concentration. Any fluid element can contain either A or B plus product (and inert diluent).

Let the reactant concentrations be $\gamma_A$ and $\gamma_B$. Then the variable $m \equiv \gamma_A - \gamma_B$ will equal $\gamma_A$ whenever it is positive (since $\gamma_B$ cannot be present with $\gamma_A$) and it equals $-\gamma_B$ when it is negative. Furthermore, if one mole of A reacts with one mole of B, then the reaction decreasing $\gamma_A$ by one will also decrease $\gamma_B$ by one and $m$ will be unchanged by the reaction. Hence the variable $m$ behaves as a *non-reactive scalar* field, and $m$ will be affected by turbulent stirring: for instance, regions of positive and negative $m$ will be mixed together to form intermediate concentrations. In fact $m$ can be thought of as a substance, like dye, that is stirred by the turbulent fluid motion. At any point in the fluid, that substance will have a concentration that varies randomly in time and has a PDF that can be constructed as in Figure 2.2.

The mean concentration of A is just the average of $m$ over positive values

$$\overline{\gamma_A} = \int_0^\infty m P(m)\, dm. \tag{2.2.7}$$

When $m > 0$ it equals $\gamma_A$, so it is readily verified that

$$\overline{\gamma_A} = \int_0^\infty m P(m)\, dm = \int_0^\infty \gamma'_A P(\gamma'_A)\, d\gamma'_A.$$

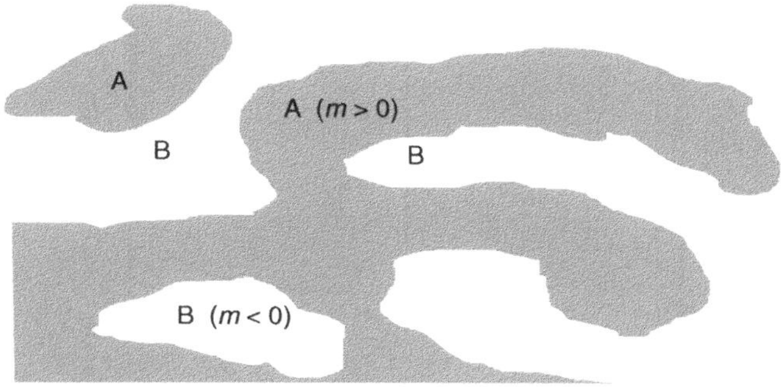


**Figure 2.3** A statistically homogeneous mixture of reactants. An infinitely fast reaction is assumed to be taking place at the interface between the zones of A and B. Turbulence contorts the interface and stirs the reactants. Product forms as the reaction proceeds and the concentrations of A and B decrease.

In the literature on turbulent combustion, Eq. (2.2.7) is referred to as Toor's analogy. If the evolution of $P(m)$ in turbulent flow can be modeled, then this provides a nice theory for turbulent reactions. But $P(m)$ is a property of a non-reactive contaminant. The intriguing aspect of this analysis is that the mean rate of chemical reactant consumption can be inferred from an analysis of how a non-reacting contaminant is mixed.

Let us see, qualitatively, how $P(m)$ evolves with mixing. Suppose the initial state is unmixed patches of concentration $\gamma_A^0$ and $\gamma_B^0$. Then initially $m$ takes the values of either $\gamma_A^0$ or $-\gamma_B^0$; these are the only two possible concentrations. The corresponding initial probability density consists of two spikes, $P(m) = \frac{1}{2}[\delta(m - \gamma_A^0) + \delta(m + \gamma_B^0)]$. The spikes are of equal probability if there are equal amounts of A and B. However, turbulent stirring and molecular diffusion will immediately produce intermediate concentrations. The PDF fills in, as shown by the solid curve in Figure 2.4. As time progresses, the concentration becomes increasingly uniform; the PDF starts to peak around the midpoint $m = \frac{1}{2}(\gamma_A^0 - \gamma_B^0)$. Ultimately the contaminant will be uniformly mixed and the only possible concentration will be this average value; in other words, the PDF collapses to a spike at the average concentration $P(m) \to \delta[m - \frac{1}{2}(\gamma_A^0 - \gamma_B^0)]$. In a sense this is a process of reversed diffusion in concentration space: instead of spreading with time, the PDF contracts into an increasingly narrow band around the mean concentration.

One approach to modeling this evolution of the PDF is to assume that $P(m)$ has the "beta" form described in Exercise 2.6. The $\beta$ distribution has two parameters, $a$ and $b$, that are related to the mean and variance by

$$\frac{a-b}{a+b} = \overline{m}, \qquad \frac{1 + (a+b)\overline{m}^2}{a+b+1} = \overline{m^2}.$$

These formulas can be derived from the PDF defined in Exercise 2.6. As $a$ and $b$ vary, PDFs representative of scalar mixing are obtained. Figure 2.5 is a comparison between

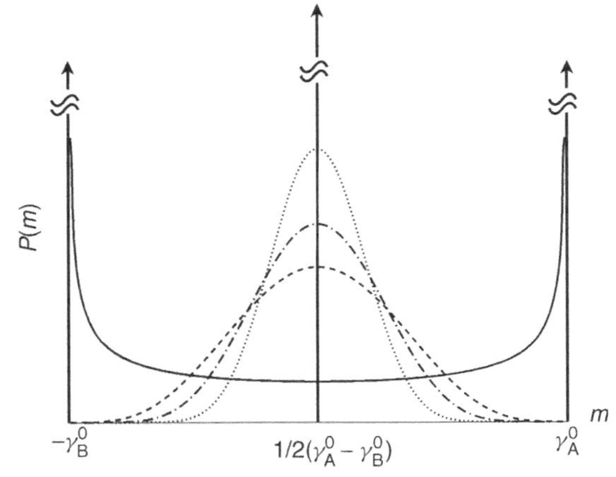


**Figure 2.4** Evolution of the probability density of a non-reactive scalar. Initially, either $m = \gamma_A^0$ or $-\gamma_B^0$ and the corresponding PDF consists of two spikes. With time, intermediate concentrations are produced by mixing and the PDF evolves toward a spike at the average concentration.

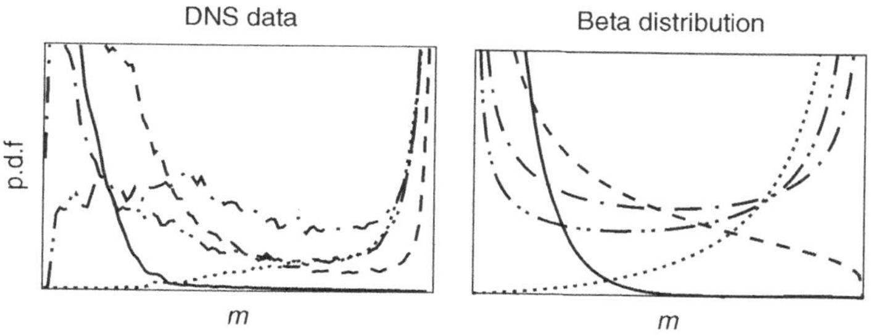


**Figure 2.5** Comparison of profiles of the PDF of $m$ obtained by direct numerical simulations to the $\beta$ PDF: $\overline{m} = -0.9$ (——), $-0.5$ (- - - -), $0.0$ (— · —), $0.5$ (— ·· —), and $0.9$ (·········). The data were obtained in a turbulent flame front. From Mantel and Bilger (1994).

the $\beta$ distribution and data from a numerical simulation of a turbulent flame. Each curve corresponds to different values of $\overline{m}$ and $\overline{m^2}$. It can be seen that the functional form of the $\beta$ PDF does a reasonable job of mimicking the data.

Once the shape of $P(m)$ has been assumed, the problem of reaction in turbulent flow reduces to modeling how $\overline{m}$ and $\overline{m^2}$ evolve in consequence of turbulent mixing. That is a problem in single-point, moment closure, which is the topic of the second part of this book.

Toor's analogy, with this assumed form for the PDF, permits reactant consumption to be computed via (2.2.7): this is called the "assumed PDF method." The simplifications consequent to Toor's analogy with an assumed PDF reduce reacting flow to a problem in turbulent dispersion. However, it should be emphasized that this is only true when the reaction rate can be considered to be infinitely fast compared to the rate of turbulent mixing. The ratio of turbulent time-scale to chemical time-scale is called the Dahmköhler number. Toor's analogy is a large Dahmköhler number approximation. It serves as a nice, although brief, introduction to concepts of turbulent combustion. Pope (1985) is a comprehensive reference on this application of probability densities.

## 2.2.2 Correlations

Correlations between random variables play a central role in turbulence modeling. The random variables in fluid flow are fields, such as the velocity field. The correlations are also fields, although they are statistics, and hence are deterministic. Correlations can be functions of position and time, or of relative position in the case of two-point correlations. The type of models used in engineering computational fluid dynamics are for single-point correlations. It will become apparent in Chapter 3 on the Reynolds averaged Navier–Stokes equation why prediction methods for engineering flows are based solely on single-point correlations. For now, it can be rationalized by noting that, in a three-dimensional geometry, single-point correlations are functions of the three space dimensions, while two-point correlations are functions of all pairs of points, or three plus three dimensions – imagine having to construct a computational grid in six dimensions! However, this section does discuss two time correlations, and two spatial point correlations are quite important in the theory of homogeneous turbulence described in Part III.

If the total, instantaneous turbulent velocity is denoted $\tilde{u}$ and the mean velocity, $U$, is defined as $U \equiv \overline{\tilde{u}}$, then the fluctuating velocity, $u$, is defined by $u = \tilde{u} - U$; in other words, the total velocity is decomposed into its mean and a fluctuation, $\tilde{u} = U + u$, with $\overline{u} = 0$. The fluctuation $u$ is usually referred to as the turbulence and $U$ as the mean flow. Of course, the velocity is a vector with components $u_i$, $i = 1, 2, 3$, so $\tilde{u}_i = U_i + u_i$ for each component.

The average of the products of the fluctuation velocity components is a second-order tensor, or a matrix in any particular coordinate system (see Section 2.3), with components $\overline{u_i u_j}$:

$$\begin{pmatrix} \overline{u_1 u_1} & \overline{u_1 u_2} & \overline{u_1 u_3} \\ \overline{u_2 u_1} & \overline{u_2 u_2} & \overline{u_2 u_3} \\ \overline{u_3 u_1} & \overline{u_3 u_2} & \overline{u_3 u_3} \end{pmatrix}. \tag{2.2.8}$$

This is a rather important matrix in engineering turbulence modeling. It is called the *Reynolds stress tensor*. The Reynolds stress tensor is symmetric, with six unique components, since $\overline{u_1 u_3} = \overline{u_3 u_1}$ etc.

The average of the product of two random variables is called their *covariance*. If the covariance is normalized by the variances, it becomes a *correlation coefficient*,

$$r_{ij} = \overline{u_i u_j} \Big/ \sqrt{\overline{u_i^2}\, \overline{u_j^2}}$$

(where there is no summation on $i$ and $j$). The correlation coefficient is not a tensor. Tensors will be defined in Section 2.3. At present it is sufficient to note that, if this were a tensor, then the convention of summation on repeated indices would be in effect on the right-hand side – but it is not. The definition of $r_{ij}$ requires a qualification that there is no summation implied. For instance, $r_{12} = \overline{u_1 u_2} \Big/ \sqrt{\overline{u_1^2}\, \overline{u_2^2}}$.

The Schwartz inequality $|r_{ij}| \leq 1$ follows from $\overline{(x \pm y)^2} \geq 0$ upon setting $x = u_i \Big/ \sqrt{\overline{u_i^2}}$ and $y = u_j \Big/ \sqrt{\overline{u_j^2}}$. The $\overline{u_i u_j}$ obtained from a turbulence model usually are not the average of a random variable; they are computed from an evolution equation for the statistic, $\overline{u_i u_j}$, itself. Then the condition $|r_{ij}| \leq 1$ might not be met by the model. If the model is such that the condition is always met, it is said to be *realizable*. In a broader sense, a realizable model is one that predicts statistics that could be those of a random process; in a narrower sense, it is one that ensures Schwartz's inequality.

As its name implies, the correlation coefficient is a measure of how closely related two random variables are. If they are equal, that is, $u_i = u_j$, then $r_{ij} = 1$. It was noted below Eq. (2.2.3) that generally $\overline{u_i u_j} \neq \overline{u}_i\, \overline{u}_j$. If, for some reason, such as a symmetry condition, this were an equality, then the two random variables $u_i$ and $u_j$ are described as *uncorrelated*. In such a case $\overline{u_i u_j} = \overline{u}_i\, \overline{u}_j = 0$ so $r_{ij} = 0$, using the fact that $\overline{u}_i = 0$.

As an example, it will be shown that the average (2.2.1) converges to the mean like $1/\sqrt{N}$, as $N \rightarrow \infty$, if the $x_i$ are uncorrelated samples. Consider the mean-squared error in estimating the true mean, $\overline{x}$, by the average:

$$\begin{aligned}
\overline{(x_{\mathrm{av}} - \overline{x})^2} &= \overline{\left[\left(\frac{1}{N}\sum_{i=1}^{N} x_i\right) - \overline{x}\right]^2} = \overline{\frac{1}{N}\sum_{i=1}^{N}(x_i - \overline{x}) \times \frac{1}{N}\sum_{j=1}^{N}(x_j - \overline{x})} \\
&= \frac{1}{N^2}\sum_{i=1}^{N}\sum_{j=1}^{N}\overline{(x_i - \overline{x})(x_j - \overline{x})} \qquad (2.2.9) \\
&= \frac{1}{N^2}\sum_{i=1}^{N}\overline{(x_i - \overline{x})(x_i - \overline{x})} = \frac{\overline{(x - \overline{x})^2}}{N}.
\end{aligned}$$

The step from the second line to the third uses the fact that the samples are uncorrelated, in the form $\overline{(x_i - \overline{x})(x_j - \overline{x})} = 0$, $i \neq j$. The $j$ sum contributes only when $i = j$. The factor $\overline{(x - \overline{x})^2}$ is the variance of the random process. Denoting it by $\sigma^2$ gives the root-mean-square error $\sqrt{\overline{(x_{\mathrm{av}} - \overline{x})^2}} = \sigma/\sqrt{N}$. So the average converges to the mean as $1/\sqrt{N}$.

The covariance $\overline{u_i u_j}$ is the average product of two components of the velocity fluctuation. Another interesting covariance is between a velocity component and itself at two different times, $\overline{u(t + \Delta t)u(t)}$ – this is called an *autocovariance*, for obvious reasons. The turbulent velocity is not a random variable like a coin toss; it is a stochastic process that evolves continuously in time, so the velocity at any time is related to that at any previous time. If the separation between these times is very small, the velocities will be essentially the same: $\lim_{\Delta t \to 0} u(t + \Delta t) = u(t)$, so $\lim_{\Delta t \to 0} \overline{u(t + \Delta t)u(t)} = \overline{u(t)^2}$. If the turbulence is stationary (Section 2.2.1), such as in a wind tunnel in steady operation, then $\overline{u^2}$ is independent of time. Furthermore, a statistically stationary process has the property that the correlation $\overline{u(t + \Delta t)u(t)}$ is a function only of the magnitude of the time separation $|\Delta t|$ but not of time itself. This follows from the definition of stationarity as translational invariance in time. A shift in the time origin, $t \to t + t_0$, leaves statistics unchanged:

$$\overline{u(t + \Delta t)u(t)} = \overline{u(t + t_0 + \Delta t)u(t + t_0)}.$$

Setting $t_0 = -t$ shows that the correlation is a function only of $\Delta t$; setting $t_0 = -t - \Delta_t$ shows it to be only a function of $|\Delta t|$. If the above covariance is normalized by $\overline{u^2}$ it becomes a *correlation function*. This function will be denoted $R(\Delta t)$: it is defined by

$$R(\Delta t) \equiv \frac{\overline{u(t + \Delta t)u(t)}}{\overline{u^2}}.$$

The following example illustrates properties of stationarity via a basic stochastic process. It also provides a framework for some important concepts in turbulence modeling.

#### 2.2.2.1 Lagrangian theory for turbulent mixing

Taylor (1921) first introduced the mathematical concept of a correlation function in connection with his famous study of *turbulent dispersion* ("turbulent dispersion" refers to mixing of a scalar, such as concentration or heat). In his paper, Taylor described a simple, instructive model for the random velocity of a fluid element. His study is the starting point for many recent developments in stochastic modeling of particle dispersion by turbulence.

Consider a discrete random process that represents the fluid velocity at time intervals of $\Delta t$:

$$u(t + \Delta t) = ru(t) + s\xi(t). \tag{2.2.10}$$

Here $r$ and $s$ are coefficients to be determined. This might be considered the velocity of a fluid element; in that sense it is a Lagrangian description of turbulence.* In this stochastic model, $\xi$ is a random variable. It has zero mean and unit variance ($\overline{\xi} = 0$, $\overline{\xi^2} = 1$), and its successive values $\{\xi(t), \xi(t + \Delta t), \xi(t + 2\Delta t), \ldots\}$ are chosen independently of one another. It could be a set of coin tosses, but in practice it would be synthesized by a computer random number generator. At each time, $u(t)$ is known, $\xi(t)$ is selected independently of this value, and the next $u(t + \Delta t)$ is computed by (2.2.10). Since $\xi(t)$ is chosen completely independent of $u(t)$, they are uncorrelated: $\overline{\xi(t)u(t)} = \overline{\xi(t)}\,\overline{u(t)} = 0 \times 0$. This property of independent increments makes (2.2.10) an example of a Markov chain. We want to find the correlation function,

$$R(\tau) \equiv \overline{u(t + \tau)u(t)}/\overline{u^2}, \tag{2.2.11}$$

for this stochastic process.

First consider how to choose the coefficients $r$ and $s$ in (2.2.10). If $u(t)$ is a statistically stationary process, then its variance should not depend on time; in particular $\overline{u(t + \Delta t)^2} = \overline{u(t)^2} = \overline{u^2}$. Squaring both sides of (2.2.10) and averaging gives

$$\overline{u(t + \Delta t)^2} = r^2\overline{u(t)^2} + s^2\overline{\xi^2} + 2rs\,\overline{u(t)\xi(t)}. \tag{2.2.12}$$

The last term is zero, as already explained, so the conditions $\overline{u(t + \Delta t)^2} = \overline{u(t)^2} = \overline{u^2}$ and $\overline{\xi^2} = 1$ give

$$s = \sqrt{(1 - r^2)\overline{u^2}}.$$

If (2.2.10) is multiplied by $u(t)$ and averaged, an equation for the autocovariance at the small time separation $\Delta t$ is obtained:

$$\overline{u(t + \Delta t)u(t)} = r\overline{u^2}. \tag{2.2.13}$$

It can be seen from this that $r$ is the correlation coefficient of $u$ with itself at two times. If $u(t)$ is statistically stationary, then the correlation between times $t$ and $t + \Delta t$ should not depend on the time, $t$, at which the correlation is measured; hence, in (2.2.13) $r$ is only a function of $\Delta t$. The precise form of $R(\tau)$ is determined by the function chosen for $r$.

---

* The Lagrangian approach is to describe fluid motion in a frame moving with fluid elements; the more familiar Eulerian approach adopts a fixed frame.

We would like Eq. (2.2.10) to become a differential equation as the time increment becomes infinitesimal, $\Delta t \to \mathrm{d}t$. If $r(\Delta t)$ is expanded in a Taylor series, it has the form

$$r = 1 + r'(0)\Delta t + \cdots \tag{2.2.14}$$

(see Exercise 2.7). In this expansion, $r'(0)$ has dimensions of $t^{-1}$, so it defines a correlation time-scale, $r'(0) \equiv 1/T_\mathrm{L}$. We will choose $r$ in (2.2.14) to be

$$r = 1 - \mathrm{d}t/T_\mathrm{L} \tag{2.2.15}$$

as $\Delta t \to \mathrm{d}t$. Correspondingly, $s = \sqrt{(1-r^2)\overline{u^2}} \approx \sqrt{2\,\mathrm{d}t\,\overline{u^2}/T_\mathrm{L}}$.

If (2.2.10) with this expression for $r$ is multiplied by† $u(t-\tau)$ it becomes

$$\overline{u(t+\mathrm{d}t)u(t-\tau)} = (1 - \mathrm{d}t/T_\mathrm{L})\overline{u(t)u(t-\tau)}$$

or

$$R(\mathrm{d}t + \tau) = (1 - \mathrm{d}t/T_\mathrm{L})R(\tau)$$

(recall that, for a stationary process, $\overline{u(t)u(t')} = \overline{u^2}R(|t-t'|)$). As $\mathrm{d}t \to 0$ this becomes

$$\frac{\mathrm{d}R(\tau)}{\mathrm{d}\tau} = -\frac{R(\tau)}{T_\mathrm{L}}. \tag{2.2.16}$$

Finally, solving this with $R(0) = 1$ gives the autocorrelation for the random process (2.2.10):

$$R(t) = \mathrm{e}^{-|\tau|/T_\mathrm{L}}$$

as was sought. $T_\mathrm{L}$ can also be defined as an integral time-scale

$$T_\mathrm{L} = \int_0^\infty R(t')\,\mathrm{d}t'. \tag{2.2.17}$$

In the theory of turbulent dispersion, this is called the *Lagrangian integral time-scale* because (2.2.10) is considered to be the velocity of fluid elements.

This example can be used to introduce the idea of an eddy diffusivity. In fact, the concepts used here to derive the eddy diffusivity are the essence of all attempts to systematically justify eddy transport coefficients, whether they are for material (diffusivity) or momentum (viscosity). Eddy viscosity is the basic idea for many practical mean flow prediction methods.

Consider a particle that is convected by the velocity (2.2.10). Its position, $X(t)$, is found by integrating $\mathrm{d}_t X = u(t)$ with the initial condition $X(0) = 0$:

$$X(t) = \int_0^t u(t')\,\mathrm{d}t'. \tag{2.2.18}$$

† A subtlety: it is necessary to multiply by $u(t-\tau)$ because that is uncorrelated with the future random variable $\xi(t)$, so $\overline{u(t-\tau)\xi(t)} = 0$. If $u(t+\tau)$ were used, this correlation would not be 0.

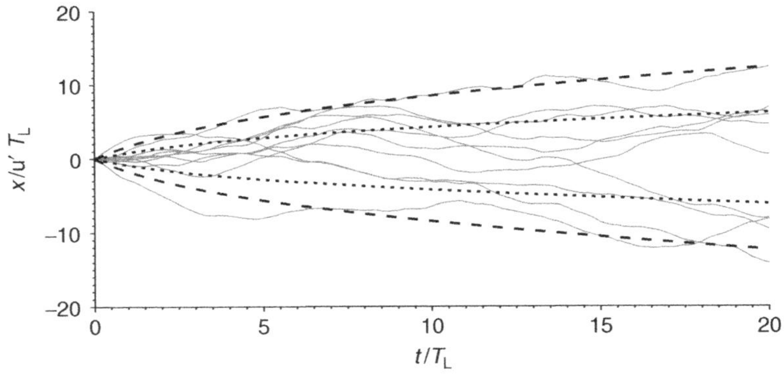


**Figure 2.6** Some trajectories and the root-mean-square spreading described by the Langevin equation: $\sqrt{\overline{X^2}}$ (·········) and $2\sqrt{\overline{X^2}}$ (- - - -).

This is a random trajectory for a given random function $u(t)$. An ensemble of $u$ values generates an ensemble of $X$ values, each of which starts at $X = 0$. A set of such trajectories is illustrated in Figure 2.6 These solutions for $X(t)$ were obtained quite simply by numerically simulating the random equation (2.2.10) along with $X(t + \Delta t) = X(t) + u(t)\Delta t$. The dashed and dotted curves show an exact solution for $\sqrt{\overline{X^2}}$ and $2\sqrt{\overline{X^2}}$.

At any given time, the trajectories constitute a set of random positions, $\{X^{(i)}(t)\}$. These have an associated probability distribution at that time. Suppose that the PDF of $X$ evolves in time according to the diffusion equation

$$\partial_t P(x; t) = \alpha_T \, \partial_x^2 P(x; t), \tag{2.2.19}$$

where $\alpha_T$ is called the *eddy diffusivity*. Now, $\overline{X^2}$ is defined as $\int_{-\infty}^{\infty} P(x)x^2 \, dx$, so an evolution equation for $\overline{X^2}$ can be obtained by multiplying both sides of (2.2.19) by $x^2$ and integrating. The right-hand side is integrated by parts assuming that $P(x) \to 0$ as $x \to \pm\infty$. This gives

$$\frac{d\overline{X^2}}{dt} = 2\alpha_T. \tag{2.2.20}$$

This result shows that the diffusivity is one-half the rate of mean-square spreading of a set of fluid trajectories that start at a common origin. A connection to the correlation function can now be drawn.

Since

$$\alpha_T = \frac{1}{2} \, d_t \overline{X(t)^2} = \overline{X(t) d_t X(t)} = \overline{u(t) X(t)},$$

Eq. (2.2.18) multiplied by $u$ and averaged gives

$$\begin{aligned}\alpha_T &= \overline{u(t)\int_0^t u(t')\,dt'} = \int_0^t \overline{u(t)u(t')}\,dt' \\ &= \overline{u^2}\int_0^t R(t-t')\,dt' = \overline{u^2}\int_0^t R(\tau)\,d\tau \\ &= \overline{u^2}\int_0^t e^{-\tau/T_L}\,d\tau = \overline{u^2}T_L(1-e^{-t/T_L}).\end{aligned} \tag{2.2.21}$$

The second line is valid for any $R(\tau)$; the third was evaluated with the present model. The variance, $\overline{X^2}$, is a simple consequence of (2.2.21): substituting (2.2.21) into (2.2.20) and integrating gives

$$\overline{X^2} = 2\overline{u^2}T_L^2(t/T_L - 1 + e^{-t/T_L}).$$

As $t/T_L \to 0$ this becomes $\overline{X^2} = t^2\overline{u^2}$; that is obviously the correct behavior, because at short times $X \approx u(0)t$ so $\overline{X^2} = \overline{(ut)^2} = t^2\overline{u^2}$.

As $t \to \infty$, (2.2.21) asymptotes to

$$\alpha_T = \overline{u^2}T_L. \tag{2.2.22}$$

This is a very useful formula for turbulence modeling: it is a formula for representing turbulent transport by an effective diffusion coefficient. Its elements are a mean-square turbulent velocity and a correlation time-scale. To use this formula, an operational definition of $\overline{u^2}$ and $T_L$ must be selected, and a model developed for their prediction in general flows. For instance, the $k$–$\varepsilon$ model (Section 6.2) selects $k$ for the velocity scale and $k/\varepsilon$ for the time-scale.

The limit $t \gg T_L$, in which the eddy diffusion approximation applies, is called the Markov limit. Eddy diffusion is formally a Markovian approximation for the random trajectory of fluid elements. In a non-homogeneous turbulent flow, this rationale is translated into a criterion that the scale of non-homogeneity, say $\delta$, should be large compared to the integral scale of the turbulence, $\delta \gg L$. In practical applications, this is not usually satisfied, but eddy transport coefficients can still be effective representations of turbulent mixing.

The concept of the dispersive property of turbulence being represented by an eddy diffusivity dates to the late 19th century; in one form or another, it remains the most widely used turbulence model. The dashed lines in Figure 2.6 show how the eddy diffusivity represents the *averaged* effect of the turbulence, not the *instantaneous* trajectories. The subject of turbulence modeling must be understood in this ensemble-averaged sense. For instance, when alluding to $T$ as an "eddy" time-scale, loose terminology is being used. This is not really an allusion to the instantaneous eddying motion. It is more correct to describe $T$ as a correlation time-scale for two-time statistics. Instantaneous turbulent eddies have only a tangential connection to the idea of an eddy diffusivity; it actually represents the average tendency of material to be dispersed by random, turbulent convection.

We conclude this section with a brief mention of the relation of this material to the subject of stochastic differential equations. Equation (2.2.10) becomes a differential equation upon substituting $r = 1 - \mathrm{d}t/T_\mathrm{L}$, $s = \sqrt{(1-r^2)}\sigma \approx \sqrt{2\,\mathrm{d}t}\sigma$, and letting $\Delta t \to \mathrm{d}t$:

$$u(t+\mathrm{d}t) = (1 - \mathrm{d}t/T_\mathrm{L})u(t) + \sigma\sqrt{\frac{2\,\mathrm{d}t}{T_\mathrm{L}}}\,\xi(t)$$

from (2.2.10). This can be rewritten in terms of differentials by setting $\mathrm{d}u = u(t+\mathrm{d}t) - u(t)$:

$$\mathrm{d}u(t) = \frac{-u(t)\mathrm{d}t}{T_\mathrm{L}} + \sqrt{\frac{2\overline{u^2}}{T_\mathrm{L}}}\,\mathrm{d}W(t), \tag{2.2.23}$$

where $\mathrm{d}W(t) \equiv \sqrt{\mathrm{d}t}\,\xi(t)$. It is standard practice not to divide this by $\mathrm{d}t$ and write the left-hand side as $\mathrm{d}u/\mathrm{d}t$, because $\mathrm{d}W/\mathrm{d}t$ is infinite (with probability 1). In the above, $\mathrm{d}W(t)$ is a white-noise process, which can be understood simply to mean that it represents a continuous function having the properties of a set of independent random variables chosen at intervals of $\mathrm{d}t$ and having variance $\mathrm{d}t$. It is formally defined as the limit of this when $\mathrm{d}t \to 0$. Since its variance tends to zero, it is a continuous function; but since $\sqrt{\overline{\mathrm{d}W^2}}/\mathrm{d}t$ tends to infinity, it is non-differentiable at any time. Equation (2.2.23) is a simple case of an Ito-type stochastic differential equation. It is often referred to as the *Langevin* equation. A general theory of such equations can be found in texts on stochastic differential equations.

#### 2.2.2.2 Spectrum of the correlation function

The idea of the energy spectrum has already been met in Section 2.1. The spectrum is formally the Fourier transform of the correlation function. So if the discussion in Section 2.1 seemed a bit vague, now that the correlation function has been defined, that discussion can be made more formal. However, the material in this section is still quite basic. The spectrum of turbulence plays a central role in Part III of this book.

In (2.1.1) the spatial wavenumber $\kappa$ was used to characterize the length scale of turbulent eddies. As mentioned previously, the term "eddies" is being used in a loose sense, to refer to properties of the statistical correlation function. The formal definition of the spectrum will be applied to the Lagrangian correlation function (2.2.11). Its spectrum, $S(\omega)$, is defined by

$$S(\omega) \equiv \frac{1}{\pi}\int_0^\infty \overline{u(t+\tau)u(t)}\cos(\omega\tau)\,\mathrm{d}\tau = \frac{\overline{u^2}}{2\pi}\int_{-\infty}^\infty R(\tau)\mathrm{e}^{-\mathrm{i}\omega\tau}\,\mathrm{d}\tau. \tag{2.2.24}$$

In the special case $\omega = 0$ this gives

$$S(0) = \frac{1}{\pi}\int_0^\infty \overline{u^2}R(\tau)\,\mathrm{d}\tau = \overline{u^2}T_\mathrm{L}/\pi.$$

Hence, the integral time-scale is related to the spectrum at zero frequency by $T_\mathrm{L} = \pi S(0)/\overline{u^2}$.

The Lagrangian spectrum (2.2.24) has the dimensions $S \sim \ell^2/t$. Invoking Kolmogoroff's reasoning about inertial subrange scaling, the spectrum should follow a $-2$ law: $S \propto \varepsilon\omega^{-2}$. This should not be confused with the Eulerian frequency spectrum, measured with a fixed probe: that corresponds to a spatial spectrum through Taylor's hypothesis (Section 2.1.1) and hence has the inertial-range behavior $\varepsilon^{2/3}\omega^{-5/3}$.

For the stochastic process described previously in this section, $R(\tau) = \mathrm{e}^{-\tau/T_\mathrm{L}}$. Its spectrum is

$$S(\omega) = \frac{\overline{u^2}}{\pi}\int_0^\infty \mathrm{e}^{-\tau/T_\mathrm{L}} \cos(\omega\tau)\,\mathrm{d}\tau = \frac{\overline{u^2}T_\mathrm{L}}{\pi(1+\omega^2T_\mathrm{L}^2)}.$$

The relation $\pi S(0) = \overline{u^2}T_\mathrm{L}$ is satisfied by this example, as it must be. At large $\omega$, $S \to \overline{u^2}/\pi T_\mathrm{L}\omega^2$. Comparing this to the inertial-range scaling shows that $\overline{u^2}/T_\mathrm{L} \propto \varepsilon$. This gives a theoretical justification for the observation that $k/\varepsilon$ is a suitable estimate of the turbulence time-scale. To this end the velocity scale $\overline{u^2} \sim k$ can be used to conclude that Taylor's integral time-scale and Kolmogoroff's inertial-range scaling are consistent if $T_\mathrm{L} \propto k/\varepsilon$. Many of the closure models described in Part II of this book invoke $k/\varepsilon$ as a time-scale.

Inverting the relation (2.2.24) shows that the correlation function is the transform of the spectrum:

$$\overline{u^2}R(\tau) = \int_{-\infty}^\infty S(\omega)\cos(\omega\tau)\,\mathrm{d}\omega. \tag{2.2.25}$$

For $\tau = 0$ this becomes $\overline{u^2} = \int_{-\infty}^\infty S(\omega)\,\mathrm{d}\omega$, where $S(\omega)d\omega$ represents the portion of the variance lying in the differential frequency band between $\omega + \frac{1}{2}\,\mathrm{d}\omega$ and $\omega - \frac{1}{2}\,\mathrm{d}\omega$; so actually, $S(\omega)$ is the *spectral energy density*.

We are now in a position to explain the representation (2.1.1) more formally. The velocity of a fluid element consists of contributions from large and small scales. Consider two particles separated by a small distance. A large eddy will convect the two particles with approximately the same velocity. It causes very little relative motion. Small-scale eddies, on the other hand, do contribute a relative velocity. The contribution of small-scale eddies is isolated by evaluating the velocity difference between two points separated by a small distance: $u_{\text{small-scale}} = u(x + r) - u(x)$ when $r \ll L$. The small-scale intensity is represented by

$$\tfrac{1}{2}\overline{[u(x+r) - u(x)]^2} = \tfrac{1}{2}[\overline{u(x+r)^2} + \overline{u(x)^2}] - \overline{u(x+r)u(x)} = \overline{u^2}[1 - R(r)],$$

using homogeneity to substitute $\overline{u(x+r)^2} = \overline{u(x)^2} = \overline{u^2}$. By analogy to (2.2.25)

$$\overline{u^2} - \overline{u^2}R(r) = \int_{-\infty}^\infty E(\kappa)\,\mathrm{d}\kappa - \int_{-\infty}^\infty E(\kappa)\cos(\kappa r)\,\mathrm{d}\kappa = 2\int_0^\infty E(\kappa)[1 - \cos(\kappa r)]\,\mathrm{d}\kappa.$$

This was used in (2.1.1). Again, it can be seen that allusions to small-scale eddies are actually a loose reference to a statistical property of the turbulence. In this case it is the mean square of the velocity difference between two nearby points. When the separation is in the inertial range, $\overline{[u(x+r) - u(x)]^2} \propto \varepsilon^{1/3}r^{2/3}$ according to inertial-range scaling.

## 2.3 Cartesian tensors

Various levels of constitutive modeling are used in turbulent flow. The simplest is to assume that the stress tensor is proportional to the rate-of-strain tensor. This is the linear eddy viscosity model; more correctly, it is a tensorally linear relation. Mathematically, if $\tau_{ij}$ is the stress and $S_{ij}$ is the rate of strain, then the linear constitutive model is $\tau_{ij} - \frac{1}{3}\delta_{ij}\tau_{kk} = \nu_T S_{ij}$, where $\nu_T$ is the eddy viscosity. Tensoral linearity means the free indices, $i, j$, on the right-hand side are subscripts of a single tensor, not of a matrix product. The functional dependence of the coefficient of eddy viscosity might require a solution of highly nonlinear equations, or it might be a function of the magnitude of the rate of strain: the stress–strain relation would still be referred to as (tensorally) linear. As long as it is only $S_{ij}$ and not $S_{ik}S_{kj}$ on the right, this is a linear constitutive model.

The convention of implied summation on repeated indices is used throughout this book. For instance, $S_{ik}S_{kj}$ is understood to equal $\sum_{k=1}^{3} S_{ik}S_{kj}$, which is the index form of the matrix product $\boldsymbol{S} \cdot \boldsymbol{S}$. Any time an index is repeated in a product it is understood to be summed. By this convention $S_{ij}S_{ji}$ is summed on both $i$ and $j$:

$$S_{ij}S_{ji} \equiv \sum_{j=1}^{3}\sum_{i=1}^{3} S_{ij}S_{ji}.$$

This is the index form for the trace of the product: $S_{ij}S_{ji} = \text{trace}[\boldsymbol{S} \cdot \boldsymbol{S}]$.

Tensor analysis provides a mathematical framework for constitutive modeling and for other aspects of single-point closure modeling. Indeed, the matrix of velocity correlations, given in (2.2.8), has already been alluded to as the Reynolds stress tensor. The full power of tensor analysis is described in the literature on continuum mechanics (Eringen, 1980). There it is used to develop constitutive relations for material properties. A turbulence model represents properties of the fluid *motion*, not material properties of the fluid. In the present application, tensor analysis is used to restrict the possible forms of the model, rather than to develop mechanical laws. It is an analytical tool. The governing laws are the Navier–Stokes equations.

A tensor consists of an indexed array and a coordinate system. Although the indexed array, such as $\overline{u_i u_j}$, is usually referred to as a tensor, the elements of this array are actually the components of the tensor in a particular coordinate system.

Consider a set of unit vectors, $\boldsymbol{e}^{(i)}$, that point along the axes of a Cartesian coordinate system. For example, the standard $x,y,z$ coordinate system is described by the unit vectors $\boldsymbol{e}^{(1)} = (1, 0, 0)$, $\boldsymbol{e}^{(2)} = (0, 1, 0)$, and $\boldsymbol{e}^{(3)} = (0, 0, 1)$. The superscripts are in brackets to indicate that they denote a member of the set of vectors, not the exponentiation of $\boldsymbol{e}$. Then a second-order tensor in three dimensions is defined by

$$\boldsymbol{T} = \sum_{j=1}^{3}\sum_{i=1}^{3} T_{ij}\boldsymbol{e}^{(i)}\boldsymbol{e}^{(j)} \equiv T_{ij}\boldsymbol{e}^{(i)}\boldsymbol{e}^{(j)}. \tag{2.3.1}$$

The complete expression (2.3.1) defines the tensor. This illustrates the respect in which $T_{ij}$ is simply the $i,j$ component of the tensor, and the respect in which a complete characterization requires both the components and the coordinate system to be stated.

A zeroth-order tensor, or a scalar, does not depend on the coordinate system; an example is pressure, which has no associated direction. The *contraction* (or trace) of

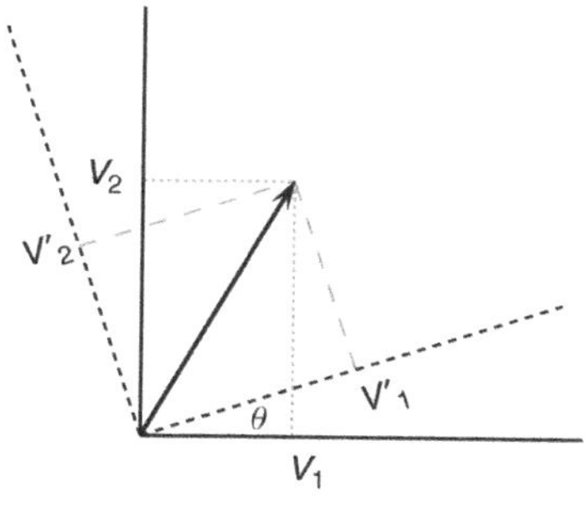


**Figure 2.7** The components of a fixed vector in two different coordinate systems.

a second order tensor is also a scalar; an example is $T_{ii} = T_{11} + T_{22} + T_{33}$. Tensors whose components do not depend on the coordinate system are termed *invariant* tensors; their components are invariant under coordinate transformations. Zeroth-order tensors are always invariant. A vector is a first-order tensor, $\boldsymbol{V} = V_i \boldsymbol{e}^{(i)}$. Its components depend on the coordinate system (see Figure 2.7).

In (2.3.1) $\boldsymbol{T}$ is the tensor, but it is common to allude to the component notation $T_{ij}$ as a second-order tensor. The order corresponds to the number of free subscripts. The tensor itself is the same in all coordinate systems, but its components will be different in different coordinate systems. This is illustrated in Figure 2.7 for the case of a first-order tensor. The figure shows how the coordinates of a fixed vector change relative to two coordinate systems at an angle of $\theta$ to one another. If the vector has components $(V_1, V_2)$ in the first system, then it has components $(V_1', V_2') = (V_1 \cos\theta + V_2 \sin\theta, V_2 \cos\theta - V_1 \sin\theta)$ in the rotated system. That is the transformation that keeps $\boldsymbol{V}$ (the arrow in the figure) unchanged; the tensor, $\boldsymbol{V}$, is independent of the coordinate system; its components are not.

To put this algebraically, consider a unit vector in each of the two frames, $\boldsymbol{e}$ and $\tilde{\boldsymbol{e}}$. The two coordinate frames are related by a linear transformation: $\boldsymbol{e}^{(i)} = E_{ij}\tilde{\boldsymbol{e}}^{(j)}$, where $\boldsymbol{E}$ is the rotation matrix:

$$\boldsymbol{E}(\theta) = \begin{pmatrix} \cos\theta & -\sin\theta & 0 \\ \sin\theta & \cos\theta & 0 \\ 0 & 0 & 1 \end{pmatrix}. \tag{2.3.2}$$

The components of the vector can be found in these coordinate systems by applying the transformation $\boldsymbol{V} = V_i \boldsymbol{e}^{(i)} = V_i E_{ij} \tilde{\boldsymbol{e}}^{(j)}$. Hence $\tilde{V}_j = V_i E_{ij}$ are the components of the given vector in the $\sim$ coordinate system.

Since $\boldsymbol{E}$ represents rotation through an angle, $\theta$, the inverse of $\boldsymbol{E}(\theta)$ is rotation through the angle $-\theta$, $\boldsymbol{E}(-\theta)$. But (2.3.2) shows that $\boldsymbol{E}(-\theta)$ is just the transpose of $\boldsymbol{E}(\theta)$: $E_{ij}(-\theta) = E_{ji}(\theta)$. Thus $E_{ij}(\theta)E_{jk}(-\theta) = E_{ij}(\theta)E_{kj}(\theta) = \delta_{ik}$ where $\delta_{ik}$ represents the components of the identity matrix:

$$\delta_{ik} = 1, \text{ if } i = k; \qquad \delta_{ik} = 0, \text{ if } i \neq k.$$

In other words, $V_i = E_{ij}(E_{kj}V_k) = \delta_{ik}V_k = V_i$ represents rotation through an angle $\theta$, then back through $-\theta$, leaving $V_i$ unchanged.

A similar transformation rule applies to a general tensor. For instance, a second-order tensor transforms according to

$$\boldsymbol{T} = T_{ij}\boldsymbol{e}^{(i)}\boldsymbol{e}^{(j)} = T_{ij}E_{ik}E_{jl}\tilde{\boldsymbol{e}}^{(k)}\tilde{\boldsymbol{e}}^{(l)}, \tag{2.3.3}$$

so the components of $\boldsymbol{T}$ change to $\tilde{T}_{ij} = T_{kl}E_{ki}E_{lj}$ in the $\sim$ coordinate system. (Note that $i, j, k, l$ are dummy indices, so this expression is identical to $\tilde{T}_{kl} = T_{ij}E_{ik}E_{jl}$.) This is the transformation rule for second-order Cartesian tensors. The requirement that the components of the tensor change in a prescribed way under coordinate transformation is an embodiment of the condition that the tensor itself be independent of the coordinate system.

The laws of physics are the same in any coordinate system, so it is a sensible principle to require that any model for turbulent statistics should comply with the requirement of coordinate independence. A corollary is that the model should be of a tensorally correct form. To pursue this further requires discussion of tensor functions of tensors, which will be provided shortly.

### 2.3.1 Isotropic tensors

The notion of isotropy plays a major role in turbulence theory. Isotropic turbulence is the idealized state of turbulent motion in which the statistics are independent of the coordinate orientation. Although exactly isotropic turbulence cannot be produced in an experiment, it is approximated by the turbulence downstream of a grid placed into a uniform stream (Figure 1.2, page 6); exactly isotropic turbulence can be simulated on a computer.

Isotropic tensors have the exceptional property that their components are the same in all coordinate systems. Figure 2.7 makes it clear that there is no such thing as an isotropic vector; the components of $\boldsymbol{V}$ change with the angle $\theta$. However, any multiple of $\delta_{ij}$ is an isotropic second-order tensor because of the transformation law (2.3.3) for the second-order tensors. Transforming $\boldsymbol{\delta}$ gives $\tilde{\delta}_{ij} = \delta_{kl}E_{ki}E_{lj} = E_{li}E_{lj} = \delta_{ij}$; the identity tensor is unchanged by coordinate transformation. An obvious corollary is that $A\delta_{ij}$, for any scalar $A$, is isotropic.

The physical interpretation of an isotropic tensor is that it describes a process that has no directional preferences, and hence its components are unchanged when viewed in a rotated system of coordinate axes. For instance, rotation through 90° transforms $x \to y$ and $y \to -x$. Velocities are transformed correspondingly: $u \to v$ and $v \to -u$. The velocity covariance $\overline{uv}$ is thereby transformed to $\overline{-vu}$. Thus, invariance of statistics under 90° rotation demands that $-\overline{vu} = \overline{uv}$. Adding $\overline{uv}$ to both sides of this shows that $\overline{uv} = 0$ if the turbulence is isotropic. Similarly, $\overline{u^2} = \overline{v^2}$. Following this argument for rotations about other axes leads to the conclusion that the isotropic Reynolds stress tensor (2.2.8) must be proportional to the identity tensor, $\boldsymbol{\delta}$.

Since there is no such thing as an isotropic vector, the correlation between pressure, which is a zeroth-order tensor, and the velocity vector must vanish: $\overline{pu_i} = 0$ for all $i$. For instance, a rotation through 180° transforms $u$ to $-u$, so isotropy would require that $\overline{pu} = -\overline{pu} = 0$.

In turbulence theory it is common to add the condition that the components of an isotropic tensor are unchanged by reflections,‡ as well as by rotations. If this additional

‡ The transformation matrix for reflection in the $y$–$z$ plane is $\boldsymbol{E} = \mathrm{diag}[-1, 1, 1]$.

requirement is made, then there are no isotropic third-order tensors. To see why, suppose there were an isotropic tensor

$$\mathcal{I} = \mathcal{I}_{ijk}\boldsymbol{e}^{(i)}\boldsymbol{e}^{(j)}\boldsymbol{e}^{(k)}.$$

Reflection in a coordinate plane transforms $\boldsymbol{e}^{(\alpha)}$ to $-\boldsymbol{e}^{(\alpha)}$, where $\alpha$ is the direction normal to the plane of reflection. This transformation gives $\tilde{\mathcal{I}}_{\alpha jk} = -\mathcal{I}_{\alpha jk}$ if $j, k \neq \alpha$ or if both $j$ and $k$ equal $\alpha$. For such components, isotropy, $\tilde{\mathcal{I}}_{\alpha jk} = \mathcal{I}_{\alpha jk}$, then requires that $\mathcal{I}_{\alpha jk} = -\mathcal{I}_{\alpha jk}$; hence, such components must vanish. This reflection would not prevent a non-zero component $\mathcal{I}_{\alpha\alpha k}$ if $k \neq \alpha$. However, reflection in the plane perpendicular to the $k$ direction demands that this component vanish as well. It follows that there are no odd-order tensors that are invariant under reflection. So the only tensors with reflectionally invariant components are of even order.§

Isotropy of the fourth-order tensor

$$\mathcal{I} = \mathcal{I}_{ijkl}\boldsymbol{e}^{(i)}\boldsymbol{e}^{(j)}\boldsymbol{e}^{(k)}\boldsymbol{e}^{(l)}$$

requires that the only non-zero components of $\mathcal{I}_{ijkl}$ be those for which the indices are pairwise equal. This is expressed by the $\delta_{ij}$ tensor as

$$\mathcal{I}_{ijkl} = A\delta_{ij}\delta_{kl} + B\delta_{ik}\delta_{jl} + C\delta_{il}\delta_{jk}. \tag{2.3.4}$$

In general, even-order isotropic tensors consist of a sum of all distinct combinations of the $\delta_{ij}$ that give the requisite number of indices (Goodbody, 1982).

An exchange of directions, such as $x \leftrightarrow y$, is a combination of rotation and reflection. An isotropic tensor must be unchanged by such exchanges: for example, $\mathcal{I}_{1122} = \mathcal{I}_{1133} = \mathcal{I}_{2233} = \cdots$. In the representation (2.3.4), these are all equal to $A$. There is no need for $\mathcal{I}_{1122}$ to equal $\mathcal{I}_{1212}$; but $\mathcal{I}_{1212} = \mathcal{I}_{1313} = \mathcal{I}_{2323} = \cdots$, all of which equal $B$ in (2.3.4).

### 2.3.2 Tensor functions of tensors; Cayley–Hamilton theorem

Second-moment closure modeling requires that certain unknown tensors be represented by functions of anisotropy and identity tensors. In constitutive modeling and in equilibrium analysis, the Reynolds stress is a tensor function of the rate-of-strain and rate-of-rotation tensors. In other words, tensor-valued functions of tensor arguments arise: $\phi_{ij} = F_{ij}(a_{kl}, \delta_{kl})$ and $\tau_{ij} = G_{ij}(S_{kl}, \Omega_{kl})$. Tensoral consistency demands that the free indices be $i, j$ on both sides of these equations.

The question that arises is this: What constraints can be placed on possible forms of the functions $F_{ij}$ and $G_{ij}$? The simple answer is that, if there are no hidden arguments, then they must be isotropic functions of their arguments. A hidden argument might be a preferential direction, such as that of gravity. In such cases, it is only necessary that the distinguished direction be made explicit; then the function must be isotropic in its arguments.

The definition of an isotropic Cartesian tensor function is that it is covariant with rotation. For a second-order tensor this is expressed by

$$\tilde{F}_{ij} = F_{ij}(\tilde{a}_{kl}, \delta_{kl}); \tag{2.3.5}$$

§ If the reflection invariance is not added, then the skew symmetric tensor $\varepsilon_{ijk}$ is isotropic. This tensor is defined by $\varepsilon_{ijk} = 0$ if any of $i, j, k$ are equal, $\varepsilon_{ijk} = -\varepsilon_{jik}$ and $\varepsilon_{123} = 1$.

or, in terms of the transformation matrix, by

$$E_{ik}E_{jl}F_{kl}(a_{kl}, \delta_{kl}) = F_{ij}(E_{km}E_{ln}a_{mn}, \delta_{kl}).$$

This is the mathematical statement that the functional form must be the same in any coordinate system: for example, if $\boldsymbol{F} = C\boldsymbol{a}^2$, then $\tilde{\boldsymbol{F}} = C\tilde{\boldsymbol{a}}^2$. The requirement (2.3.5) is that the laws of physics be independent of coordinate orientation.

Note that any matrix product of the $a$ satisfies (2.3.5). For instance, if $F(\boldsymbol{a}) = \boldsymbol{a}^2$, or in component form $F_{ij}(a_{kl}) = a_{in}a_{nj}$, then

$$\begin{aligned}\tilde{F}_{ij}(\boldsymbol{a}) &= E_{ik}E_{jl}F_{kl} = E_{ik}a_{km}a_{ml}E_{jl}\\ &= E_{ik}a_{km}E_{nm}E_{np}a_{pl}E_{jl} = (E_{ik}E_{nm}a_{km})(E_{np}E_{jl}a_{pl})\\ &= \tilde{a}_{in}\tilde{a}_{nj} = F_{ij}(\tilde{\boldsymbol{a}}),\end{aligned}$$

where $E_{nm}E_{np} = \delta_{mp}$ was used in the second line. The same reasoning applies to any power of $\boldsymbol{a}$. This suggests that an isotropic function of a tensor might be written as a linear combination of products of powers of $a_{ij}$ and $\delta_{ij}$. As an example

$$\tilde{F}_{ij} = Aa_{ij} + Ba^2_{ij} + ca^4_{ij} \tag{2.3.6}$$

would qualify. Here $a^4_{ij}$ is shorthand for the matrix product $a_{ik}a_{kl}a_{lm}a_{mj}$. A combination of all powers of $\boldsymbol{a}$ provides the most general tensor function, with the qualification that coefficients like $A$, $B$, and $C$ in the above equation can be functions of the *invariants* of the tensor $\boldsymbol{a}$ – invariants will be defined shortly. Fortunately, this power series terminates after a few terms because higher powers of $\boldsymbol{a}$ can be written as linear combinations of lower powers: that is the content of the *Cayley–Hamilton theorem*.

The Cayley–Hamilton theorem (Goodbody, 1982) says that, in $n$ dimensions, $a^n_{ij}$ is a linear combination of lower powers of $a$. In three dimensions, the explicit formula is

$$a^3_{ij} = III_a\delta_{ij} - II_a a_{ij} + I_a a^2_{ij}, \tag{2.3.7}$$

where $III_a$, $II_a$, and $I_a$ are the invariants of $\boldsymbol{a}$. Note that, if (2.3.7) is multiplied by $\boldsymbol{a}$ and then used to eliminate $\boldsymbol{a}^3$, a formula relating $\boldsymbol{a}^4$ to $\boldsymbol{a}^2$, $\boldsymbol{a}$, and $\boldsymbol{\delta}$ is obtained:

$$\boldsymbol{a}^4 = III_a\boldsymbol{a} - II_a\boldsymbol{a}^2 + I_a\boldsymbol{a}^3 = III_a\boldsymbol{a} - II_a\boldsymbol{a}^2 + I_a(III_a\boldsymbol{\delta} - II_a\boldsymbol{a} + I_a\boldsymbol{a}^2).$$

The same reasoning applies to all powers higher than 3; hence it follows from (2.3.7) that the most general form of (2.3.6) is a linear combination of multiples of $\delta_{ij}$, $a_{ij}$, and $a^2_{ij}$, with no higher powers of $\boldsymbol{a}$:

$$\tilde{F}_{ij} = A\delta_{kl} + Ba_{ij} + Ca^2_{ij}. \tag{2.3.8}$$

This represents the most general tensoral dependence for an isotropic second-order function of a second-order tensor. The coefficients of the sum can be arbitrary functions of scalar parameters, such as the invariants of $\boldsymbol{a}$.

As another corollary to (2.3.7), the inverse of the matrix $\boldsymbol{a}$ is given by

$$a^{-1}_{ij} = (a^2_{ij} + II_a\delta_{ij} - I_a a_{ij})/III_a.$$

This is proved by multiplying (2.3.7) by $\boldsymbol{a}^{-1}$.

To prove (2.3.7), assume that the matrix $\boldsymbol{a}$ has three independent eigenvectors with eigenvalues $\lambda_1$, $\lambda_2$, and $\lambda_3$. Then any vector $\mathbf{x}$ can be written as a linear combination of the three eigenvectors $\boldsymbol{x} = a\boldsymbol{\xi}_1 + b\boldsymbol{\xi}_2 + c\boldsymbol{\xi}_3$. It follows from this that the product

$$(\boldsymbol{a} - \lambda_1\boldsymbol{\delta}) \cdot (\boldsymbol{a} - \lambda_2\boldsymbol{\delta}) \cdot (\boldsymbol{a} - \lambda_3\boldsymbol{\delta}) \cdot \boldsymbol{x} = 0$$

vanishes for any $\boldsymbol{x}$ because one of the bracketed terms annihilates each of the $\boldsymbol{\xi}_i$. Because this is true for an arbitrary vector, the product of matrices itself must vanish:

$$(\boldsymbol{a} - \lambda_1\boldsymbol{\delta}) \cdot (\boldsymbol{a} - \lambda_2\boldsymbol{\delta}) \cdot (\boldsymbol{a} - \lambda_3\boldsymbol{\delta}) = 0.$$

Expanding this product gives

$$a^3_{ij} = III_a\delta_{ij} - II_a a_{ij} + I_a a^2_{ij}$$

in index notation, where $III_a = \lambda_1\lambda_2\lambda_3$, $II_a = \lambda_1\lambda_2 + \lambda_1\lambda_3 + \lambda_2\lambda_3$, and $I_a = \lambda_1 + \lambda_2 + \lambda_3$. This proves (2.3.7).

Note that $I_a = a_{kk}$ because the sum of the eigenvalues equals the trace of the matrix. The other invariants can also be related to matrix traces. Taking the trace of Eq. (2.3.7) shows that $3III_a = a^3_{kk} + II_a I_a - a^2_{kk} I_a$. Expanding the equation $(\lambda_1 + \lambda_2 + \lambda_3)^2 = I_a^2$ shows that $2II_a = I_a^2 - a^2_{kk}$, after noting that the sum of squares of the eigenvalues equals the trace of the matrix squared: $a^2_{kk} = \lambda_1^2 + \lambda_2^2 + \lambda_3^2$. So the invariants of $\boldsymbol{a}$ can be written directly in terms of traces of powers of the matrix as:

$$I_a = a_{kk}, \qquad II_a = -\tfrac{1}{2}(a^2_{kk} - I_a^2), \qquad (a^3_{kk} + II_a I_a - a^2_{kk} I_a). \tag{2.3.9}$$

The $I_a$, $II_a$, and $III_a$ are called the first, second, and third *principal invariants* of the matrix. As the name implies, they are the same in any coordinate system.

The special case of (2.3.7) for a trace-free matrix, $\boldsymbol{b}$, arises in second-moment closure modeling. The trace-free condition is $I_b = 0$ so

$$\boldsymbol{b}^3 = \tfrac{1}{3}\, b^3_{kk}\boldsymbol{\delta} + \tfrac{1}{2}\, b^2_{kk}\boldsymbol{b}, \tag{2.3.10}$$

where the principal invariants $II_b = -\frac{1}{2}\, b^2_{kk}$ and $III_b = \frac{1}{3}\, b^3_{kk}$ have been substituted.

A quite useful observation is that the Cayley–Hamilton theorem is a device to solve matrix equations (Pope, 1975). The approach can be illustrated by considering the problem of solving an equation for a tensor function of a given tensor, $\boldsymbol{S}$. Let the dependent variable be $a_{ij}$ and suppose we are asked to solve the equation

$$\boldsymbol{a} = \boldsymbol{a} \cdot \boldsymbol{S} + \boldsymbol{S} \cdot \boldsymbol{a} + \boldsymbol{S} \tag{2.3.11}$$

for $\boldsymbol{a}$. The right-hand side of this equation contains both left and right matrix multiplications, so it cannot be solved directly by matrix inversion. The Cayley–Hamilton theorem proves that the most general function $\boldsymbol{a}(\boldsymbol{S})$ is of the form $\boldsymbol{a} = A\boldsymbol{\delta} + B\boldsymbol{S} + C\boldsymbol{S}^2$, where $A$, $B$, and $C$ are functions of the invariants of $\boldsymbol{S}$. Substituting this form into the example (2.3.11) gives

$$\begin{aligned} A\boldsymbol{\delta} + B\boldsymbol{S} + C\boldsymbol{S}^2 &= 2(A\boldsymbol{S} + B\boldsymbol{S}^2 + C\boldsymbol{S}^3) + \boldsymbol{S} \\ &= 2[A\boldsymbol{S} + B\boldsymbol{S}^2 + C(III_s\boldsymbol{\delta} - II_s\boldsymbol{S} + I_s\boldsymbol{S}^2)] + \boldsymbol{S}, \end{aligned}$$

where (2.3.7) was applied to $S^3$. Equating separately the coefficients of $S$, $S^2$, and $\delta$ on both sides of this equation gives

$$A = 2III_s C, \qquad B = 2A - 2II_s C + 1, \qquad C = 2B/(1 - 2I_s).$$

Solving for $A$, $B$, and $C$ gives

$$\boldsymbol{a} = \frac{4III_s \boldsymbol{\delta} + (1 - 2I_s)\boldsymbol{S} + 2\boldsymbol{S}^2}{1 - 2I_s - 8III_s + 4II_s}$$

as the solution to (2.3.11) – provided the solvability condition $1 - 2I_s - 8III_s + 4II_s \neq 0$ is met. The solvability condition is analogous to the determinant condition for matrix inversion.

#### 2.3.2.1 Tensor functions of two tensors

In constitutive modeling the stress tensor can be a function of two tensors, $\boldsymbol{S}$ and $\boldsymbol{\Omega}$, in addition to $\boldsymbol{\delta}$. The principle of material frame indifference demands that *material properties* are independent of the rate-of-rotation tensor $\boldsymbol{\Omega}$, but *turbulent stresses* are most definitely affected by rotation. This is because they are not material properties, but are properties of the flow.

The development of representation theorems in three dimensions is quite cumbersome, so we will first address the two-dimensional case. Although it is not obvious at this point, the two-dimensional formulas are useful in three-dimensional turbulent flow; indeed, they are far more helpful than are the three-dimensional formulas. The two-dimensional Cayley–Hamilton theorem is

$$a_{ij}^2 = I_a a_{ij} - II_a \delta_{ij}, \tag{2.3.12}$$

where $I_a = a_{kk}$ and $II_a = -\frac{1}{2}(a_{kk}^2 - I_a^2)$, with $i = 1, 2$ and $j = 1, 2$. It follows that the square of a trace-free tensor is proportional to the identity:

$$\boldsymbol{a}^2 = \tfrac{1}{2}\, a_{kk}^2 \boldsymbol{\delta} \tag{2.3.13}$$

if $I_a = 0$.

Let a tensor be a function of two trace-free tensors: $\boldsymbol{\tau} = \mathcal{F}(\boldsymbol{A}, \boldsymbol{B})$. By the Cayley–Hamilton theorem (2.3.12), the most general isotropic function can depend tensorally on $\boldsymbol{A}$, $\boldsymbol{B}$, and $\boldsymbol{A}^2$ (and their transposes), where the last is used in place of the identity, in accord with (2.3.13). However, it can also depend on products of these. It can be shown that the most general such dependence is on only $\boldsymbol{A} \cdot \boldsymbol{B}$. To show this, it must be demonstrated that higher products, such as $\boldsymbol{A} \cdot \boldsymbol{B} \cdot \boldsymbol{A} \cdot \boldsymbol{B}$, can be reduced to products of at most two tensors. The number of tensors in the product is called its extension; so the claim is that the most general tensor product is of extension 2 or less. The proof is by constructing a sequence of products of increasing extension. At each level any tensor of lower extension can be ignored because it has already been included.

Substituting $\boldsymbol{a} = \boldsymbol{A} + \boldsymbol{B}$ into (2.3.13) and using the fact that both $\boldsymbol{A}$ and $\boldsymbol{B}$ satisfy (2.3.13) gives

$$\boldsymbol{B} \cdot \boldsymbol{A} + \boldsymbol{A} \cdot \boldsymbol{B} = (A_{kj} B_{jk})\boldsymbol{\delta}.$$

Since $\boldsymbol{\delta}$ is a product of extension 1, this can be stated as

$$\boldsymbol{B} \cdot \boldsymbol{A} = -\boldsymbol{A} \cdot \boldsymbol{B} + \text{l.o.e.}, \tag{2.3.14}$$

where "l.o.e." stands for lower-order extension. All possible tensors of extension 2 are: $\boldsymbol{A}^2$, $\boldsymbol{B}^2$, $\boldsymbol{A} \cdot \boldsymbol{B}$, and $\boldsymbol{B} \cdot \boldsymbol{A}$. The first two are proportional to $\boldsymbol{\delta}$ and so are of extension 1, while the last is proportional to the third, plus a lower-order extension, by (2.3.14). Hence $\boldsymbol{A} \cdot \boldsymbol{B}$ is the only unique tensor of extension 2.

A tensor of extension 3 cannot contain $\boldsymbol{A}^2$ or $\boldsymbol{B}^2$, because those can be replaced by the identity, so it must be something like $\boldsymbol{A} \cdot \boldsymbol{B} \cdot \boldsymbol{A}$, which equals

$$-\boldsymbol{A} \cdot \boldsymbol{A} \cdot \boldsymbol{B} + \text{l.o.e.} = II_A \boldsymbol{B} + \text{l.o.e.} = \text{l.o.e.}$$

by (2.3.14) and (2.3.13); in other words, it is not a tensor of extension 3. By this reasoning, all products of more than two tensors can be seen to be reducible to extensions of order 2 or lower. Essentially, this is because (2.3.14) permits permutation of the tensors (+ l.o.e.) and thereby a reduction of the extension.

The complete set of independent tensors ($\boldsymbol{A}, \boldsymbol{B}, \boldsymbol{A} \cdot \boldsymbol{B}, \boldsymbol{A}^2$) is termed the integrity basis (Spencer and Rivlin, 1959). Even though only one of $\boldsymbol{A} \cdot \boldsymbol{B}$ or $\boldsymbol{B} \cdot \boldsymbol{A}$ is needed, the representation theorem for a symmetric tensor is usually written as

$$\tau = c_1 \boldsymbol{A} + c_2 \boldsymbol{B} + c_3 [\boldsymbol{A} \cdot \boldsymbol{B} - \boldsymbol{B} \cdot \boldsymbol{A}] + c_4 \boldsymbol{A}^2. \tag{2.3.15}$$

Symmetry means $\tau_{ij} = \tau_{ji}$. This form shows that the $c_3$ term vanishes if $\boldsymbol{A}$ and $\boldsymbol{B}$ are symmetric. But the case of interest here is where $\boldsymbol{A}$ is symmetric (the rate of strain, $\boldsymbol{S}$) and $\boldsymbol{B}$ is antisymmetric (the rate of rotation, $\boldsymbol{\Omega}$); then $c_2$ must be zero. The use of representation (2.3.15) in solving tensor equations is illustrated by Exercise 2.11.

In constitutive modeling for three-dimensional turbulent stresses in a two-dimensional mean flow, (2.3.15) represents stresses in the plane of the flow. The third dimension is added to the stress tensor by including the three-dimensional identity tensor:

$$\tau = a_1 \boldsymbol{S} + a_2 [\boldsymbol{S} \cdot \boldsymbol{\Omega} - \boldsymbol{\Omega} \cdot \boldsymbol{S}] + a_3 \boldsymbol{S}^2 + a_4 \boldsymbol{\delta}. \tag{2.3.16}$$

This idea will be developed in context later in the book (Section 8.3).

A three-dimensional analog to (2.3.15) can be developed by the same method of examining successive extensions and finding those that are unique up to l.o.e. However, the three-dimensional formula is less useful to turbulence modeling than is (2.3.15). A derivation of the requisite representation theorems can be found in Spencer and Rivlin (1959); we simply quote the result for the case of interest here. If $\boldsymbol{S}$ is symmetric and $\boldsymbol{\Omega}$ is antisymmetric, then the most general symmetric isotropic tensor function is

$$\begin{aligned} \tau = {} & c_0 \boldsymbol{\delta} + c_1 \boldsymbol{S} + c_2 \boldsymbol{S}^2 + c_3 \boldsymbol{\Omega}^2 + c_4 (\boldsymbol{S\Omega} - \boldsymbol{\Omega S}) + c_5 (\boldsymbol{S}^2 \boldsymbol{\Omega} - \boldsymbol{\Omega S}^2) \\ & + c_6 (\boldsymbol{S\Omega}^2 + \boldsymbol{\Omega}^2 \boldsymbol{S}) + c_7 (\boldsymbol{\Omega}^2 \boldsymbol{S}^2 + \boldsymbol{S}^2 \boldsymbol{\Omega}^2) + c_8 (\boldsymbol{S\Omega S}^2 - \boldsymbol{S}^2 \boldsymbol{\Omega S}) \\ & + c_9 (\boldsymbol{\Omega S \Omega}^2 - \boldsymbol{\Omega}^2 \boldsymbol{S\Omega}) + c_{10} (\boldsymbol{\Omega S}^2 \boldsymbol{\Omega}^2 - \boldsymbol{\Omega}^2 \boldsymbol{S}^2 \boldsymbol{\Omega}) \end{aligned} \tag{2.3.17}$$

in three dimensions. Each term of this sum was formed by adding an element of the integrity basis to its transpose, eliminating any that result in l.o.e. The irreducible representation has 11 terms. The reason why (2.3.17) is of limited use in turbulence modeling is because of its length and nonlinearity. When used in numerical computations, such a constitutive model would be expensive to evaluate and would cause numerical stiffness. Sometimes truncated versions of (2.3.17) have been proposed as constitutive models.

# Exercises

**Exercise 2.1.** *Dissipation-range scaling.* In the legend of Figure 2.1 $R_\lambda$ is the Reynolds number based on Taylor's microscale (see Exercise 2.5). Let $R_\lambda = R_T^{1/2}$. Assuming that the energetic range begins where the data leave the $-5/3$ line, do these data roughly confirm the $R_T$ scaling of $\eta/L$? (Note that the scaling and estimates give proportionalities, not equalities.)

**Exercise 2.2.** *DNS.* One application of dimensional analysis is to estimating the computer requirements for direct numerical simulation of turbulence. The computational mesh must be fine enough to resolve the smallest eddies, and the computational domain large enough to resolve the largest. Explain why this implies that the number of grid points, $N$, scales as $N \propto (L/\eta)^3$ in three dimensions. Obtain the exponent in $N \sim R_T^n$. Estimate the number of grid points needed when $R_T = 10^4$.

**Exercise 2.3.** *Relative dispersion.* The inertial-range velocity $(\varepsilon r)^{1/3}$ can be described as the velocity at which two fluid elements that are separated by distance $r$ move apart (provided their separation is in the inertial range, $\eta \ll r \ll L$). Deduce the power law for the time dependence of the mean-square separation $\overline{r^2}(t)$. Use dimensional analysis. Also infer the result by integrating an ordinary differential equation. The scaling $\overline{v^2} \propto r^{2/3}$ is often called Richardson's 2/3 law.

**Exercise 2.4.** *Averaging via PDFs.* Let the PDF of a random variable be given by the function

$$P(x) = Ax^2e^{-x}, \qquad \text{for } \infty > x \geq 0,$$

$$P(x) = 0, \qquad \text{for } x < 0.$$

What is the value of the normalization constant $A$? Evaluate $\overline{\sin(ax)}$ where $a$ is a constant.

**Exercise 2.5.** *Taylor microscale.* Let $u$ be a statistically homogeneous function of $x$. A microscale, $\lambda$, is defined by

$$\lim_{\xi \to 0} d_\xi^2 \left[ \overline{u(x)u(x+\xi)} \right] = -\frac{\overline{u^2}}{\lambda^2},$$

if the correlation function is twice differentiable. Show that $\overline{u^2}/\lambda^2 = \overline{(d_x u)^2}$. Assume that $\overline{(d_x u)^2}$ follows dissipation-range scaling and obtain the $R_T$ dependence of $\lambda/L$.

**Exercise 2.6.** *Toor's analogy, Eq. (2.2.7).* To what does the term "analogy" refer?
The PDF

$$P(m) = (1+m)^{a-1}(1-m)^{b-1}/B(a,b), \qquad -1 < m < 1,$$

$$P(m) = 0, \qquad |m| \geq 1,$$

is referred to as the "beta" probability density. The normalization coefficient is $B(a, b) = 2^{a+b-1}\Gamma(a)\Gamma(b)/\Gamma(a+b)$, where $\Gamma$ is the factorial function $\Gamma(a) = (a-1)!$, extended to non-integer arguments.

The beta distribution is a popular model for the mixture fraction PDF in reacting flows. Given that $\Gamma(1/2) = \sqrt{\pi}$, evaluate $\overline{\gamma_{\mathrm{A}}}$ for $a = b = 1/2$, $a = b = 2$, and $a = b = 4$. Plot the $\beta$ distribution for these same three cases. Based on these cases, describe how the PDF evolves as mixing and reaction proceed.

**Exercise 2.7.** *Langevin equation.* Why does the expansion (2.2.14) begin with 1 and why is the next term negative?

In (2.2.10) and subsequent equations let $s^2 = (1-r^2)\sigma^2$. Derive the equation

$$\tfrac{1}{2}\,\mathrm{d}_t\overline{u^2} = -\frac{\overline{u^2}}{T_{\mathrm{L}}} + \frac{\sigma^2}{T_{\mathrm{L}}}$$

from (2.2.12). Solve for $\overline{u^2}(t)$ with initial condition $\overline{u^2}(0) = 0$, $\sigma$ and $T_{\mathrm{L}}$ being constants. Is $u$ a statistically stationary random variable? The exact solution for the variance that is derived here will be used to test a Monte Carlo simulation in the next exercise.

**Exercise 2.8.** *More on stochastic processes.* Most computer libraries have a random number algorithm that generates values in the range $\{0, 1\}$ with equal probability, that is, $P(\tilde{u}) = 1,\ 0 \le \tilde{u} \le 1$.

Show that $\overline{\tilde{u}} = \frac{1}{2}$ and that $\overline{u^2} = \frac{1}{12}$, where $u$ is the fluctuation $\tilde{u} - \overline{\tilde{u}}$. Deduce that

$$\xi \equiv \sqrt{12}\,(\tilde{u} - \tfrac{1}{2})$$

has $\overline{\xi} = 0$ and $\overline{\xi^2} = 1$, as is needed in the model (2.2.10). A Gaussian random variable can be approximated by summing $N$ values $\xi_i$ and normalizing by $\sqrt{N}$. For $N = 16$,

$$\xi_g = \frac{1}{4}\sum_{1}^{16} \xi_i.$$

Program this and verify by averaging a large number (10 000 say) of values that $\overline{\xi_g} \approx 0$ and $\overline{\xi_g^2} \approx 1$. Let $s$ in (2.2.10) be as in the previous exercise. Normalize $t$ by $T_{\mathrm{L}}$ and $u$ by $\sigma$ and solve (2.2.10) numerically, starting with $u = 0$ and integrating to $t/T_{\mathrm{L}} = 10$ by steps of 0.05. Compute $\overline{u^2}(t)$ by averaging 100 such solutions, by averaging 1000 such solutions, and by averaging 4000 such solutions. Plot these estimates of $\overline{u^2}$ versus $t$ and compare to the exact result found in Exercise 2.7. Does it look as if the average is converging like $1/\sqrt{N}$?

**Exercise 2.9.** *Isotropy.* Verify that (2.3.4) is the most general fourth-order, isotropic tensor.

**Exercise 2.10.** *Solving equations via Cayley–Hamilton.* Use the Cayley–Hamilton theorem to solve

$$b_{ij} = S_{ik}b_{kj} - \tfrac{1}{3}\,\delta_{ij}S_{kl}b_{lk} + S_{ij},$$

in which $\boldsymbol{S}$ is a given, trace-free matrix ($S_{kk} = 0$) and $\boldsymbol{b}$ is the unknown. Why is $\boldsymbol{b}$ also trace-free? Your solution is only valid if $\boldsymbol{S}$ is such that the solution is not infinite. What is this solvability criterion?

**Exercise 2.11.** *Solving equations via generalized Cayley–Hamilton.* Suppose that $\boldsymbol{\tau}$ is the solution to

$$\boldsymbol{\tau} = \boldsymbol{\tau} \cdot \boldsymbol{\Omega} - \boldsymbol{\Omega} \cdot \boldsymbol{\tau} + \mathrm{Trace}(\boldsymbol{S} \cdot \boldsymbol{\tau})\delta - \boldsymbol{S}$$

in which $\boldsymbol{S}$ is a given symmetric, trace-free tensor and $\boldsymbol{\Omega}$ is a given antisymmetric, trace-free tensor. Show that $\boldsymbol{\tau}$ is symmetric. Use (2.3.15) to solve this equation in two dimensions for $\boldsymbol{\tau}(\boldsymbol{S}, \boldsymbol{\Omega})$.

# 3

# Reynolds averaged Navier–Stokes equations

> The labours of others have raised for us an immense reservoir of important facts. We merely lay them on, and communicate them, in a clear and gentle stream ...
>
> – Charles Dickens

Turbulent flow is governed by the Navier–Stokes momentum equations, the continuity equation, and, in compressible flow, energy and state equations. Here we consider only incompressible, constant-density flow. Derivations of the Navier–Stokes equations can be found in many books on viscous flow, such as Batchelor (1967) or White (1991). The reader is assumed to have familiarity with laminar viscous flow, or should consult such references.

The turbulence problem, as presently formulated, is to describe the statistics of the velocity field, without access to realizations of the random flow. It seems sensible to start by attempting to derive equations for statistics. Toward this end, averages of the Navier–Stokes equations can be formed, in hopes of finding equations that govern the mean velocity, the Reynolds stresses, and so on. Regrettably, the averaged equations are unclosed. In that sense they fall short of any ambition to arrive at governing laws for statistics. However, the Reynolds averaged equations give an insight into the factors that govern the evolution of mean flow and Reynolds stresses. In the present chapter, we will be contented with physical interpretations of terms in the unclosed equations; how to close the equations is the subject of Part II of the book.

*Statistical Theory and Modeling for Turbulent Flows, Second Edition* P. A. Durbin and B. A. Pettersson Reif

## 3.1 Background to the equations

The equations governing viscous incompressible flow, whether turbulent or laminar, are

$$\partial_t \tilde{u}_i + \tilde{u}_j \, \partial_j \tilde{u}_i = -\frac{1}{\rho} \, \partial_i \tilde{p} + \nu \nabla^2 \tilde{u}_i,$$
$$\partial_i \tilde{u}_i = 0. \qquad (3.1.1)$$

The first expresses conservation of momentum. The second expresses the incompressibility of fluid volumes, which is equivalent to mass conservation in the present case. Throughout this text the shorthand $\partial_i$ is used for the partial derivative $\partial/\partial x_i$, and $(u_1, u_2, u_3) = (u, v, w)$; for example, $\partial_1 u_2 = \partial v/\partial x$.

Stokes showed that Poiseuille pipe flow is a solution to these equations. Poiseuille's formula for the pressure drop versus flow rate agreed with experiments up to some speed, after which the flow rate was underpredicted. Stokes argued that Eqs. (3.1.1) were not at fault. He knew that the fluid motion developed unsteadiness at high speeds. He suggested that the equations remained valid, but that an unsteady solution was needed. In 1883 Osborne Reynolds performed an experiment to illustrate that the drop in flow rate (or increase in drag coefficient) did indeed correspond to the occurrence of unsteadiness. He injected a thin stream of dyed water into a clear pipe through which the main stream flowed. As long as the dye filament remained thin through the length of the pipe, Poiseuille's law was valid. When the drag began to increase above the laminar, Poiseuille value, the dye filaments showed undulations, eventually leading to erratic motion that dispersed the filament into a puff of dyed fluid. Figure 3.1 is his schematic of this process. A transition was occurring from flow in lamina, to more disorderly flow.

While Reynolds' observations confirmed Stokes' conjecture concerning unsteadiness, they did not prove that Eqs. (3.1.1) describe the unsteady flow. To support that conjecture, Reynolds invented what is now called "Reynolds averaging." Methods to solve the full, unsteady equations did not exist in his time, so he considered the equations, averaged so as to remove the unsteadiness. He showed that, above a certain value of the Reynolds number,* a finite-amplitude, unsteady motion was consistent with the averaged, governing equations. Actually, his investigation was the forerunner of energy stability theory, but the idea of studying averaged equations laid the foundation for the statistical theory of turbulence.

Nowadays, we would say that the pipe flow in Reynolds' experiment became unstable and made a transition to turbulence above a critical Reynolds number (Figure 3.2). Stokes and Reynolds were contending that this process was within the province of the Navier–Stokes equations. As a result of years of research, including computer simulations of turbulence, it has been convincingly proved that Eqs. (3.1.1) are the equations of turbulent fluid flow. The only doubts that have arisen were over whether large accelerations could occur that would violate the linear stress–strain rate relation for Newtonian fluids, or even violate the continuum assumption. Those doubts have been laid to rest.

* The terminology "Reynolds number" was not introduced until 1908, 25 years after Reynolds' famous publication.

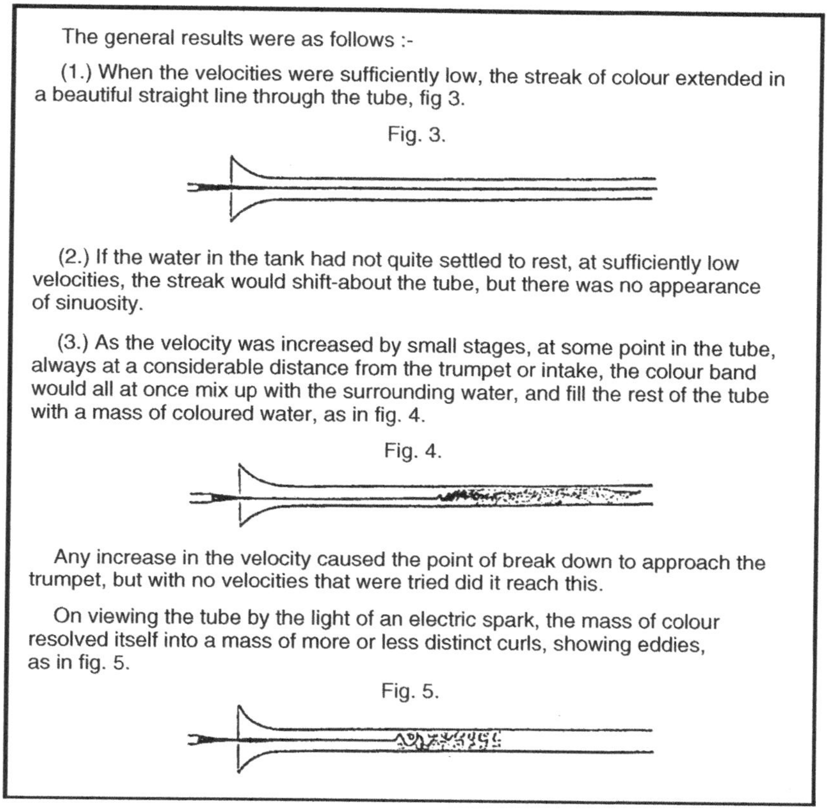

The general results were as follows :-

(1.) When the velocities were sufficiently low, the streak of colour extended in a beautiful straight line through the tube, fig 3.

Fig. 3.

(2.) If the water in the tank had not quite settled to rest, at sufficiently low velocities, the streak would shift-about the tube, but there was no appearance of sinuosity.

(3.) As the velocity was increased by small stages, at some point in the tube, always at a considerable distance from the trumpet or intake, the colour band would all at once mix up with the surrounding water, and fill the rest of the tube with a mass of coloured water, as in fig. 4.

Fig. 4.

Any increase in the velocity caused the point of break down to approach the trumpet, but with no velocities that were tried did it reach this.

On viewing the tube by the light of an electric spark, the mass of colour resolved itself into a mass of more or less distinct curls, showing eddies, as in fig. 5.

Fig. 5.

**Figure 3.1** Schematic of a dye filament in a flow undergoing transition to turbulence, as drawn by Reynolds (1883).

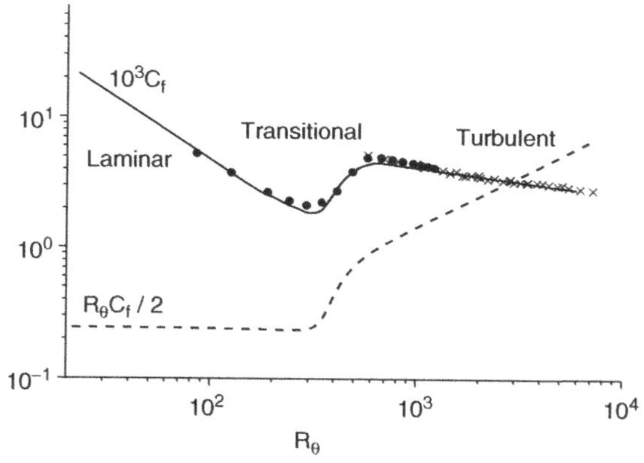


**Figure 3.2** Friction coefficient versus momentum thickness Reynolds number $(U_\infty \theta/\nu)$ in laminar–turbulent transition. The skin friction is normalized in two ways, by $\rho U_\infty^2/2$ (——), and by $\mu U_\infty/\theta$ (- - - -). Some experimental data are shown by crosses and full circles.

## 3.2 Reynolds averaged equations

The Navier–Stokes equations (3.1.1) govern fluid turbulence. If that were the solution to the "problem of turbulence," there would be no need for this book. The snag is that the phenomenon of turbulence is the complete solution to these equations – a chaotic, spatially, and temporally complex solution. Such solutions are not easily obtained, even on massively parallel supercomputers. A much simpler level of description is needed: this calls for a statistical approach. There are no closed equations for the statistics of turbulent flow. The equations obtained by averaging the exact laws (3.1.1) contain more unknowns than the number of equations, as will be seen.

In Section 2.2.2 the total velocity was decomposed into a sum of its mean and a fluctuation, $\tilde{u}(\mathbf{x}, t) = U(\mathbf{x}, t) + u(\mathbf{x}, t)$, where $U \equiv \overline{\tilde{u}}$. If this decomposition is substituted into (3.1.1) they become

$$\begin{aligned} \partial_t(U_i + u_i) + (U_j + u_j)\,\partial_j(U_i + u_i) &= -\frac{1}{\rho}\,\partial_i(P + p) + \nu\nabla^2(U_i + u_i), \\ \partial_i(U_i + u_i) &= 0. \end{aligned} \tag{3.2.1}$$

The average of these equations is obtained by drawing a bar over each term, noting the rules $\overline{U} = U$ and $\overline{u} = 0$ (see Eq. 2.2.3):

$$\begin{aligned} \partial_t U_i + U_j\,\partial_j U_i &= -\frac{1}{\rho}\,\partial_i P + \nu\nabla^2 U_i \underbrace{-\partial_j\overline{u_j u_i}}, \\ \partial_i U_i &= 0. \end{aligned} \tag{3.2.2}$$

These are the Reynolds averaged Navier–Stokes (RANS) equations. Equations (3.2.2) for the mean velocity are the same as Eqs. (3.1.1) for the total instantaneous velocity, except for the last term of the momentum equation, highlighted with the brace. This term is a derivative of the Reynolds stress tensor (2.2.8). It comes from the convective derivative, after invoking continuity, to write $u_i\,\partial_i u_j = \partial_i(u_i u_j)$. So, strictly, the Reynolds stresses are not stresses at all – they are the averaged effect of turbulent convection. But we know from the example of the Langevin equation, Section 2.2.2, especially Eq. (2.2.20), that the ensemble-averaged effect of convection can be diffusive; here it is momentum that is being diffused. At the molecular level, momentum is diffused by viscosity, and appears in the governing equations as the viscous stress; the analogous diffusive nature of $\overline{u_i u_j}$ is the origin of it being termed the "Reynolds stress." This also helps to explain why Reynolds stresses are often modeled by an eddy viscosity: $-\overline{u_i u_j} \approx \nu_T[\partial_j U_i + \partial_i U_j]$.

The mean flow equations (3.2.2) are unclosed because they are a set of four equations ($i = 1, 2, 3$) with 10 unknowns ($P$; $U_i$, $i = 1, 2, 3$; and $\overline{u_i u_j}$, $i = 1, 2, 3$, $j \leq i$). The extra six unknowns are the components of the Reynolds stress tensor. The statistical problem (3.2.2) for the mean, or first moment, requires knowledge of the covariance, or second moment. This is because the Navier–Stokes equations have a quadratic nonlinearity. Any nonlinearity causes moment equations to be unclosed; here the first-moment equation contains second moments, the second-moment equation will contain third moments, and so on up the hierarchy. Only the second level of the hierarchy is considered in this book. That is the highest level used directly in single-point moment closure modeling.

Dynamical equations for the Reynolds stress tensor can be derived from the equation for the fluctuating velocity. After subtracting (3.2.2) from (3.2.1), the result

$$\partial_t u_i + U_k\,\partial_k u_i + u_k\,\partial_k U_i + \partial_k(u_k u_i - \overline{u_k u_i}) = -\frac{1}{\rho}\,\partial_i p + \nu\nabla^2 u_i \tag{3.2.3}$$

is obtained. Multiplying this by $u_j$, averaging and adding the result to the same equation with $i$ and $j$ reversed gives the equations for the Reynolds stress tensor. It is left for the reader to work through the details (Exercise 3.1). We obtain

$$\begin{aligned}\partial_t\overline{u_i u_j} + U_k\,\partial_k\overline{u_i u_j} = &-\underbrace{\frac{1}{\rho}\,\overline{(u_j\,\partial_i p + u_i\,\partial_j p)}}_{\text{redistribution}} - \underbrace{2\nu\,\overline{\partial_k u_i\,\partial_k u_j}}_{\text{dissipation}}\\ &\underbrace{-\,\partial_k\overline{u_k u_i u_j}}_{\text{turbulent transport}}\ \underbrace{-\,\overline{u_j u_k}\,\partial_k U_i - \overline{u_i u_k}\,\partial_k U_j}_{\text{production}} + \nu\nabla^2\overline{u_i u_j}.\end{aligned} \tag{3.2.4}$$

The cumbersome equation (3.2.4) is referred to as the *Reynolds stress transport equation*. Obviously, it is not a closed equation for the second moment. The usual terminology for the various terms on the right is indicated; it will be discussed below. The final term is not annotated: it is simply the molecular viscous transport; this and production are closed terms because they contain only the dependent variable $\overline{u_i u_j}$ and mean flow gradients.

The turbulent kinetic energy equation, for the present case of constant-density flow, is one-half of the trace (set $i = j$ and sum over $i$) of (3.2.4):

$$\begin{aligned}\partial_t k + U_k\,\partial_k k = &-\underbrace{\frac{1}{\rho}\,\partial_i\overline{u_i p}}_{\text{pressure-diffusion}} - \underbrace{\nu\,\overline{\partial_k u_i\,\partial_k u_i}}_{\text{dissipation}}\\ &\underbrace{-\,\frac{1}{2}\,\partial_k\overline{u_k u_i u_i}}_{\text{turbulent transport}}\ \underbrace{-\,\overline{u_i u_k}\,\partial_k U_i}_{\text{production}} + \nu\nabla^2 k,\end{aligned} \tag{3.2.5}$$

where $k \equiv \frac{1}{2}\overline{u_i u_i}$ is actually the kinetic energy per unit mass. The change in terminology from "redistribution" in (3.2.4) to "pressure-diffusion" in (3.2.5) will be explained in due course.

## 3.3 Terms of kinetic energy and Reynolds stress budgets

The annotations in Eqs. (3.2.4) and (3.2.5) indicate physical interpretations of the individual terms. The term marked "dissipation" is preceded by a negative sign and represents decay of turbulence. It is usually represented as $\varepsilon_{ij} \equiv 2\nu\,\overline{\partial_k u_i\,\partial_k u_j}$. One half of the trace of the dissipation, $\nu\,\overline{\partial_k u_i\,\partial_k u_i} = \nu\,\overline{|\nabla \boldsymbol{u}|^2}$, is usually denoted succinctly as $\varepsilon$; $\varepsilon$ is the rate of dissipation of turbulent kinetic energy. Clearly $\varepsilon \geq 0$. The components of $\varepsilon_{ij}$ permit each component of the Reynolds stress tensor to dissipate at a different rate; $\varepsilon_{ij}$ is a positive definite matrix.

The "transport" term is so termed because it redistributes energy in space without creating or destroying it. In other words, the transport term is conservative. Conservative terms are of the form of the divergence of a flux. By the divergence theorem, the integral of such terms over a fluid volume equals the flux through the surface of the volume:

$$\int_V \partial_k \overline{u_k u_i u_i}\, \mathrm{d}V = \int_S \hat{n}_k\, \overline{u_k u_i u_i}\, \mathrm{d}S,$$

showing mathematically that this term neither produces nor destroys energy within the volume. The "pressure-diffusion" term is so-called because it too is of conservative form. However, this is a rather peculiar terminology because pressure effects are nonlocal and instantaneous (in incompressible flow), while diffusion occurs slowly, down local gradients. Fortunately the pressure-diffusion term is usually small compared to the others in Eq. (3.2.5).

The physical effect of transport terms is to spread the Reynolds stresses in space. It is generally assumed that they drive the spatial distribution toward uniformity, in analogy to gradient diffusion by molecular processes. While that is not necessarily true, it is a good understanding in the vast majority of flows.

The rate of turbulent energy production is $\mathcal{P} = -\overline{u_i u_k}\partial_k U_i$. Note that the negative sign is included in the definition of $\mathcal{P}$. Despite its name, this term does not represent net production of energy. It represents the rate at which energy is transferred from mean flow to turbulent fluctuations. An equal and opposite term appears in the mean flow energy equation (Exercise 3.2). A turbulence model must respect the conservation of net energy. This is easy to check: the equation for total energy, $\frac{1}{2}|U|^2 + k$, should not contain a production term.

In most cases turbulent energy is generated from mean shear. This might be thought of as an instability process, such as illustrated in Figure 3.1, that takes place continually within the flow. An additional process is that the turbulent vortices are stretched and intensified by the mean rate of strain (Figure 1.6). Equation (3.2.5) alludes only to the averaged effect of such processes. Intuitively one expects the averaged flow of energy to be from the orderly mean flow to the disorderly turbulence. However, there is no guarantee that $\mathcal{P} \geq 0$: in fact, $\mathcal{P}$ can be negative in flows subject to strongly stabilizing forces, such as centrifugal acceleration. But in most fully turbulent flows it is non-negative.

To see why this is so, consider the following "mixing length" type of argument (Figure 3.3). Let the mean flow be a parallel shear layer, $U(y)$, in the $x$ direction. The only non-zero mean velocity gradient is $\partial_2 U_1 = \partial_y U$, so $\mathcal{P} = -\overline{uv}\,\partial_y U$. A fluid element initially located at $y = Y_0$ will be convected by the turbulent velocity, and at time $t$ it will end up at the cross-stream position $Y(t) = Y_0 + \int_0^t v(t')\,\mathrm{d}t'$. Suppose that

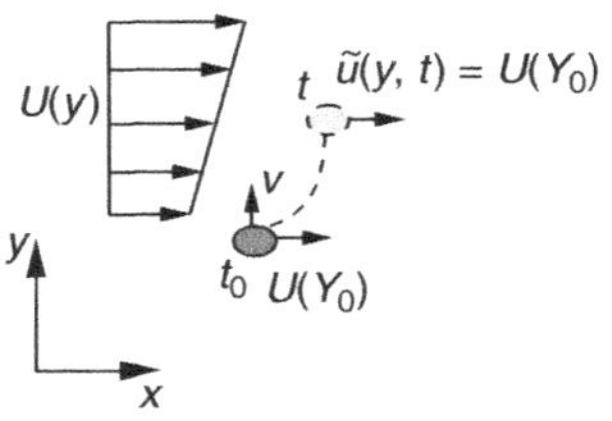


**Figure 3.3** Schematic of the mixing length rationale.

it retains its initial $x$ velocity $U(Y_0)$, ignoring $u$ in comparison to $U$. When it arrives at $Y(t) = y$, the instantaneous velocity will be the particle's velocity $\tilde{u} = U(Y_0)$. Since $Y_0 = y - \int_0^t v(t')\,dt'$, this velocity is equivalent to $\tilde{u} = U(y - \int_0^t v(t')\,dt')$. By definition, the velocity fluctuation is $u = \tilde{u} - U$. In the present line of reasoning, $u$ equals

$$u = U\left(y - \int_0^t v(t')\,dt'\right) - U(y) \approx -\int_0^t v(t')\,dt'\,\partial_y U. \tag{3.3.1}$$

The first term in a Taylor series was used for the approximation. This approximation is formally justified if the displacement, $\ell_m \equiv \int_0^t v(t')\,dt'$, is small compared to the scale of mean flow variation, $\delta$, say,[†] so that higher derivatives in the Taylor series can be neglected.

The *Reynolds shear stress*, $\overline{uv}$, is found by multiplying (3.3.1) by $v(t)$ and averaging:

$$\overline{uv} = -\int_0^t \overline{v(t)v(t')}\,dt'\,\partial_y U. \tag{3.3.2}$$

We have already come across this integral in (2.2.21): there it was the eddy diffusivity; here it is an "eddy viscosity." Thus

$$\overline{uv} = -\nu_T\,\partial_y U. \tag{3.3.3}$$

It is now clear why a negative sign was included in the definition of production:

$$\mathcal{P} = -\overline{uv}\,\partial_y U = \nu_T(\partial_y U)^2 \tag{3.3.4}$$

is non-negative if the eddy viscosity $\nu_T$ is non-negative. In situations where $\mathcal{P} < 0$, the eddy viscosity model (3.3.3) would not be valid because negative viscosity is not acceptable.

If $\partial_y U > 0$ then (3.3.1) shows that a positive $v$ correlates with a negative $u$. This is an intuitive understanding of why $\overline{uv}$ tends to have the opposite sign of the mean flow gradient in parallel shear flow, independent of a specific eddy viscosity assumption. Upward motion carries fluid parcels with lower mean velocity into a region of higher velocity, where they are a negative $u$ fluctuation. The general tendency for $\overline{uv}$ to be negative (if $\partial_y U > 0$) in shear flow is why $\mathcal{P}$ tends to be positive.

The derivation of (3.3.3) assumed that the fluid element conserves its initial value of $U(Y_0)$. That assumption is better justified when applied to the concentration of a passive contaminant, $\tilde{c} = C + c$, such as a small amount of heat or dye (Exercise 3.3). In the absence of molecular diffusion, concentration is carried by fluid elements, $D_t(C + c) = 0$, so the only approximations needed are $c \ll C$ and $\ell_m \ll \delta$. By contrast, momentum is affected by pressure gradients, $\rho D_t \tilde{u} = -\nabla\tilde{p}$. The mixing length reasoning has been called into question on the ground that it neglects pressure gradients, but as an empirical model the eddy viscosity formula (3.3.3) is often not bad.

A consequence of the caveat that pressure gradients affect momentum transport is that the eddy viscosity will not equal the eddy diffusivity in general, even though the derivations (2.2.21) and (3.3.2) make them identical. The ratio $\nu_T/\alpha_T$ is called the *turbulent Prandtl number*; see Section 4.4. Experimental measurements suggest that a value

[†] Usually the strict condition $\ell_m \ll \delta$ is not satisfied, but this mixing length formalism is a reasonable conceptual model.

of $\mathrm{Pr_T} = 0.9$ can be used in boundary layers while $\mathrm{Pr_T} = 0.7$ is often more suitable in free-shear flows. It should be emphasized, however, that $\mathrm{Pr_T}$ is not a material property and it can depend on many factors that influence the flow field. Turbulent transport of momentum and concentration is a complicated process; eddy viscosities and turbulent Prandtl numbers represent pragmatic simplifications that do not always work!

The distribution of the terms in (3.2.5) across a boundary layer is plotted in Figure 3.4. The wall is at $y = 0$ and $\delta$ is the 99% thickness of the boundary layer. These data are from a direct numerical simulation (DNS) at $R_\delta = 5300$, which is a rather low Reynolds number compared to most experiments. The virtue of data generated by computer simulation is that all terms in the budget are known accurately. It is very difficult to measure pressure-diffusion in the laboratory. The DNS data confirm that it is tiny.

Figure 3.4 shows how production, $\mathcal{P}$, and dissipation, $\varepsilon$, are approximately equal over most of the flow. However, they attain their largest magnitude and are unequal near to the wall. The production becomes large at small $y/\delta$ because the shear, $\mathrm{d}U/\mathrm{d}y$, is large near the wall. A velocity profile, $U(y/\delta)$, is included to the right of the figure so that the distribution of shear can be seen. The velocity rises steeply from the wall to $y/\delta \approx 0.15$. This is the region of maximum production.

The no-slip boundary condition at $y = 0$ requires that $u_i = 0$, so $\mathcal{P}$ must vanish at the wall. However, $\varepsilon = \nu\, \overline{\partial_y u_i\, \partial_y u_i}$ at $y = 0$, which need not vanish (near-wall asymptotics are covered in Section 7.3.3). Indeed, $\varepsilon$ has a sharp maximum at the wall.

The terms of the Reynolds stress budget (3.2.4) have similar interpretations to those of the kinetic energy budget (3.2.5). The production tensor, from (3.2.4), is

$$\mathcal{P}_{ij} = -\overline{u_j u_k}\partial_k U_i - \overline{u_i u_k}\partial_k U_j. \tag{3.3.5}$$

In parallel shear flow the velocity is $U_1(x_2)$ in tensor notation. Using this in (3.3.5) gives $\mathcal{P}_{11} = -2\overline{u_1 u_2}\partial_2 U_1 = -2\overline{uv}\,\partial_y U$. This is twice (3.3.4). Energy flows from the mean shear into the streamwise $\overline{u_1^2}$ component. On the other hand, $\mathcal{P}_{22} = 0 = \mathcal{P}_{33}$ in parallel shear flow; there is no production of $\overline{u_2^2}$ or $\overline{u_3^2}$. The only way energy can get to these

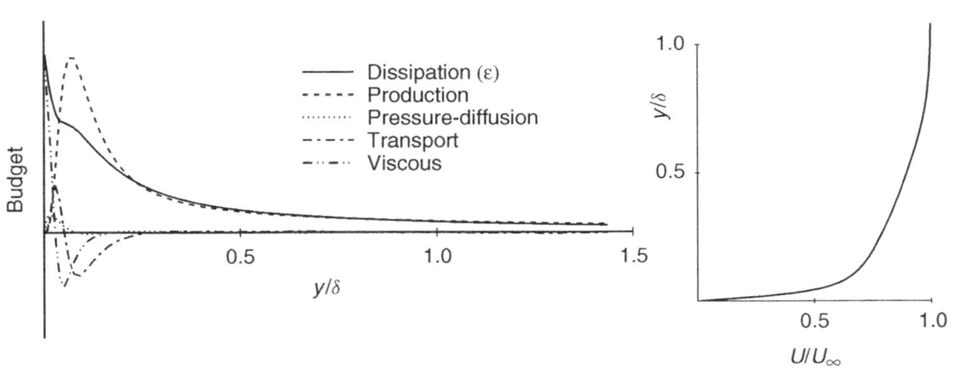


**Figure 3.4** Turbulent kinetic energy budget in a boundary layer; the curves are DNS data from Spalart (1988). The mean velocity profile is shown at the right.

components is for it to be fed from the streamwise component into the other normal stress components.

The only new term in the Reynolds stress budget (3.2.4) is that denoted "redistribution." It plays precisely the role of feeding variance from one Reynolds stress component into others. The terminology, *redistribution*, implies that variance is shifted between components of $\overline{u_i u_j}$ without altering the total energy, $\frac{1}{2}\overline{u_k u_k}$. The qualitative effect of redistribution is usually to shift energy from the larger components of the Reynolds stress tensor into the smaller components. Strictly speaking, if redistribution occurs with no generation of net energy, the trace of the redistribution term should vanish. The trace of the term denoted "redistribution" in (3.2.4) does not quite vanish; hence it does not exactly define the redistribution tensor. It might be more accurate to simply call it the velocity–pressure gradient correlation. It will be denoted by the symbol $\phi_{ij}$:

$$\phi_{ij} \equiv \frac{1}{\rho}\left(\overline{u_j\,\partial_i p} + \overline{u_i\,\partial_j p}\right). \tag{3.3.6}$$

This has trace $\phi_{kk} = (2/\rho)\partial_k \overline{u_k p}$, which is only strictly zero if the turbulence is homogeneous in space. (Recall that homogeneity means that statistics are independent of position in space; homogeneity in a direction, $x_k$, implies that derivatives of statistics vanish in that direction, $\partial_k = 0$.)

The formula

$$\Pi_{ij} = \phi_{ij} - \tfrac{1}{3}\phi_{kk}\delta_{ij} \tag{3.3.7}$$

is a more proper definition of the redistribution tensor. This trace-free tensor is (3.3.6) minus 1/3 of its trace times the identity tensor. Another commonly used approach is to avoid an aggregate redistribution term, but to write

$$-\overline{u_j\,\partial_i p} - \overline{u_i\,\partial_j p} = -(\partial_i \overline{u_j p} + \partial_j \overline{u_i p}) + \overline{p(\partial_i u_j + \partial_j u_i)}$$

and to refer to the first term on the right as "pressure-diffusion" and the second as "pressure–strain." Pressure–strain is redistributive because it has zero trace. To a large extent, the difference between separating out pressure–strain, or retaining a redistribution tensor, is a matter of semantics. It has little bearing on closure modeling, except in near-wall treatments (Section 7.3.4); in that context it is preferable to retain $\Pi_{ij}$ because this vanishes at no-slip walls. In homogeneous turbulence, pressure–strain and redistribution are identical.

Some insight into the operation of the redistribution term is gained by anticipating the isotropization of production (IP) model of Section 7.1.4. Let $\phi_{ij}$ be represented by the formula

$$\phi_{ij} = C(\mathcal{P}_{ij} - \frac{2}{3}\delta_{ij}\mathcal{P}),$$

where $C$ is a constant, equal to 3/5 in the IP model. Technically, this is a "rapid" redistribution model. In parallel shear flow $\phi_{11} = \frac{4}{3}C\mathcal{P}$ and $\phi_{22} = \phi_{33} = -\frac{2}{3}C\mathcal{P}$; $\phi_{11}$ is positive and the other two are negative. They sum to zero, which makes them redistributive: energy is drawn from $\overline{u_1^2}$ into $\overline{u_2^2}$ and $\overline{u_3^2}$. That is the mandatory behavior of redistribution in parallel shear flow.

To conclude this section, and for future reference, the kinetic energy and Reynolds stress transport equations for homogeneous turbulence will be cited. Again, homogeneity implies that spatial derivatives of all fluctuation statistics are zero. However, derivatives of the mean flow, $\partial_j U_i$, need not be zero; they need only be independent of position, $\boldsymbol{x}$. That is because homogeneity requires that the coefficients in equations for $\overline{u_i u_j}$ be constant, and those coefficients involve derivatives of $U$, but not $U$ itself. Thus, the general mean flow for which the turbulence can be homogeneous is of the form $U_{(\mathbf{x},t)} = A_{ij}(t)x_j + B_i(t)$. Here $\boldsymbol{A}$ is a matrix that determines the mean rate of strain and rate of rotation:

$$\begin{aligned} S_{ij} &\equiv \tfrac{1}{2}(\partial_i U_j + \partial_j U_i) = \tfrac{1}{2}(A_{ji} + A_{ij}), \\ \Omega_{ij} &\equiv \tfrac{1}{2}(\partial_i U_j - \partial_j U_i) = \tfrac{1}{2}(A_{ji} - A_{ij}). \end{aligned} \tag{3.3.8}$$

The equations of homogeneous turbulence are obtained by setting the spatial derivatives of statistics to zero in (3.2.4) and (3.2.5). The turbulent kinetic energy equation becomes

$$\partial_t k = \mathcal{P} - \varepsilon. \tag{3.3.9}$$

Note that $\varepsilon$ is not zero in homogeneous turbulence; it is a statistic *of* derivatives, not a derivative *of* statistics. Turbulent kinetic energy, $k$, evolves by (3.3.9) in consequence of imbalance between production, $\mathcal{P} = \frac{1}{2}\mathcal{P}_{kk}$, and dissipation, $\varepsilon$.

Setting spatial derivatives to zero in (3.2.4) gives

$$\partial_t \overline{u_i u_j} = -\phi_{ij} + \mathcal{P}_{ij} - \varepsilon_{ij} \tag{3.3.10}$$

in notation introduced previously for the redistribution, production, and dissipation tensors. The mean flow gradient appears in the definition of $\mathcal{P}_{ij}$, which is why the turbulence cannot be homogeneous unless the velocity gradients are independent of position.

## 3.4 Passive contaminant transport

Turbulence transports passive contaminants in much the same way as it transports momentum. A passive contaminant is defined as a transported substance, $c$, that does not affect the dynamical equation (3.1.1). The mean flow field can be analyzed without regard to the passive scalar field. The scalar field can subsequently be computed *a posteriori*. The contaminant may be a species concentration in a reacting flow, or a pollutant in the atmosphere. In many instances, heat can be regarded as a passive scalar; exceptions occur when buoyancy forces are significant, or when temperature variations cause fluid properties, such as viscosity or density, to vary noticeably. In those cases, the contaminant is referred to as "active" and the set of governing equations are fully coupled.

The Reynolds averaged equations governing the concentration, $c$, of a passive contaminant can be developed in the same manner as in the previous sections. The starting point is the convection–diffusion equation for the concentration of a scalar:

$$\partial_t \tilde{c} + \tilde{u}_j\, \partial_j \tilde{c} = \alpha \nabla^2 \tilde{c}. \tag{3.4.1}$$

Decomposing the concentration as $\tilde{c}(\mathbf{x},t) = C(\mathbf{x},t) + c(\mathbf{x},t)$ and averaging results in the equation

$$\partial_t C + U_j\,\partial_j C = \alpha\nabla^2 C - \partial_i \overline{cu_i} \tag{3.4.2}$$

for the mean scalar concentration $C$. The last term on the right-hand side comes from the convective derivative $u_j\,\partial_j c$ by invoking the continuity constraint $\partial_j u_j = 0$. The new term, $\overline{cu_i}$, referred to as the *scalar flux*, is unknown *a priori* and needs to be modeled in order to close the set of governing equations.

The equation for the scalar flux is obtained analogously to the transport equations governing the kinematic Reynolds stress tensor (3.2.4): subtract (3.4.2) from (3.4.1), then multiply by $u_i$, and add the result to the $i$ component of the momentum equation, multiplied by $c$. The result is the transport equation for the *Reynolds flux*:

$$\begin{aligned}
&\partial_t \overline{cu_i} + U_j\,\partial_j \overline{cu_i} \\
&\quad = -\frac{1}{\rho}\overline{c\,\partial_i p} + \frac{1}{2}(\alpha-\nu)\,\partial_j(\overline{u_i\,\partial_j c} - \overline{c\,\partial_j u_i}) + \frac{1}{2}(\nu+\alpha)\nabla^2\overline{cu_i} \\
&\qquad \underbrace{-\,(\alpha+\nu)\,\overline{\partial_j u_i\,\partial_j c}}_{\text{dissipation}} \;\underbrace{-\,\partial_k\overline{u_k u_i c}}_{\text{transport}} \;\underbrace{-\,\overline{u_i u_j}\,\partial_j C - \overline{cu_j}\,\partial_j U_i}_{\text{production}}\,.
\end{aligned} \tag{3.4.3}$$

The terms on the left-hand side correspond to the time rate of change and advection of turbulent flux. The scalar pressure-gradient correlation $-(1/\rho)\overline{c\,\partial_i p}$ plays a redistributive role, although the trace-free constraint is not applicable because this is an equation for a vector. The last two terms on the second line in (3.4.3) can alternatively be expressed as $\partial_j(\alpha\,\overline{u_i\,\partial_j c} + \nu\,\overline{c\,\partial_j u_i})$ and represent diffusion. The first term on the last line is the rate of dissipation of turbulent fluxes, whereas the second redistributes fluxes in space. The terms $-\overline{cu_j}\,\partial_j U_i$ and $-\overline{u_i u_j}\,\partial_j C$ are the rate of turbulence flux production due to mean flow and scalar gradients, respectively. The Reynolds flux equation contains four unknown terms that need to be modeled.

The convection–diffusion equation (3.4.1) is a linear equation for the passive scalar concentration. Hence the superposition principle applies. For instance, if $c(\boldsymbol{x}; z_1) \equiv c_1$ is the concentration field produced by a source located at $z_1$ and $c(\boldsymbol{x}; z_2) \equiv c_2$ is the concentration field produced by a source at $z_2$, then $c_1 + c_2$, is the concentration produced by both sources being turned on simultaneously. The concentration fields simply add together. Since averaging is a linear operation, this superposition method applies to $C$ and $\overline{cu_i}$ as well. Any closure model should preserve the superposition property of the mean field and of the scalar flux, $\overline{cu_i}$.

Models that introduce the variance $\overline{c^2}$ into the equation for the mean concentration of a passive scalar violate superposition and must be avoided. The variance due to a single source is $\overline{c_1^2}$; the variance due to a pair of sources is $\overline{(c_1+c_2)^2} = \overline{c_1^2} + \overline{c_2^2} + 2\overline{c_1c_2}$. Hence the variance of two sources is not simply the sum of the variances of the individual sources. Superposition does not apply to the square, or to any other nonlinear function of the concentration.

# Exercises

**Exercise 3.1.** *Derivation of Reynolds stress transport equation.* This exercise may seem laborious, but it is a good introduction to the use of Reynolds averaging, and to the Reynolds averaged Navier–Stokes (RANS) equations.

Derive (3.2.3) from (3.2.2) and (3.2.1) and then obtain (3.2.4). Symbolically the steps are

$$\mathrm{RS}_{ij} = \overline{u_j[\mathrm{NS}(U+u)_i - \overline{\mathrm{NS}(U+u)_i}]} + \overline{u_i[\mathrm{NS}(U+u)_j - \overline{\mathrm{NS}(U+u)_j}]},$$

where "NS" represents the Navier–Stokes equations. How many equations and how many unknowns are there? Why not also average the product of $u_i$ with the continuity equation?

**Exercise 3.2.** *Production of k.* Using (3.2.2) show that the rate at which mean energy (per unit mass) $\frac{1}{2}U_iU_i$ is lost to the turbulence is $\overline{u_iu_k}\,\partial_k U_i = -\mathcal{P}$. (Conservation terms are not an "energy loss.") This demonstrates that the term "production" is actually referring to the transfer of energy from the mean flow to the turbulence, and not to a net source of energy.

**Exercise 3.3.** *The mixing length rationale.* Consider the mean concentration $C(x_i)$ of a passive quantity that is convected by a turbulent velocity vector $u_j(t)$. Derive an eddy diffusion formula analogous to (3.3.3). The eddy diffusivity should come out as a second-order tensor, that is, a matrix of components. Is this tensor symmetric in general? Does the answer change when the turbulence is statistically stationary?

**Exercise 3.4.** *Anisotropy equation.* The Reynolds stress *anisotropy* tensor is defined as $b_{ij} = \overline{u_iu_j}/k - \frac{2}{3}\delta_{ij}$. Using (3.3.9) and (3.3.10) derive the evolution equation for $b_{ij}$ in homogeneous turbulence. The equation should involve $b_{ij}$, $\partial_j U_i$, $\phi_{ij}$, $\varepsilon_{ij}$, and $k$, with no explicit or implicit dependence on $\overline{u_iu_j}$.

"Isotropy" means complete lack of any directional preference. Hence the identity matrix $[\delta_{ij}]$ is isotropic because all the diagonal components are equal; or more correctly, because if the coordinate system were rotated, the identity matrix would remain unchanged. The tensor $b_{ij}$ measures the departure of $\overline{u_iu_j}$ from isotropy.

# 4

# Parallel and self-similar shear flows

> Science is the century-old endeavor to bring together by means of systematic thought the perceptible phenomena of this world into as thorough-going an association as possible.
>
> – Albert Einstein

This chapter is devoted to a description of prototypical building blocks of turbulent flows. Often we attempt to understand a complex flow field by identifying certain of these elements. The full flow field might include boundary layers on non-planar walls, mixing layers separating from edges, wakes behind bodies, and so on. An understanding can be gained by identifying such elements and studying them in isolation. In a practical application, they are not likely to be isolated; they are components of the overall flow. For instance, consider a jet impinging onto a plane wall. Prior to impingement, the generic flow element is an unconfined jet. This jet is known to entrain ambient fluid and to spread at rates that can be accurately measured in a controlled laboratory environment. After impingement, a transition from an unconfined to a wall jet takes place. This element too can be studied in a carefully controlled laboratory experiment. Certainly, the transition between these elements is not a generic flow; indeed, a driving motive to develop turbulence models is the need to analyze the entire flow, including the non-generic elements. However, the features that can be isolated and carefully studied provide a framework on which the general-purpose models can be anchored.

Practical applications invariably involve a multitude of complicating peculiarities. These might be vortices that form at the juncture between a plane wall and an appendage, or a pressure gradient transverse to the flow that skews the flow direction, or a large variety of other geometrical and fluid dynamical intricacies. Not all of them can be decomposed

*Statistical Theory and Modeling for Turbulent Flows, Second Edition* P. A. Durbin and B. A. Pettersson Reif

into simpler elements. The present chapter is meant to introduce some of the widely used building blocks that often can be identified.

## 4.1 Plane channel flow

The simplest, non-homogeneous turbulent shear flow is fully developed plane channel flow. The geometry is illustrated by Figure 4.1. The turbulence is non-homogeneous in the $y$ direction only. The choice of where the origin of $x$, $z$ or $t$ is located is immaterial; hence, derivatives of statistical quantities with respect to these directions and with respect to time are zero. The mean pressure gradient is not zero in the $x$ direction; it drives the flow through the channel.

After dropping $x$ and $z$ derivatives and invoking the two-dimensional (2D) parallel flow assumption that $U$ is a function only of $y$, the mean flow equations (3.2.2) simplify to

$$-\frac{1}{\rho}\,\partial_x P = \partial_y(\overline{uv} - \nu\,\partial_y U),$$
$$-\frac{1}{\rho}\,\partial_y P = \partial_y \overline{v^2}. \tag{4.1.1}$$

But $\partial_x \overline{v^2} = 0$, so differentiating the second of these with respect to $x$ shows that $\partial_y\,\partial_x P = 0$, or that $\partial_x P$ is constant, independent of $y$. This constant pressure gradient can be related to the skin friction via the overall momentum balance: integrating the first of (4.1.1) from 0 to $2H$ gives

$$-\frac{2H}{\rho}\,\partial_x P = 2\nu\,\partial_y U(0) \equiv 2\frac{\tau_w}{\rho} \equiv 2u_*^2. \tag{4.1.2}$$

The no-slip boundary condition has been used to set $\overline{uv}(0) = 0 = \overline{uv}(2H)$. In (4.1.2) $\tau_w$ is the frictional force acting per unit area on the surface; $u_* \equiv \sqrt{|\tau_w/\rho|}$ is referred to as the *friction velocity*.

Before proceeding, let us consider the symmetry of the flow upon reflection across the centerline $y = H$. Reflection in $y = H$ is equivalent to the replacements $y \to 2H - y$ and $x \to x$. Differentiating these and dividing by d$t$ gives the corresponding replacements $v \to -v$ and $u \to u$. Thus, the requirements of symmetry under reflection are $U(y) = U(2H - y)$, $\overline{uv}(y) = -\overline{uv}(2H - y)$, and $\mathrm{d}_y U(y) = -\mathrm{d}_y U(2H - y)$. Thus, $U(y)$ is symmetric, as shown in Figure 4.1, and $\mathrm{d}_y U(y)$ and $\overline{uv}$ are antisymmetric. By the latter antisymmetry, at $y = H$, $\overline{uv}(H) = -\overline{uv}(H)$ and hence the Reynolds stress vanishes at the channel centerline, $\overline{uv}(H) = 0$.

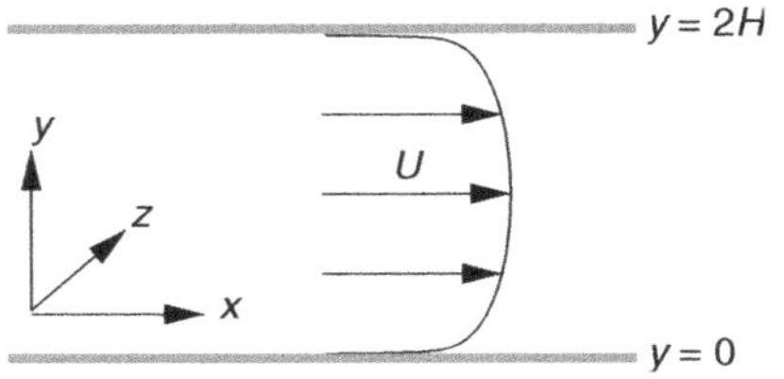


**Figure 4.1** Schematic of channel geometry and flow.

Substituting (4.1.2) into (4.1.1) gives

$$u_*^2 = H\,\partial_y(\overline{uv} - \nu\,\partial_y U). \tag{4.1.3}$$

This is to be solved subject to the no-slip condition $U(0) = 0 = \overline{uv}(0)$ and the symmetry condition $\mathrm{d}_y U(H) = 0 = \overline{uv}(H)$. Non-dimensional variables, referred to as "plus units," are introduced by

$$y_+ = yu_*/\nu, \qquad U_+ = U/u_*, \qquad \overline{uv}_+ = \overline{uv}/u_*^2, \qquad \text{and} \qquad R_\tau = u_* H/\nu.$$

Integrating (4.1.3) with respect to $y$, subject to the boundary conditions, and introducing these non-dimensional variables gives

$$\mathrm{d}_{y_+} U_+ - \overline{uv}_+ = 1 - \frac{y_+}{R_\tau}. \tag{4.1.4}$$

The left-hand side is just the sum of the viscous and Reynolds shear stresses. This total stress varies linearly with $y_+$. At any but the very lowest Reynolds numbers, the viscous term, $\mathrm{d}_{y_+} U_+$, is only important near to boundaries – say when $y_+ < 40$. The relative contributions of viscous and Reynolds stresses to the total is illustrated in Figure 4.2. The viscous contribution is practically zero over most of the channel.

Present interest is in high Reynolds number, turbulent flow with $R_\tau \gg 1$: turbulence cannot be maintained when $R_\tau \lesssim 100$. We will first consider (4.1.4) for small $y_+$, then move on to the log region at larger $y_+$.

By its definition, $y_+$ falls in the range $0 \le y_+ \le 2R_\tau$. Near to the lower wall, $y_+$ takes values of order one. In this region, called the *viscous sublayer*, $y_+/R_\tau$ is small, given that $R_\tau \gg 1$ and $y_+ = O(1)$. Hence, (4.1.4) is approximately $\mathrm{d}_{y_+} U_+ - \overline{uv}_+ = 1$; in dimensional terms, the total stress is approximately constant and equal to the surface stress $\tau_w$.

The channel geometry enters only through $R_\tau$, so within the constant-stress region there is no explicit dependence on the geometry. For this reason the constant-stress layer

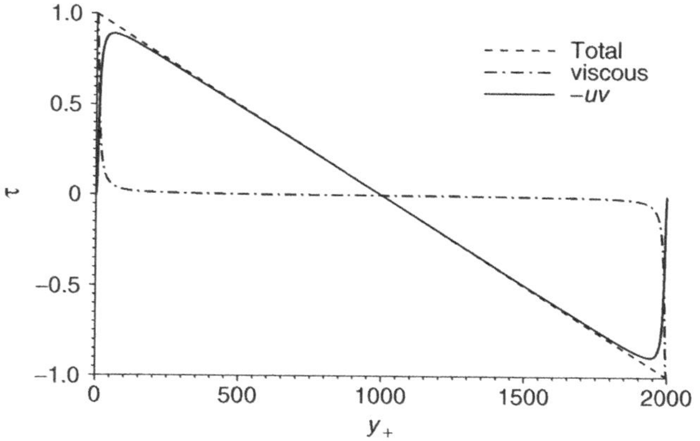


**Figure 4.2** Viscous and Reynolds shear stresses in plane channel flow: turbulent (———), viscous (— · —), and total (- - - -). The illustration shows that viscous stress is only important near to boundaries.

is referred to as a "universal equilibrium" layer, meaning that it is relatively insensitive to the details of the flow farther from the wall. A corollary to universality is that the present considerations are applicable to boundary layers, pipe flow, and other situations in which the region proximate to the boundary can be treated as a quasi-equilibrium, constant-stress layer. The term *quasi*-equilibrium is used because the outer flow is not irrelevant; it influences the viscous sublayer indirectly by determining the value of $u_*$. The wall stress is a function of Reynolds number and hence of the geometry of the flow.

The no-slip conditions $u = w = 0$ at $y = 0$ suggest that $u, w \rightarrow O(y)$ as $y \rightarrow 0$. From continuity, $\partial_y v = -\partial_x u - \partial_z w$. It follows that $v \rightarrow O(y^2)$ and hence that $\overline{uv}_+ \rightarrow O(y_+^3)$ right next to the wall. Then, integrating (4.1.4),

$$U_+ = y_+ + O(y_+^4) + O(R_\tau^{-1}). \tag{4.1.5}$$

Thus, $U$ varies linearly in a region adjacent to the wall. In dimensional units, the linear profile is $U = \tau_w y/\mu$. Although the present reasoning suggests that $U_+$ varies linearly only when $y_+ < 1$, Figure 4.4 shows that this is a good approximation up to $y_+ \approx 5$. In other flows, that often remains a good rule of thumb.

Farther from the wall, but still in the constant-stress region, $y_+ \ll R_\tau$, the turbulent stresses start to become important. The rate of energy production $\mathcal{P}_+ = -\overline{uv}_+ \mathrm{d}_{y_+} U_+$ (see Section 3.3) equals $-\overline{uv}_+(1 + \overline{uv}_+)$, if (4.1.4) is used to substitute $\mathrm{d}_{y_+} U_+ \approx \overline{uv}_+ + 1$. Hence $\mathcal{P}_+$ has a maximum value of $\frac{1}{4}$ when $-\overline{uv}_+ = \frac{1}{2} = \mathrm{d}_{y_+} U_+$. The equality of $-\overline{uv}_+$ and $\mathrm{d}_{y_+} U_+$ means that, at this point of maximum production, the Reynolds shear stress is equal to the viscous stress. Experiments and numerical simulations give a value of $y_+ \approx 10$–$15$ for the position of maximum production. Beyond this distance from the wall, the turbulent stress rapidly becomes larger than the viscous stress. By $y_+ \approx 40$ the so-called *logarithmic layer* is entered. The log layer is discussed in the next section.

Figure 4.3 shows the distribution of $\overline{u_i u_j}$ across a channel at a relatively modest Reynolds number $R_\tau = 590$. Beyond the near-wall region, the viscous shear stress is small and $-\overline{uv} \approx 1 - y_+/R_\tau$. The streamwise intensity, $\overline{u^2}$, is produced by mean shear (Eq. (3.3.5)) and has a peak where $\mathcal{P}$ peaks, at about $y_+ = 15$. Other normal stress

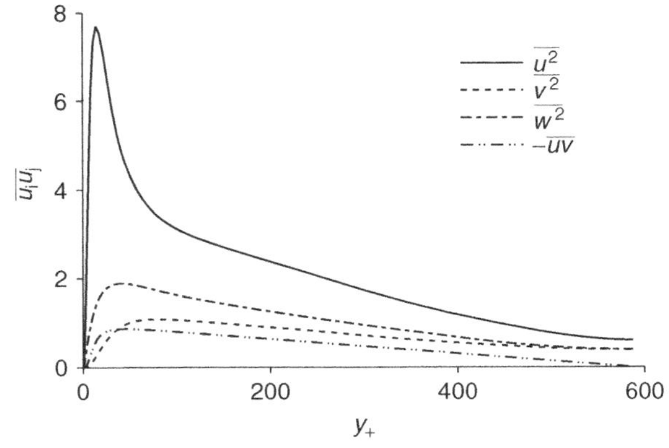


**Figure 4.3** Reynolds stress distribution across a channel at $R_\tau = 590$. Direct numerical simulation data of Moser *et al.* (1999). $y^+ = 590$ represents the center of the channel.

components are fed by redistribution of intensity from $\overline{u^2}$ (see Chapter 3); hence, they fall below $\overline{u^2}$. An enlarged plot in the region $y_+ < 5$ would show the behaviors $\overline{u^2} \sim O(y_+^2)$, $\overline{w^2} \sim O(y_+^2)$, $\overline{v^2} \sim O(y_+^4)$, and $\overline{uv} \sim O(y_+^3)$, which follow from no-slip and continuity, as described above and in Section 7.3.3.

### 4.1.1 Logarithmic layer

The log layer and log law are somewhat analogous to the inertial subrange and Kolmogoroff $-5/3$ law that were discussed in Section 2.1. As in that case, there is an inner viscous region, and an outer region that is affected by geometry. The two are connected by an intermediate region whose form is found by dimensional reasoning. The inner viscous region was discussed in the preceding section. The length scale for that region is $\ell_+ = \nu/u_*$, which defines "plus units." Toward the central region of the channel the length scale $\ell_H = H$ becomes relevant. In a boundary layer, the corresponding scale would be the overall boundary-layer thickness, $\delta$.

The mean momentum is highest in the central region of the channel. In the near-wall, viscous region, momentum is diffused to the wall and lost by viscous action. At high Reynolds number, there must be an intermediate region in which momentum is transferred toward the wall, but in which viscous stresses are not directly important. This is analogous to the reasoning in Section 2.1 that energy flows from large to small scales across an inertial subrange.

When $1 \ll y_+ \ll R_\tau$ the viscous term in (4.1.1) will be small. Then, from Eq. (4.1.4), $-\overline{uv}_+ \approx 1$, or $-\overline{uv} \approx u_*^2$ in dimensional form. We can continue the analogy to the inertial subrange. In the intermediate region neither $\ell_+$ nor $H$ can be the appropriate length scale. The former is unsuitable because in this region $y/\ell_+ \to \infty$; the latter is unsuitable because $y/H \to 0$ (this follows because the assumption $y_+ \ll R_\tau$ is equivalent to $y/H \ll 1$). In either scaling, the non-dimensionalized distance $y$ becomes a constant, while we want it to be an independent variable. As there are no remaining parameters with dimension of length, the coordinate $y$ itself is the only relevant length scale. From this, it will be inferred that the mean velocity must vary logarithmically. There are a variety of ways to deduce the log law; since the eddy viscosity has already been discussed in Chapters 2 and 3, that route will be taken.

The eddy viscosity has dimensions of $\ell^2/t$. In the constant-stress layer an appropriate velocity scale is $u_*$. Using $y$ for the length gives $\nu_T = \kappa u_* y$. Here $\kappa$ is a coefficient of proportionality called the Von Karman constant, after Theodore Von Karman, who first deduced the log law. It is considered to be a fundamental constant – though certainly not a constant of nature. Substituting this $\nu_T$ into (3.3.3) with $-\overline{uv} = u_*^2$ gives the famous result:

$$\mathrm{d}_y U = \frac{u_*}{\kappa y}. \tag{4.1.6}$$

This is a statement of the log law. That terminology makes sense after this is integrated to

$$U = \frac{u_*}{\kappa} \log(y_+) + B u_*, \tag{4.1.7}$$

where $B$ is an integration constant. Here and elsewhere log is the natural logarithm. From experiments, the constants are found to be $\kappa \approx 0.41$ and $B \approx 5.1$. The dashed

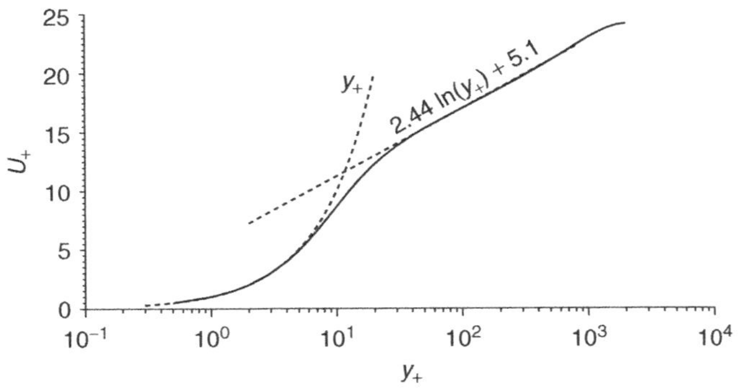


**Figure 4.4** Mean velocity in log–linear coordinates. The dashed lines show the near-wall asymptote $U \to y_+$ and the log law.

lines in Figure 4.4 show this formula and the near-wall behavior (4.1.5). The two formulas intersect at $y_+ \approx 11$. The portion of the velocity profile that interpolates between the viscous sublayer and the log layer is often referred to as the *buffer layer*. This is usually about $5 < y_+ < 40$.

It might seem that there is a sleight of hand here because $y$ was made non-dimensional by $\ell_+ = \nu/u_*$ in (4.1.7), while it had previously been argued that the viscous scale should be irrelevant in this intermediate layer. However, the scaling of the eddy viscosity and of (4.1.6) was consistent, so it is better to regard the necessary reappearance of $\ell_+$ in (4.1.7) as a theoretical deduction, rather than as an inconsistency. A consequence of this deduction is that (4.1.7) should be regarded as an approximation that is formally justified for large $\log(R_\tau)$. This will be explained shortly.

First, let us turn the apparent inconsistency to our advantage. One way to circumvent the ambiguity would be to make it irrelevant whether $\ell_+$ or $H$ is used in the argument of the logarithm. With $y$ normalized by $H$, Eq. (4.1.7) becomes

$$U(y) = \frac{u_*}{\kappa}[\log(y/H) + \log(R_\tau)] + Bu_*.$$

Let us devise a solution that is valid across the central region of the channel and that reduces to this in the log layer. Symmetry suggests using an eddy viscosity of the form $\kappa u_* y(1 - y/2H)$, which extends across the central region and varies linearly in the log layer near either wall. The solution to $\nu_T \partial_y U = u_*^2(1 - y/H)$ (see Eq. (4.1.4)) for the mean velocity is then

$$U(y) = \frac{u_*}{\kappa}\{\log[(y/H)(1 - y/2H)] + \log(R_\tau)\} + Bu_*. \tag{4.1.8}$$

This becomes the log-layer formula near either wall, when $y/2H \ll 1$ or when $1 - y/2H \ll 1$. Evaluating (4.1.8) at $y = H$ shows that $u_*$ and the centerline mean velocity, $U(H)$, are related by

$$u_* = \frac{U(H)\kappa}{\log(R_\tau/2) + B\kappa}. \tag{4.1.9}$$

Instead of working with $u_*$, let the non-dimensional skin friction coefficient be defined by

$$C_f \equiv \frac{2u_*^2}{U_H^2}. \tag{4.1.10}$$

Equation (4.1.9) can be written as the implicit equation

$$\sqrt{\frac{2}{C_f}} = \frac{1}{\kappa}\log\left(\frac{R_H}{2}\sqrt{\frac{C_f}{2}}\right) + B \tag{4.1.11}$$

for the friction coefficient – note that (4.1.11) is an implicit relation for $C_f$. The Reynolds number based on centerline velocity is $R_H = U(H)H/\nu$.

So a rather interesting result has come out of pursuing the dilemma: the friction velocity is related to the centerline velocity by a logarithmic drag law. The attempt to resolve the apparent inconsistency in the rationale behind the logarithmic velocity profile has led to a formula that requires the friction velocity to depend on $\log(R_H)$. The reasoning is more subtle than first appears. At the outset $u_*$ and $\nu$ were taken as independent parameters. Now, it seems that $u_*$ depends on $\nu$ and $H$ through Eq. (4.1.9). However, the whole theory can be shown to be formally consistent at high Reynolds number. Typically $C_f = O(10^{-3})$, so $u_*/U_H = O(10^{-1})$ or smaller (see Exercise 4.2). By Eq. (4.1.9), $u_*$ becomes small like $1/\log R_\tau$ at high Reynolds number. Such observations lead to an asymptotic justification of the log-layer theory (Lundgren, 2007).

Formula (4.1.11) is an example of a semi-theoretical turbulent drag law. Various other empirical formulas for skin friction as a function of $R_H$ have been proposed for channel flow and boundary layers (Schlichting, 1968). Some will be met later in this book. Figure 3.2 on page 47 shows the variation of $C_f$ with momentum thickness Reynolds number in a zero pressure-gradient boundary layer. The slow decrease of $C_f$ with $R_\theta$ on the turbulent part of the solid curve is qualitatively consistent with (4.1.9) because the logarithmic dependence represents a mildly decreasing function. On the laminar part of that curve, $C_f \propto R_\theta^{-1}$ can be derived from the Blasius profile; this has slope $-1$ on the log–log plot. As can be seen, in the transitional region of Figure 3.2 there is considerable sensitivity of $C_f$ to $R_\theta$. However, the turbulent $C_f$ is considerably less sensitive to Reynolds number. This weak sensitivity to Reynolds number is characteristic of fully developed turbulence. The turbulent friction coefficient is often fit explicitly by $C_f \propto R_\theta^{-1/4}$ (see Eq. (4.2.8)) instead of by an implicit expression like (4.1.11).

## 4.1.2 Roughness

Surface roughness can have a profound effect on the transfer of momentum or heat between the fluid and wall. The viscous sublayer adjacent to a smooth wall presents a high impedance to transport to and from the surface; protrusions that penetrate the viscous layer increase transfer rates between the surface and the fluid. They do so by generating irregular, turbulent motion and by extending the surface into the flow.

Intuitively, asperities on the surface will increase the drag force exerted by the wall on the flow. In a channel flow with given pressure drop, the increased drag would decrease the mass flux and the centerline velocity. The additive constant, $B$, in (4.1.7) should therefore be decreased. Let $r$ be a scale for the size of the roughness. Assume that it is of the random, sand grain, variety. The log law can be rewritten with $y$ normalized by $r$ and with $r_+$ defined as $ru_*/\nu$:

$$U = u_* \left[ \frac{1}{\kappa} \log(y/r) + \frac{1}{\kappa} \log(r_+) + B_r(r_+) \right]$$
$$= u_* \left[ \frac{1}{\kappa} \log(y/r) + \mathcal{B} \right]. \qquad (4.1.12)$$

The function $B_r(r_+)$ represents the alteration of the additive constant by roughness.

The new additive term $\mathcal{B} \equiv (1/\kappa)\log(r_+) + B_r(r_+)$ has been measured experimentally. Ligrani and Moffat (1986) fit the curve

$$\begin{aligned} B_r &= B, && r_+ < 2.25, \\ B_r &= \xi[8.5 - \log(r_+)/\kappa - B] + B, && 2.25 \le r_+ \le 90, \\ B_r &= 8.5 - (1/\kappa)\log(r_+), && r_+ > 90, \end{aligned} \qquad (4.1.13)$$

through such measurements. This formula is broken into three regions: effectively smooth, transitionally rough, and fully rough. The interpolation function $\xi$ in (4.1.13) is

$$\xi = \sin\left[\frac{\frac{1}{2}\pi \log(r_+/2.25)}{\log(90/2.25)}\right],$$

which increases from 0 to 1 through the transitionally rough range $2.25 \le r_+ \le 90$.

The rough-wall law (4.1.12) is sometimes represented by

$$U = \frac{u_*}{\kappa} \log(y/z_0),$$

where $z_0 \equiv r e^{-\kappa\mathcal{B}}$ can be called the hydrodynamic roughness length. In the fully rough regime, $r_+ > 90$, Eqs. (4.1.13) become $\mathcal{B} = 8.5$, giving $z_0 = 0.031r$. The hydrodynamic roughness length is a small fraction of the geometrical roughness size. In the smooth-wall regime, $r_+ < 2.5$, $z_0 = e^{-\kappa B}\nu/u_*$, which recovers the original formula (4.1.7).

If the roughness elements are not random, or are not densely placed, the formula (4.1.13) alters its form. The changes are primarily in the intermediate range of $r_+$. The particular functional form for $B_r(r_+)$ depends on the nature of the roughness. Ligrani and Moffat (1986) provide a formula for close-packed hemispherical elements, in addition to the formula (4.1.13) for sand grain roughness.

Sometimes two-dimensional, spanwise ribs appended to the wall are referred to as roughness elements. The terminology "$d$-type" roughness is used for closely placed ribs, while more widely spaced ribs are of "$k$-type." The $k$-type behave like the random, three-dimensional roughness treated above. If the rib spacing is less than approximately its height, the $d$-type roughness can trap pockets of fluid between the ribs, lowering the drag on the surface compared to $k$-type behavior. Rib roughness is especially important to heat-transfer applications. Often, it is best treated as flow over a deterministic geometry, and not lumped with the random roughness considered here. In other words, in a numerical analysis, the features of the geometry should be resolved by the computational grid.

## 4.2 Boundary layer

Ideas concerning the log law and the viscous wall region are equally applicable to boundary layers. The relevance is illustrated by experimental data plotted in Figure 4.5. Near the wall the boundary-layer profiles in log–linear coordinates, and "plus" variables, are similar to those for channel flow: adjacent to the surface $U_+ \propto y_+$, and when $y_+ > 40$, the form $U_+ \propto \log y_+$ is again seen.

However, there is a difference farther from the surface. The mean boundary-layer velocity must become constant in the free stream, $U \to U_\infty$ as $y \to \infty$. For some purposes the profile might be cut off at $y = \delta_{99}$ (defined as the elevation where $U = 0.99U_\infty$) to make the flow analogous to that in half of a channel. However, the boundary-layer profile has a tendency to rise above the logarithmic line, especially in the adverse pressure-gradient (APG) data of Figure 4.5. This area where the profile rises above the log line is called the "law-of-the-wake" region. Thus boundary-layer structure can be separated into two regions: the *law-of-the-wall* region lies near the surface, where the velocity falls below the log line in Figure 4.5; the *law-of-the-wake* region lies in the outer part of the flow, where the velocity usually rises above the log line. The log line itself is the common asymptote of the two regions and is considered to be part of both zones. These regions are labelled and demarcated by cross-hatching in the left-hand side plot of Figure 4.5.

In a zero pressure-gradient boundary layer, the law of the wall occupies the interval $0 < y \lesssim 0.2\delta_{99}$; the log layer occupies the portion $40\nu/u_* \lesssim y \lesssim 0.2\delta_{99}$; and the law of the wake occupies $40\nu/u_* \lesssim y < \infty$. Of course the upper limit for the wall region and the lower limit for the wake region could have been selected anywhere within the log layer;

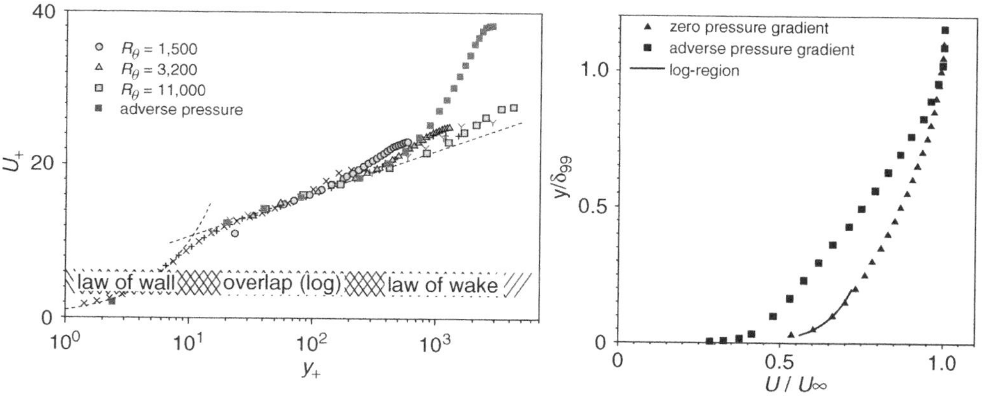


**Figure 4.5** Mean velocity in zero and adverse pressure-gradient boundary layers. The log–linear plot illustrates the law-of-the-wall and the law-of-the-wake regions. The APG data illustrate how the wake region grows when the boundary layer is retarded. The linear plot at the right shows how the APG erodes the boundary-layer profile. This figure also shows the log region $40\nu/u_* < y < 0.2\delta_{99}$ by a solid curve. The symbols are experimental data.

the above values simply include the entire log layer in both the inner and the outer regions. The specific values $40\nu/u_*$ and $0.2\delta_{99}$ should be treated as representative, not rigorous.

It is instructive to show the log layer on a linear plot. That is done at the right of Figure 4.5. In linear coordinates the log line appears rather mundane. It is a short region lying above the layer of steep shear adjacent to the surface.

The law-of-the-wake region is analogous to the outer region in the channel flow if $\delta_{99}$ is used in place of $H$ for the length scale. There are two ways to represent the mean velocity in this region, corresponding to the log–linear and linear–linear plots in Figure 4.5. Corresponding to the linear–linear plot, the velocity profile is written in the defect form

$$U = U_\infty - u_* f(y/\delta_{99}). \tag{4.2.1}$$

In this formula, $u_* f(y/\delta_{99})$ equals the departure of $U$ from its free-stream value; it is the amount by which $U(y)$ lies to the left of $U_\infty$ in the right half of Figure 4.5. Asymptotically $U \to U_\infty$ at large $y$ and $f \to 0$.

This representation of the wake function will be used to illustrate how the overlap requirement leads to a drag law. Various functions have been proposed for $f(y/\delta_{99})$. Irrespective of its behavior at arbitrary $y$, when $y \ll \delta_{99}$ it must tend to a logarithmic form. The form

$$f(y/\delta_{99}) = -\frac{1}{\kappa}\log(y/\delta_{99}) + C$$

can be assumed when $y \ll \delta_{99}e^{\kappa C}$, where $C$ is a constant that depends on the flow. $C \approx 2.3$ is the experimental value for the zero pressure-gradient boundary layer; lower values are found for channel flow because it has a smaller law-of-the-wake defect.

If it is required that, as $y/\delta_{99} \to 0$, the wake form (4.2.1) matches to the wall form (4.1.7), then

$$U_\infty + \frac{u_*}{\kappa}\log(y/\delta_{99}) - Cu_* = \frac{u_*}{\kappa}\log(yu_*/\nu) + Bu_*$$

in their region of common validity. This gives the skin friction law

$$\frac{U_\infty}{u_*} = \frac{1}{\kappa}\log(\delta_{99}u_*/\nu) + B + C. \tag{4.2.2}$$

For instance, given the free-stream velocity and the Reynolds number, this is a formula to predict the friction velocity $u_*$.

Corresponding to the log–linear plot on the left-hand side of Figure 4.5, the wake law is written as

$$\frac{U}{u_*} = \frac{1}{\kappa}\log(yu_*/\nu) + B + \frac{\Pi}{\kappa}w(y/\delta_{99}) \tag{4.2.3}$$

instead of (4.2.1). The wake parameter, $\Pi$, depends on the imposed pressure gradient. Comparing (4.2.3) to (4.1.7), $\Pi w(y/\delta_{99})/\kappa$ is just the amount by which the velocity rises above the log law in Figure 4.5. The letter $w$ stands for "wake" function. Coles (see Kline *et al.*, 1968) defined $w(1)$ to be 2, so that $2\Pi/\kappa$ is defined as the velocity excess above the log law at $y = \delta_{99}$. It is readily evaluated from log–linear plots of experimental data.

The wake parameter, $\Pi$, must increase with increasingly adverse pressure gradient in order to mimic the effect seen in Figure 4.5. An empirical formula that has been fit to data is

$$\Pi = 0.8(\beta + 0.5)^{3/4}, \tag{4.2.4}$$

where the pressure-gradient parameter is defined as $\beta = (\delta^*/\tau_w)d_x P_\infty$. This formula ceases to be valid when $\beta = -0.5$; for stronger favorable pressure gradients $\Pi$ becomes negative. Coles used the form $w(y/\delta_{99}) = 2\sin^2(\pi y/2\delta_{99})$ as an estimate of the wake function. This is constrained by $w(1) = 2$ and $w(0) = 0$, with the derivative with respect to $y$ vanishing at both endpoints. The effect of pressure gradient on skin friction can be inferred from (4.2.3) as in Exercise 4.6.

In addition to $\delta_{99}$, other measures of boundary-layer thickness are the momentum thickness $\theta$, and displacement thickness $\delta^*$, defined by

$$U_\infty^2\theta = \int_0^\infty U(y)[U_\infty - U(y)]\,dy,$$
$$U_\infty\delta^* = \int_0^\infty U_\infty - U(y)\,dy. \tag{4.2.5}$$

These measure the momentum and mass flux deficits from the free-stream flow caused by the presence of the boundary layer. The ratio $H = \delta^*/\theta$ is called the "form factor." In a zero pressure-gradient layer it is typically 1.4–1.3, decreasing with increasing Reynolds number. The corresponding value for a laminar, Blasius boundary layer is 2.5.

The form factor $H$ characterizes the fullness of the profile. As the deficit fills in, $H$ decreases toward 1; $H$ would be almost unity for the nearly flat profile, $U(y) \approx U_\infty$ for most $y$. Turbulent eddies stir high-velocity fluid toward the wall, filling out the profile and decreasing $H$ below its laminar value. The zero pressure-gradient profile in Figure 4.5 displays this filled-out form. It has a very steep shear near the wall: that is why $C_f$ is much larger than it would be in a laminar boundary layer at the same Reynolds number.

An adverse pressure gradient increases $H$ and decreases $C_f$. A sufficiently persistent decelerating pressure gradient will cause the boundary layer to separate when $H$ reaches about 3. These integrated properties provide some insight into the development of turbulent boundary layers. They will be discussed below and in Section 6.1.1.

The boundary-layer approximation (White, 1991) is to neglect the $x$ derivative of viscous and Reynolds stresses in comparison to $y$ derivatives. Also, the mean pressure gradient is that imposed by the free stream. This is a valid approximation for any thin shear layer. In the boundary layer the Reynolds averaged mean flow equations (3.2.2) become

$$U\,\partial_x U + V\,\partial_y U = -\partial_x P_\infty + \partial_y(\nu\,\partial_y U - \overline{uv}),$$
$$\partial_x U + \partial_y V = 0. \tag{4.2.6}$$

The pressure gradient can be eliminated by letting $y$ tend to the free stream. In the free stream, viscous and turbulent stresses are negligible and so is the $y$ derivative. Hence (4.2.6) asymptotes to $\partial_x P_\infty = -U_\infty\,\partial_x U_\infty$. After substituting this, the first of (4.2.6) can be rearranged to

$$\partial_x[U(U - U_\infty)] + \partial_y[V(U - U_\infty)] + (U - U_\infty)\,\partial_x U_\infty = \partial_y(\nu\partial_y U - \overline{uv}).$$

Integrating this equation with respect to $y$, from the wall to $\infty$, gives the boundary-layer momentum integral equation (Exercise 4.5). In the case, $\partial_x U_\infty = 0$, of zero pressure gradient, the momentum integral equation is

$$U_\infty^2 \, \mathrm{d}_x \theta = \tau_\mathrm{w}/\rho \equiv u_*^2, \tag{4.2.7}$$

which is a rather simple formula for the growth of the boundary layer. It can also be written as $\mathrm{d}_x \theta = C_\mathrm{f}/2$. Since $C_\mathrm{f} = O(10^{-3})$, Eq. (4.2.7) shows that the boundary-layer thickness grows very slowly. Note that (4.2.7) is valid in laminar or turbulent flow; but the rate of boundary-layer growth is greater in turbulent flow. What is the difference? The difference arises because (4.2.7) is an unclosed equation. A relation between $\tau_\mathrm{w}$ and the mean flow is needed to close the equation. That relation is rather different in laminar or turbulent flow.

Equation (4.2.7) can be converted into a closed equation for boundary-layer growth if a functional dependence of the form $C_\mathrm{f}(R_\theta)$ is prescribed. The previously mentioned data correlation

$$C_\mathrm{f} = a R_\theta^{-1/4} \tag{4.2.8}$$

can be used in turbulent flow, where $a$ is a constant found in experiments to be approximately equal to 0.025. With this formula for $C_\mathrm{f}$, the momentum integral (4.2.7) becomes

$$\mathrm{d}_x \theta = \frac{a}{2} \left( \frac{\nu}{U_\infty \theta} \right)^{1/4} \tag{4.2.9}$$

and has solution

$$\frac{\theta}{x} = \left( \frac{5a}{8} \right)^{4/5} \left( \frac{U_\infty x}{\nu} \right)^{-1/5}. \tag{4.2.10}$$

The thickness grows a bit less than linearly, as $x$ to the power 4/5, with the slope decreasing as the $x$ Reynolds number to the power $-1/5$. This can be compared to the laminar boundary layer, for which the exponent on the right-hand side of (4.2.10) is $(U_\infty x/\nu)^{-1/2}$ and $\theta$ grows as $x^{1/2}$. The slope of the turbulent boundary-layer thickness is closer to linear than that of the laminar layer.

How pressure gradients alter the profiles of Reynolds stresses can be deduced from their effect on the mean flow. Consider velocity profiles like those in Figure 4.5. The adverse pressure gradient reduces the wall stress and generates shear farther out in the boundary layer. This alteration to the mean flow will move the production of turbulence away from the wall, into the central portion of the layer. The near-wall peak of turbulent kinetic energy then diminishes and a broad maximum develops in the outer part of the boundary layer. This is a generic behavior observed in adverse pressure-gradient boundary layers. Experimental data and a Reynolds stress computation in Figure 4.6 illustrate the evolution of turbulent stresses upon moving into an increasingly adverse pressure gradient. The computation was carried out with one of the closure models developed in Chapter 7. Measurements near the wall are difficult. The diminution of the $k$ peak is not evident in the data; it is illustrated clearly by the computation.

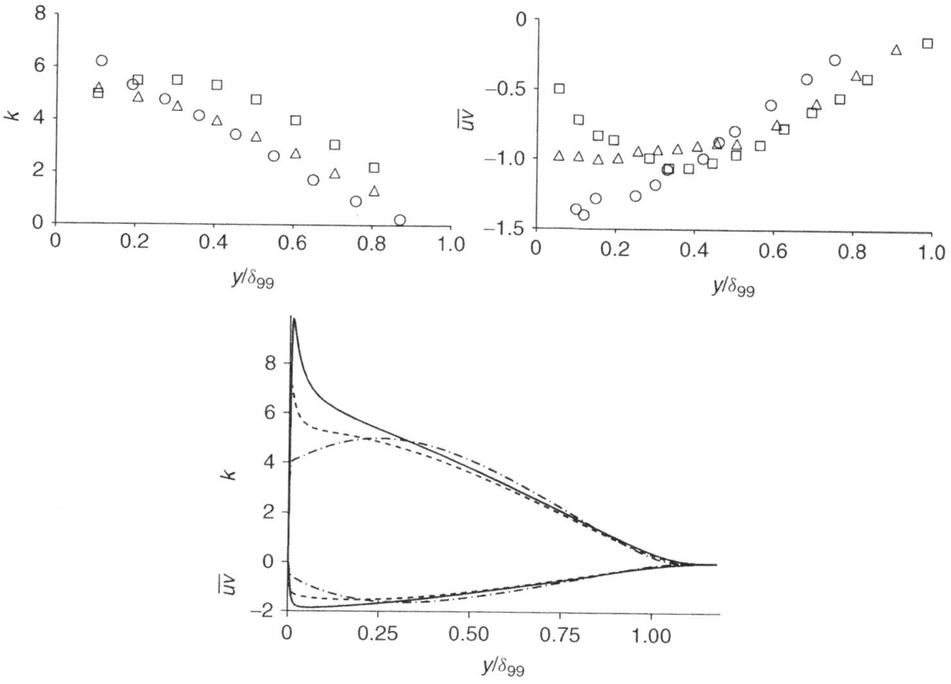


**Figure 4.6** The evolution of Reynolds stresses in an adverse pressure gradient. Top, experimental data from Samuel and Joubert (1974): initial, zero pressure gradient (○); downstream, adverse pressure gradient (△); farther downstream (□). Bottom, a calculation with a Reynolds stress model: zero pressure gradient (——); downstream (- - - -); farther downstream (— · —).

## 4.2.1 Entrainment

As the boundary layer grows in thickness, more fluid becomes turbulent. In this sense, the boundary layer is *entraining* free-stream, laminar fluid. The rate of entrainment is analyzed as follows. At the top of the boundary layer, the mean velocity is approximately $U_\infty$ in the $x$ direction and $V_{99}$ in the $y$ direction, the non-zero value of $V$ being the displacement effect of the growing boundary layer. Hence, the velocity vector is $\boldsymbol{U}_{99} = (U_\infty, V_{99})$, to an accuracy of $O(\mathrm{d}\delta_{99}/\mathrm{d}x)^2$. Taking the 99% thickness, $\delta_{99}(x)$, as a measure of the "edge" of the boundary layer, the outward normal to this edge is $\hat{n} = (-\mathrm{d}_x\delta_{99}, 1)$, to the same order of accuracy. The entrainment velocity is defined as the velocity normal to the edge. It is taken as positive into the boundary layer, so it is given by $V_E = -\hat{n} \cdot \boldsymbol{U}_{99}$, or

$$V_E = U_\infty\,\mathrm{d}_x\delta_{99} - V_{99}.$$

Integrating the continuity equation of (4.2.6) across the boundary layer gives

$$V_{99} = -\int_0^{\delta_{99}} \partial_x U \,\mathrm{d}y = \int_0^{\delta_{99}} \partial_x (U_\infty - U)\,\mathrm{d}y - \delta_{99}\,\mathrm{d}_x U_\infty$$
$$\approx \mathrm{d}_x \int_0^{\infty} (U_\infty - U)\,\mathrm{d}y - \delta_{99}\,\mathrm{d}_x U_\infty = \mathrm{d}_x(U_\infty \delta^*) - \delta_{99}\mathrm{d}_x U_\infty.$$

Substituting this into the expression for $V_E$ gives

$$V_E = \mathrm{d}_x[U_\infty(\delta_{99} - \delta^*)] \tag{4.2.11}$$

for the velocity with which free-stream fluid enters the boundary layer. For the purposes of description, entering the boundary layer is defined here as crossing the 99% thickness. This quantifies the notion of entrainment of free-stream fluid by the boundary layer.

Typically $\delta_{99}/\delta^*$ might be about 6–8, so (4.2.11) represents a velocity into the boundary layer of about $6\,\mathrm{d}_x(U_\infty \delta^*)$. If $U_\infty$ and the form factor $H$ are constant, then this can be written $V_E/U_\infty \approx 6H\,\mathrm{d}_x\theta = 3HC_f$. Inserting typical values of $H = 1.35$ and $C_f = 3 \times 10^{-3}$ gives the order-of-magnitude estimate $V_E/U_\infty \sim 10^{-2}$.

Entrainment is considered to be associated with large eddies inside the boundary layer engulfing irrotational, free-stream fluid. Coherent, large eddies are the subject of Chapter 5. Smaller eddies are ultimately responsible for mixing the irrotational fluid into the vortical, turbulent region. Figure 5.11 (on page 101) shows an experimental view of this process.

## 4.3 Free-shear layers

Turbulent shear layers that are not adjacent to boundaries are referred to as free-shear layers. Examples are wakes, jets, and mixing layers. These are illustrated in Figures 1.8, 1.7, and 4.7. Such flows are characterized by a distinct vortical region, with a

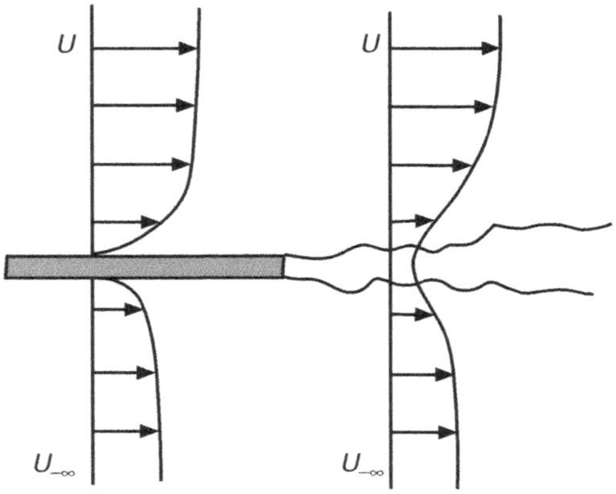


**Figure 4.7** Schematic of a two-stream mixing layer.

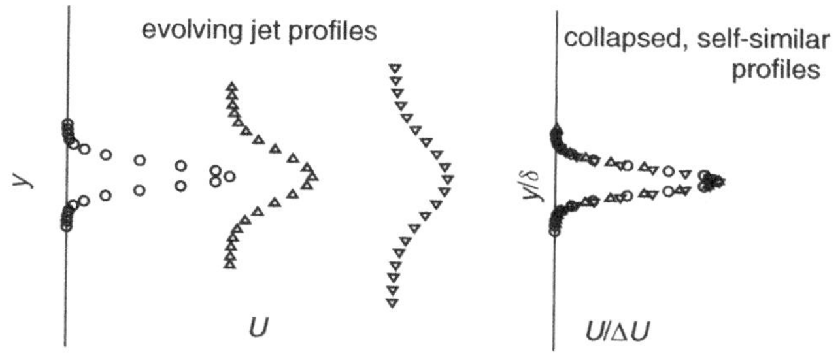


**Figure 4.8** Illustration of evolving jet profiles and collapse of the data by plotting $U/\Delta U(x)$ versus $y/\delta(x)$.

characteristic thickness, say $\delta$. As the shear layer proceeds downstream, the thickness of the vortical region grows via entrainment. If the profiles of velocity (and Reynolds stresses) at different distances can be collapsed by rescaling the velocity and transverse coordinate, then the flow is *self-similar*. The reference scale with which the coordinate is non-dimensionalized is the local shear-layer thickness, $\delta(x)$. The reference velocity is a characteristic difference, $\Delta U(x)$. The left portion of the schematic in Figure 4.8 shows how a turbulent jet spreads and how its centerline velocity diminishes with downstream distance. But if the $y$ coordinates of these synthetic data are divided by the jet thickness, $\delta(x)$, defined here as the point where $U$ is half of its centerline value, and if the velocity data are divided by the centerline value, $\Delta U = U_{\mathbb{C}}(x)$, then these data collapse onto a single curve as shown to the right of the figure.

Velocity profiles and turbulence data from experiments on self-similar mixing layers and wakes are shown in Figures 4.9 and 4.10. They have been collapsed by the appropriate similarity scaling. The appropriate scaling will be developed below.

In most engineering applications, shear layers are not self-similar. For example, immediately after a round jet exits a nozzle, it consists of a "potential core" surrounded by a shear layer, the latter being the downstream continuation of the boundary layer inside the nozzle. The free-shear layers spread, eventually merging in the center at about five jet diameters downstream. Only after about 15 diameters does the jet become self-similar. In aeroacoustic applications, the non-similar potential core region is where noise is produced; in impingement cooling applications the jets rarely, if ever, attain self-similarity before they reach the surface. Nevertheless, our basic understanding of turbulent shear layers comes from studying fully developed flows because they are amenable to theoretical attack and can be studied by generic experiments.

The two-dimensional, thin-layer equations (4.2.6) apply to free-shear layers as well as to boundary layers. In high Reynolds number flow, away from any surface, the viscous term can be dropped – this is because it is not needed for imposing the no-slip condition. We will also consider only flows without a free-stream pressure gradient. Then the streamwise momentum and continuity equations become

$$U\,\partial_x U + V\,\partial_y U = -\partial_y \overline{uv},$$
$$\partial_x U + \partial_y V = 0. \tag{4.3.1}$$

In order to derive a momentum integral for the two-dimensional jet, the continuity equation is used to rewrite the first of (4.3.1) as

$$\partial_x(UU) + \partial_y(VU) = -\partial_y\overline{uv}.$$

Integrating this between $y = \pm\infty$, with $U \to 0$ as $y \to \pm\infty$, gives

$$\mathrm{d}_x \int_{-\infty}^{\infty} U^2\, \mathrm{d}y = 0.$$

When the self-similarity assumption $U = \Delta U f(y/\delta)$ is substituted, this becomes

$$\mathrm{d}_x \left[ \delta \Delta U^2 \int_{-\infty}^{\infty} f^2(y/\delta)\, \mathrm{d}(y/\delta) \right] = 0.$$

The term $\int_{-\infty}^{\infty} f^2(y/\delta)\, \mathrm{d}(y/\delta)$ is a number that is independent of $x$. It depends on the velocity profile, $f$, but that is assumed to be of a universal, self-similar form. Here, "universal" means that it does not depend on the particularities of the nozzle from which the jet emerged. It follows that

$$\delta \Delta U^2 = \text{constant.} \tag{4.3.2}$$

and hence that $\Delta U \propto 1/\sqrt{\delta}$. This is the relation between the decay of the centerline velocity and the spread of the jet in Figure 4.8. Equation (4.3.2) expresses the constancy of the momentum flux in the jet; the constant on the right-hand side has dimensions of $\ell^3/t^2$. The momentum flux must be constant because there are no forces that would change it acting on the free jet.

For an axisymmetric jet the momentum and continuity equations are

$$\partial_x UU + \frac{1}{r}\,\partial_r rVU = -\frac{1}{r}\,\partial_r r\overline{uv},$$
$$\partial_x U + \frac{1}{r}\,\partial_r rV = 0. \tag{4.3.3}$$

The momentum integral is now

$$\mathrm{d}_x(\delta^2 \Delta U^2) \int_0^{\infty} f^2(r/\delta)(r/\delta)\, \mathrm{d}(r/\delta) = 0,$$

showing that

$$\delta^2 \Delta U^2 = \text{constant.} \tag{4.3.2a}$$

Hence $\Delta U \propto 1/\delta$ in axisymmetric flow.

The two other generic free-shear flows are wakes and mixing layers. The mixing layer has a high-speed side, with velocity $U_\infty$, and a low-speed side, with velocity $U_{-\infty}$, as in Figure 4.7. A point of terminology: when $U_{-\infty} = 0$, this is referred to as a "single-stream" mixing layer; if $U_{-\infty} > 0$, it is a "two-stream" mixing layer. A scaled mean flow profile is shown in Figure 4.9. The velocity scale

$$\Delta U = U_\infty - U_{-\infty}$$

is independent of $x$.

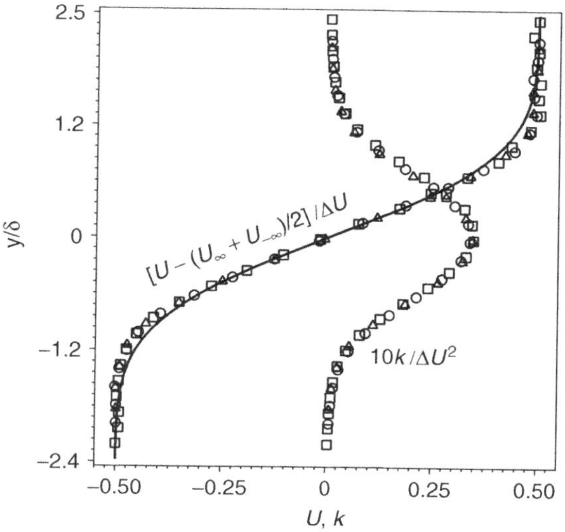


**Figure 4.9** Mean velocity and turbulent energy profiles in a mixing layer. Velocities are scaled as indicated. The $y$ coordinate is scaled on the thickness obtained by fitting an error function curve through the mean flow data; the Erf fit is shown by the solid curve. Experiments of Bell and Mehta (1990).

For the 2D, or plane, wake with a uniform free-stream velocity, the momentum equation is written as

$$\partial_x[U(U - U_\infty)] + \partial_y[V(U - U_\infty)] = -\partial_y \overline{uv}. \tag{4.3.4}$$

The wake velocity profile is of the form $U_\infty - \Delta U f(y/\delta)$. The scaled profile, $f(y/\delta)$, is plotted in Figure 4.10. A formal analysis shows that the wake cannot be strictly self-similar. Loosely speaking, this is because $U_\infty$ is independent of $x$ while $\Delta U$ decreases as the wake spreads, so there are two velocity scales with different $x$ dependence. That prevents self-similarity. For the same reason, a jet in co-flow cannot be self-similar. Such flows can only be approximately self-similar. The approximately self-similar state is reached when $\Delta U \ll U_\infty$. This means that the self-similar region is far downstream of the body that produces the wake.

The momentum integral of (4.3.4) is

$$\mathrm{d}_x \int_{-\infty}^{\infty} U(U - U_\infty)\,\mathrm{d}y = \mathrm{d}_x \left\{ \delta \Delta U \int_{-\infty}^{\infty} [U_\infty - \Delta U f(y/\delta)] f(y/\delta)\,\mathrm{d}(y/\delta) \right\} = 0, \tag{4.3.5}$$

assuming that the profile of $U$ can be represented by $U_\infty - \Delta U f(y/\delta)$. This integral is not just a number: it is of the form $AU_\infty + B\Delta U$, which depends on $x$. Again, this suggests that the wake is not self-similar unless $U_\infty \gg \Delta U$. In that limit the integrand of (4.3.5) reduces to $U_\infty f(y/\delta)$ so that

$$\delta \Delta U U_\infty = \text{constant}. \tag{4.3.6}$$

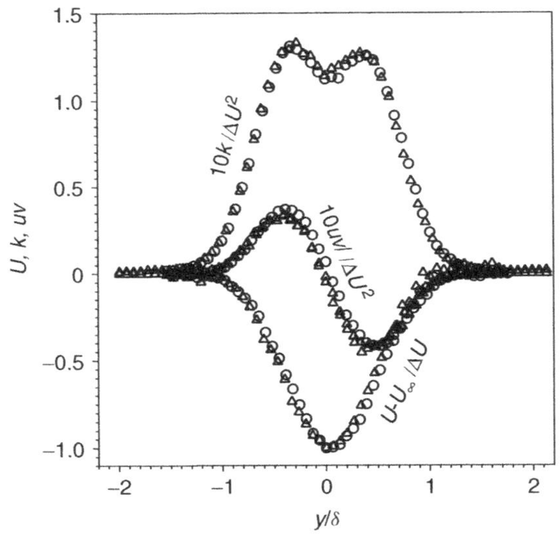


**Figure 4.10** Wake defect, turbulent energy, and shear stress in a plane wake. The $y$ coordinate is scaled on the half-width of the wake. Experiments of Weygandt and Mehta (1995).

In this limit the momentum deficit is constant and is convected with velocity $U_\infty$, so the momentum equation would simply be $U_\infty\, \partial_x U = -\partial_y \overline{uv}$. In the axisymmetric case

$$\delta^2 \Delta U U_\infty = \text{constant}. \tag{4.3.6a}$$

The constant in (4.3.6) or (4.3.6a) is proportional to the drag on the body that produced the wake. A self-propelled body produces a zero-momentum deficit wake because the propulsive thrust must equal the resistive drag. The constants in (4.3.6) and (4.3.6a) are then zero and other considerations are required to obtain the self-similar scaling (Exercise 4.10).

Equations (4.3.2), (4.3.2a), (4.3.6), and (4.3.6a) relate the reference velocity to the thickness of the shear layer for jets and wakes. In summary

$$\begin{aligned}
&\Delta U = U_{\mathbb{C}} \propto 1/\sqrt{\delta}, && \text{2D jet,}\\
&\Delta U = U_{\mathbb{C}} \propto 1/\delta, && \text{axisymmetric jet,}\\
&\Delta U \propto 1/\delta, && \text{2D wake,}\\
&\Delta U \propto 1/\delta^2, && \text{axisymmetric wake,}\\
&\Delta U = U_\infty - U_{-\infty}, && \text{mixing layer.}
\end{aligned} \tag{4.3.7}$$

But how do these shear layers grow with $x$? This can be answered by dimensional reasoning and the dispersion formula (2.2.20). (A more formal approach is described in Exercise 4.9.)

Equation (2.2.20), page 30, can be applied to dispersion in the $y$ direction across thin shear layers:

$$\frac{D\overline{Y^2}}{Dt} = 2\alpha_T. \tag{4.3.8}$$

Although dispersion is not the present topic, the qualitative reasoning behind that formula suits the present purpose of simple scaling analysis. To this end, $D\overline{Y^2}/Dt$ can be regarded as proportional to $U_c\, d_x\delta^2$, where $U_c$ is a convection velocity. This uses the idea that $\overline{Y^2}$ is a measure of shear-layer thickness, and introduces a scaling estimate for the convection velocity. For the jet, an appropriate velocity scale is the centerline velocity; for the small-deficit wake, it is $U_\infty$; for the mixing layer, is it the average velocity $\frac{1}{2}(U_\infty + U_{-\infty})$. Using these and the replacement $D\overline{Y^2}/Dt \to U_c\, d_x\delta^2$ in (4.3.8) gives

$$\begin{aligned} U_{\mathbb{C}}\, d_x\delta &\propto \alpha_T/\delta, && \text{jet,} \\ U_\infty\, d_x\delta &\propto \alpha_T/\delta, && \text{wake,} \\ \tfrac{1}{2}(U_\infty + U_{-\infty})\, d_x\delta &\propto \alpha_T/\delta, && \text{mixing layer.} \end{aligned} \tag{4.3.9}$$

An estimate for $\alpha_T$ is needed to obtain the growth rate of $\delta$ with $x$.

The diffusivity $\alpha_T$ has dimensions $\ell^2/t$. The relevant length scale for the large eddies, that do most of the mixing, is $\delta$. Their velocity scale is $\Delta U$ because they are produced by the mean shear. Hence, $\alpha_T \propto \delta\Delta U$: in fact this scaling is demanded if the flow is to be self-similar. An estimate that the constant of proportionality is $O(10^{-2})$ is provided by Exercise 4.11. Substituting $\alpha_T/\delta \propto \Delta U$ into (4.3.9) and then using (4.3.7) produces equations for $d\delta/dx$. They integrate to

$$\begin{aligned} &\delta \propto x, && \text{jet,} \\ &\delta \propto x^{1/2}, && \text{2D wake,} \\ &\delta \propto x^{1/3}, && \text{axisymmetric wake,} \\ &\delta \propto \frac{U_\infty - U_{-\infty}}{U_\infty + U_{-\infty}}\, x, && \text{mixing layer.} \end{aligned} \tag{4.3.10}$$

Note that jets and mixing layers spread linearly. The wakes do not: in a wake, the mean vorticity is convected downstream at the constant speed $U_\infty$, but it is spread by an increasingly less intense turbulence. In the jet, both the convection speed and turbulence intensity diminish with downstream distance, leading to linear growth. The mixing-layer growth rate scales on the parameter $(U_\infty - U_{-\infty})/(U_\infty + U_{-\infty})$, which is a measure of the strength of the mean shear.

Self-similarity can be applied to the Reynolds stresses as well as to the mean flow. For instance, the turbulent kinetic energy assumes the form $k = \Delta U^2 h(y/\delta)$. A formal approach to similarity analysis is developed in Exercise 4.9. The equations given in that exercise show that the turbulent stresses must collapse if strict similarity is obeyed. The experimental data for $k$ provided in Figure 4.9 do indeed collapse, as do the wake data on $k$ and $\overline{uv}$ provided in Figure 4.10.

### 4.3.1 Spreading rates

The mixing-layer thickness is defined by fitting the measured velocity profile to

$$\frac{U - \frac{1}{2}(U_\infty + U_{-\infty})}{U_\infty - U_{-\infty}} = \tfrac{1}{2}\,\mathrm{Erf}(y/\delta(x)).$$

The error function fit is shown in Figure 4.9. At $y = \delta$ the scaled velocity equals $\frac{1}{2}\,\mathrm{Erf}(1) = 0.42$. In a two-stream mixing layer, with the velocity ratio $U_{-\infty}/U_\infty = 0.6$, Bell and Mehta (1990) measured $\mathrm{d}\delta/\mathrm{d}x = 0.02$. The mixing-layer spreading rate is sometimes quoted as twice this, corresponding to the distance between the scaled velocity equaling $\pm 0.42$. The reproducibility of this spreading rate is on the order of 10%. It is rather more difficult to obtain a self-similar state in a single-stream mixing layer; consequently the spreading rates are less reproducible than in two-stream layers. A reasonable fit to data on spreading rate as a function of velocity ratio is

$$\frac{\mathrm{d}\delta}{\mathrm{d}x} = 0.084\,\frac{1 - U_{-\infty}/U_\infty}{1 + U_{-\infty}/U_\infty}. \tag{4.3.11}$$

Let the thickness of the round jet be defined as the radius at which the velocity is one-half of the centerline value: $U = \frac{1}{2}U_{℄}$ at $y = \delta(x)$. Hussein *et al.* (1994) obtained $\mathrm{d}\delta_{1/2}/\mathrm{d}x = 0.094$. They observed that previous measurements were lower due to wind tunnel confinement of the jet; a discrepancy of 10% was cited. Additionally, stationary hot wire measurements were high by 10% due to errors in measuring flow angles. The lessons to learn from this are that spreading rates are not likely to be accurate to more than 10%, even in carefully controlled experiments – and that the ideal jet exhausting into still air is not likely to be met in practical applications.

The plane jet spreading rate is also about $\mathrm{d}\delta_{1/2}/\mathrm{d}x = 0.1$. Many turbulence models predict unequal rates of spreading for round and plane jets. The $k$–$\varepsilon$ model (Section 6.2) predicts the round jet to spread about 9% faster than the plane jet. This has been called the "round jet–plane jet anomaly," alluding to the erroneous model prediction, not to a physical anomaly.

A plane wake conserves the momentum flux deficit

$$U_\infty^2 \Delta = \int_{-\infty}^{\infty} U(U_\infty - U)\,\mathrm{d}y,$$

where $\Delta$ is a constant determined by the drag on the body that produces the wake. A thickness can be defined by the half-width, $[U_\infty - U(y)]/(U_\infty - U_{℄}) = \frac{1}{2}$ when $y = \pm\delta(x)$. Approximate similarity in the far wake implies that

$$\delta = c\sqrt{x\Delta}. \tag{4.3.12}$$

The constant of proportionality has been measured to be $c = 0.29$ for a wide variety of wake generators by Wygnanski *et al.* (1986). The value of $c$ shows a variability of about $\pm 0.02$ for bluff-body wakes.

### 4.3.2 Remarks on self-similar boundary layers

The two-layer (law of the wall, law of the wake) structure of boundary layers precludes complete self-similarity. To obtain similarity there must be a single length scale in the problem. This means that the wall region scale $\nu/u_*$ has to be proportional to the wake region scale $\delta$, independently of $x$. But the ratio of scales, $u_*\delta/\nu$, is a Reynolds number that increases with $x$: using the previous estimate (4.2.10), $\delta \propto x^{4/5}$, and

$$u_*/U_\infty \propto \sqrt{C_f} \propto \sqrt{R_\theta^{-1/4}}$$

gives $u_*\delta/\nu \propto x^{7/10}$ in the zero pressure-gradient boundary layer. That $x$ dependence prevents similarity in the boundary layer as a whole. Similarity also demands there to be a single velocity scale, which means that $U_\infty/u_*$ should be independent of $x$, and this is not satisfied either.

However, the wake region is only weakly affected by the wall region through the log layer. To a first approximation, similarity can be assumed independently in each of the two regions, ignoring the coupling across their overlap layer. The nature of the approximation is determined by the matching condition on the mean velocity. Matching was employed to derive (4.2.2). That formula implies that the error is due to an $x$ dependence of order $\log(u_*\delta/\nu) \sim \log(x)$ in the wake region. This is a very mild $x$ dependence. Regional self-similarity seems a reasonable approximation, and it is indeed found to be so in experiments.

The region-by-region similarity has already been covered, in essence, in Section 4.2. The law-of-the-wall scaling collapses the inner region and the law-of-the-wake collapses the outer.

Clauser (1956) showed experimentally that a family of self-similar law-of-the-wake functions could be produced in boundary layers subject to pressure gradients. His family is analogous to the Falkner–Skan profiles in laminar boundary layers. In turbulent flow, the criterion for a self-similar wake region is that the pressure-gradient parameter $\beta = (\delta^*/\tau_w)\, d_x P_\infty$ should be constant, independently of $x$. This can be seen by reference to (4.2.3). Subtracting that formula evaluated at $y = \delta_{99}$ from the same formula at $y$ gives

$$U = U_{99} + \frac{u_*}{\kappa}\{\Pi(\beta)[w(y/\delta_{99}) - w(1)] + 1/\kappa \log(y/\delta_{99})\}.$$

This is of the self-similar wake form, $U = U_\infty - u_* f(y/\delta)$, provided that $U_\infty/u_*$ and $\beta$ are (approximately) independent of $x$. Clauser devised an experiment in which suction through the upper wall of his wind tunnel could be adjusted to produce a pressure gradient with constant $\beta$; this is a non-trivial problem because $\beta$ depends on the flow field through the factor $\delta^*/\tau_w$. With some ingenuity, he was able to succeed, producing several instances of self-similar turbulent boundary layers.

## 4.4 Heat and mass transfer

Consider the transport of a passive contaminant by a turbulent flow. Once introduced, the contaminant is mixed by fluid motions. Large-scale convective mixing is affected by the turbulence and is not immediately a function of the particular material being mixed. Hence, this is a problem in turbulent convection: the ability of turbulent motion to mix a passive contaminant is a property of the flow field, not of the contaminant. Representations of turbulent mixing are, therefore, independent of molecular Prandtl number, to a first approximation.

However, the molecular Prandtl number is not irrelevant to heat and mass transfer. It is only the turbulent part of the mixing that, to a suitable degree of approximation, is independent of molecular properties. Concentration is also diffused by molecular motion. Some dependence of heat and mass transfer on molecular Prandtl number is to be expected due to the presence of these two modes of mixing. At low Reynolds number, the molecular mode becomes substantial. In wall-bounded flow, molecular effects become important in the near-wall region rather than at low Reynolds number *per se*. The near-wall region is the origin of the dependence of heat and mass transfer on molecular properties.

For the sake of correctness, the qualification that only large-scale mixing is under consideration in this section should be reiterated. Molecular diffusion is always important at small scales. A more diffusive contaminant will erase fine structure created by turbulent eddies; a less diffusive contaminant will develop fine-grained structure. Such influences of molecular diffusion are relevant to the spectrum of concentration fluctuations; we are concerned here only with the mean concentration.

### 4.4.1 Parallel flow and boundary layers

The view that passive contaminant dispersion is affected by turbulence, independently of the specific identity of the contaminant, suggests that a prediction method might be construed as a relationship between a turbulent scalar flux and the Reynolds stresses responsible for that flux. The simplest model for contaminant transport is to assume it to be analogous to momentum transport. There are two versions of this analogy, one called the turbulent Prandtl number, the other called Reynolds analogy. The turbulent Prandtl number is a constant of proportionality between an eddy diffusivity and eddy viscosity, as was mentioned in Section 3.3.2. The Reynolds analogy is an equivalence between momentum and concentration profiles.

The equation governing the mean concentration of a convected scalar is (3.4.2). In parallel flow, the left-hand side of that equation is zero. If the flow is planar, parallel to a wall at $y = 0$, the mean concentration equation simplifies to $\alpha\, \partial^2 C/\partial y^2 - \partial\overline{vc}/\partial y = 0$ or

$$\alpha \frac{\partial C}{\partial y} - \overline{vc} = Q_{\mathrm{w}}, \tag{4.4.1}$$

where $Q_{\mathrm{w}}$ is an integration constant, equal to the flux to or from the wall. A good deal of insight can be obtained by invoking an eddy diffusion model, $-\overline{vc} \approx \alpha_{\mathrm{T}}\, \partial C/\partial y$. Then Eq. (4.4.1) can be rearranged to

$$C - C_0 = \int_0^y \frac{Q_{\mathrm{w}}}{\alpha + \alpha_{\mathrm{T}}(y')}\, \mathrm{d}y'. \tag{4.4.2}$$

If, for instance, $C$ represents temperature, and a channel is heated to temperature $C_H$ on the upper wall, and held at temperature $C_0$ on the lower wall, then

$$Q_{\mathrm{w}} = \frac{C_H - C_0}{\int_0^H [1/(\alpha + \alpha_{\mathrm{T}})]\, \mathrm{d}y'}.$$

The denominator highlights the role of $1/\alpha_{\mathrm{T}}$ as an impedance to heat transfer. Under typical turbulent conditions, $\alpha \ll \alpha_{\mathrm{T}}$ outside of the viscous wall layer. Hence, regions where $\alpha_T$ is low make the largest contribution to the integral. They thereby control the magnitude of heat flux, $Q_w$.

The usual region of low $\alpha_T$ is adjacent to no-slip walls, at which $\alpha_T$ tends to zero. The near-wall layer exerts disproportionate control on heat transfer to the underlying surface. The behavior of turbulence in the viscous zone proximate to a wall is discussed at length in Section 7.3.3. It is sufficient to note that, as $y \to 0$, $v \to O(y^2)$. As a constant-temperature boundary is approached, $c$ tends to zero in proportion to $y$, giving $\overline{vc} = O(y^3)$. The eddy viscosity also tends to zero as $y^3$. In this zone $|\overline{vc}| \ll |\alpha\, \partial_y C|$, or with an eddy diffusion approximation, $\alpha_T \ll \alpha$. Not only does the near-wall layer exert a disproportionate influence on heat transfer, it also introduces a dependence on molecular properties through the value of $\alpha$.

Since the eddy viscosity is a property of the turbulent velocity, not of the contaminant concentration, it is appropriate to normalize it on the molecular viscosity. Hence the viscous region scaling cited above can be written as $\alpha_T \sim \nu y_+^3$ as $y_+ \to 0$. Then the region where molecular diffusion dominates turbulent transport is $\nu y_+^3 \ll \alpha$, or

$$y_+ \ll \mathrm{Pr}^{-1/3}, \tag{4.4.3}$$

where $\mathrm{Pr} = \nu/\alpha$ is the molecular Prandtl number. This inequality assumes that $y_+$ is in the viscous layer $y_+ \lesssim 5$. For fluids with high Prandtl number, that is, with low diffusivity, turbulent transport dominates in most of the viscous layer, according to the estimate (4.4.3). Molecular diffusion is unable to make up for the suppression of turbulence by the wall, and a high-impedance region exists. For fluids with low Prandtl number, the high diffusivity compensates for reduced turbulent transport near the wall. In fact, in the limit of very large $\alpha$, as occurs for liquid metals, the contaminant transport is by molecular diffusion throughout the fluid.

Section 4.2 presented ideas about the mean velocity profile. It is of interest to develop analogous results for the mean scalar profile. A useful approach is to derive the scalar profile from that of the velocity. Both profiles are heavily influenced by turbulent mixing, so they might be expected to be somewhat similar if their boundary conditions are analogous. In order to compare the two profiles, they must be non-dimensionalized. To this end, $q_* = Q_w/u_*$ defines a friction scale for the contaminant concentration. The comparison will be developed in "plus" units. The difference between the scalar and velocity profiles in a constant-stress layer can be evaluated using Eq. (4.4.2) as

$$\frac{C - C_0}{q_*} - \frac{U}{u_*} = \int_0^{y_+} \frac{1}{\mathrm{Pr}^{-1} + \alpha_T^+}\, \mathrm{d}y'_+ - \int_0^{y_+} \frac{1}{1 + \nu_T^+}\, \mathrm{d}y'_+, \tag{4.4.4}$$

where $\alpha_T^+ = \alpha_T/\nu$. If $\alpha = \nu$ and $\alpha_T = \nu_T$, then the right-hand side is zero and $C - C_0 = q_* U/u*$. This is an embodiment of the *Reynolds analogy* between contaminant and momentum transport. When suitably normalized, the two have the same profile.

Reynolds analogy is more often stated in terms of surface transfer coefficients. The Stanton number is defined as a normalized surface heat flux

$$\mathrm{St} = \frac{Q_w}{(C_\infty - C_0)U_\infty}.$$

A corollary to the analogy $C - C_0 = q_* U/u*$ is that

$$\mathrm{St} = \frac{\alpha}{U_\infty} \frac{\partial_y C(0)}{(C_\infty - C_0)} = \frac{\nu}{U_\infty} \frac{\partial_y U(0)}{U_\infty} = \frac{u_*^2}{U_\infty^2} = \frac{1}{2} C_{\mathrm{f}},$$

again under the condition $\alpha = \nu$, or $\mathrm{Pr} = 1$. The strict equality $\mathrm{St} = \frac{1}{2} C_{\mathrm{f}}$ does not apply if $\mathrm{Pr} \neq 1$. The ratio $2\,\mathrm{St}/C_{\mathrm{f}}$ is sometimes termed the Reynolds analogy factor. Measured values often are near to, but not equal to, unity; for heat transfer in air, it has been measured to be about 1.2 in fully turbulent boundary layers. The Reynolds analogy only makes sense if the transport of the contaminant is primarily a consequence of the ambient turbulence, and is only weakly a function of the contaminant diffusivity.

A turbulent Prandtl number can be defined by $\mathrm{Pr_T} = \nu_{\mathrm{T}}/\alpha_{\mathrm{T}}$. In a parallel shear flow, the Reynolds shear stress is modeled by $\overline{uv} = -\nu_{\mathrm{T}}\, \partial_y U$, according to formula (3.3.3) on page 51. Similarly, the Reynolds flux of a contaminant is modeled as $\overline{vc} = -\alpha_{\mathrm{T}}\, \partial_y C$. The turbulent Prandtl number can be found experimentally by measuring terms in these definitions. It is obtained directly by measuring the ratio on the right-hand side of

$$\mathrm{Pr_T} = \frac{\nu_{\mathrm{T}}}{\alpha_{\mathrm{T}}} = \frac{\overline{uv}}{\partial_y U} \frac{\partial_y C}{\overline{vc}}.$$

In a boundary layer, the empirical value $\mathrm{Pr_T} = 0.85$ is a good estimate of available data (Kays, 1994).

Inserting $\mathrm{Pr_T}$ in front of $U$ in Eq. (4.4.4) and substituting $\alpha_{\mathrm{T}} = \nu_{\mathrm{T}}/\mathrm{Pr_T}$ give

$$\frac{C - C_0}{q_*} - \mathrm{Pr_T} \frac{U}{u_*} = \int_0^{y_+} \frac{\mathrm{Pr_T}}{\mathrm{Pr_T}\,\mathrm{Pr}^{-1} + \nu_{\mathrm{T}}^+} - \frac{\mathrm{Pr_T}}{1 + \nu_{\mathrm{T}}^+}\, \mathrm{d}y'_+. \tag{4.4.5}$$

In the log layer, $\nu_{\mathrm{T}}^+ \gg 1$ (Figure 6.7). If $\mathrm{Pr_T}/\mathrm{Pr}$ is $O(1)$, so that $\nu_{\mathrm{T}}^+ \gg \mathrm{Pr_T}\,\mathrm{Pr}^{-1}$ as well, then the integrand of Eq. (4.4.5) is vanishingly small in the log layer. Then for large $y_+$ the range of integration can be taken from 0 to $\infty$, so that the temperature and velocity profiles are related by

$$\frac{C - C_0}{q_*} = \mathrm{Pr_T} \frac{U}{u_*} + B_C(\mathrm{Pr}, \mathrm{Pr_T}), \tag{4.4.6}$$

where

$$B_C = \int_0^{\infty} \frac{\mathrm{Pr_T}}{\mathrm{Pr_T}\,\mathrm{Pr}^{-1} + \nu_{\mathrm{T}}^+} - \frac{\mathrm{Pr_T}}{1 + \nu_{\mathrm{T}}^+}\, \mathrm{d}y'_+.$$

$B_C$ is an additive coefficient for the temperature profile. It is clear that this coefficient depends on the molecular Prandtl number. With expression (4.1.7) for the velocity, Eq. (4.4.6) becomes the log law for concentration,

$$\frac{C - C_0}{q_*} = \frac{\mathrm{Pr_T}}{\kappa} \log(y_+) + \mathrm{Pr_T} B + B_C. \tag{4.4.7}$$

Kader (1981) found that the formula

$$\mathrm{Pr_T} B + B_C = (3.85\,\mathrm{Pr}^{1/3} - 1.3)^2 + 2.12 \log(\mathrm{Pr}) \tag{4.4.8}$$

fit boundary-layer data over the range $6 \times 10^{-3} < \mathrm{Pr} < 4 \times 10^4$. He also gave the value $\mathrm{Pr_T}/\kappa = 2.12$: with a Von Karman constant of $\kappa = 0.41$, this implies a value $\mathrm{Pr_T} = 0.87$.

Further insight into the Prandtl number dependence of the additive coefficient, $B_C$, can be gained by a heuristic argument proposed by Kays and Crawford (1994). If Pr is not extremely small, then next to the surface there is a high-impedance layer in which $\alpha > \alpha_T$. Define a top to that layer at $y = y^h$. Conceptually, the flow might be broken into two zones: where $y_+ < y_+^h$, turbulent transport is negligible; where $y_+ > y_+^h$, turbulent transport dominates. Then Eq. (4.4.2) can be approximated as

$$\frac{C - C_0}{q_*} = y_+^h \mathrm{Pr} + \int_{y_+^h}^{y_+} \frac{\mathrm{Pr_T}\, dy'_+}{\nu_T^+}, \tag{4.4.9}$$

when $y_+ > y_+^h$. For moderate Prandtl numbers, the inequality (4.4.3) is motivation to select $y_+^h = a\mathrm{Pr}^{-1/3}$, where $a$ is a constant. In the log region, $\nu_T^+ = \kappa y_+$ and the estimate (4.4.9) becomes

$$\frac{C - C_0}{q_*} = \frac{\mathrm{Pr_T}}{\kappa} \log(y_+) - \frac{\mathrm{Pr_T}}{\kappa} \log(a\mathrm{Pr}^{-1/3}) + a\mathrm{Pr}^{2/3}. \tag{4.4.10}$$

This is of the same form as (4.4.7). The empirical formula (4.4.8) should be used in practice, rather than the above heuristic estimate of $B_C$.

The additive term, $B_C$, shifts the intercept of the log–linear plot of $C$ versus $\log(y_+)$ with respect to the $C$ axis. The dependence of this shift on molecular Prandtl number originates in the high-impedance layer: the thinner the layer, the larger is $C$ at the intercept, and hence the greater is the shift. In other words, the smaller $\alpha$ is in comparison to $\nu$, the thinner is the layer. Another way to see this is to consider a hot stream and a cold wall. The smaller is $\alpha$, the closer will the hot fluid approach the wall; hence, the higher will the temperature be in the log layer.

A heat-transfer formula analogous to the skin friction formula (4.1.11) can be found from (4.4.6). Assume the latter equation to be valid to the edge of the boundary layer. Then as $y \to \infty$, it becomes

$$\frac{C_\infty - C_0}{q_*} = \mathrm{Pr_T}\frac{U_\infty}{u_*} + B_C(\mathrm{Pr}, \mathrm{Pr_T}).$$

Substituting the definitions $q_* = Q_w/u_*$, $\mathrm{St} = Q_w/(C_\infty - C_0)U_\infty$, and $C_f = 2u_*^2/U_\infty^2$ gives

$$\frac{2\,\mathrm{St}}{C_f} = \frac{1}{\mathrm{Pr_T} + B_C(\mathrm{Pr}, \mathrm{Pr_T})\sqrt{C_f/2}} \tag{4.4.11}$$

after some rearrangement. Equation (4.4.11) can be used in conjunction with the skin friction formulas in Chapters 4 or 6 to estimate heat-transfer rates.

For a turbulent boundary layer in air, $\mathrm{Pr} = 0.71$, and using $B = 5.0$ in formula (4.4.8) gives $B_C = -0.42$. With $\mathrm{Pr_T} = 0.85$, the Reynolds analogy factor is $2\,\mathrm{St}/C_f = 1.18/(1 - 0.5\sqrt{C_f/2})$. For a typical value of $\frac{1}{2}C_f \sim 10^{-3}$, this factor is 1.2. Heat is transported a bit more effectively than momentum, to the extent that $\mathrm{St} > \frac{1}{2}C_f$. This might be due to a tendency of fluctuating pressure gradients to reduce momentum transport.

Pursuing further the parallel between momentum and concentration boundary layers, a thickness of the concentration boundary layer can be defined by analogy to the momentum thickness:

$$\Delta = \int_0^\infty \frac{U}{U_\infty} \frac{C - C_0}{C_\infty - C_0} \, dy. \tag{4.4.12}$$

When $C$ represents heat, this is called the enthalpy thickness. Kays and Crawford (1993) suggest that formula (4.4.11) for the Stanton number can be approximated by the simpler data correlation $\mathrm{St} = 0.0125 R_\Delta^{-0.25} \mathrm{Pr}^{-0.5}$ when $0.5 < \mathrm{Pr} < 1.0$ and when $R_\theta$ is in the range $10^3$ to $8 \times 10^3$. Here $R_\theta$ and $R_\Delta$ are the Reynolds numbers based on momentum and enthalpy thickness.

The equations governing momentum and contaminant transport would be the same if $\alpha = \nu$ in Eq. (3.4.2), provided the turbulent fluxes of momentum and concentration were proportionate *and provided* $\partial_i P = 0$ in Eq. (3.2.2). This last condition implies that any analogy between heat and momentum transfer will break down in flows with non-zero pressure gradient. For example, an accelerating free stream can reduce the turbulence in a boundary layer, decreasing St. If the reduction of the turbulence is not too great, then the tendency for $C_\mathrm{f}$ to increase in an accelerated flow will not be reversed. Then St can fall, while $C_\mathrm{f}$ rises. As another example, $C_\mathrm{f}$ must vanish at a stagnation point, while St need not.

Although the use of an overall analogy between heat transfer and skin friction becomes problematic in most flows of practical interest, the proportionality between $\nu_\mathrm{T}$ and $\alpha_\mathrm{T}$, as embodied in the turbulent Prandtl number, remains a useful assumption. Kays (1994) cites data which suggest that pressure gradients might affect the value of $\mathrm{Pr}_\mathrm{T}$, but concludes that such effects are small. In many practical flows, heat transfer can be predicted with the eddy viscosity closure models discussed in Chapter 6 and with a turbulent Prandtl number of 0.85. Complex flow effects are assumed to be embodied in the predictions of eddy viscosity that those models make.

### 4.4.2 Dispersion from elevated sources

Transport to surfaces is not always the motivation to study turbulent mixing. Various applications involve mixing of a contaminant with ambient fluid internally to the flow. An example is the dispersion of pollutants in the atmosphere.

Consider a source of contaminant that is well away from any boundary. The ratio of convective to molecular transport is characterized by the Peclet number, $\mathrm{Pe} = UL/\kappa$. If both the Peclet and Reynolds numbers are high, then to a first approximation it should be possible to neglect molecular diffusion in the equation for the *mean* concentration. (Note that molecular diffusion cannot be ignored in the equation for the concentration *variance*, because it is the origin of dissipation.) Let the source of contaminant have a characteristic scale $L$, and let the scale for both the turbulent velocity and the mean velocity variations be $\hat{u}$. Non-dimensionalizing Eq. (3.4.2) with these scales gives

$$\partial_t C + U_j \, \partial_j C = \mathrm{Pe}^{-1} \nabla^2 C - \partial_i \overline{c u_i},$$

where $\mathrm{Pe} = \hat{u} L/\alpha$ is the turbulent Peclet number. For $\mathrm{Pe} \gg 1$,

$$\partial_t C + U_j \partial_j C = -\partial_i \overline{c u_i}.$$

This is the Reynolds average of the random convection equation

$$\partial_t \tilde{c} + \tilde{u}_j \, \partial_j \tilde{c} = 0. \tag{4.4.13}$$

The present convention of using a tilde to denote the total of mean plus fluctuation should be recalled. Equation (4.4.13) provides a simple view of mean scalar transport at high Peclet number: mixing can be regarded as the ensemble-averaged effect of random convection alone. Again, it should be warned that the simplification (4.4.13) would be quite erroneous for scalar variance: dissipation of scalar variance by molecular diffusion cannot be neglected, even at high Peclet number.

Consider a convected fluid particle. Its trajectory solves the nonlinear equation $\mathrm{d}_t X_i = \tilde{u}_i(\boldsymbol{X})$ with initial position $X_i = x_{0_i}$. If the initial scalar profile is a given function $C_0(\boldsymbol{x})$, then the solution to Eq. (4.4.13) is

$$\tilde{c}(\boldsymbol{X}) = C_0(\boldsymbol{x}_0)$$

because $\tilde{c}$ is constant along trajectories and equal to its initial value. This is a Lagrangian solution because it is phrased in terms of the fluid trajectories. It can be termed the "formula for forward dispersion." Often reversed dispersion provides a more intuitive perspective; indeed, the mixing length argument of Section 3.2 was phrased in terms of reversed dispersion.

The reversed view is obtained by restricting attention to particles with the current position $\boldsymbol{X} = \boldsymbol{x}$ and following their trajectory back in time. The initial position of such particles is $\boldsymbol{x}_0 = \boldsymbol{X}^{-1}(t, \boldsymbol{x})$: in other words, they are found by tracing trajectories backward from a given final position, $\boldsymbol{x}$, through the field of turbulence, to a random initial position, $\boldsymbol{x}_0$. The solution now is written

$$\tilde{c}(\boldsymbol{x}) = C_0(\boldsymbol{x}_0(t, \boldsymbol{x})).$$

Since $\boldsymbol{x}_0$ is random, its probability density can be introduced for use in averaging. The transition probability for a particle at $\boldsymbol{x}$, at time $t$, to have originated at $\boldsymbol{x}_0$ is denoted $P(\boldsymbol{x}, t; \boldsymbol{x}_0)$. With this, the average concentration is found to be

$$C(\boldsymbol{x}, t) = \iiint C_0(\boldsymbol{x}_0) P(\boldsymbol{x}, t; \boldsymbol{x}_0) \, \mathrm{d}^3 \boldsymbol{x}_0, \tag{4.4.14}$$

invoking the definition (2.2.5) of an average.

A case that arises in atmospheric dispersion is the concentrated source. Assume that $C_0 = Q/V$ is constant in a small region of volume $V$ around the origin, and is zero elsewhere. Then formula (4.4.14) is approximately

$$C(\boldsymbol{x}) = Q P(\boldsymbol{x}, t; 0).$$

For this initial condition, the PDF gives the scalar concentration directly. For example, if the PDF is Gaussian, then, in principal axes of the $\overline{X_i X_j}$ correlation,

$$C = \frac{Q}{\sqrt{(2\pi)^3 \overline{X^2}\, \overline{Y^2}\, \overline{Z^2}}} \exp\left[ \frac{x^2}{2\overline{X^2}} + \frac{y^2}{2\overline{Y^2}} + \frac{z^2}{2\overline{Z^2}} \right]. \tag{4.4.15}$$

The analysis of dispersion reduces to obtaining $\overline{X_i X_j}$ in the case of a Gaussian cloud. This will be illustrated by a discussion of the phenomenon of shear dispersion.

An insight into the interaction between turbulent dispersion and a mean shear is gained by simplifying the analysis to two dimensions, corresponding to a "line source." Also the mean flow is in the $x$ direction, so that $\tilde{u} = U + u$ and $\tilde{v} = v$. We invoke a thin-layer approximation and neglect $u$ in comparison to $U$, so that the trajectory equations are simply

$$\mathrm{d}_t X = U(Y),$$

$$\mathrm{d}_t Y = v,$$

with $X(0) = 0$ and $Y(0) = 0$. Furthermore, it will be assumed that the turbulence can be treated as stationary, despite the mean shear. The virtue of these various assumptions is that they lead to explicit formulas, illustrating the phenomenon of shear dispersion.

From $Y = \int_0^t v\,\mathrm{d}t$, the mean of $Y$ is seen to be zero because $\overline{v} = 0$. If $v$ is statistically stationary, the variance of $Y$ is

$$\begin{aligned}\overline{Y^2} &= \overline{\int_0^t v(t')\,\mathrm{d}t' \int_0^t v(t'')\,\mathrm{d}t''} = \int_0^t \int_0^t \overline{v(t')v(t'')}\,\mathrm{d}t''\,\mathrm{d}t' \\ &= \overline{v^2} \int_0^t \int_0^t R(t'' - t')\,\mathrm{d}t''\,\mathrm{d}t' = 2\overline{v^2} \int_0^t \int_0^{t'} R(t'' - t')\,\mathrm{d}t''\,\mathrm{d}t' \\ &= 2\overline{v^2} \int_0^t \int_0^{t'} R(\tau)\,\mathrm{d}\tau\,\mathrm{d}t' = 2\overline{v^2} \int_0^t (t - \tau)R(\tau)\,\mathrm{d}\tau\end{aligned}$$

by analogy to Eq. (2.2.21). For an exponential correlation function, the value of the integral is given below Eq. (2.2.21):

$$\overline{Y^2} = 2\overline{v^2}\, T_{\mathrm{L}}^2 (t/T_{\mathrm{L}} - 1 + \mathrm{e}^{-t/T_{\mathrm{L}}}). \tag{4.4.16}$$

where $T_L$ denotes the Lagrangian integral time scale. A new element, termed *shear dispersion*, arises when the variance of $X$ is computed. The mean of $X$ could in general be computed from the PDF of $y$, invoking $\mathrm{d}_t\overline{X} = \int P(y, t)U(y)\,\mathrm{d}y$. However, the essential elements of shear dispersion are present in the simple, homogeneous shear, $U = \mathcal{S}Y$, where $\mathcal{S}$ is a constant rate of shear. The mean of $\mathrm{d}_t\overline{X}$ equals $\mathcal{S}\overline{Y}$, which is zero. Thus, without loss of generality, $\overline{X} = 0$. The variance is

$$\overline{X^2} = 2\mathcal{S}^2 \int_0^t \int_0^{t'} \overline{Y(t')Y(t'')}\,\mathrm{d}t''\,\mathrm{d}t'. \tag{4.4.17}$$

This is not zero, even though no turbulence was included in the $x$ direction: that is the essence of shear dispersion, as will be explained shortly.

Evaluation of $\overline{X^2}$ requires two-time statistics of $Y$. Note that $Y$ is *not* statistically stationary; in fact, its variance has the time dependence (4.4.16). The two-time correlation is given by

$$\overline{Y(t')Y(t'')} = \int_0^{t'}\int_0^{t''} \overline{v(s')v(s'')}\,\mathrm{d}s''\,\mathrm{d}s'$$
$$= \overline{v^2}\int_0^{t'}\int_0^{t''} R(s'-s'')\,\mathrm{d}s''\,\mathrm{d}s'. \qquad (4.4.18)$$

For the exponential correlation function, $R(\tau) = \mathrm{e}^{-|\tau|/T_\mathrm{L}}$, straightforward integration gives

$$\overline{Y(t')Y(t'')} = \overline{v^2}\,T_\mathrm{L}^2\left(\mathrm{e}^{-t'/T_\mathrm{L}} + \mathrm{e}^{-t''/T_\mathrm{L}} - \mathrm{e}^{-(t'-t'')/T_\mathrm{L}} - 1 + \frac{2t''}{T_\mathrm{L}}\right)$$

when $t'' < t'$. Substituting this into Eq. (4.4.17) and integrating gives

$$\overline{X^2} = 2\overline{v^2}\mathcal{S}^2 T_\mathrm{L}^4\left[1 - \frac{t^2}{2T_\mathrm{L}^2} + \frac{t^3}{3T_\mathrm{L}^3} - \left(1 + \frac{t}{T_\mathrm{L}}\right)\mathrm{e}^{-t/T_\mathrm{L}}\right]. \qquad (4.4.19)$$

The short-time behavior, $\overline{X^2} \to \frac{1}{4}\overline{v^2}\mathcal{S}^2 t^4$, is readily seen to be correct. It is a consequence of $Y \to vt$; for then $\mathrm{d}_t X = \mathcal{S}Y = v\mathcal{S}t$, so that $X = \frac{1}{2}v\mathcal{S}t^2$ and $\overline{X^2} = \frac{1}{4}\overline{v^2}\mathcal{S}^2 t^4$.

The long-time behavior of expression (4.4.19) is its most interesting aspect. The third term in the bracket dominates, giving $\overline{X^2} \to \frac{2}{3}\overline{v^2}\mathcal{S}^2 T_\mathrm{L} t^3$ as $t/T_\mathrm{L} \to \infty$; this can also be found directly by substituting the long-time behavior

$$\overline{Y(t'')Y(t')} \to 2T_\mathrm{L}\overline{v^2}\min(t', t'')$$

into the integral (4.4.17). The streamwise variance grows like $t^3$, in contrast to the $t$ growth of $\overline{Y^2}$. At long times, the streamwise dispersion is substantially greater than the cross-stream dispersion, even though it is affected by turbulence in the latter direction. This is an intriguing conclusion, and obviously of some practical importance.

The mechanism of streamwise shear dispersion was first explored by Taylor (1954). Consider a strip of contaminant that is initially vertical. The strip will be sheared by the mean flow, producing a gradient in the $y$ direction (Figure 4.11). Shearing the strip does not cause mixing with the ambient fluid. However, when the gradient caused by mean shear is acted on by the $v$ component of turbulence, as in Figure 4.11, concentration is mixed into the ambient, producing a dispersed profile in the $x$ direction.

Note that the covariance tensor

$$\begin{pmatrix} \overline{X^2} & \overline{XY} \\ \overline{XY} & \overline{Y^2} \end{pmatrix}$$

is not diagonal. The off-diagonal term is found to be

$$\overline{XY} = \mathcal{S}\int_0^t \overline{Y(t)Y(t')}\,\mathrm{d}t' = \overline{v^2}\mathcal{S}T_\mathrm{L}^2 t\left(\frac{t}{2T_\mathrm{L}} - 1 + \mathrm{e}^{-t/T_\mathrm{L}}\right), \qquad (4.4.20)$$

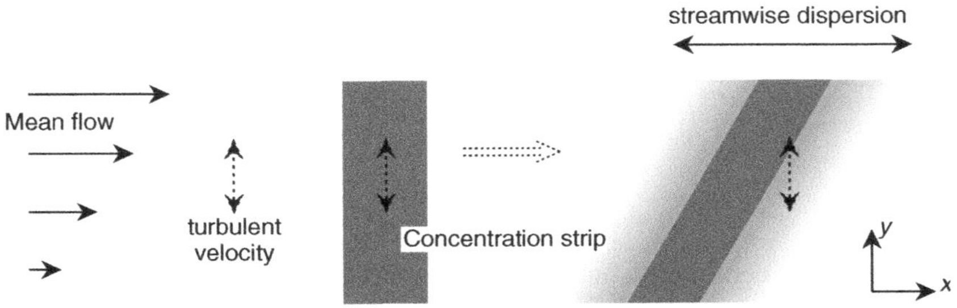


**Figure 4.11** The combination of turbulent mixing in $y$ and mean shear in $x$ causes dispersion in the streamwise direction.

again, for an exponential velocity autocorrelation. This off-diagonal correlation causes the constant-probability contours of the Gaussian to be ellipses with principal axes angled relative to the $x, y$ axes. For a two-dimensional, Gaussian cloud, the centerline concentration varies as one over the determinant of the dispersion tensor (see Eq. (4.4.15)). Specifically,

$$C(0,0) = \frac{Q}{2\pi(\overline{X^2}\,\overline{Y^2} - \overline{XY}^2)}.$$

As $t \to \infty$ the denominator becomes $O(t^4)$; hence, the centerline concentration falls as $t^{-4}$ in consequence of shear dispersion. Without shear, turbulent dispersion in $x$ and $y$ would cause $C(0,0)$ to fall only as $t^{-2}$. The presence of shear can promote mixing quite dramatically.

# Exercises

**Exercise 4.1.** *Turbulent kinetic energy equation in the channel.* Write the form of the turbulent kinetic energy equation (3.2.5) for the special case of parallel flow: $U(y)$, homogeneity in $x, z$, and stationarity in $t$. Show that

$$\int_0^{2H} \mathcal{P}\,\mathrm{d}y = \int_0^{2H} \varepsilon\,\mathrm{d}y.$$

What boundary condition on $k$ did you use, and why? Recall that $\mathcal{P}$ is the rate at which turbulent energy is produced from the mean shear. This exercise suggests that, in shear layers, production and dissipation are similar in magnitude.

**Exercise 4.2.** *Magnitude of $C_f$.* Let the channel half-height Reynolds number be $R_H = 10^4$. Use formula (4.1.11) to estimate the value of $C_f$.

**Exercise 4.3.** *Pressure drop across a channel.* Because of the favorable pressure gradient, the log law can be considered valid practically to the center of a fully developed, plane channel flow. The centerline velocity, $U_{\mathbb{C}}$, and volume flux, $Q$, are related by

$$Q \approx 0.87 U_{\mathbb{C}} A,$$

where $A$ is the cross-sectional area of the channel. Its height is $2H$. The channel is long and narrow, so the sidewalls can be ignored.

A certain volume flux, $Q$, of airflow is desired. Write a formula that can be used to estimate the necessary pressure drop across a channel of length $L$ – do not try to solve it for $\Delta P$; that could be done numerically.

**Exercise 4.4.** *Zones of the boundary layer.* Make a sketch of the mean velocity distribution, $U(y)$, as in Figure 4.5. Indicate the law-of-the-wall and law-of-the-wake regions in both log–linear and linear–linear coordinates. In zero pressure gradient, what is the approximate magnitude of $U/U_\infty$ at the top of the log layer? What is it for the adverse pressure-gradient data?

**Exercise 4.5.** *The momentum integral.* Derive the momentum integral equation analogous to (4.2.7) for the case where $\mathrm{d}_x U_\infty \neq 0$. Substitute $\delta^* = H\theta$. Will a favorable pressure gradient ($\mathrm{d}_x U_\infty > 0$ and $\mathrm{d}_x P < 0$) increase or decrease the growth rate of $\theta$? Answer the same question for an adverse pressure gradient.

**Exercise 4.6.** *Law of the wake.* Infer a drag law by evaluating (4.2.3) at $y = \delta_{99}$. Use Coles' wake function for $w(y)$. Set $U_{99} = U_\infty$ and compare this to (4.2.2). Comment on the effect of pressure gradient on skin friction.

**Exercise 4.7.** *Drag law.* The power-law form $U = U_d(y/d)^a$, $y < d$, is sometimes used to fit the mean flow profile in the outer region; $d$ is the boundary-layer edge, so $U_d = U_\infty$. Above the boundary layer, $U = U_\infty$, $y \geq d$. The exponent $a$ is a small number, about 1/7, although it varies with Reynolds number, as you will show. Calculate $\theta$, $\delta^*$, and $H = \delta^*/\theta$ for this profile. For small $a$, $(y/d)^a = \mathrm{e}^{a \log(y/d)} \approx 1 + a \log(y/d)$. If this is matched to the log law (4.1.7), how is $a$ related to $u_*/U_\infty$? What is the corresponding skin friction law, written like (4.1.11) in terms of friction coefficient, $C_\mathrm{f}$, and momentum thickness Reynolds number, $R_\theta = U_d\theta/\nu$?

**Exercise 4.8.** *Measurement of friction velocity.* Measurements in a particular turbulent boundary layer give a thickness $\delta_{99} = 3.2$ cm. The velocity $U$ is measured at three heights with the results: $U = 40.1\ \mathrm{m\,s^{-1}}$ at $y = 2.05$ cm, $U = 34.7\ \mathrm{m\,s^{-1}}$ at $y = 0.59$ cm, and $U = 30.7\ \mathrm{m\,s^{-1}}$ at $y = 0.22$ cm.

(i) Estimate the friction velocity $u_*$.

(ii) Let the fluid be air with $\nu = 0.15\ \mathrm{cm^2\,s^{-1}}$. Is this an adverse or favorable pressure-gradient boundary layer?

**Exercise 4.9.** *Governing equations for self-similar flow.* Substitute the assumed forms

$$U = \Delta U(x) f(y/\delta(x)), \qquad V = \Delta U(x) g(y/\delta(x)), \qquad \overline{uv} = \Delta U^2 h(y/\delta)$$

for the case of a 2D jet into (4.3.1). Use (4.3.7) but ignore (4.3.10). Let $\zeta = y/\delta$. Rewrite the equations that you derived as ordinary differential equations involving $f(\zeta)$, $g(\zeta)$, and $h(\zeta)$. In order for the independent variable to be $\zeta$ alone, the coefficients in this equation cannot be functions of $x$ or $y$ – if they were, then the solution would be of the form $f(\zeta, x)$ or $f(\zeta, y)$, which are not self-similar. What condition must be imposed on $\Delta U$ and $d_x\delta$ for the equations to permit self-similarity? Do you think that all the results (4.3.10) can be derived by this approach?

**Exercise 4.10.** *The momentumless wake.* For a zero momentum-deficit, plane wake, the small-deficit approximation implies $U_\infty \int_{-\infty}^{\infty}(U - U_\infty)\,dy = 0$. Hence, the constant in (4.3.6) is zero and the similarity analysis for wakes does not hold.

Use the eddy viscosity approximation $-\overline{uv} = \nu_T\,\partial_y U$. Assume $\nu_T$ to be constant, use the small-defect approximation for the convection velocity, and form the second moment of (4.3.4) with respect to $y^2$, after making these approximations. Thus, find the similarity scaling in this case. ("Form the second moment" means multiply by $y^2$, then integrate between $\pm\infty$.)

**Exercise 4.11.** *Plane self-similar jet via eddy viscosity.* For a plane jet, Eqs. (4.3.10) and (4.3.7) can be written $\delta = ax$ and $U_{\mathbb{C}} = b/\sqrt{x}$, where $a$ and $b$ are constants. Generally, the eddy viscosity (3.3.3) will depend on $y$, but a reasonable estimate of self-similar profiles can be obtained by setting $\nu_T = cU_{\mathbb{C}}\delta$, where $c$ is an empirical constant. Show that the 2D, thin shear-layer equations have a self-similar solution of the form $U = U_{\mathbb{C}} f(\eta)$, $V = U_{\mathbb{C}} g(\eta)$, with $\eta = y/\delta$. Solve for $f$. [Hint: The form of the solution is the same as in laminar flow.]

Let $a$ be defined by the 50% thickness of the jet: in other words $f(1) = 0.5$. Find the constant of proportionality between the "Reynolds number" $U_{\mathbb{C}}\delta/\nu_T = 1/c$ and the spreading rate $d\delta/dx = a$. Use $a = 0.05$ to obtain an estimate of the model constant $c$. This exercise illustrates how the turbulent viscosity controls the jet spreading rate. Conversely, an experimental measurement of $a$ permits the model constant to be calibrated.

**Exercise 4.12.** *Enthalpy thickness.* Show that the enthalpy thickness evolves according to

$$\frac{d\Delta}{dx} = \mathrm{St}$$

in analogy to the evolution equation (4.2.7) for momentum thickness. Compare the evolution of $\Delta(x)$ to that of $\Theta(x)$ for a zero pressure-gradient boundary layer in air over the range $10^3 < R_\theta < 10^4$. The comparison can be in the form of computed curves using a data correlation for $C_f$.

**Exercise 4.13.** *Reflected plume model.* Because the equation of a passive scalar is linear, superposition is applicable. The mean concentration of a sum of sources is the sum of their individual mean concentrations. The Gaussian cloud

$$C = \frac{Q}{\sqrt{(2\pi)^2\overline{X^2}\,\overline{Y^2}}}\exp\left[-\frac{x^2}{2\overline{X^2}} - \frac{(y - y_s)^2}{2\overline{Y^2}}\right]$$

is the mean concentration of a line source located at $x = 0$, $y = y_s$. If a no-flux wall exists at $y = 0$, a fictitious source can be added at $y = -y_s$ to satisfy the boundary condition

$\partial_y C = 0$ on $y = 0$. Write the concentration distribution for this "reflected plume" model. Show that the concentration integrated across the physical domain, which is $y > 0$, $-\infty < x < \infty$, is constant in time.

There is no mean flow, just turbulence, so $\mathrm{d}_t X = u$ and $\mathrm{d}_t Y = v$. The height of the plume centerline is defined as

$$Y_{\mathrm{L}}(t) = \int_{-\infty}^{\infty}\int_{0}^{\infty} yC(y)\,\mathrm{d}y\,\mathrm{d}x \Bigg/ \int_{-\infty}^{\infty}\int_{0}^{\infty} C(y)\,\mathrm{d}y\,\mathrm{d}x.$$

Show that this increases with time, even though there is no mean velocity in the $y$ direction. For the exponential correlation analyzed in the text, find the "Lagrangian mean velocity," $\mathrm{d}_t Y_{\mathrm{L}}$ when $y_{\mathrm{s}} = 0$. Why is the Lagrangian mean velocity not zero, even though there is no Eulerian mean velocity?

# 5

# Vorticity and vortical structures

Big whorls have little whorls,
  which feed on their velocity;
and little whorls have smaller whorls,
  and so on to viscosity (in the molecular sense).

– Lewis Fry Richardson

The above is an oft-quoted rhyme from Richardson's prophetic book on numerical weather prediction (Richardson, 1922). In a 1937 paper, G. I. Taylor wrote, in prose: "... They represent the fundamental processes in turbulent flow, namely the grinding down of eddies produced by solid obstructions (and on a scale comparable with these obstructions) into smaller and smaller eddies until these eddies are of so small a scale that they die away owing to viscosity ..." (Taylor and Green, 1937). This chapter is an interlude on "whorls" and "eddies." Some of the material originates in the 1920s and 1930s with Richardson and Taylor; some is more recent vintage; all of it fills in between the lines of this poetry and prose.

The bulk of this text uses the term "eddy" as a vague conception of the fluid motions that are ultimately responsible for mixing and dissipative properties of turbulence. The precise definition of an eddy is largely irrelevant. As long as the complexity of the fluid motions admits a statistical description, only ensemble averages of the random flow field arise. Sometimes it is informative to look behind the statistics, to get a sense of the instantaneous eddying motion. High Reynolds number turbulence is certainly disorderly, but identifiable vortical structures embedded in the irregular flow are occasionally glimpsed. The purpose of the present chapter is to introduce the concept of coherent and vortical structures. It also serves as a bridge to some of the material in Parts II, III, and IV.

Cases arise in which notions of eddy structure have been put to use. Some ideas on how to control turbulence have been based on enhancing or destroying the recognizable structure: breaking up long-lived, large eddies can reduce drag by shortening the range

*Statistical Theory and Modeling for Turbulent Flows, Second Edition* P. A. Durbin and B. A. Pettersson Reif

over which momentum is stirred. Aeroacoustic noise is associated with large structures that form on the jet potential core. Fine-scale mixing of chemical reactants is increased when large structures disintegrate into highly three-dimensional eddies.

However, the most intense vorticity in turbulent flows is associated with the smallest scales of motion. This small-scale vorticity is thought to be nearly isotropic in its structure. The reasoning is that, as large eddies become distorted by other eddies and by instabilities, they spawn smaller, less oriented eddies, which in turn spawn smaller eddies, and so on, with the smallest scales having lost all directional preference.* This is the usual conception of the energy cascade, discussed in Section 2.1, in Chapter 10, and in the above quotations of Richardson and Taylor. The concept of universality at small scale could be interpreted as a statement that the smallest scales are nondescript. Such is not the case for the largest eddies. Coherence, if it exists, is to be found in the large scales.

The largest eddies do have directional preferences, and their shapes are characteristic of the particular mean flow. For instance, an axisymmetric jet may show evidence of vortex rings, while a mixing layer may show evidence of long rolls (see Figures 5.1 and 5.3). These recognizable eddies are called "coherent structures." The allusion is to human ability to recognize the forms, rather than to statistical concepts of coherence. This section on coherent structures is a taxonomy of eddy shapes. In some cases there is debate over the nature of the dominant structures in a particular flow. The present chapter is meant as an initial exposure to this topic, not as an exhaustive compilation.

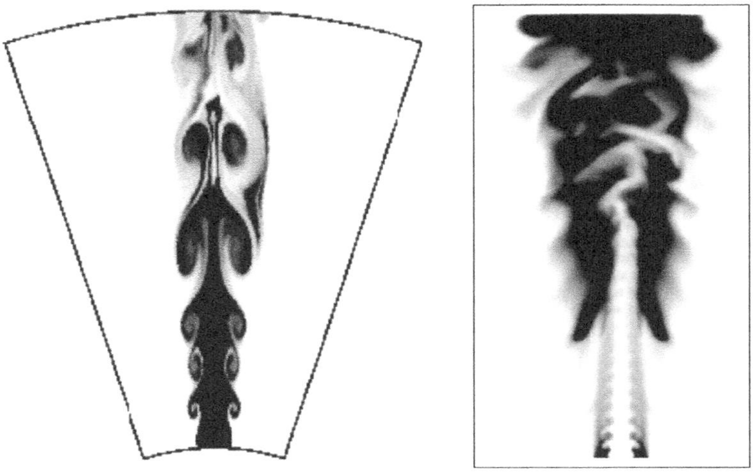

**Figure 5.1** Jets undergoing transition to turbulence, visualized via a scalar concentration field. From numerical simulations by Steiner and Busche (1998) (left) and Danaila and Boersma (1998) (right).

* Actually, the extent to which the smallest scales are isotropic and unaffected by mean straining remains a subject of debate and investigation.

## 5.1 Structures

Coherent structures are features of the turbulence field. The meaning of the word "structure" depends on how they are visualized. Often, smoke or particle displacements are used to visualize the velocity field. Structures are then flow patterns, recognized atop the more disorderly motions. In numerical simulations, contours of vorticity or streamlines might be utilized. Structures are then patterns seen in level sets of the quantity being plotted.

The basic concept is that a coherent structure is a recognizable form seen amidst the disorderly motion. Its value is to give an insight into properties of turbulence via concepts like vortex kinematics and dynamics. It associates a concrete form to the term "eddy." However, the extent to which recognizable forms can be identified, and the value of doing so, have been a subject of debate. Ideas about coherent structures have not had a substantial impact on statistical turbulence modeling, but they have contributed to the general understanding of turbulent shear flows.

Large eddies are associated with energetic scales of motion. The designation "large eddy" includes coherent vortices, but it also includes less definitive patterns of fluid movement. The issue is clouded even further by the fact that coherent eddies occur irregularly in space and time, so they often can be considered as just a part of the spectrum of random turbulent motions. Attempts to distinguish coherent structures from the sea of large eddies have included pattern recognition (Mumford, 1982) and education (Hussain, 1978) techniques. The first approach is to objectively seek recurring flow patterns, irrespective of where they occur; the second is to stimulate the pattern with external forcing, so that it occurs at a fixed location.

The structures that occur in fully turbulent free-shear flow are generally thought to be qualitatively the same as those seen in the late stages of transition. Many of the illustrations in this chapter are from transitional flow, where the structures are seen clearly. In fully developed turbulence, they are obscured by small scales and by intensely irregular flow; indeed, some of the structures may cease to be present in high Reynolds number turbulence.

### 5.1.1 Free-shear layers

The mixing layer is a generic element of many shear flows. A provocative study of coherent structures in mixing layers by Brown and Roshko (1974) served as a stimulus to research on coherent structures in the 1970s.

Mixing-layer coherent eddies consist of spanwise rolls, with streamwise ribs superimposed (Figures 5.2 and 5.3). The spanwise rolls correspond to the vortices seen in two-dimensional Kelvin–Helmholtz instability. The gap between the rolls is called the "braid region." The dominant structures in the braid region are termed "ribs." Ribs are the legs of vortex loops that span the gap between successive rolls, as in Figure 5.3. Hence they occur as alternating positive and negative streamwise vortices. Intense straining in the braid region amplifies the streamwise vorticity.

In a mixing layer streamwise-oriented vortices make a large contribution to entrainment into the layer; they are probably synonymous with what Townsend (1976) called

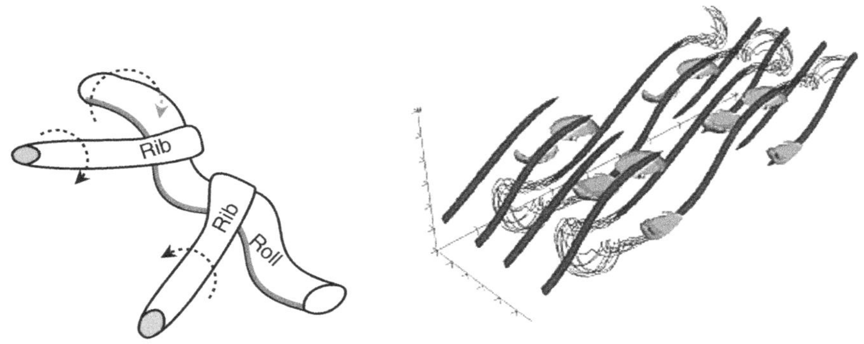


**Figure 5.2** Left: schematic of ribs wrapping around and distorting rolls. Right: ribs in a transitional mixing layer, as seen via contours of $\sqrt{\omega_x^2 + \omega_y^2}$; reproduced from Moser and Rogers (1991), with permission.

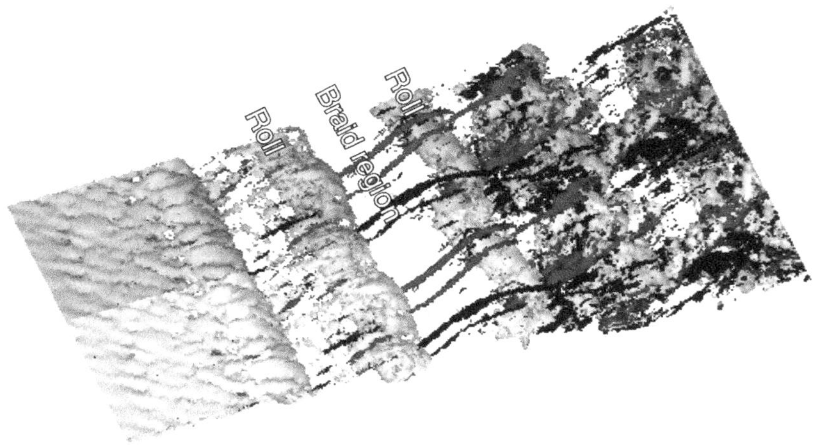


**Figure 5.3** Rolls and ribs in a transitional mixing layer: contours of streamwise vorticity, $\omega_x$. The gray and black streamwise contours, seen between the rolls, are the edges of positive and negative vortices of the braid region. Figure courtesy of P. Comte and P. Bégou (Comte *et al.* 1998).

entrainment eddies. In fully turbulent flow, the braid region contains a good deal of small-scale turbulence. The ribs are far less pronounced than the clear-cut structure that is seen in the transitional mixing layer of Figure 5.3.

Townsend and co-workers in the 1950s inferred a vortical structure from measurements of two-point correlations (Townsend, 1976). The structure is common to many shear layers. The term "double roller" was used to describe its form: Figure 5.4 shows this conception. The method of inference was to construct a pattern of eddies that was consistent with measured correlation functions. Mumford (1982) subsequently obtained

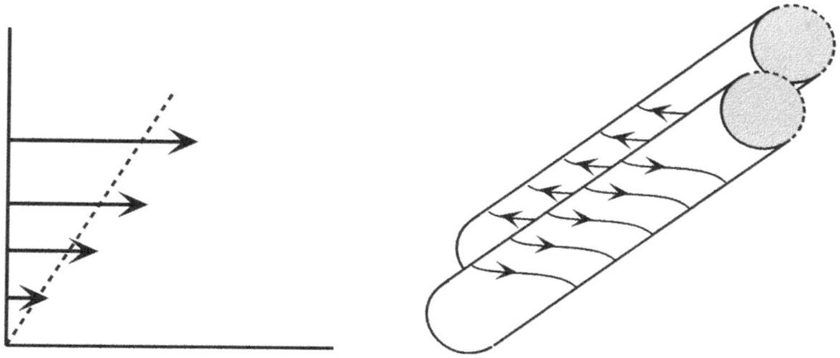

**Figure 5.4** Double rollers in free-shear flows, after Townsend (1976).

evidence that this pattern was representative of an actual geometry. He developed a statistical pattern recognition method to extract a representative form from instantaneous velocity fields. In this body of research, rollers were inferred to be present in planar jets, wakes, and mixing layers. They align at $\sim 40^\circ$ to the shear, irrespective of its direction. The schematic Figure 5.4 shows their alignment in a mixing layer. In a plane jet, two sets of rollers are inferred, forming a horizontal "V" shape, such that each set is aligned to the shear in the upper and lower parts of the flow.

Streamwise rib vortices distort the spanwise rolls (Figure 5.2), producing corrugations and secondary instabilities. Likewise, the rolls stretch and distort the streamwise vortices. Rolls contribute to mixing-layer growth in the transitional stage, but it should not be thought that two-dimensional vortices are the coherent structures of fully turbulent mixing layers. Streamwise, sloping entrainment eddies, as in Figure 5.4, are probably more important to turbulent mixing-layer dynamics. However, external forcing can be used to stimulate rolls, as illustrated by Figure 5.5.

The upper panel of Figure 5.5 is a natural mixing layer. It shows little evidence of rolls. The other panels show the development of forced mixing layers, as simulated numerically by Rogers and Moser (1994). When the shear layer is stimulated by periodic forcing of suitable frequency, spanwise structure coagulates. Toward the middle of the lowest panel, two structures are seen in the act of pairing. As time evolves, they roll around each other and amalgamate into a larger feature. The outcome is an increased mixing-layer thickness. External forcing can increase the growth rate of a mixing layer by the mechanism of vortex pairing.

One engineering significance of turbulent coherent eddies is that they entrain free-stream fluid into a growing shear layer. Shear-layer entrainment in an averaged sense was discussed in conjunction with Eq. (4.2.11). An instantaneous view is that the external fluid is engulfed into the shear layer by swirling round large entrainment eddies. The engulfed fluid then acquires vortical turbulence as it is mixed by smaller-scale eddies. This is the origin of the term "entrainment eddies," which has been applied to the double roller. Evidence is that double rollers are a statistical view of real eddies; but instantaneous structures are not very distinct in high Reynolds number shear layers.

Another reason that ideas about coherent structures are of engineering interest is that they are concrete flow patterns that might be controlled if they are stimulated by periodic forcing, as in the numerical simulation of Figure 5.5. In physical experiments, the

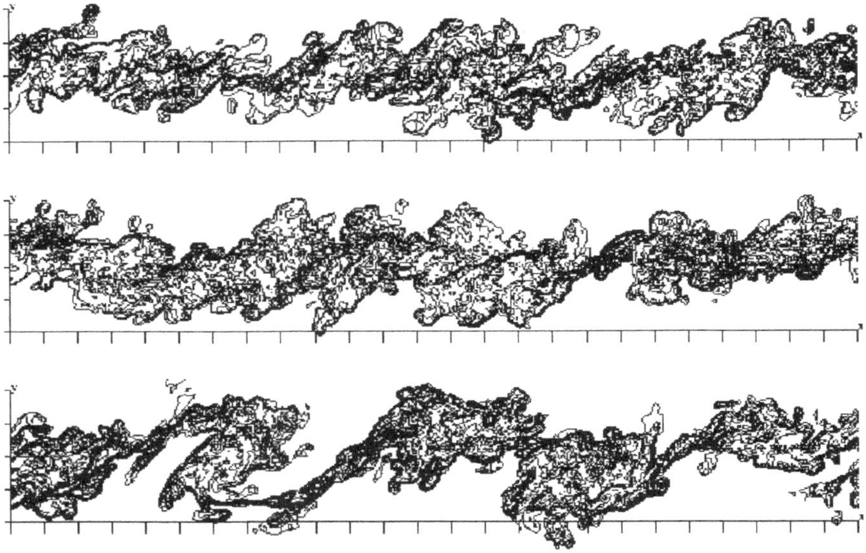

**Figure 5.5** Contours of a passive tracer delineate mixing-layer eddies. Top: a natural mixing layer; middle: a mixing layer subject to moderate, periodic forcing; bottom: a mixing layer with strong forcing. From Rogers and Moser (1994); reproduced with permission.

spanwise rolls in a mixing layer can be locked into step by forcing them with a loudspeaker or vibrating vane. This may be a mechanism for controlling the development of a shear layer – but turbulence control is a topic outside our present scope.

Processes similar to those of the plane mixing layer occur in the early stages of turbulent jet development (Figure 5.1). In this case the large vortices are ring-shaped. They can become corrugated and also can move off-axis to produce a field of eddies that occur irregularly in space and time. In both the plane mixing layer and circular jet, the vortices can intersect and join one another. This can be an amalgamation of eddies, referred to as vortex pairing (as in the lower part of Figure 5.5); or it can be local joining, referred to as cut and connect (Figure 5.6). Cut and connect is a topological change. Vortex segments with opposite signed vorticity merge, the vorticity of opposite signs diffuse together and cancel, and a new configuration of vortex lines emerges.

A jet nozzle has a sharp edge at which the flow separates. The fixed, circular separation line tends to impose axisymmetry on the initial large-scale eddies. This suggests that the axisymmetry can be broken by corrugating the lip of the nozzle, which is the case. The corrugations hasten the breakup of axisymmetric vortices into smaller, irregular eddies. This can occur through fission of vortex rings into smaller rings, as illustrated in Figure 5.6. Small-scale mixing can be promoted by encouraging the breakup of large structures.

While a similarity might be expected between the coherent eddies in the wake of a sphere and those downstream of the exit of a jet, that is not the case. Even though the mean flow is axisymmetric, the sphere does not shed vortex rings. Flow measurements

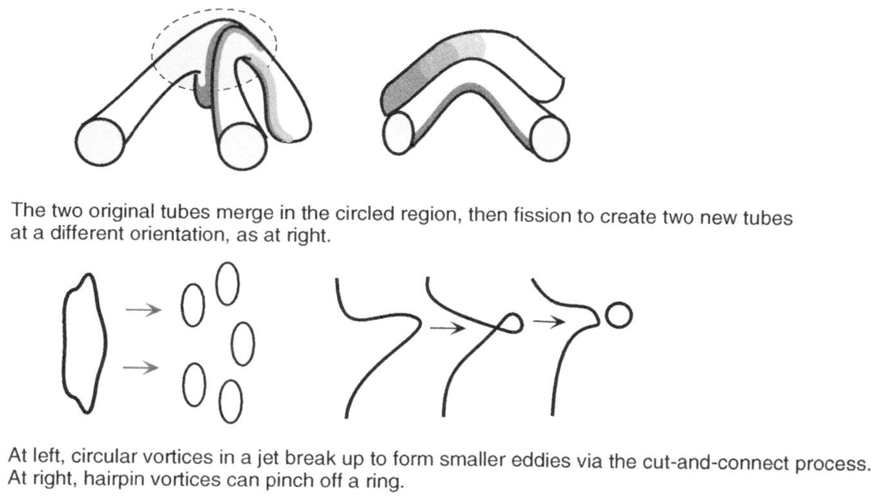

The two original tubes merge in the circled region, then fission to create two new tubes at a different orientation, as at right.
At left, circular vortices in a jet break up to form smaller eddies via the cut-and-connect process. At right, hairpin vortices can pinch off a ring.


**Figure 5.6** Reorganization of vortices: cut and reconnection of vortex tubes, following Hussain (1978).

in the wake of a sphere have long been known to suggest a helical structure in the wake. The helix can be left- or right-handed, either of which destroys the cylindrical symmetry. The way to restore axisymmetry on the average is for the helix to vary randomly between left and right polarization; indeed, this is what is observed. It took a long time before the structure of the helix was successfully visualized, first in experiments, then in numerical simulations. It consists of loops of vorticity that are shed from the sphere at locations that circle the perimeter of the sphere, creating a helical signature in the wake (Figure 5.7). It should be emphasized that this "coherent" structure of the sphere wake is rather hard to detect in high Reynolds number experiments or computations. Hot wire anemometers separated by 180° around the circumference show strong coherence, evidencing the presence of some orderliness, but laboratory visualizations have been only at laminar or transitional Reynolds numbers.

### 5.1.2 Boundary layers

The coherent eddies of a flat-plate boundary layer may also be in the form of vortex loops, in this case alluded to as "horseshoe" or "hairpin" vortices. However, analysis of two-point correlations and pattern recognition programs often show only the legs of the hairpin vortices. This has led to the boundary-layer coherent structure being described as "double cones" aligned at about 45° to the wall and nearly attached to the surface (Townsend, 1976). Figure 5.8 is a schematic of the form of attached eddies that was inferred by statistical analysis of eddy structure.

Smoke visualizations of the instantaneous flow field in a turbulent boundary layer by Head and Bandyopadhyay (1981) showed evidence of the upper loop of hairpin vortices. The attached eddies may well be the lower portion of such vortices. At very high Reynolds

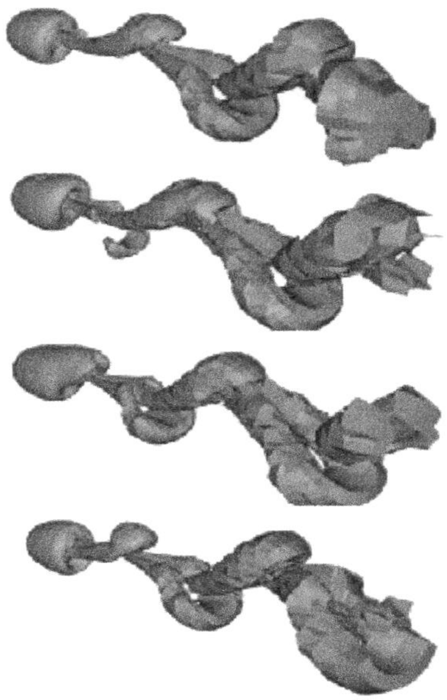

**Figure 5.7** Vortical structures downstream of a sphere. The surfaces enclose regions where rate of rotation is greater than rate of strain. The loops rotate direction to create a helical signature. From Johnson and Patel (1999), reproduced with permission.

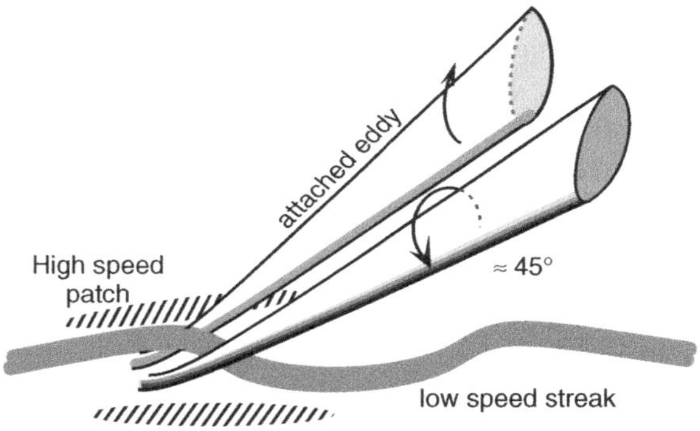


**Figure 5.8** Attached, conical eddies in a boundary layer, as proposed by Townsend (1976).

numbers, the hairpin vortices might occur in the outer region of the boundary layer and not be seen as attached to the wall; a precise image does not currently exist. There is evidence that sometimes the hairpins occur as a packet of several vortices, forming a sort of tunnel. Figure 5.9 was constructed by Zhou *et al.* (1999) from low Reynolds number direct numerical simulation (DNS) data. It shows a sequence of vortex loops. These horseshoe-shaped loops are only seen at low Reynolds number; at higher Reynolds number the legs are straighter, as sketched in the lower part of Figure 5.11. The low

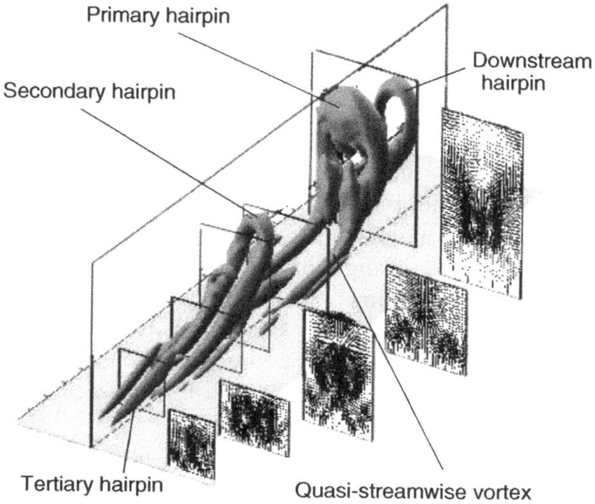


**Figure 5.9** Hairpin eddies in low Reynolds number channel flow. Courtesy of R. Adrian, J. Zhou, and S. Balachandar, reproduced with permission.

Reynolds number simulation in Figure 5.9 suggests a process by which self-generation of vortex loops might lead to a packet. As loops lift from the surface, they are stretched by the mean shear, ultimately becoming hairpins, with long, nearly straight legs, lying at $\sim 45^\circ$ to the flow. These legs induce a distortion in the underlying boundary layer that locally intensifies the spanwise vorticity. The local patch of high vorticity becomes kinked into another horseshoe, which lifts from the surface. This process has been termed "burst and sweep," the lift-up being called the burst and the regeneration being initiated by the sweep (Offen and Kline 1975).

In a parallel shear layer the rate-of-strain tensor is

$$\begin{bmatrix} 0 & \frac{1}{2}\partial_y U \\ \frac{1}{2}\partial_y U & 0 \end{bmatrix}.$$

This has eigenvectors proportional to $(1, 1)$ and $(-1, 1)$. These are vectors at $\pm 45^\circ$ to the wall. The $\sim 45^\circ$ orientation of attached eddies tends to align them with the principal directions of strain. This suggests that these large eddies efficiently extract energy from the mean flow. The circulation round vortices lying at $45^\circ$ to the flow lies in a plane at $-45^\circ$ to the flow. That orientation produces an anticorrelation between $u$ and $v$, consistently with the Reynolds shear stress $\overline{uv}$ being negative.

The angled alignment of structures requires a balance of forces. Mean shear would rotate a material line until it became parallel to the wall. However, a horseshoe vortex has a self-induced velocity that lifts it away from the wall. To achieve an orientation of $\approx 45^\circ$, an approximate balance between rotation by the mean shear and self-induced lifting is required. This would suggest that only structures with sufficiently large circulation can maintain themselves.

As the horseshoe vortex lifts away from the wall, the mean rate of strain will intensify the vorticity in its legs, making them a dominant feature of the flow. The legs of lifted vortices may be the origin of the conical eddy structure (Figure 5.8) inferred from statistical correlations.

Between the conical eddies, *low-speed streaks* are seen near the surface, as portrayed in Figure 5.8. This means that the instantaneous velocity $\tilde{u}$ is less than the average velocity $U$ in these regions. The flow between the eddies is upward from the surface, so it convects low-momentum fluid away from the surface to produce these low-speed streaks. The appeal here is to a qualitative "mixing length" argument, according to which $u$ momentum is conserved during a displacement. That is a reasonable concept if the displaced fluid is long in the $x$ direction. In that case $\partial_x p$ is small for the displaced element and $u$ momentum is approximately conserved. It is probably not coincidental that the streaks are indeed elongated in the $x$ direction. Consisting of predominantly $u$ velocity, and being elongated in $x$, gives streaks a jet-like character. A theoretical reason for this is discussed in Chapter 11.

Typically the streaks are spaced about $100\nu/u*$ apart in the spanwise, $z$ direction; in other words, $\Delta z_+ \approx 100$ on average. The streaks themselves have a width of $\Delta z_+ \approx 30$ and a length that can be more than $\Delta x_+ \approx 2000$. The elongation is consistent with evidence from pattern recognition analysis.

The legs of the horseshoe vortices in Figure 5.9 curve up from the surface. The portion along the surface might be considered to produce the highly elongated, low-speed streaks (see Figure 5.8). However, there is a different perspective, discussed in Chapter 11, by which the streaks are streamwise-elongated, jet-like disturbances that are selectively amplified by mean shear. Then the long vortices near the surface might simply be the boundaries of the jet-like structure.

On the outside of the pair of conical eddies, *high-speed streaks* are seen. They are produced by convection of high-momentum fluid toward the wall. High-speed streamwise velocity contours are shorter than the low-speed streaks and can also be seen as patches of high surface shear stress. The concept of high- and low-speed streaks evolved through the analysis of many flow visualizations, and is well substantiated. Instantaneous snapshots of turbulent boundary layers give an unambiguous impression of the streaky structure, but it is certainly not as clear-cut as the cartoon in Figure 5.8 implies; the cartoon represents a conditionally averaged flow structure. An instantaneous velocity field in a plane parallel to the wall is contained in Figure 5.10. A long streak of negative $u$ can be identified in the middle of the figure. The connection between attached eddies and the elongated velocity contours in Figure 5.10 is not entirely certain. To the extent that the present discussion of coherent structure is a taxonomy, there is no need to provide a definitive connection between the various features that have been observed in boundary layers. At the present time it is not clearly understood.

The upper edge of the boundary layer, sometimes called the superlayer, shows large-scale undulations. Their length is of order 1–2 times the boundary-layer thickness and they convect downstream at about $0.85U_\infty$ (Cantwell, 1981). Smaller irregularities in the interface may be caused by the upper portion of the hairpin vortices, as well as by the general, disorderly turbulent motion. The photograph at the top of Figure 5.11 provides a clear impression that the interface is perturbed by small vortices. It has been proposed that large bulges are sometimes created by an envelope of hairpin vortices.

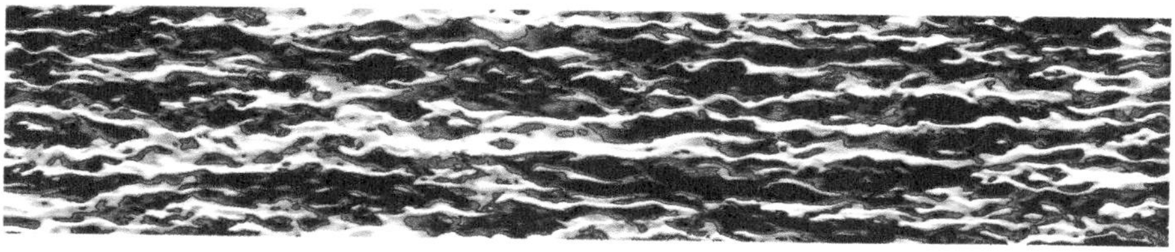

**Figure 5.10** Streaks in a turbulent boundary layer (Jacobs and Durbin, 2000). These are contours of the instantaneous $u$ component in a plane near the wall. Light regions are $u < 0$, dark are $u > 0$. A long streak of negative $u$ is seen spanning the middle of the figure. White lines are overlaid to show the $u = 0$ contours.

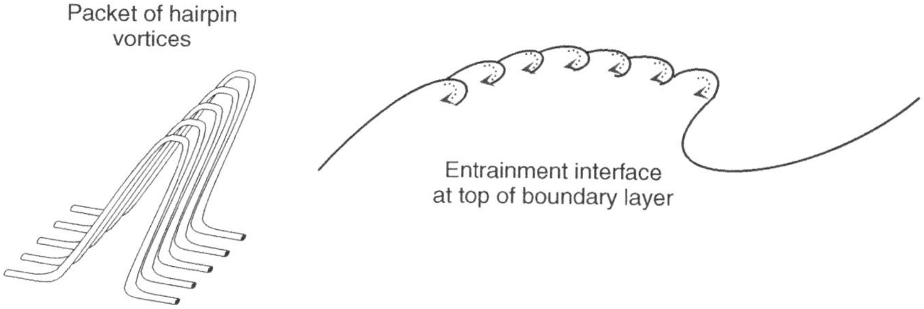


**Figure 5.11** Upper: Smoke wire visualization of the instantaneous eddying motion in a turbulent boundary layer. Courtesy of T. Corke and H. Nagib. Lower: Schematic of hairpin vortices and entrainment interface. After Head and Bandyopadhyay (1981).

Entrainment occurs via engulfing into the bulges, in conjunction with mixing by small eddies. Deep intrusions of laminar fluid are seen between turbulent zones in Figure 5.11. The smaller eddies along the entrainment interface ingest portions of laminar fluid. This two-fold process advances the interface between the turbulent boundary layer and the free stream.

Three types of structure seen in boundary layers are widely accepted: very long, narrow, low-speed streaks near the surface; hairpin vortices, or conical eddies, in the central portion; and an undular interface with the free stream. These form the basis

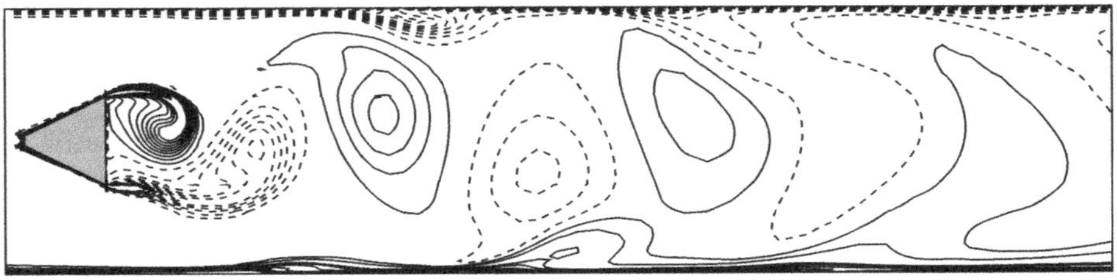

**Figure 5.12** Vorticity contours behind a triangular cylinder in a channel show unsteady vortex shedding. This illustrates that the ensemble-averaged, mean flow can be unsteady; nevertheless, these are contours of averaged vorticity and not of the instantaneous, random flow.

for a qualitative understanding of mixing, entrainment, pressure fluctuations, and other properties of turbulent boundary layers.

### 5.1.3 Non-random vortices

Many other characteristic structures have been seen in various turbulent shear flows. Quite often they are reminiscent of vortices seen more clearly in laminar flow. Examples are streamwise eddies on concave walls or swept wings, associated with Görtler and cross-flow vortices, and toroidal eddies in circular Couette flow, corresponding to Taylor vortices. These will not be discussed here, since these and other examples are best dealt with in the context of hydrodynamic stability.

Vortex streets, reminiscent of the laminar Von Karman vortex street, can usually be detected behind bluff bodies even in a fully turbulent regime. This is illustrated in Figure 5.12 by an unsteady Reynolds averaged computation using a turbulence model. If truly non-random eddies exist, they should be distinguished from the turbulent motions. In an experiment the non-random, periodic shedding appears as a spike in the frequency spectrum that is clearly distinguished from the broadband, random background. The decomposition into mean and fluctuation, $\tilde{u} = U + u$, used in Chapter 3, includes the periodic component in the mean flow $U(t)$. The decomposition could be extended by dividing $U$ into time-averaged and periodic parts, $U(t) = \overline{U} + U'(t)$. However, this division is not at all useful in the context of unsteady Reynolds averaged computation. $U(t)$ is the ensemble-averaged velocity that appears in the equations of Chapter 3; when it is time-dependent due to vortex shedding, ensemble averaging cannot be replaced by time averaging. In an experiment, averages can be computed at a fixed phase of the oscillation and along a direction of homogeneity. Randomness and ensemble averaging cause the vortex street to decay rather more rapidly than in laminar flow. This example is discussed further in Section 7.4.2.

## 5.2 Vorticity and dissipation

Vortical structures occur in the large, energetic scales of motion. Although large structures are vortical, the small-scale vorticity is far more intense. At high Reynolds number, coherent structures can be difficult to distinguish in plots of vorticity contours because the most pronounced feature is small-scale irregularity. The dominance of small-scale vorticity can be deduced by applying the scaling arguments outlined in Section 2.1.

The length scale of the large eddies is denoted $L$ and is on the order of the shear-layer thickness $\delta$; their velocity scale is on the order of the turbulent intensity $\sqrt{k}$. The length scale of the smallest eddies is the dissipative scale $\eta = (\nu^3/\varepsilon)^{1/4}$; their velocity is of order $(\varepsilon\nu)^{1/4}$. In the inertial range, the size of the eddies is $r$, the distance between two points; their velocity is of order $(\varepsilon r)^{1/3}$. Vorticity has dimensions of velocity/length. The scalings of length, velocity, and, consequently, vorticity are summarized in Table 5.1. Within the inertial range, as $r \to 0$ the energy decreases, but the vorticity increases; vorticity is dominated by small eddies, velocity is dominated by large scales.

The requirement that the inertial subrange matches to the energetic range as $r \to L$, and to the dissipation range as $r \to \eta$, has an instructive consequence. Consider the vorticity scaling in Table 5.1: matching the inertial-range formula to the energetic range formula as $r \to L$ gives $(\varepsilon/r^2)^{1/3} \to \sqrt{k}/L$. This can be restated as $(\varepsilon/L^2)^{1/3} \sim \sqrt{k}/L$, or as

$$\varepsilon \sim k^{3/2}/L. \tag{5.2.1}$$

This estimate for the rate of dissipation is the basis of the $k$–$\ell$ turbulence model discussed in Section 6.2.2.2.

Similarly, the requirement that the inertial-range formula matches to the dissipation-range formula as $r \to \eta$ is $(\varepsilon/r^2)^{1/3} \to (\varepsilon\nu)^{1/2}$. This simply recovers the scaling $\eta \sim (\nu^3/\varepsilon)^{1/4}$. The interesting perspective given by this asymptotic analysis is that expression (5.2.1) and the formula for $\eta$ are requisites for the inertial range to interpolate between the energetic and dissipative scales.

Comparing the relative magnitudes of vorticity in Table 5.1:

$$\frac{\omega_{\text{energetic}}}{\omega_{\text{inertial}}} = \left(\frac{r}{L}\right)^{2/3}, \qquad \frac{\omega_{\text{energetic}}}{\omega_{\text{dissipation}}} = \left(\frac{\eta}{L}\right)^{2/3},$$

where Eq. (5.2.1) has been invoked. In the inertial range $\eta \ll r \ll L$, so the relative magnitudes of vorticity are $\omega_e \ll \omega_i \ll \omega_d$. In fact $\omega_e/\omega_d$ is of order $R_T^{-1/2}$, where $R_T = \sqrt{k}L/\nu$, as in Section 2.1. This shows that at high Reynolds number the small-scale vorticity is much more intense than the large-scale vorticity. The reason is essentially that the velocity gradient is inversely proportional to the length scale, so the small scales are associated with large velocity gradients.

As the vorticity is most intense at small scales, it might seem intuitively that vorticity has some connection to the rate of turbulent energy dissipation. The vorticity vector is

**Table 5.1** Scalings of length, velocity, and vorticity.

| Range | Energetic $r \sim L$ | Inertial $\eta \ll r \ll L$ | Dissipative $r \sim \eta$ |
|---|---|---|---|
| length | $L$ | $r$ | $(\nu^3/\varepsilon)^{1/4}$ |
| velocity | $\sqrt{k}$ | $(\varepsilon r)^{1/3}$ | $(\varepsilon\nu)^{1/4}$ |
| vorticity | $\sqrt{k}/L$ | $(\varepsilon/r^2)^{1/3}$ | $(\varepsilon/\nu)^{1/2}$ |

defined as† $\omega_i = \varepsilon_{ijk}\,\partial_j u_k$. The mean-square magnitude of the vorticity is therefore

$$\overline{\omega^2} = \overline{\omega_i\,\omega_i} = \overline{\varepsilon_{ijk}\,\partial_j u_k \varepsilon_{ilm}\,\partial_l u_{\mathrm{m}}} = (\delta_{jl}\delta_{km} - \delta_{jm}\delta_{kl})\,\overline{\partial_j u_k\,\partial_l u_{\mathrm{m}}}$$
$$= \overline{\partial_j u_k\,\partial_j u_k} - \overline{\partial_j u_k\,\partial_k u_j}.$$

The dissipation rate is defined by $\varepsilon = \nu\,\overline{\partial_j u_k\,\partial_j u_k}$. For incompressible flow $\partial_j u_j = 0$ and the above formula for $\overline{\omega^2}$ can be rewritten as

$$\nu\,\overline{\omega^2} = \varepsilon - \nu\,\partial_j\partial_k\overline{u_k u_j}. \tag{5.2.2}$$

The last term of this equation vanishes in homogeneous turbulence, in which case $\varepsilon = \nu\,\overline{\omega^2}$. Usually the small scales are approximately homogeneous and this relation between mean-squared vorticity and dissipation holds to a good degree of accuracy (Exercise 5.2).

## 5.2.1 Vortex stretching and relative dispersion

The question of how vorticity accumulates at small sizes, and what sets that size, can be answered by considering the dynamics of vorticity. The essential elements are stretching and diffusion.

A notable property of turbulence is its ability to mix contaminants. A concentration blob will be distorted into a filamentary shape by the turbulence, as illustrated in Figure 5.13. When molecular diffusion comes into play, the filaments will ultimately mix with the surrounding fluid. But the stirring by fluid turbulence is the first and more important step.

Consider two points, A and B, in the dye blob (Figure 5.13). A corollary to the observation that turbulence mixes the dye is that on average these marked elements will separate; this is Richardson's idea of relative dispersion (Exercise 2.3). There is nothing special about the particle pairs being in the dye: any pair of particles tends to separate in consequence of random convection. If the particle pairs are connected by a vortex tube, then a corollary to relative dispersion is that vortex tubes tend to be stretched. As

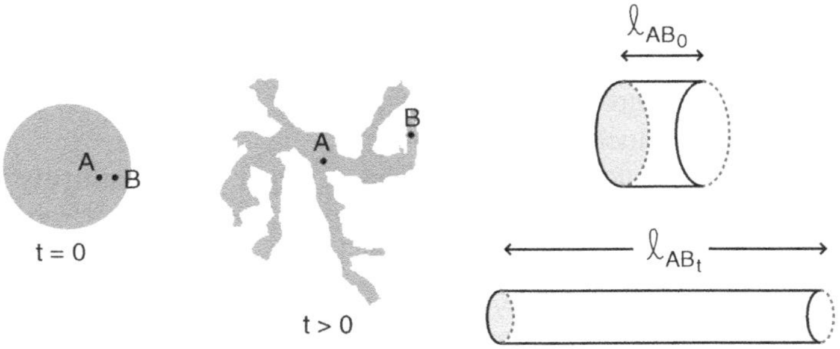


**Figure 5.13** Schematic of blob dispersion and implied vortex stretching. Turbulent dispersion of particle pairs lies behind mixing of contaminants and elongation of vortex tubes.

† The skew symmetric tensor can be defined by $\varepsilon_{ijk} = 1$ if $ijk$ are any three successive integers in 123123 and $\varepsilon_{ijk} = -1$ if $ijk$ are any three successive integers in 132132.

a vortex tube is stretched, its vorticity increases. This provides an understanding of how the small-scale vorticity is maintained: turbulent convection tends to stretch vortex lines.

Consider a cylindrical vortex tube of cross-sectional area $\mathcal{A}_0$ and initial length $\ell_{\mathrm{AB}_0}$. Its volume is $\mathcal{A}_0 \ell_{\mathrm{AB}_0}$. If the flow is incompressible, the volume of fluid elements must be constant. At a later time, $\mathcal{A}/\mathcal{A}_0 = \ell_{\mathrm{AB}_0}/\ell_{\mathrm{AB}}$. By definition, the circulation $\omega\mathcal{A}$ around a vortex tube is constant, so

$$\omega = \omega_0 \frac{\mathcal{A}_0}{\mathcal{A}} = \omega_0 \frac{\ell_{\mathrm{AB}}}{\ell_{\mathrm{AB}_0}}. \tag{5.2.3}$$

As two fluid particles separate, the length of the material line connecting them grows and the vorticity increases in proportion to the length of the material line. The cross-section grows smaller, and the vorticity amplifies; hence vorticity intensifies at small scale.

Ultimately, the vorticity amplification by stretching will be counteracted by molecular diffusion. An exact solution to the Navier–Stokes equations that illustrates the balance between stretching and diffusion is the Burgers vortex, illustrated by Figure 5.14. The axial component of the steady, axisymmetric vorticity equation is

$$U_r\, \partial_r \omega_x = \omega_x\, \partial_x U_x + \frac{\nu}{r}\, \partial_r r\, \partial_r \omega_x. \tag{5.2.4}$$

If the $x$ and $r$ components of velocity are given by $U_x = \alpha x$ and $U_r = -\frac{1}{2}\alpha r$, then the flow is an incompressible, axisymmetric rate of strain along the axis of the vortex. Substituting this into (5.2.4) gives

$$-\tfrac{1}{2}\alpha\, \partial_r (r^2 \omega_x) = \nu\, \partial_r r\, \partial_r \omega_x.$$

A solution to this equation is

$$\omega_x = \omega_{\mathbb{C}}\, e^{-\alpha r^2/4\nu}, \tag{5.2.5}$$

where $\omega_{\mathbb{C}}$ is a constant. The solution (5.2.5) is a steady vorticity distribution, for which axial stretching and radial diffusion balance exactly. The thickness of the vortex is $O(\sqrt{\nu/\alpha})$. The circulation of the vortex, $\Gamma = \int_0^\infty \omega\, \mathrm{d}(\pi r^2)$, is constant during the stretching and equal to $4\pi\nu\omega_{\mathbb{C}}/\alpha$.

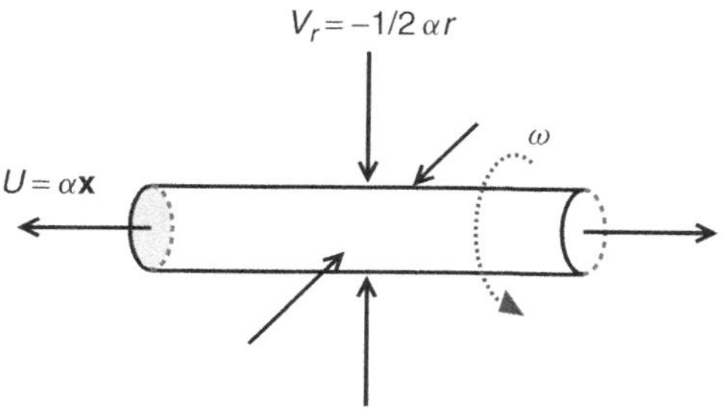


**Figure 5.14** Straining flow and vorticity configuration for a cylindrical Burgers vortex.

Suppose there is a number $N_b/A$ of Burgers vortices per unit area of fluid. Then the averaged rate of energy dissipation is

$$\varepsilon = \frac{N_b}{A}\nu \int_0^\infty \omega^2 \, d(\pi r^2) = \frac{N_b \alpha \Gamma^2}{8\pi A}$$

(see Eq. (5.2.2)). This illustrates that the balance between stretching and diffusion results in a dissipation rate that is independent of molecular viscosity. The Burgers vortex is suggestive of the idea that at high Reynolds number $\varepsilon$ is independent of $\nu$, even though dissipation is a viscous process.

Recall the estimate (5.2.1), $\varepsilon \sim k^{3/2}/L$, that relates dissipation to the energetic scales of motion. It might have seemed curious that this estimate is independent of molecular viscosity. The physical justification is that just given; that at high Reynolds number the balance between stretching and diffusion of small-scale vorticity causes the rate of energy dissipation to be independent of viscosity. The length scale of the vortices for which this balance is achieved is $\sqrt{\nu/\alpha} \sim (\nu^3/\varepsilon)^{1/4} = \eta$ if the rate of strain, $\alpha$, is of order $\sqrt{\varepsilon/\nu}$. Intensification of small-scale eddies by self-straining stops when the Kolmogoroff scale is reached and a balance with viscous diffusion is struck.

### 5.2.2 Mean-squared vorticity equation

The concepts illustrated by the Burgers vortex are manifested in the full vorticity equation. A transport equation for the mean-squared vorticity can be derived by the same procedure as was used to derive the kinetic energy equation in Chapter 3: the equation for the fluctuating vorticity vector is multiplied by the vorticity vector and averaged. The result for $\overline{|\omega|^2} \equiv \overline{\omega_i \omega_i}$ is

$$\begin{aligned}\partial_t \overline{|\omega|^2} + U_j \, \partial_j \overline{|\omega|^2} &= -2\overline{u_j \omega_i} \, \partial_j \Omega_i - \partial_j \overline{u_j \omega_i \omega_i} + 2\overline{\omega_i \omega_j} \, \partial_j U_i \\ &\quad + 2\Omega_j \, \partial_i \overline{\omega_i u_j} + \nu \nabla^2 \overline{|\omega|^2} + 2\overline{\omega_i \omega_j \, \partial_j u_i} - 2\nu \, \overline{\partial_j \omega_i \, \partial_j \omega_i},\end{aligned}$$

where $\Omega$ is the mean vorticity. On restriction to homogeneous turbulence, the simpler equation

$$\tfrac{1}{2} \partial_t \overline{|\omega|^2} = \overline{\omega_i \omega_j} \, \partial_j U_i + \overline{\omega_i \omega_j \, \partial_j u_i} - \nu \, \overline{\partial_j \omega_i \, \partial_j \omega_i} \qquad (5.2.6)$$

is obtained. Oftentimes, $|\omega|^2$ is referred to as the fluctuating enstrophy.

It has already been deduced that the largest contribution to the vorticity is from the dissipative range, where $\omega \sim (\varepsilon/\nu)^{1/2}$. For the purpose of order-of-magnitude estimation, assume that production and dissipation are comparable. Then $-\overline{u_i u_j} \, \partial_j U_i \sim \varepsilon \sim k^{3/2}/L$. With the order-of-magnitude scaling $\overline{u_i u_j} \sim k$, this gives the mean flow gradient magnitude of $k^{1/2}/L$. Turbulent gradients are of order $\omega$. Consider the relative magnitudes of the terms in Eq. (5.2.6): for instance

$$\frac{\overline{\omega_i \omega_j \, \partial_j u_i}}{\overline{\omega_i \omega_j} \, \partial_j U_i} \sim \frac{\omega L}{k^{1/2}} \sim \sqrt{\frac{\varepsilon L^2}{\nu k}} \sim \sqrt{\frac{k^{1/2} L}{\nu}} = R_T^{1/2}.$$

This type of reasoning shows that, at high turbulent Reynolds number, the terms containing only turbulent quantities dominate (5.2.6). The lowest-order balance is between the last two terms

$$\overline{\omega_i \omega_j \, \partial_j u_i} \approx \nu \, \overline{\partial_j \omega_i \, \partial_j \omega_i} \tag{5.2.7}$$

(see Exercise 5.3). This represents a balance between stretching of turbulent vorticity by the turbulent velocity gradients and dissipation of mean-squared vorticity by molecular diffusion. The Burgers vortex described in the previous section is a model for this balance if $\alpha$ represents the small-scale rate of strain.

The self-stretching can be related to a product of velocity derivatives. It can be shown that

$$2\omega_i \omega_j \, \partial_j u_i = \partial_i \partial_j \partial_k (u_i u_j u_k) - 2\partial_k u_i \, \partial_k u_j \, \partial_i u_j - 3\partial_j (u_j \, \partial_k u_i \, \partial_i u_k).$$

If the turbulence is homogeneous, averaging this formula gives

$$\overline{\omega_i \omega_j \, \partial_j u_i} = -\overline{\partial_k u_i \, \partial_k u_j \, \partial_i u_j}.$$

Vortex stretching occurs when the covariance on the right-hand side is not zero. For a Gaussian random variable, third moments vanish and so would this self-stretching term. Non-Gaussianity in the small-scale statistics is critical to the maintenance of turbulence.

The third moment defines the *skewness* of a probability distribution. The *velocity derivative skewness* is defined by

$$S_{\mathrm{d}} \equiv -\overline{\partial_k u_i \, \partial_k u_j \, \partial_i u_j} / (\overline{\partial_j u_i \, \partial_j u_i})^{3/2}.$$

In isotropic turbulence this can be shown (Batchelor and Townsend, 1947) to be equal to

$$S_{\mathrm{d}} = -\frac{7}{6\sqrt{15}} \frac{\overline{(\partial_1 u_1)^3}}{\overline{(\partial_1 u_1)^2}^{3/2}}.$$

The gradient $\partial_1 u_1$ is something that can be measured by a fixed probe in a wind tunnel, upon invoking Taylor's hypothesis (Section 2.1.1) to relate $\partial/\partial x_1$ to $\partial/\partial t$. The velocity derivative skewness is indeed measured to be non-zero (Mydlarski and Warhaft, 1996); experimental values are in the range 0.5–1.0.

Usually $S_{\mathrm{d}}$ is defined as $\overline{(\partial_1 u_1)^3} / \overline{(\partial_1 u_1)^2}^{3/2}$, without the numerical factor, but the above definition suits the present purpose. It follows from the skewness formulas and (5.2.2) that (5.2.6) can be written as

$$\tfrac{1}{2}\, \partial_t \overline{|\omega|^2} = \overline{\omega_i \omega_j} \, \partial_j U_i + S_{\mathrm{d}} (\overline{|\omega|^2})^{3/2} - \nu \, \overline{\partial_j \omega_i \, \partial_j \omega_i}. \tag{5.2.8}$$

In order for there to be self-stretching of the vorticity, $S_{\mathrm{d}}$ must be non-zero. In the absence of viscosity and mean flow gradients, only the middle term on the right-hand side of (5.2.8) remains. Then it can formally be integrated to

$$\overline{|\omega|^2} = \left( \frac{|\omega|_0}{1 - |\omega|_0 \int_0^t S_{\mathrm{d}} \, \mathrm{d}t} \right)^2,$$

where the subscript 0 denotes the initial value. The fact that the denominator could vanish at some time raises the possibility that the vorticity could become infinite in a finite time. The time to singularity would seem to be $O(1/|\omega|_0 S_\mathrm{d})$. The matter is actually more subtle, since it depends on the vorticity remaining aligned with the straining. There is ongoing debate among mathematicians over whether or not the inviscid equations do produce a finite-time singularity. Even if it does occur in an ideal fluid, molecular viscosity will prevent this singularity. However, it can be seen that self-stretching is a powerful vorticity amplifier.

## Exercises

**Exercise 5.1.** *Useful mathematical relation.* Prove that

$$\varepsilon_{ilp}\varepsilon_{jnp} = (\delta_{ij}\delta_{ln} - \delta_{in}\delta_{lj}).$$

The brute force approach is to write out all 81 components. However, all but 12 are zero. Use this identity to show that $\nu\overline{|\omega|^2} = \varepsilon$ in homogeneous, incompressible turbulence. The vorticity vector is $\omega_i = \varepsilon_{ijk}\,\partial_j u_k$.

**Exercise 5.2.** *Vorticity and dissipation.* Suppose that the scale of non-homogeneity is $\delta \sim L$. Show that the magnitude of the first term on the right-hand side of (5.2.2) relative to the second is $O(R_\mathrm{T})$.

**Exercise 5.3.** *The balance between stretching and diffusion.* Show by dimensional analysis of the stretching $(\omega_j\,\partial_j u_i)$ and diffusion $(\nu\nabla^2\omega_i)$ terms of the vorticity equation that a balance is reached when the vortex radius is $O(\eta)$. Note that the rate of strain is of order $\sqrt{\varepsilon/\nu}$ at small scales.

**Exercise 5.4.** *Introduction to rapid distortion theory.* Find a solution of the form

$$\omega_3 = A(t)\,\mathrm{e}^{\mathrm{i}(k_1(t)x_1 + k_2x_2)}$$

to the inviscid vorticity equation if the velocity is of the form $u_3 = \alpha x_3$, $u_2 = v_2$, and $u_1 = v_1 - \alpha x_1$, and $\nabla\cdot v = 0$. The initial condition is $\omega_3(0) = \omega_0\,\mathrm{e}^{\mathrm{i}(\kappa_1x_1 + \kappa_2x_2)}$. Because $v$ is the velocity generated by the vorticity, it too is proportional to $\mathrm{e}^{\mathrm{i}[k_1(t)x_1 + k_2x_2]}$. [Hint: Don't forget the continuity equation.]

**Exercise 5.5.** *Burgers vortex.* Solve the two-dimensional vorticity equation, corresponding to (5.2.4), for $\omega_x(y)$ when the straining flow is $U = \alpha x$ and $V = -\alpha y$. What is $W(y)$? Why is this an exact solution to the Navier–Stokes equations?

# Part II

# SINGLE-POINT CLOSURE MODELING

# 6

# Models with scalar variables

> The pictures and vocabulary are the key to physical intuition, and the intuition makes possible the sudden leaps of insight ... mathematics, of course is the ultimate arbiter; only it can say whether the leaps of insight are correct.
>
> – Richard Price and Kip Thorn

The equations for the statistics of turbulent flow are fewer in number than the unknowns appearing in them: they do not form a closed set of predictive equations. The purpose of closure modeling is to formulate further equations such that a soluble set is obtained. When the purpose is to predict non-homogeneous flow, possibly in complex engineering geometries, then these closure equations must contain empiricism. It is generally preferable for empiricism to enter via a fairly small number of experimentally determined model constants. However, sometimes functions are introduced to fit experimental curves: this is especially true in the simplest models, such as integral closure or mixing lengths. By adopting functional forms, it is implicitly assumed that turbulent flows exhibit a universal behavior of some sort. Unfortunately, universality does not apply to complex engineering flows – flows that closure models are formulated to predict. The more elaborate, transport equation models solve partial differential equations in order to minimize the need to specify functional forms.

Consider the problem of predicting the mean flow field. The four equations (3.2.2) for the mean contain 10 unknowns: $U_i$, $P$, and $\overline{u_i u_j}$, with $i = 1–3$, $j = 1–3$. To close these equations, a semi-empirical formulation to predict the $\overline{u_i u_j}$ is needed. For instance, the Reynolds stress might be explicitly related to the mean flow by a Newtonian constitutive equation with an eddy viscosity: $\overline{u_i u_j} = -2\nu_T S_{ij} + \frac{2}{3}\delta_{ij}k$. The term "semi-empirical" means that aspects of the model are obtained from a combination of analysis, fluid dynamics, and experimental data. We have already seen examples of this: to obtain the log law, the relation $-\overline{uv} = u_*\kappa y\, \partial_y U$ was deduced as a closure formula by dimensional reasoning, but the constant $\kappa$ had to be determined from experiments.

*Statistical Theory and Modeling for Turbulent Flows, Second Edition* P. A. Durbin and B. A. Pettersson Reif

It has sometimes mistakenly been thought that empirical coefficients are a bane of turbulence modeling. Actually, the opposite is true. Closure models enable one to predict statistics of the very complex phenomenon of turbulent flow by solving a remarkably simple set of equations (at least for the majority of models, such as those discussed herein). If no empirical constants were required, the implication would be that the closure represented either exact laws of fluid dynamics, or a systematically derived approximation. That degree of exactitude is impossible in predictive models for engineering flows. The alternative is to use equations that are not systematically derived and that have an empirical component. The model then becomes a method to use data measured in simple flows that can be studied in reproducible experiments, and therefrom to predict far more complex flows. The creation and analysis of these models proves to be a challenging field of practical, applied mathematics.

Having pondered the problem of predicting the mean, $U_i$, and recognized that its exact evolution equation contains $\overline{u_i u_j}$, one might think to derive the exact equation for $\overline{u_i u_j}$ (which is Eq. (3.2.4)) in hopes of achieving closure. To one's dismay, the problem has got worse: not only has a further moment $\overline{u_i u_j u_k}$ arisen, but a velocity–pressure gradient correlation, $\overline{u_i \, \partial_j p}$, also appears. There are now two sources for the lack of closure: nonlinearity and nonlocality. Quadratic nonlinearity in the Navier–Stokes equations results in a moment hierarchy in which the $n$th-moment equation contains the $(n+1)$th velocity moment. The formation of successively higher moment equations does not achieve closure, and rapidly becomes impracticable. The second lack of closure arises from two (or more) point statistics entering the single-point moment equations. Pressure is a nonlocal effect of turbulent velocity fluctuations (pressure is governed by an elliptic, or *nonlocal*, Poisson equation); the velocity–pressure gradient correlation $\overline{u_i \, \partial_j p}$ implicitly introduces the two-point velocity correlation. In this case one could consider a hierarchy of multi-point correlations, but already at two points it has become impracticable. In a three-dimensional geometry, the two-point correlation $\overline{u_i(\boldsymbol{x}) u_j(\boldsymbol{x}')}$ is a function of six coordinates, $(\boldsymbol{x}, \boldsymbol{x}')$. Modeling in three dimensions is already a challenge; it would be very difficult to make a case for modeling in six spatial dimensions. Only models for single-point, first- or second-moment statistics are of interest here.

The type of modeling discussed in this book is that widely used in Reynolds averaged computational fluid dynamics (CFD). This is the type of CFD that is used in virtually all industrial applications of turbulent flow computation. The models have a strong motivation of filling the needs of computer-aided analysis. But the emphasis here is not on pragmatics; the intent is to cover how models are developed: Where does mathematics guide the process? Where does empiricism enter?

This book is not a survey of the many models that can be found in the literature. However, it does cover the range of approaches, from the simplest, integral closure, to the complex Reynolds stress transport methods. The material selected for this text is meant to cover fundamental techniques of modeling and the particulars of a few models. Thereby the groundwork is laid for the full range of models that can be found in the literature.

## 6.1 Boundary-layer methods

The evolution of the field of turbulence modeling has been heavily influenced by advances in computer technology. The earliest models were for boundary-layer flows because parabolic, boundary-layer equations could be solved on the computing equipment of that time. In fact, the earliest computationally tractable models were integral closures, which only require integration of ordinary differential equations (for 2D boundary layers). Closure models for boundary-layer flow are still of some value, but the state of computer technology has enabled full Reynolds averaged Navier–Stokes (RANS) computations to be performed routinely. For such applications, general-purpose models are needed – they are the subject of later sections; we commence with the more limited methods.

### 6.1.1 Integral boundary-layer methods

Integral boundary-layer methods were an early approach to turbulent flow prediction, one that has been replaced by more elaborate methods in most applied computational fluid dynamics computer codes. These methods were developed to estimate boundary-layer effects in streamlined flow. Although integrated equations are no longer an active area of research in turbulence modeling, they are still sometimes an effective level of closure. In some applications, such as to aerodynamics, the primary concern is with streamlined, potential flow around a body. If boundary layers are thin and attached, the role of the closure model may be to estimate the skin friction, or to compute a small displacement effect that modifies the potential flow, or to assess whether the boundary layer is likely to remain attached under a particular potential flow. Integral models may be adequate for such purposes. They have the advantage of being extremely simple and computationally inexpensive, requiring only the solution of ordinary differential equations.

In the boundary-layer approximation, the free-stream velocity is found by an independent potential flow computation. Given $U_\infty(x)$ from such a computation, predictions of momentum thickness, $\theta$, skin friction, $C_\mathrm{f}$, and displacement thickness, $\delta^*$, are wanted. The integral of the boundary-layer momentum equation (4.2.7), on page 68, is unclosed because it contains three unknowns: $\theta$, $\delta^*$, and $C_\mathrm{f}$. The coefficient of $\delta^*$ vanishes in zero pressure gradient. In that case the empirical formula $C_\mathrm{f} = aR_\theta^{-1/4}$ was used in (4.2.10) to close the differential equation. This is a simple example of an integral equation method.

The momentum integral can be written

$$\mathrm{d}_x\theta + (\delta^* + 2\theta)\frac{\mathrm{d}_x U_\infty}{U_\infty} = \frac{C_\mathrm{f}}{2} \tag{6.1.1}$$

when there is a free-stream pressure gradient, as was derived in Exercise 4.5. A common practice is to compute $\delta^*$ from the entrainment law (4.2.11)

$$\mathrm{d}_x[U_\infty(\delta_{99} - \delta^*)] = V_\mathrm{E}. \tag{6.1.2}$$

Equations (6.1.1) and (6.1.2) are a pair of evolution equations for $\delta^*$ and $\theta$ that contain five unknowns. They can be closed by prescribing three functions of the form $C_\mathrm{f} = F(R_\theta, H)$, $\delta_{99} = \delta^* G(H)$, and $V_\mathrm{E} = U_\infty E(H)$, where $H \equiv \delta^*/\theta$. This brings the total number of equations to five. The role of the form factor, $H$, is to incorporate effects of pressure gradient. The definition

$$H = \frac{U_\infty \int_0^\infty U_\infty - U(y)\,\mathrm{d}y}{\int_0^\infty U(y)[U_\infty - U(y)]\,\mathrm{d}y}$$

shows that $H > 1$ if $U(y)/U_\infty$ is monotonically increasing. The larger the region where $U/U_\infty$ is on the order of $\frac{1}{2}$, the larger $H$ will be. Adverse pressure gradients produce substantial deficits in the velocity profile, which increase $H$. So a role of the form factor, $H$, is to parameterize the impact of pressure gradient on the velocity profile. As $H$ – and hence the velocity deficit – increases, the skin friction decreases. This effect will enter the empirical function for $C_f$ if its arguments include both $R_\theta$ and $H$.

The non-dimensional ratios $\delta_{99}/\delta^*$ and $V_E/U_\infty$ are outer region properties. At high Reynolds number they are assumed to be mildly influenced by viscosity, and to first approximation independent of Reynolds number. Therefore they are represented as functions of $H$ alone.

A set of semi-empirical formulas that has been used by some investigators is

$$
\begin{aligned}
C_f &= 0.246 R_\theta^{-0.268} 10^{-0.678H}, \\
V_E &= \frac{0.031 U_\infty}{[0.15 + 1.72/(H-1) - 0.01(H-1)^2]^{0.617}}, \\
\delta_{99} - \delta^* &= \theta \left[ 3.15 + \frac{1.72}{H-1} - 0.01(H-1)^2 \right].
\end{aligned}
\qquad (6.1.3)
$$

The constants and exponents in (6.1.3) were obtained simply by fitting formulas to experimental data. Closure via a formula for $V_E$ is called Head's entrainment method. Other sorts of integral closure can be found in Kline *et al.* (1968) and Bradshaw (1976).

The formula in (6.1.3) for $C_f$ is the Ludwig-Tillman data correlation. It does not permit $C_f$ to reach zero and hence does not apply near a point of separation. An alternative formula, such as

$$
C_f = \min[0.246 \times 10^{-0.678H}, 0.046 \log(3/H)] R_\theta^{-0.268}, \qquad H \le 3, \qquad (6.1.3a)
$$

would permit the skin friction to reach zero. This assumes $H = 3$ to be the separation criterion. The boundary-layer equations cannot be solved beyond separation, so there is really no need for the modification (6.1.3a); a computation using the Ludwig–Tillman correlation (6.1.3) can simply be stopped when $H = 3$, with a declaration that separation has occurred.

In conjunction, (6.1.1), (6.1.2), and (6.1.3) constitute five equations in five unknowns. Substituting the first of (6.1.3) on the right-hand side of (6.1.1) and the second on the right-hand side of (6.1.2) gives a pair of ordinary differential equations that can be solved for the boundary-layer growth. If the third equation is used to replace $\delta_{99} - \delta^*$ in (6.1.2), a pair of equations for $H$ and $\theta$ is obtained. The specifics of (6.1.3) are not important; the gist is to close integrated balance equations by semi-empirical drag and entrainment laws.

In a decelerated boundary layer, $d_x U_\infty < 0$ and $H$ increases with $x$ (Exercise 4.5). The first of (6.1.3) shows that the effect of increasing $H$ is to cause $C_f$ to fall more rapidly than in a zero pressure-gradient boundary layer. In a strong pressure gradient $C_f$ can fall to zero as the boundary layer separates. This is illustrated by Figure 6.1, which shows how the adverse pressure gradient also produces a rapid thickening of the boundary layer. This figure was computed with (6.1.1), (6.1.2), and (6.1.3).

Although a sufficiently decelerating pressure gradient can drive the skin friction to zero and cause the boundary layer to separate, turbulent boundary layers are far more

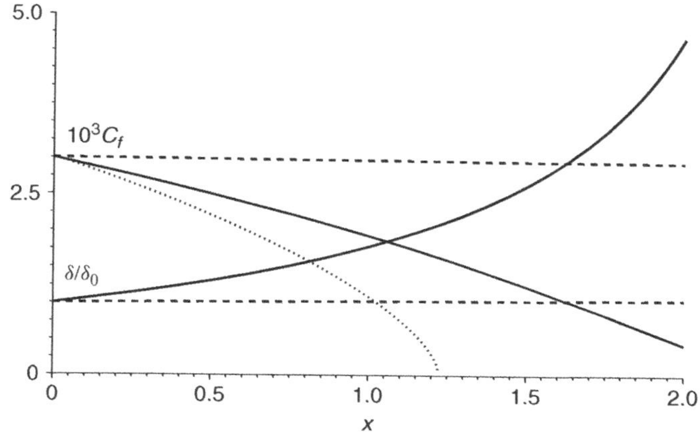


**Figure 6.1** Skin friction coefficient beneath zero and adverse pressure-gradient boundary layers: zero pressure gradient (– – – –); adverse pressure gradient (———); and laminar, adverse pressure gradient (··········).

resistant to separation than are laminar layers. A laminar curve is included in Figure 6.1. In the given pressure gradient, the laminar layer separates, while the turbulent layer does not. The mechanism of separation is a competition between the decelerating pressure gradient promoting flow reversal near the surface and transport processes resisting it by carrying forward momentum from the free stream toward the surface. Turbulent entrainment and Reynolds stresses are more efficient mean momentum transporters than is molecular viscosity. Hence they are more able to resist the tendency toward flow reversal.

An intriguing example of the ability of turbulence to delay separation is the "drag crisis." If the boundary layer on a cylinder or sphere is laminar, it will separate shortly before the topmost point. But if the boundary layer is turbulent, the flow will remain attached beyond the top and separate to the rear of the object. Momentum transport by turbulent mixing resists the tendency to separate. The pressure drag on the cylinder is due to separation: on the front of the cylinder the pressure is high due to stagnation; in the separated zone on the rear it is low. As the separation point moves to the rear, less surface area is in the separated zone, so there is less drag. Starting with a laminar boundary layer, as the Reynolds number is increased beyond a critical value ($Ud/\nu \approx 10^5$ for a cylinder), the boundary layer will make a transition to turbulence (see Figures 3.2 and 3.1), at which point the separation shifts rearward and the drag suddenly *decreases*. Figure 6.2 illustrates aspects of the drag crisis. Beyond the critical Reynolds number the drag starts to increase slowly with Reynolds number, as the adverse pressure gradient becomes more effective relative to the eddy viscosity.

### 6.1.2 Mixing length model

Prandtl introduced the mixing length model in 1925 as an analogy to the kinetic theory of gases. The analogy is quite loose and does not bear scrutiny. Kinetic theory relates macroscopic transport coefficients to molecular motion. There is a great disparity in length scales that justifies a weak non-equilibrium approximation. By contrast, the length

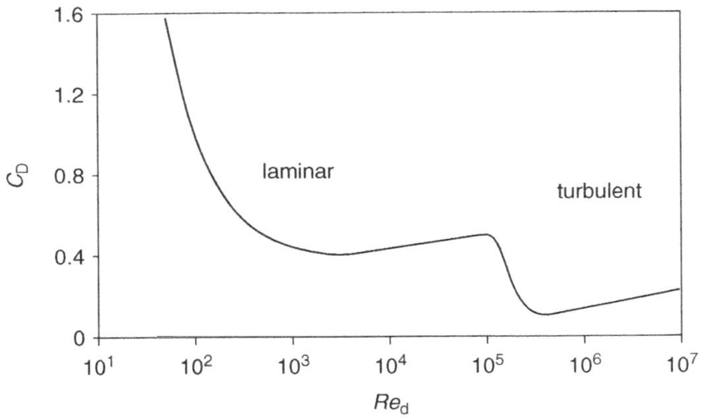


**Figure 6.2** The drag crisis in flow round a sphere. At $R_d = Ud/\nu \approx 10^5$, the drag coefficient, $C_D$, suddenly drops as the boundary layer on the sphere becomes turbulent.

scale of eddying turbulent motion is *not* disparate to the scale of mean flow variation. In kinetic theory, transport coefficients are derived in the transformation from a statistical mechanical to a thermodynamic description. In turbulent flow, the transformation is from fluctuating fluid dynamics to averaged fluid dynamics; the change is not from fine-grained to coarse-grained description. The analogy to kinetic theory is not justifiable, but the mixing length concept has proved a quite useful empirical model, irrespective of its rationale.

Here is a quite heuristic rationalization that follows from (3.3.1). In many thin shear layers, it is found experimentally that

$$-\overline{uv} \approx 0.3k\,\text{sign}(\partial_y U). \tag{6.1.4}$$

A formula relating Reynolds shear stress to the mean flow can be obtained by estimating the turbulent kinetic energy, $k$. Assume that turbulent fluctuations are produced by displacing mean momentum. This idea was developed below Eq. (3.3.1). If $\ell'$ is a random $y$ displacement, the velocity fluctuation it causes is $u = U(y - \ell') - U(y) \approx -\ell'\,\partial_y U$. This can be used to estimate the turbulent energy,

$$k = \tfrac{1}{2}\,\overline{u_i u_i} \propto \overline{\ell'^2}\,|\partial_y U|^2. \tag{6.1.5}$$

Letting $\ell_{\mathrm{m}}^2 = \overline{\ell'^2}$, Eq. (6.1.4) becomes the mixing length model

$$-\overline{uv} = \ell_{\mathrm{m}}^2 |\partial_y U| \partial_y U. \tag{6.1.6}$$

Here $\ell_{\mathrm{m}}$ is regarded as an empirical length scale and constants of proportionality are simply absorbed into its definition. A formula to prescribe $\ell_{\mathrm{m}}$ as a function of distance to the wall – and maybe of other parameters – must be provided, as discussed below and in Exercise 6.3.

An early criticism of Prandtl's mixing length reasoning was that it ignored pressure gradients. The reasoning around Eq. (3.3.1) is equivalent to assuming that the streamwise momentum equation can be approximated by $D_t(U + u) = 0$. Then tracking a Lagrangian trajectory back from $y$ to a random initial position $y_0$ gives

$$u(y) = U(y_0) - U(y)$$

if the initial velocity fluctuation vanishes, $u(y_0) = 0$. Squaring and averaging over the random initial position gives

$$\overline{u^2}(y) = \overline{[U(y_0) - U(y)]^2}.$$

The mixing length could be defined by $\ell_m \propto \overline{[U(y_0) - U(y)]^2}^{1/2} / \partial_y U$. It is a length scale for momentum dispersion.

There is a possible counter to the criticism that Prandtl's rationale ignores the term $\partial_x p$ in the fluctuation streamwise momentum equation. It is found theoretically, and observed experimentally, that strong shear amplifies eddying motion, causing $u$ to become approximately independent of $x$. For instance, near to a wall in a turbulent boundary layer, streamwise streaks are observed in flow visualizations. These are seen in smoke or hydrogen bubble visualizations. The turbulence is organized into jet-like motions, parallel to the surface. Numerical simulations show the same pattern in contours of $u(x, z)$ in a plane near the wall (Figure 5.10, page 101). Streaky, jet-like motion, with small $\partial_x u$, also has small $\partial_x p$. Displacement of momentum therefore is approximately a valid concept for such structures. This rationale is of comfort, even if it has no immediate bearing on the model. The mixing length is rather an operational method, based on selecting formulas for $\ell_m(y)$ to produce velocity profiles that agree with experimental data.

Equation (6.1.6) can be written in eddy viscosity form, $-\overline{uv} = \nu_T \, \partial_y U$ with

$$\nu_T = \ell_m^2 |\partial_y U|. \tag{6.1.7}$$

Using this, the boundary-layer momentum equation (4.2.6) with (6.1.6) becomes the closed equation

$$U \, \partial_x U + V \, \partial_y U = -\partial_x P_\infty + \partial_y [(\nu + \ell_m^2 |\partial_y U|) \, \partial_y U] \tag{6.1.8}$$

for $U$. The mean flow evolution can be computed by solving this, instead of its integrated form (6.1.1). Formulas for $\ell_m$ relate it to the wall-normal coordinate, $y$, pressure gradient, transpiration, and so on (Kays and Crawford, 1993). A variety of such formulations have been devised in the course of time. We will consider only the most basic formulation in terms of the inner and outer structure of turbulent boundary layers. This aspect is common to all mixing length models.

The two asymptotic zones of the turbulent boundary layer are the law-of-the-wall and law-of-the-wake regions (Section 4.2). Dimensional analysis gives the functional forms

$$\begin{aligned} \ell_m &= \nu/u_* F(yu_*/\nu), && \text{law of wall}, && y \lesssim 0.2\delta_{99}, \\ \ell_m &= \delta_{99} G(y/\delta_{99}), && \text{law of wake}, && y \gtrsim 40\nu/u_*. \end{aligned} \tag{6.1.9}$$

In the logarithmic overlap region, these must be equal. From the first of (6.1.9), $\ell_m$ cannot depend on $\delta_{99}$ in the overlap region; from the second it cannot depend on $\nu/u_*$. So all that is left is for $\ell_m$ to be proportional to $y$. More formally, let $R = u_*\delta_{99}/\nu$ and let

$Y = y/\delta_{99}$. Then in the logarithmic region both of the forms (6.1.9) are simultaneously valid. Equating them gives

$$G(Y) = \frac{F(RY)}{R}.$$

The only way for the functional dependences to be equivalent is $F = cRY$ and $G = cY$, where $c$ is a constant that is readily shown to be the Von Karman constant. It follows because in the log layer $\nu_T = u_*\kappa y$ and $|\partial_y U| = u_*/\kappa y$. Then from Eq. (6.1.7), $\ell_m = \kappa y$. So $\ell_m$ varies linearly in the overlap region, but its behavior in the rest of the inner and outer regions remains to be specified.

Van Driest suggested an extension of the mixing length to the wall. His approach is referred to as a "viscous damping function." He multiplied $\kappa y$ by an exponential damping function to obtain a formula for the inner region. Van Driest's argument was that viscous friction would reduce turbulent mixing near the wall. A more important effect is that the mixing normal to the wall is suppressed because the wall is impermeable. Although Van Driest justified an exponential damping function by appeal to the oscillatory Stokes boundary layer, in practice it is simply a convenient form. Thus the inner region formula is

$$\ell_m = \kappa y(1 - e^{-y_+/A_+}). \tag{6.1.10}$$

The empirical constant $A_+$ is determined by the additive constant $B$ in the log-layer velocity profile (4.1.7): the value $A_+ = 26$ gives $B = 5.3$. As usual, $\kappa = 0.41$ is the value for the Von Karman constant.

In the outer part of the boundary layer the mixing length is usually made proportional to the boundary-layer thickness; stated otherwise, $G(y/\delta_{99}) =$ constant in (6.1.9). Assuming that the outer region begins at $0.2\delta_{99}$ and patching on to the linear variation in the log layer gives

$$\begin{aligned} \ell_m &= \kappa y, & y &< 0.2\delta_{99}, \\ \ell_m &= 0.2\kappa\delta_{99}, & y &\geq 0.2\delta_{99}. \end{aligned} \tag{6.1.11}$$

Note that $\delta_{99}$ (as well as $\delta^*$ and $\theta$) can be evaluated when (6.1.8) is solved for $U(y)$, so no model like (6.1.3) is needed. In this method the shape and evolution of the boundary-layer profile $U(y; x)$ are computed via the partial differential equation (6.1.8).

But, as Figure 6.3 shows, the constant mixing length is not in agreement with data in the outer region. A much better approximation is to use a constant eddy viscosity,

$$\nu_T = 0.2\kappa u_*\delta_{99}, \qquad y \geq 0.2\delta_{99}. \tag{6.1.12}$$

Figure 6.3 also shows the formula (6.1.12) for a boundary layer with $R_\theta = 1410$ along with DNS data of Spalart (1988). Higher Reynolds number laboratory data (DeGraaff and Eaton, 2000) are also shown. The data are plotted in inner layer variables. Formula (6.1.10) for the inner region is valid at all Reynolds numbers in this scaling, and agrees with all of the data.

In inner variables, Eq. (6.1.12) is $\nu_T^+ = 0.2\kappa\delta_{99}^+$. In these variables the outer eddy viscosity increases with $R_\theta$; only the case $R_\theta = 1440$ is included in the figure. At higher Reynolds number the horizontal line shifts upward, mimicking the data. Often the eddy viscosity is multiplied by an "intermittency factor" $1/[1 + 5.5(0.3y/\delta_{99})^6]$ to make it taper off in the free stream, the way the data does.

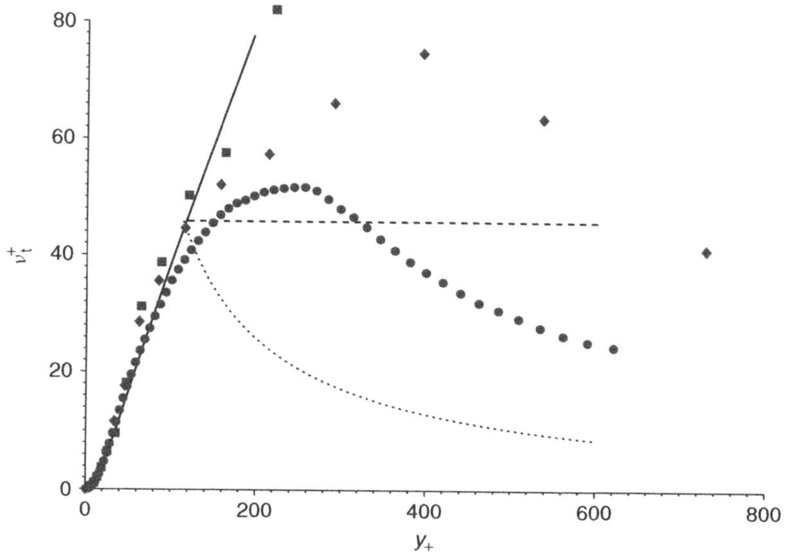


**Figure 6.3** Eddy viscosity as predicted by the mixing length model, and determined from data. Symbols are DNS and experimental data. The solid curve is the inner region formula. The dotted curve is obtained with a constant mixing length in the outer region, and the dashed line is a constant outer region eddy viscosity, both for $R_\theta = 1440$. Data at $R_\theta = 1440$ (●), 2900 (♦), and 13 310 (■).

If the boundary layer flows along a curvilinear wall, the normal and tangential directions vary with position. Then the dominant velocity derivative will vary with position; it will not always be $\partial_y U$. A coordinate-independent form must replace the velocity gradient, $\partial_y U$ (see Section 8.1.1.1). It can be either the rate-of-strain or the rate-of-rotation invariant. These are $|S|^2 = S_{ij}S_{ji}$ and $|\Omega|^2 = -\Omega_{ij}\Omega_{ji} = \frac{1}{2}|\omega|^2$, where

$$S_{ij} = \tfrac{1}{2}\,[\partial_i U_j + \partial_j U_i], \qquad \Omega_{ij} = \tfrac{1}{2}\,[\partial_i U_j - \partial_j U_i] = \tfrac{1}{2}\,\varepsilon_{ijk}\omega_k$$

and $\boldsymbol{\omega}$ is the vorticity vector. In parallel shear flow, $2|S|^2 = |\omega|^2 = |\partial_y U|^2$, so the mixing length model could be generalized either as $\nu_T = \sqrt{2}\,\ell_m^2|S|$ or as $\nu_T = \ell_m^2|\omega|$; both reduce to $\ell_m^2|\partial_y U|$ in parallel shear flow.

The second statement, $\nu_T = \ell_m^2|\omega|$, was adopted by Baldwin and Lomax (1978). In their form, the mixing length idea became a mainstay of computational aerodynamics. In early implementations, this model was used in conjunction with the thin-layer Navier–Stokes equations (Tannehill *et al.*, 1997).

Flows over airfoils have a large, irrotational free stream. The use of $|\omega|$ ensures that the eddy viscosity vanishes in the free stream. However, in rotating flows, such as occur in turbomachinery or in stirring tanks, a spurious eddy viscosity can occur. This is so because $|\Omega|^2$ must be replaced by the absolute vorticity invariant $|\Omega^A|^2$ in a rotating frame of reference in order to preserve the frame invariance of the model: this issue is discussed in detail in Section 8.1.1.1. In such cases, $|S|$ would be a more suitable invariant of the velocity gradient to use when defining $\nu_T$.

The mixing length model also becomes questionable in separated flow; for instance, this model may well predict attached flow when the real flow is separated. Indeed, it was largely the failure to predict trailing edge separation and stall that initially led to the use of more advanced models in the field of computational aerodynamics. However, the Baldwin–Lomax model can be put to good use, as long as these limitations are recognized.

With some elaboration, the mixing length type of model can be used for mildly separated boundary layers. Thus it is rather more flexible than integral equation models, but that flexibility is achieved at the expense of having to solve a partial differential equation, rather than an ordinary differential equation, for the mean flow. An obvious limitation to this approach is that prescription of the mixing length becomes problematic in flows that are not approximately parallel, thin shear layers. For instance, if two shear layers intersect at some angle, how should $\delta_{99}$ be defined? What should $0.99U_\infty$ be when there is significant streamline curvature (so that velocity gradients exist in the free stream)? What about highly non-parallel flows? Flow over a backward-facing step is a common example of a modestly complex flow (Figure 6.4). A shear layer detaches from an upstream step and reattaches to the surface about six step heights downstream. In such a flow, there are two shear layers at any $x$ within the separated region, the detached mixing layer and the bottom-wall boundary layer. It is not clear how $\ell_m(\boldsymbol{x})$ would be prescribed in such a flow.

With such questions in mind, Baldwin and Lomax (1978) proposed a length scale estimate that has more flexibility than $\delta_{99}$, so that mixing lengths could be used in mildly complex geometries. They proposed the outer region viscosity

$$\nu_T = 0.1312 \min(d_{max} F_{max};\ \Delta U^2 d_{max}/F_{max}) \tag{6.1.13}$$

instead of (6.1.12). Here

$$F(d) \equiv \kappa d(1 - e^{-d_+/26})|\omega| \qquad \text{and} \qquad F_{max} = \max_d F(d) = F(d_{max}),$$

where $d$ is the shortest distance from the wall and $\Delta U^2 = |\max_d U(d) - \min_d U(d)|^2$.

In a flat-plate boundary layer, $d_{max} \approx \delta_{95}$ and formula (6.1.13) is very close to (6.1.12). In a detached shear layer, $d_{max} \to \infty$ and the second term in (6.1.13) comes into play: if $\omega_{max} \equiv F_{max}/d_{max}$, then, as $d_{max} \to \infty$, $\nu_T \to 0.1312\Delta U^2/\omega_{max}$. The transition from

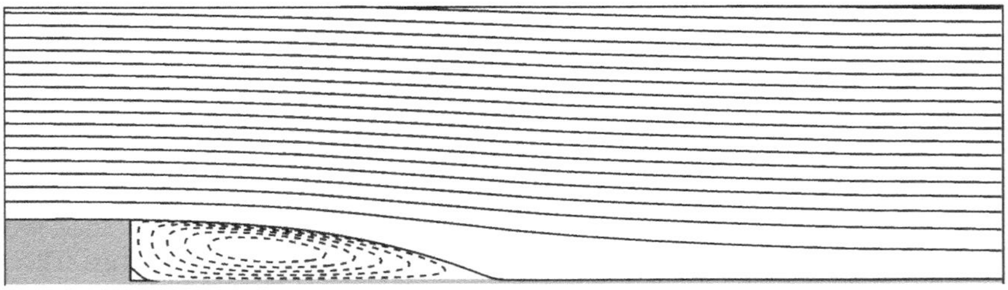

**Figure 6.4** Streamlines of flow over a backward-facing step. This is a widely used test case for turbulence models. It illustrates the need for a model that can be applied to non-parallel flows, including separated regions.

the eddy viscosity (6.1.7) with (6.1.10) to the formula (6.1.13) takes place at the value of $d$ where they are equal.

A fundamental difficulty here is that the only natural length scales are $\nu/u_*$ and the distance to the wall, $d$, which either is used directly or weighted by vorticity. However, the viscous length scale is only dominant near the surface, $d_+ \lesssim 40$, and the wall distance does not characterize turbulent eddies far from the surface. Similarly, the only clearly defined velocity scale for the turbulence is the friction velocity, $u_*$, and this is only valid near the surface. Using a mean velocity difference, $\Delta U$ in Eq. (6.1.13), requires a search along an arbitrary direction: in a thin-layer Navier–Stokes code it would be along a grid line, which introduces an inexorable grid dependence into all computations.

These considerations suggest that it would be better to obtain turbulence scales by solving differential equations that govern their evolution and spatial distribution, rather than trying to prescribe them *a priori*. Then the eddy viscosity can be computed at each point of a general geometry grid and will adapt to the particular flow configuration. This is the motivation for turbulent transport models.

## 6.2 The $k-\varepsilon$ model

The $k-\varepsilon$ model is the most widely used general-purpose turbulence transport model. The current form was initially developed by Jones and Launder (1972). What is now called the "standard" $k-\varepsilon$ model is the Jones–Launder form, without wall damping functions, and with the empirical constants given by Launder and Sharma (1974). Although an enormous number of variations on the $k-\varepsilon$ model have been proposed, the standard model is still the basis for most applications, with the caveat that some form of fix is needed near to solid boundaries, as will be discussed later. In part, the popularity of $k-\varepsilon$ is for historical reasons: it was the first two-equation model used in applied computational fluid dynamics. Other models may be more accurate, or more computationally robust in certain applications, but $k-\varepsilon$ is the most widely used.

As the applications of computational fluid dynamics have grown, various other models have been developed or adapted to their particular needs. It is not the purpose of this book to compile a handbook of models and their most propitious applications. The general concepts of scalar transport models can be developed by a detailed look at $k-\varepsilon$ as a prototype.

The purpose of models of the $k-\varepsilon$ type is to predict an eddy viscosity. The connection of turbulent kinetic energy and its dissipation to eddy viscosity could be rationalized as follows. At high Reynolds number the rate of dissipation and production are of similar order of magnitude (see Figure 3.4). Thus, we estimate $\varepsilon \approx \mathcal{P}$. Multiplying this by $\nu_T$ gives

$$\nu_T\varepsilon \approx \nu_T\mathcal{P} = \nu_T(-\overline{uv}\,\partial_y U) = (\overline{uv})^2 \approx 0.09k^2, \tag{6.2.1}$$

using $-\overline{uv} = \nu_T\,\partial_y U$. The last, approximate, equality stems from the experimental observation that the stress-intensity ratio $\overline{uv}/k \approx 0.3$ in the log layer (see (6.1.4)). Rearranging (6.2.1), $\nu_T \approx 0.09k^2/\varepsilon$. This formula is usually written

$$\nu_T = C_\mu k^2/\varepsilon, \tag{6.2.2}$$

and the standard value of $C_\mu$ is 0.09.

Alternatively, it can simply be argued by dimensional analysis that the turbulence correlation time-scale is $T \sim k/\varepsilon$ and the velocity scale squared is $k$. Then by analogy to (2.2.22), $\nu_T \sim u^2 T = C_\mu k^2/\varepsilon$. By either rationale, one sees that formulas to parameterize turbulent mixing can be evaluated from a model that predicts $k$ and $\varepsilon$.

The mean flow is computed from the scalar eddy viscosity and the constitutive relation

$$-\overline{u_i u_j} = 2\nu_T S_{ij} - \tfrac{2}{3} k\delta_{ij}, \qquad (6.2.3)$$

where $S_{ij}$ is the mean rate-of-strain tensor (3.3.8). The constitutive equation (6.2.3) is a linear stress–strain relation, as for a Newtonian fluid. It inherently assumes an equilibrium between Reynolds stress and mean rate of strain. This may be violated in some flows, such as strongly three-dimensional boundary layers, where the Reynolds stress is not proportional to the mean rate of strain, but it works surprisingly well in a wide variety of flows. The problem addressed in the $k$–$\varepsilon$ transport model is how to robustly predict $\nu_T$. The formula (6.2.2) reduces this to predicting the spatial and temporal distribution of $k$ and $\varepsilon$.

Equation (3.2.5), page 49, is the exact, but unclosed, evolution equation for $k$. To "close" it, the transport and pressure-diffusion terms together are replaced by a gradient transport model:

$$-\partial_j\left(\frac{1}{2}\overline{u_j u_i u_i} - \frac{1}{\rho}\overline{u_j p}\right) \approx \partial_j(\nu_T\, \partial_j k). \qquad (6.2.4)$$

This closure preserves the conservation form of the unclosed term on the left. It is based on the notion that the third velocity moment represents random convection of turbulent kinetic energy and this can be modeled by diffusion. Invoking the notation in (3.3.9), the transport equation for $k$ with (6.2.4) substituted is

$$\partial_t k + U_j\, \partial_j k = \mathcal{P} - \varepsilon + \partial_j((\nu + \nu_T)\, \partial_j k). \qquad (6.2.5)$$

It is common to add a parameter $\sigma_k$ as a denominator of $\nu_T$, analogously to $\sigma_\varepsilon$ in (6.2.6); but the usual value is $\sigma_k = 1$. Note that $\mathcal{P} \equiv -\overline{u_i u_j}\partial_j U_i$ is related to $S_{ij}$ by (6.2.3):

$$\mathcal{P} = 2\nu_T S_{ij}\, \partial_j U_i - \tfrac{2}{3} k\, \partial_i U_i = 2\nu_T S_{ij} S_{ji} - \tfrac{2}{3} k\, \partial_i U_i = 2\nu_T |S|^2 - \tfrac{2}{3} k \nabla \cdot \boldsymbol{U}.$$

For incompressible flow this reduces to $\mathcal{P} = 2\nu_T|S|^2$.

The modeled transport equation for $\varepsilon$ cannot be derived systematically. Essentially it is a dimensionally consistent analogy to the above $k$ equation:

$$\partial_t \varepsilon + U_j\, \partial_j \varepsilon = \frac{C_{\varepsilon 1}\mathcal{P} - C_{\varepsilon 2}\varepsilon}{T} + \partial_j\left(\left(\nu + \frac{\nu_T}{\sigma_\varepsilon}\right)\partial_j \varepsilon\right). \qquad (6.2.6)$$

The time-scale $T = k/\varepsilon$ makes this dimensionally consistent. Equation (6.2.6) is analogous to (6.2.5), except that empirical constants $C_{\varepsilon 1}$, $C_{\varepsilon 2}$, and $\sigma_\varepsilon$ have been added because the $\varepsilon$ equation is just an assumed form. The terms on the right-hand side can be referred to as "production of dissipation," "dissipation of dissipation," and "diffusion of dissipation."

The empirical coefficients are chosen in order to impose certain experimental constraints. These will be discussed subsequently. The "standard" values for the constants are

$$C_\mu = 0.09, \qquad C_{\varepsilon 1} = 1.44, \qquad C_{\varepsilon 2} = 1.92, \qquad \sigma_\varepsilon = 1.3. \tag{6.2.7}$$

These constants were chosen some time ago. More recent data suggest that slightly different values might be suitable, but the standard constants give reasonable engineering results in many situations. It is not likely that minor adjustments would significantly affect the predictive accuracy.

Substituting the constitutive model (6.2.3) closes (3.2.2):

$$\partial_t U_i + U_j\,\partial_j U_i = -\frac{1}{\rho}\,\partial_i\left(P + \frac{2}{3}\rho k\right) + \partial_j[(\nu + \nu_{\mathrm{T}})(\partial_j U_i + \partial_i U_j)]. \tag{6.2.8}$$

This is solved with (6.2.5), (6.2.6), and (6.2.2) in general geometries, provided suitable boundary conditions can be formulated, which is not a trivial issue. Before discussing that issue, some properties of this model will be explored.

### 6.2.1 Analytical solutions to the $k$–$\varepsilon$ model

It should be possible to determine the empirical constants of the $k$–$\varepsilon$ model from simple measurements that isolate each term. If the model were exact, then a set of four suitable measurements would give the values (6.2.7). In practice, these are not constants of nature, so their values would depend somewhat on the particular data used to evaluate them. Nevertheless, it is informative to consider how closed-form solutions to (6.2.5) and (6.2.6) enable estimates of these constants. The following are three such solutions.

#### 6.2.1.1 Decaying homogeneous, isotropic turbulence

The simplest turbulent flow is homogeneous isotropic turbulence – which can be approximated by grid-generated turbulence, as in Figure 1.2 on page 6. The simplifications in homogeneous turbulence were already discussed in Section 3.3; all gradients of statistics vanish and hence so do all transport terms. For isotropic turbulence there are no mean flow gradients either, so $\mathcal{P} = 0$. Equations (6.2.5) and (6.2.6) simplify to

$$\mathrm{d}_t k = -\varepsilon,$$
$$\mathrm{d}_t \varepsilon = -C_{\varepsilon 2}\frac{\varepsilon}{T}.$$

We seek a power-law solution

$$k = \frac{k_0}{(t/t_0 + 1)^n}.$$

The first equation then shows that

$$\varepsilon = \frac{n k_0}{t_0 (t/t_0 + 1)^{n+1}}.$$

With $T = k/\varepsilon = (t + t_0)/n$, the $\varepsilon$ equation becomes $n + 1 = nC_{\varepsilon 2}$ or

$$C_{\varepsilon 2} = \frac{n+1}{n}. \tag{6.2.9}$$

The decay exponent, $n$, can be found by fitting the curve $k = k_0(x/x_0 + 1)^{-n}$ through measurements of grid turbulence in a wind tunnel. Experimental data tend to fall in the range $n = 1.3 \pm 0.2$. A typical value of $n = 1.2$ would give $C_{\varepsilon 2} = 1.83$; the standard value in (6.2.7) corresponds to a low decay exponent, $n = 1.09$. While this suggests that the standard value of $C_{\varepsilon 2}$ should be revised, when the rest of the model was correspondingly recalibrated, its predictions in practical flows would be imperceptibly altered.

#### 6.2.1.2 Homogeneous shear flow

The next constant to be related to basic experimental data is $C_{\varepsilon 1}$. This is a rather important constant. For given values of $C_\mu = 0.09$ and $C_{\varepsilon 2} = 1.92$ – which we have related to very fundamental, reproducible data – the value of $C_{\varepsilon 1}$ controls the spreading rate of free-shear layers. The standard value was chosen so that the basic model would give a reasonable value for the spreading rate $\mathrm{d}_x\delta$ in a plane mixing layer (Section 4.3). Numerical computations are required to obtain a prediction of this spreading rate. For a plane, two-stream mixing layer with a velocity ratio of 0.1, the spreading rates are $\mathrm{d}_x\delta = 0.094$ for $C_{\varepsilon 1} = 1.44$, $\mathrm{d}_x\delta = 0.074$ for $C_{\varepsilon 1} = 1.54$, and $\mathrm{d}_x\delta = 0.119$ for $C_{\varepsilon 1} = 1.34$. This might seem to evidence a high degree of sensitivity to the value of $C_{\varepsilon 1}$: a 7% change causes a 20% change in $\mathrm{d}_x\delta$. However, it is more appropriate to quote the corresponding values of $C_{\varepsilon 2} - C_{\varepsilon 1}$, which are 0.48, 0.38, and 0.58. The spreading rate varies almost proportionately to these numbers. The following analysis explains why the difference between $C_{\varepsilon 2}$ and $C_{\varepsilon 1}$ is a suitable measure of the model constants.

The closed-form solution that sheds light on the evaluation of $C_{\varepsilon 1}$ is for homogeneous shear flow (although, again, this it is not how the standard value was obtained). In parallel shear flow, (6.2.3) gives $-\overline{uv} = C_\mu(k^2/\varepsilon)\,\partial_y U$. If the shear is homogeneous, then $\partial_y U$ is a constant, which will be denoted $\mathcal{S}$. In this case the $k$–$\varepsilon$ model becomes

$$\begin{aligned} \mathrm{d}_t k &= C_\mu \mathcal{S}^2 \frac{k^2}{\varepsilon} - \varepsilon, \\ \mathrm{d}_t \varepsilon &= \left( C_{\varepsilon 1} C_\mu \mathcal{S}^2 \frac{k^2}{\varepsilon} - C_{\varepsilon 2}\varepsilon \right) \frac{\varepsilon}{k}. \end{aligned} \tag{6.2.10}$$

A solution is sought in the form $k = k_0\,\mathrm{e}^{\lambda \mathcal{S} t}$, $\varepsilon = \varepsilon_0\,\mathrm{e}^{\lambda \mathcal{S} t}$. Substituting this assumption into (6.2.10) and solving gives

$$\begin{aligned} \lambda &= \frac{C_{\varepsilon 2} - C_{\varepsilon 1}}{\sqrt{(C_{\varepsilon 2} - 1)(C_{\varepsilon 1} - 1)/C_\mu}}, \\ \frac{\mathcal{P}}{\varepsilon} &= C_\mu \left( \frac{\mathcal{S}k}{\varepsilon} \right)^2 = \frac{C_{\varepsilon 2} - 1}{C_{\varepsilon 1} - 1}, \end{aligned} \tag{6.2.11}$$

for the exponential growth rate, $\lambda$, and ratio of production to dissipation. The growth rate is determined by the excess of production to dissipation:

$$\frac{\mathcal{P}}{\varepsilon} - 1 = \frac{C_{\varepsilon 2} - C_{\varepsilon 1}}{C_{\varepsilon 1} - 1}.$$

It is now clear why $C_{\varepsilon 2} - C_{\varepsilon 1}$ controls the mixing layer spreading rate.

This solution could be used to evaluate $C_{\varepsilon 1}$ from measured growth rates of homogeneously sheared turbulence, given values of the other coefficients. Experimental values of $\mathcal{P}/\varepsilon$ are around $1.6 \pm 0.2$ (Tavoularis and Karnik 1989). Substituting the standard constants (6.2.7) into (6.2.11) gives a too high value of $\mathcal{P}/\varepsilon = 0.92/0.44 = 2.1$. The upper limit of the experimental uncertainty would give $C_{\varepsilon 1} = 1.51$ if $C_{\varepsilon 2} = 1.92$. As mentioned previously, the value $C_{\varepsilon 1} = 1.44$ of (6.2.7) was selected to produce a reasonable $\mathrm{d}\delta/\mathrm{d}x$ in mixing layers.

This analysis makes clear the reason why the values of $C_{\varepsilon 2}$ and $C_{\varepsilon 1}$ are usually quoted to three decimal places: the leading "1" cancels out of $C_{\varepsilon 2} - C_{\varepsilon 1}$ and of $C_{\varepsilon 1} - 1$; the differences have only two significant decimal places.

The $C_{\varepsilon 1}$ calibration illustrates a dilemma with which one is sometimes confronted: the constants of the model do not have unique values that can be determined from a single calibration experiment. This is in contrast to the material properties, like the molecular viscosity, $\nu$, which can be measured in any viscometric experiment. The model constants are not material properties; in some respects they represent statistical properties of the flow. In principle, a closure model is only applicable to a subset of turbulent flows, determined by its mathematical formulation and by the particular calibration of the model coefficients. However, this nebulous subset can only be defined in broad terms.

#### 6.2.1.3 The logarithmic layer

A third closed-form solution is that for the log layer. This solution provides the value of $\sigma_\varepsilon$, given the previous values for the other constants. Recall that the log layer is a constant-stress layer, $-\overline{uv} = u_*^2$, and that the log law can be stated as $\partial_y U = u_*/\kappa y$. Also, molecular viscosity is small in this layer: $\nu \ll \nu_{\mathrm{T}}$. On dimensional grounds $k$ is constant; the reasoning is based on $u_*$ and $y$ being the only dimensional parameters and hence $k/u_*^2 =$ constant. If $k$ is constant, all derivatives in (6.2.5) are zero and it becomes

$$\mathcal{P} = \varepsilon; \tag{6.2.12}$$

production and dissipation are locally in balance. But then

$$\mathcal{P} = -\overline{uv}\,\partial_y U = u_*^3/(\kappa y) = \varepsilon \tag{6.2.13}$$

has to be the solution for $\varepsilon$. The eddy viscosity is given by $\nu_{\mathrm{T}} \equiv -\overline{uv}/\partial_y U = \kappa u_* y$. The $k$–$\varepsilon$ model equates this to $C_\mu k^2/\varepsilon$. With the solution just given for $\varepsilon$, this implies that $\kappa u_* y = C_\mu k^2 \kappa y/u_*^3$ or

$$k = \frac{u_*^2}{\sqrt{C_\mu}},$$

and equivalently $C_\mu = (\overline{uv}/k)^2$. The stress-intensity ratio, $|\overline{uv}/k|$, is found experimentally to be about 0.3 in many shear flows (Townsend, 1976). Hence $C_\mu = 0.09$ as in (6.2.7).

Substituting $\varepsilon = u_*^3/(\kappa y)$ and $\nu_T = \kappa u_* y$ into (6.2.6) gives

$$\kappa^2 = (C_{\varepsilon 2} - C_{\varepsilon 1})\sigma_\varepsilon \sqrt{C_\mu}\,, \tag{6.2.14}$$

after some algebra. This can be used to evaluate $\sigma_\varepsilon$ from experimental measurements of the Von Karman constant, $\kappa$. Those measurements are mostly in the range $0.41 \pm 0.2$. The value $\kappa = 0.43$ and the previous values of $C_{\varepsilon 2}$, $C_{\varepsilon 1}$, and $C_\mu$ give the standard coefficient $\sigma_\varepsilon = 1.3$.

The law-of-the-wall, law-of-the-wake, and logarithmic overlap structure of the mean flow is considered to be established above $R_\theta \approx 3000$ (Purtell *et al.*, 1981). Unfortunately, turbulent stresses do not so readily adopt this structure. An inference from the log-layer scaling is that the Reynolds stress tensor $\overline{u_i u_j}^+$ should be constant in this layer. The $k$–$\varepsilon$ solution reproduces this. Experiments do show a plateau of turbulent intensity in the region $40 \lesssim y_+ \lesssim 0.2\delta_{99}^+$, but it requires a rather higher $R_\theta$ than 3000 (Figure 6.5). The classic boundary-layer experiments did not show a clear plateau; however, the value of $k_+$ near $y+ \approx 200$ had the value of about 3.3 corresponding to $C_\mu = 0.09$. Models have largely been calibrated with these older data.

Recent experiments by DeGraaff and Eaton (2000) question the correctness of the older data. Figure 6.5 shows their measured behavior of $\overline{u_+^2}^{1/2}$ profiles for various $R_\theta$. Maybe by $R_\theta \approx 8000$ the Reynolds stresses are in agreement with the full log-layer scaling, but even then the level of the plateau is not completely insensitive to $R_\theta$.

It is a curious property of high Reynolds boundary layers that the normal component of turbulent intensity $\overline{v^2}$ develops a plateau at a much lower Reynolds number than the tangential component $\overline{u^2}$. This is illustrated by Figure 6.6. There, $R_\theta = 3000$ seems to

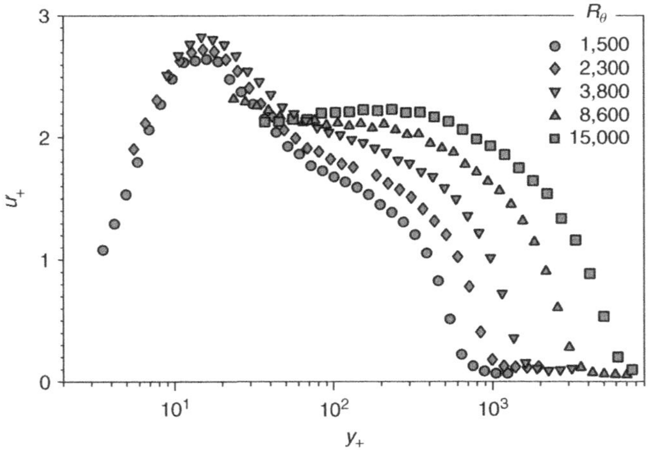


**Figure 6.5** Profiles of $u' \equiv \sqrt{\overline{u_+^2}}$ in a turbulent boundary layer. Hot wire data from DeGraaff and Eaton (2000).

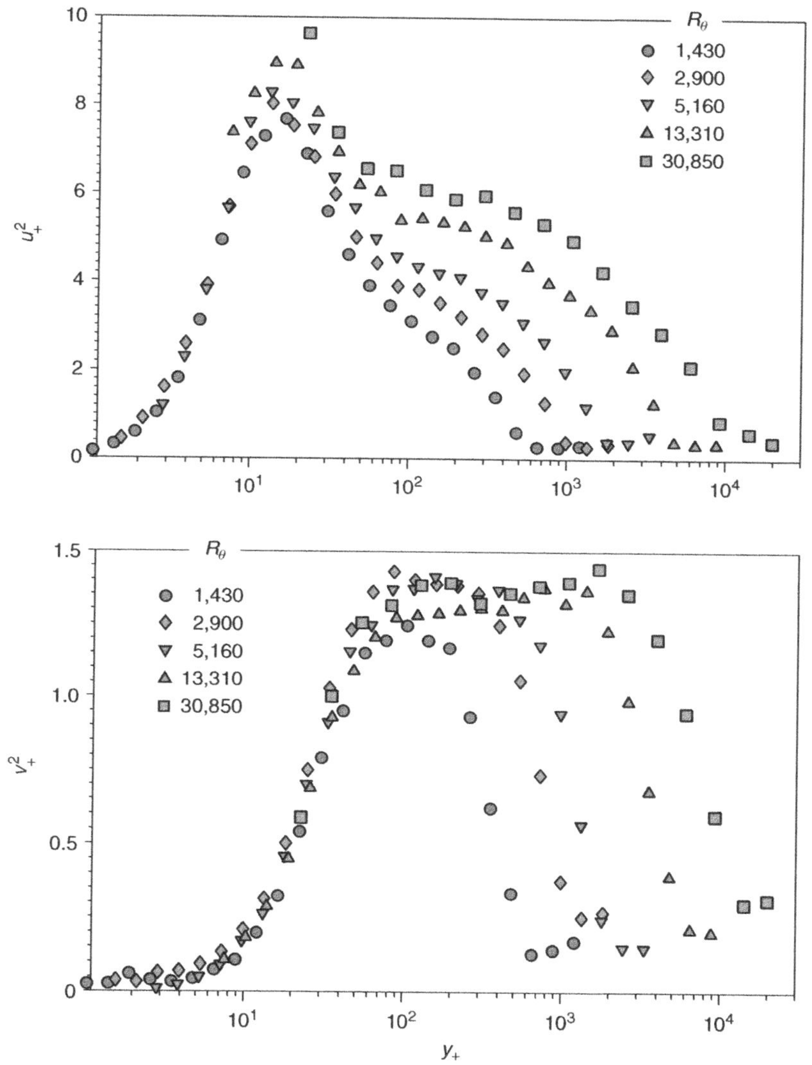


**Figure 6.6** Profiles of $\overline{u^2_+}$ and $\overline{v^2_+}$ in a turbulent boundary layer. Laser Doppler velocimetry data from DeGraaff and Eaton (2000).

be high enough for a region with log-layer scaling to be seen in $\overline{v^2}$. The fact that both $\overline{v^2}$ and $U$ achieve high Reynolds number scaling before $\overline{u^2}$ suggests that a model might be based on $\overline{v^2}$ instead of $k$ (Durbin, 1991).

The usual estimate

$$\overline{w^2} \approx \frac{1}{2}(\overline{u^2} + \overline{v^2}) \qquad \text{or} \qquad k \approx \frac{3}{4}(\overline{u^2} + \overline{v^2})$$

applied to the DeGraaff and Eaton (2000) data gives a high Reynolds number value of $k_+ \approx 5.4$. This would give $C_\mu = 1/5.4^2 = 0.034$; indeed DeGraaff and Eaton suggest that $1/k_+^2$ might continue to fall, like $C_f^{-1}$, as the Reynolds number increases. The standard $k$–$\varepsilon$ model does not predict the correct high Reynolds number behavior of $k$ for this reason. That is of some concern, but it is not devastating: the log-layer eddy viscosity is predicted to be $u_* \kappa y$, which is correct at high Reynolds number. Therefore, mean flow predictions do not fail. Introducing Reynolds number dependence into $C_\mu$ might seem necessary, but other constants would have to counteract it to preserve the known Reynolds number independence of the Von Karman constant, $\kappa$.

### 6.2.2 Boundary conditions and near-wall modifications

The boundary conditions to the $k$–$\varepsilon$ model at a no-slip wall are quite natural, but the near-wall behavior of the model is not. This is rather a serious issue; in many applications the region near the wall is crucial. Turbulent mixing is suppressed by the proximate boundary, causing a great reduction of transport across this layer (Section 4.4). Representing the suppression of mixing is critical to accurate predictions of skin friction and heat transfer. Unfortunately, the $k$–$\varepsilon$ model does not represent this effect and it breaks down catastrophically below the log layer. A variety of patches have been proposed in the course of time.

To start, consider the boundary conditions. At a no-slip surface $\boldsymbol{u} = 0$, so $k = \frac{1}{2}\overline{|u|^2}$ has a quadratic zero. Hence both $k$ and its normal derivative vanish. The natural boundary condition is

$$k = 0 \qquad \partial_n k = 0, \tag{6.2.15}$$

where $\partial_n \equiv \hat{\boldsymbol{n}} \cdot \boldsymbol{\nabla}$ is the derivative in the normal direction. For expository purposes, a coordinate system with $y$ in the normal direction will be used in the rest of this section. But it should be recognized that the considerations apply in general geometries if $\hat{n}$ is considered to be the unit normal to the wall and $y$ to be the minimum distance to a point on the wall.

Equation (6.2.15) specifies two conditions on $k$ and none on $\varepsilon$. That suffices to solve the coupled $k$–$\varepsilon$ system. However, when the model is implemented into CFD codes, it is common practice to convert these into $k = 0$ and a condition on $\varepsilon$. As the wall is approached, $\mathcal{P} \to 0$ and $\nu_T \to 0$, so that (6.2.5) has the limiting behavior

$$\varepsilon_w = \nu\, \partial_y^2 k. \tag{6.2.16}$$

The wall value of dissipation, $\varepsilon_w$, is not zero. Indeed, from its definition, $\varepsilon = \nu\, \overline{\partial_i u_j\, \partial_i u_j}$, and from $\boldsymbol{u} = 0$ on $y = 0$, it follows that $\varepsilon_w = \nu[\overline{(\partial_y u)^2} + \overline{(\partial_y w)^2}] \neq 0$. (Note that $\partial_y v = 0$ by continuity and the no-slip condition.) Integrating (6.2.16) gives

$$k \to A + By + \frac{\varepsilon_w y^2}{2\nu},$$

where $A$ and $B$ are integration constants. By (6.2.15), $A = B = 0$. So the wall value of dissipation is

$$\varepsilon_w = \lim_{y \to 0} \frac{2\nu k}{y^2}. \tag{6.2.17}$$

In CFD codes, this or something equivalent often is used as a means to impose the no-slip condition (6.2.15). The right-hand side of (6.2.17) might be evaluated at the first computational node above the wall.

Unfortunately, deriving the correct no-slip boundary conditions is neither the only, nor the most important, issue in near-wall modeling. A second need is to prevent a singularity in the $\varepsilon$ equation (6.2.6). If the time-scale $T = k/\varepsilon$ is used, then $T \to O(y^2)$ as $y \to 0$ and the right-hand side of (6.2.6) becomes singular like $\varepsilon^2/k$. In fact this would preclude the behavior $\varepsilon = O(1)$ at the wall. The fault is that the correlation time-scale should not vanish at the wall. Wall scaling, and evidence from DNS (Antonia and Kim 1994), show that the Kolmogoroff scale, $\sqrt{\nu/\varepsilon}$, is appropriate near the surface. The formula

$$T = \max(k/\varepsilon,\ 6\sqrt{\nu/\varepsilon}) \tag{6.2.18}$$

was proposed by Durbin (1991). The coefficient of 6 in formula (6.2.18) is an empirical constant obtained by matching DNS data. This can be written more generally as

$$T = \sqrt{\nu/\varepsilon}\, F(\sqrt{k^2/\nu\varepsilon}). \tag{6.2.19}$$

The argument of the function $F$ is the square root of the turbulent Reynolds number, so this is sometimes alluded to as a *low* Reynolds number correction. However, Kolmogoroff scaling only applies to the viscous region of *high* Reynolds flow, so the terminology "low Reynolds number" is misleading: it is more correct to understand this as a near-wall modification. For instance, this scaling does not apply to very weak turbulence, or to turbulence in its final period of decay.

The function $F(x)$ in (6.2.19) is arbitrary except for the limiting behaviors $F(0) = O(1)$ and $F(x) \to x$ as $x \to \infty$. It could be specified from experimental data, but the previous formula, $F = \max(\sqrt{k^2/\nu\varepsilon}, 6)$ suffices for most purposes. Several other methods to avoid a singularity in the $\varepsilon$ equation are discussed in Patel *et al.* (1984).

But the near-wall dilemma goes further, beyond just preventing the singularity and imposing a boundary condition. The formula $\nu_T = C_\mu k^2/\varepsilon$ gives an erroneous profile of eddy viscosity even if exact values of $k(y)$ and $\varepsilon(y)$ are known. The solid line in Figure 6.7 is the viscosity constructed from DNS data (Moser *et al.*, 1999) by evaluating the exact definition $\nu_T^+ = -\overline{uv}/\nu\, \mathrm{d}_y U$. It is compared to curves constructed by substituting the *exact** $k$ and $\varepsilon$ into the $k-\varepsilon$ formula (6.2.2). The model formula is seen to be grossly in error below $y_+ \approx 50$. To emphasize that the failure is near the wall, a curve has been plotted with $C_\mu = 0.08$ to improve agreement with the exact data farther from the wall. The region of failure is where turbulent transport is low; hence it is the high-impedance region. Overpredicting the eddy viscosity in this region will greatly overpredict skin friction on the surface.

A physical understanding of the failure illustrated in Figure 6.7 is that $k$ is the wrong velocity scale for transport to and from the wall. Near the surface, the kinetic energy is dominated by the tangential components of intensity, $k \approx \frac{1}{2}(\overline{u^2} + \overline{w^2})$. However, turbulent transport to or from the surface is more closely related to the normal component $\overline{v^2}$. Figure 6.6 is consistent with this observation. If $k$ controlled surface stress, then the data below $y_+ \approx 100$ would collapse when normalized as $k_+ \equiv k/u_*^2$. They do not, but the $\overline{v_+^2}$ data do.

* This is often referred to as "*a priori* testing."

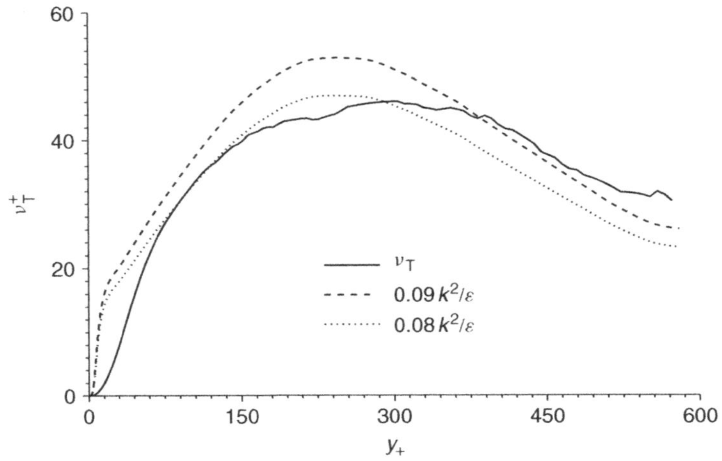


**Figure 6.7** Exact eddy viscosity compared to the $k$–$\varepsilon$ formula in channel flow at $R_\tau = 590$: curves were computed from DNS data of Moser *et al.* (1999).

One device to fix (6.2.2) consists of damping the viscosity, as in the Van Driest mixing length formula. To this end it is replaced by

$$\nu_T = f_\mu C_\mu k^2/\varepsilon. \tag{6.2.20}$$

A *damping function*, $f_\mu$, has been inserted. In the literature, this has been made a function of $yk/\nu$, or of $k^2/\varepsilon\nu$, depending on the model. For example, Launder and Sharma (1974) use

$$f_\mu = \exp\left[\frac{-3.4}{(1 + R_T/50)^2}\right], \qquad R_T = k^2/\nu\varepsilon.$$

While the argument of $f_\mu$ also could be $yu_*/\nu$, that form is of limited use because $u_*$ vanishes at a separation point. The friction velocity $u_*$ is usually replaced by $k$ by invoking the log-layer solution $u_* \to C_\mu^{1/4} k^{1/2}$.

Some of the bewildering array of wall damping schemes that have been proposed in the course of time are summarized in Patel *et al.* (1984). One could simply fit $f_\mu$ to the ratio of the exact to the $k$–$\varepsilon$ curves in Figure 6.7 – invoking the definition that the damping function is the ratio of "what you want to what you have." However, it has been found that damping functions that are fitted to zero pressure-gradient data fail to predict flows with adverse pressure gradient (Hanjalic and Launder 1980). The large number of damping schemes that have been proposed is testimony to the ineffectiveness of this approach. It also tends to suffer from numerical stiffness. The reader might consult Patel *et al.* (1984) or Chen and Jaw (1998) for tabulations of various "low Reynolds number $k$–$\varepsilon$" formulations.

It is not sufficient simply to damp the eddy viscosity. All low Reynolds number $k$–$\varepsilon$ models also modify the $\varepsilon$ equation in one way or another. As an example, Launder and

Sharma (1974) replace it with

$$\partial_t \tilde{\varepsilon} + U_j\, \partial_j \tilde{\varepsilon} = \frac{\tilde{\varepsilon}}{k}(C_{\varepsilon 1}\mathcal{P} - f_2 C_{\varepsilon 2}\tilde{\varepsilon}) + \partial_j \left( \left( \nu + \frac{\nu_T}{\sigma_\varepsilon} \right) \partial_j \tilde{\varepsilon} \right) + 2\nu\nu_T \left( \frac{\partial^2 U}{\partial y^2} \right)^2. \quad (6.2.21)$$

The last term on the right-hand side is not coordinate-independent, but it is only important near the surface, so $\hat{y}$ can be considered the wall normal direction. The dependent variable $\tilde{\varepsilon}$ is defined as $\varepsilon - \nu|\nabla k|^2/2k$. This is a device to cause $\tilde{\varepsilon}$ to be zero at the wall. The blending function $f_2 = 1 - 0.3\exp(-R_t^2)$ also was inserted, on the supposition that the decay exponent, $n$, should be about 2.5 near the wall. Actually $n = 2.5$ applies to grid turbulence in its final period of decay, not to the wall region of high Reynolds number boundary layers. (The final period of decay is discussed in Section 10.2.3.)

The benefits of adopting low Reynolds number $k$–$\varepsilon$ models in practical situations seem limited, given the apparent arbitrariness of the formulations, their numerical stiffness, and their inaccurate predictions in flows with significant pressure gradient. This formulation has little to recommend it and will not be described further.

#### 6.2.2.1 Wall functions

Another method to circumvent the erroneous predictions in the near-wall region is to abandon the $k$–$\varepsilon$ equations in a zone next to the wall and impose boundary conditions at the top of that zone. Within the zone the turbulence and mean velocity are assumed to follow prescribed profiles. This is the "wall function" method. Conceptually, the wall function is used in the law-of-the-wall region and the $k$–$\varepsilon$ model predicts the flow field farther from the surface. The two are patched in the logarithmic overlap layer. Let $y_p$ be the distance from the wall at which the patching is done. At that point the log-layer functions $dU/dy = u_*/\kappa y_p$, $k = u_*^2/\sqrt{C_\mu}$, and $\varepsilon = u_*^3/\kappa y_p$ are assumed to be valid. These are an exact solution to the standard $k$–$\varepsilon$ model in a constant-stress layer, so smooth matching is possible in principle; in practice, wall functions are used even when they are not mathematically justified, such as in a separating flow. At a two-dimensional separation point, $u_* = 0$. To avoid problems as $u_*$ changes sign, the boundary conditions are expressed in terms of $u_k \equiv (k\sqrt{C_\mu})^{1/2}$. They assume the form

$$\frac{dU}{dy} = \frac{u_*}{\kappa y_p}, \qquad \varepsilon = \frac{u_k^3}{\kappa y_p}, \qquad \frac{dk}{dy} = 0. \quad (6.2.22)$$

In practice, the conditions (6.2.22) are applied at the grid point closest to the solid boundary, $y(1)$; this point should be located above $y_+ \approx 40$ and below $y \approx 0.2\delta_{99}$. The skin friction is found from the logarithmic drag law applied at this point by solving $U(1) = u_*[\log(y(1)u_*/\nu)/\kappa + B]$ for $u_*$, given a computed $U(1)$. The tangential surface shear stress is assumed to be parallel to the direction of the mean velocity.

The wall function procedure is rationalized by appeal to the two-layer (law of the wall, law of the wake) boundary-layer structure (Section 4.2). The law of the wall is assumed to be of a universal form, unaffected by pressure gradients or flow geometry. The meaning of universality is that the flow in this region can be prescribed once and for all. Its large $y_+$ asymptote, (6.2.22), is its only connection to the non-universal part of the flow field. Through this, the skin friction can respond to external forces. However, in highly perturbed flows, including boundary layers with separation, reattachment, strong curvature, or strong lateral pressure gradients, the assumption of a universal wall layer is not consistent with experiments. These sorts of distortions upset the assumed state of quasi-equilibrium that lies behind the use of wall functions. Predictions made with wall functions then deteriorate.

As a practical matter, it is sometimes impossible to ensure that the first point of a computational grid lies in the log layer, if a log layer exists at all. The definition of $y_+$ requires $u_*$, but this is computed as part of a flow solution. It is not possible *a priori* to generate a computational mesh that will ensure that the first computational node is neither too close nor too far from the wall. If the first node lies too close to the wall, then the $k$–$\varepsilon$ model will be used in the region at small $y_+$ in Figure 6.7, in which it severely overpredicts $\nu_T$. In boundary-layer flows, the tendency will then be to overpredict surface skin friction. Accurate computation may require *a posteriori* modification of the mesh to achieve a suitable $y_+$.

In complex flows, it is likely that the wall function will be used beyond its range of justifiability. On the other hand, wall functions can significantly reduce the cost of a CFD analysis. The steepest gradient of turbulent energy occurs near the wall ($y_+ \lesssim 10$). By starting the solution above this region, the computational stiffness is reduced. Because of this stiffness, the near-wall region requires a disproportionate number of grid points; avoiding it with a wall function reduces grid requirements. Wall functions therefore are widely used for engineering prediction. However, almost all commercial CFD codes either warn the user if the $y_+$ criteria are not met, or enable the user to monitor $y_+$ values.

An alternative to the standard wall function, described above, is the *non-equilibrium wall function*. This approach partly obviates a major shortcoming of the standard formulation by introducing a mild dependence on pressure gradients. This improves performance in complex flow computations. The flow is assumed to consist of two zones: a viscous sublayer, $y_+ < y_+^v$; and a fully developed turbulent layer, $y_+ > y_+^v$; typically $y_+^v \approx 12$. If $y_+ < y_+^v$, the following boundary conditions are applied:

$$k = \left(\frac{y}{y^v}\right)^2 k_{\mathrm{P}}, \qquad \varepsilon = \frac{2\nu k}{y^2}. \tag{6.2.23}$$

If $y_+ > y_+^v$, the standard conditions (6.2.22) on $k$ and $\varepsilon$ are used. In addition to the more elaborate treatment of turbulent quantities, the drag law is also modified to include pressure-gradient effects, as was discussed in Section 4.2.

In summary, the wall function method is used in industrial CFD computations mainly because it reduces computational cost by not requiring a near-wall grid. However, regardless of specific formulation, the wall function method assumes that a universal wall law exists. This assumption can fail severely in complex flows. While the method works reasonably well when boundary layers are attached, with limited influence of pressure gradients, it is less reliable in cases that depart from this state. Reliable computations of industrial flows require proper representation of wall boundary conditions; an alternative approach is desirable.

#### 6.2.2.2 Two-layer models

A compromise between the assertion of a universal wall layer, as made by wall function methods, and a full simulation of the wall adjacent region, is to formulate a simplified model for that layer and patch it onto the full $k$–$\varepsilon$ model. The full model is solved in the outer region. The $k$–$\ell$ formulation has been used to this end, in an approach called the "two-layer $k$–$\varepsilon$ model" (Chen and Patel, 1988).

The $k$–$\ell$ model uses the $k$ equation (6.2.5), but replaces the $\varepsilon$ equation (6.2.6) by the algebraic formula

$$\varepsilon = k^{3/2}/\ell_\varepsilon. \tag{6.2.24}$$

The dissipation length $\ell_\varepsilon$ must be prescribed. Most simply, it can be made analogous to the mixing length (6.1.10)

$$\ell_\varepsilon = C_\ell y(1 - \mathrm{e}^{-y\sqrt{k}/\nu A_\varepsilon}). \tag{6.2.25}$$

The limiting form (6.2.16) of the $k$ equation near a no-slip surface now becomes

$$\nu\, \partial_y^2 k = \varepsilon = \frac{k^{3/2}}{\ell_\varepsilon}.$$

Substituting expression (6.2.25) and letting $y \to 0$ gives

$$\nu\, \partial_y^2 k \to \frac{\nu A_\varepsilon k}{C_\ell y^2}.$$

The correct behavior $k \propto y^2$, as follows from (6.2.15), is obtained if

$$A_\varepsilon = 2C_\ell. \tag{6.2.26}$$

The eddy viscosity (6.2.2) will not have the right damping if (6.2.24) is substituted. Doing so would give $\nu_\mathrm{T} = \sqrt{k}\,\ell_\varepsilon$. Therefore a separate length is used in the formula

$$\nu_\mathrm{T} = C_\mu \sqrt{k}\,\ell_\nu. \tag{6.2.27}$$

Again, the exponential Van Driest form

$$\ell_\nu = C_\ell y(1 - \mathrm{e}^{-y\sqrt{k}/\nu A_\nu}) \tag{6.2.28}$$

proves convenient and effective. Substituting the log-layer solution $k_+ = 1/\sqrt{C_\mu}$ and $\nu_\mathrm{T}^+ = \kappa y_+$ along with (6.2.28) into (6.2.27) results in

$$C_\ell = \kappa / C_\mu^{3/4} \tag{6.2.29}$$

when the limit $y\sqrt{k}/\nu \gg 1$ is invoked. Given the Von Karman constant $\kappa = 0.41$ and given that $C_\mu$ retains its value 0.09, the only new empirical constant is $A_\nu$. This can be found by selecting a value that produces $B \approx 5$ for the additive constant in the log law, as was done with the mixing length (6.1.10). A similar method of calibration is to select the value of $A_\nu$ that most closely reproduces the curve of skin friction versus Reynolds number in a flat-plate boundary layer. This is illustrated in Figure 6.8, where calibration of the above two-layer $k$–$\varepsilon$ model consists of selecting the value of $A_\nu$ that best agrees with experiment. The value $A_\nu = 62.5$ gives a good fit.

It should be emphasized that the value of $A_\nu$ depends on the specific equations of the model. The two-layer $k$–$\varepsilon$ model that is calibrated in Figure 6.8 consists of the above

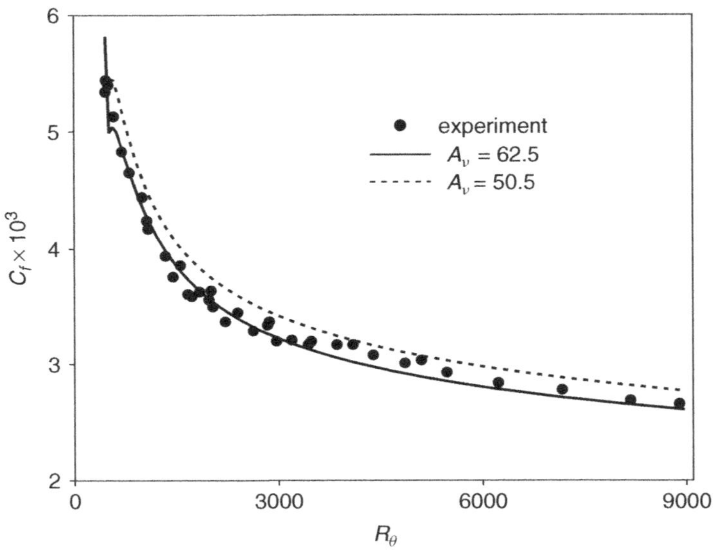


**Figure 6.8** Calibration of the constant $A_\nu$ in the two-layer $k$–$\varepsilon$ model.

$k$–$\ell$ model near the wall and the standard $k$–$\varepsilon$ model farther away. The two models were patched at the value of $y$ at which $1 - \mathrm{e}^{-y\sqrt{k}/\nu A_\nu}$ reaches 0.95; this occurs at $y_{\mathrm{sw}} = \log(20) A_\nu \nu / \sqrt{k(y_{\mathrm{sw}})}$.

Succinctly, the two-layer model solves the $k$ equation (6.2.5) at all points in the flow, but instead of (6.2.6) the $\varepsilon$ equation is represented by

$$\mathcal{L}[\varepsilon] = \mathcal{S}. \tag{6.2.30}$$

The operator $\mathcal{L}$ and source $\mathcal{S}$ are defined as

$$\mathcal{L} = \begin{cases} \partial_t + U_j\,\partial_j - \partial_j \left(\nu + \dfrac{\nu_{\mathrm{T}}}{\sigma_\varepsilon}\right)\partial_j, & y > y_{\mathrm{sw}}, \\ 1, & y \le y_{\mathrm{sw}}, \end{cases} \tag{6.2.31}$$

and

$$\mathcal{S} = \begin{cases} \dfrac{C_{\varepsilon 1}\mathcal{P} - C_{\varepsilon 2}\varepsilon}{T}, & y > y_{\mathrm{sw}}, \\ \dfrac{k^{3/2}}{\ell_\varepsilon}, & y \le y_{\mathrm{sw}}. \end{cases} \tag{6.2.32}$$

This two-layer formulation has proved quite effective in practical computations (Rodi, 1991) and usually gives better predictions than wall functions. But it does require a fine grid near to walls, and so is more expensive computationally than wall functions.

### 6.2.3 Weak solution at edges of free-shear flow; free-stream sensitivity

It was remarked by Cole (see Kline *et al.*, 1968) that such equations can develop propagating front solutions. An equation of the form

$$\partial_t \nu_T = \partial_y (\nu_T \, \partial_y \nu_T)$$

has a ramp solution $\nu_T = A(ct - y)$ for $y < ct$ and $\nu_T = 0$ for $y > ct$. Substitution into the above equation gives the front speed $c = A$. The front develops because $\nu_T$ spreads by diffusion and the diffusion coefficient tends to zero at the front. The solution cannot diffuse beyond the point where $\nu_T$ vanishes; it becomes identically zero from there onward.

The $k$–$\varepsilon$ model behaves similarly to this simple nonlinear diffusion equation (Cazalbou *et al.*, 1994). We will ignore molecular viscosity and consider a time-evolving solution. It will be supposed that, near the front, production and dissipation can be dropped to lowest order of approximation. That will be justified after the solution is obtained. The model equations that describe the propagating front are

$$\begin{aligned} \partial_t k &= \partial_y (\nu_T \, \partial_y k), \\ \partial_t \varepsilon &= \partial_y \left( \frac{\nu_T}{\sigma_\varepsilon} \partial_y \varepsilon \right), \\ \nu_T &= C_\mu k^2/\varepsilon. \end{aligned} \tag{6.2.33}$$

The possibility for a front to form exists if a consistent solution has the property $\nu_T \to 0$ as $k, \varepsilon \to 0$. Then the solution will not be able to diffuse past the point where this occurs. That such behavior is possible is verified by seeking a solution in the form

$$k = A_k (ct - y)^m, \qquad \varepsilon = A_\varepsilon (ct - y)^{2m-1}, \qquad \nu_T = C_\mu \frac{A_k^2}{A_\varepsilon}(ct - y)$$

for $y < ct$, and $k = \varepsilon = \nu_T = 0$ for $y > ct$. Substituting these into the governing equations (6.2.33) shows that

$$m = \frac{1}{2 - \sigma_\varepsilon} \qquad \text{and} \qquad c = \frac{C_\mu A_k^2}{A_\varepsilon (2 - \sigma_\varepsilon)}.$$

The standard value $\sigma_\varepsilon = 1.3$ gives $k \sim (ct - y)^{1.43}$ as $y$ approaches $ct$ from below. Therefore, $k$ approaches 0 with a singular second derivative. This propagating front solution is found near the edge of free-shear flows. It requires a very fine grid to resolve the singular solution in a computer calculation. Either numerical or molecular diffusion will smooth it.

In retrospect, it can be verified that the terms retained in (6.2.33) are of order $(ct - y)^{0.43}$ while $\varepsilon \sim (ct - y)^{1.86}$ and $\mathcal{P}$ is of the same size (Exercise 6.9). Hence it is consistent to drop production and dissipation in the local analysis of the weak solution near shear-layer edges.

It was argued by Cazalbou *et al.* (1994) that the front is a desirable property of the $k$–$\varepsilon$ model that prevents spurious sensitivity to free-stream conditions. If $\sigma_\varepsilon < 2$ then $c > 0$ and the turbulent layer propagates toward the free stream. The supposition is that the shear layer is then not unduly influenced by external disturbances. The $k$–$\omega$ model

(Section 6.3) corresponds to using $\sigma_\varepsilon = 2$, which precludes the above solution. That model was found by Menter (1994) to have a debilitating sensitivity to free-stream conditions, which makes its predictions of free-shear layers somewhat unreliable. It might seem that reducing $\sigma_\varepsilon$ (actually $\sigma_\omega$) to a value less than 2 would solve this problem, according to the above edge analysis. That is not the case: reducing $\sigma_\varepsilon$ does not remove the spurious sensitivity displayed by $k$–$\omega$ solutions. The matter of free-stream sensitivity is quite important to turbulence modeling, but its cause is not well understood mathematically. Perhaps that is because the front solution is a possible local behavior that might not arise in a complete solution. In practice, free-stream sensitivity is checked by numerical calculation of free-shear layers.

## 6.3 The $k$–$\omega$ model

Since the $\varepsilon$ equation is primarily a dimensionally consistent analog to the $k$ equation, and the variable $\varepsilon$ is in part used to define a time-scale $T = k/\varepsilon$, one might instead consider combining the $k$ equation directly with a time-scale equation. In homogeneous turbulence, it is largely irrelevant to do so: the evolution equation for $T \equiv k/\varepsilon$, derived from the standard $k$ and $\varepsilon$ equations, is

$$\begin{aligned} \mathrm{d}_t T &= \frac{\mathrm{d}_t k}{\varepsilon} - \frac{k\,\mathrm{d}_t \varepsilon}{\varepsilon^2} = (C_{\varepsilon_2} - 1) - 2C_\mu (C_{\varepsilon_1} - 1)|S|^2 T^2 \\ &= 0.92 - 0.88 C_\mu |S|^2 T^2. \end{aligned} \tag{6.3.1}$$

Solving this and $k$ is exactly the same as solving $\varepsilon$ and $k$.

In non-homogeneous flow, a turbulent diffusion term must be added; now there is a basis of distinction. If $T$ is assumed to diffuse, that term is $\partial_y(\nu_\mathrm{T}/\sigma_\mathrm{T}\,\partial_y T)$. Consider steady flow in a constant-stress, log layer. Because $\mathcal{P}/\varepsilon = 2C_\mu|S|^2T^2 = 1$ in the log layer, the homogeneous contribution to the $T$ equation (6.3.1) is 0.92–0.44, which is positive. The left-hand side of (6.3.1) is zero in steady state. Hence, the diffusion term must be negative to achieve a balance. This means that $T$ is not a good quantity to work with because the simplest model requires negative diffusion, and negative diffusion is mathematically ill-posed and computationally unstable. Instead, $1/T$, or $\omega$, can be used. This quantity is found, by the above line of reasoning, to require a positive diffusion coefficient. Replacement of the $\varepsilon$ equation by an $\omega$ equation is, indeed, viable.

Wilcox (1993) elected to define $\omega$ as $\varepsilon/C_\mu k$. His $k$–$\omega$ model can be written

$$\begin{aligned} \partial_t k + U_j\,\partial_j k &= 2\nu_\mathrm{T}|S|^2 - C_\mu k\omega + \partial_j\left(\left(\nu + \frac{\nu_\mathrm{T}}{\sigma_k}\right)\partial_j k\right), \\ \partial_t \omega + U_j\,\partial_j \omega &= 2C_{\omega 1}|S|^2 - C_{\omega 2}\omega^2 + \partial_j\left(\left(\nu + \frac{\nu_\mathrm{T}}{\sigma_\omega}\right)\partial_j \omega\right), \\ \nu_\mathrm{T} &= k/\omega. \end{aligned} \tag{6.3.2}$$

The $k$ equation is altered only by changing $\varepsilon$ to $C_\mu k\omega$. The $\omega$ equation is quite analogous to the $\varepsilon$ equation. The standard constants are $C_{\omega 1} = 5/9$, $C_{\omega 2} = 3/40$, $\sigma_\omega = \sigma_k = 2$, and $C_\mu = 0.09$. The calibration of constants is the same as for the $k$–$\varepsilon$ model: for instance, $C_\mu/C_{\omega 2} = 1.2$ is the decay exponent for grid turbulence, and the constants are related to $\kappa$ by the log-layer solution $\kappa^2 = (C_{\omega 2}/C_\mu - C_{\omega 1})\,\sigma_\omega \sqrt{C_\mu}$. The properties of the $k$–$\omega$ model model are discussed at length in Wilcox (1993).

Near a no-slip surface the dominant balance of terms in this model is

$$\partial^2_{y_+} k_+ = C_\mu k_+ \omega_+ \qquad \text{and} \qquad \partial^2_{y_+} \omega_+ = C_{\omega 2}\omega_+^2$$

in non-dimensional units. The solution is $\omega_+ = 6/(C_{\omega 2} y_+^2)$ and $k \propto y_+^m$ with $m = \frac{1}{2} + \sqrt{149/20} = 3.23$. This shows that $\omega$ is singular at no-slip boundaries and that $k$ does not behave as $y^2$. Despite these apparent drawbacks, Wilcox (1993) has shown this model to be usable near boundaries, without a requirement for wall functions, or for wall damping – that is its remarkable property. The reason for this favorable attribute is that extra dissipation is produced near walls. The source of extra dissipation can be understood by rewriting the $\omega$ equation as an $\varepsilon$ equation.

The rate of dissipation in the $k$ equation is $C_\mu k\omega = \varepsilon$. Forming the evolution equation for this product from (6.3.2) gives

$$\partial_t \varepsilon + U_j\, \partial_j \varepsilon = \frac{C_{\varepsilon 1}\mathcal{P} - C_{\varepsilon 2}\varepsilon}{T} + \partial_j \left( \left( \nu + \frac{\nu_{\mathrm{T}}}{\sigma_\varepsilon} \right) \partial_j \varepsilon \right) + S_\omega, \qquad (6.3.3)$$

with $T = 1/\omega$, $C_{\varepsilon 1} = 1 + C_{\omega 1}$, $C_{\varepsilon 2} = 1 + C_{\omega 2}/C_\mu$, and $\sigma_\varepsilon = \sigma_\omega$. This reproduces the standard $\varepsilon$ model (6.2.6) with an additional term $S_\omega$. The source term

$$S_\omega = \frac{2}{T}\left( \nu + \frac{\nu_{\mathrm{T}}}{\sigma_\omega} \right) \left[ \frac{|\nabla k|^2}{k} - \frac{\nabla k \cdot \nabla \varepsilon}{\varepsilon} \right]$$

largely distinguishes between the $k$–$\omega$ and $k$–$\varepsilon$ models. In the viscous sublayer, $k$ increases with wall distance, while $\varepsilon$ decreases; hence $S_\omega > 0$. The extra term is a source, not a sink, of dissipation. Consequently, a larger $\varepsilon$ is produced.

Figure 6.9 shows a solution to the $k$–$\omega$ model model in plane channel flow at $R_\tau = 590$. The profile of $\varepsilon = C_\mu k\omega$ has been multiplied by 50 for display. The region of

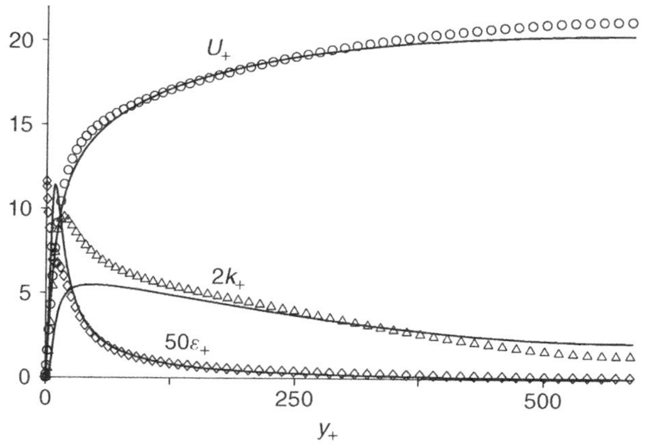


**Figure 6.9** Plane channel flow at $R_\tau = 590$. The full curves are solutions to the $k$–$\omega$ model. DNS data for $U$ (○); $2k$ (△); and $50\varepsilon$ (◇).

interest is the law-of-the-wall layer, between $y_+ = 0$ and $y_+ \approx 40$. That region is critical to transfer of heat or momentum between the fluid and the surface. In the wall layer, the $k-\omega$ predictions of $k$ and $\varepsilon$ are at odds with the data: $\varepsilon$ erroneously goes to zero at the surface and has a spurious peak near $y_+ = 10$. The consequence of the spurious peak is that $k$ is excessively dissipated near the wall. The DNS data (Moser *et al.*, 1999) for $k$ have a sharp maximum at $y_+ \approx 10$; that peak is entirely missing in the $k-\omega$ prediction. While this grossly erroneous prediction of $k$ might at first be disconcerting, in fact the underestimation of $k$ and overestimation of $\varepsilon$ are exactly what are needed to counter the overprediction of $\nu_T$ displayed by the formula $C_\mu k^2/\varepsilon$ in Figure 6.7. Both of these features, not just the deletion of the peak from the $k$ profile, are critical to obtaining a reasonable distribution of $\nu_T$. Indeed, the $U$ predictions in Figure 6.9 agree quite well with data, given that no wall corrections have been made to the model. Unfortunately, difficulties with $k-\omega$ arise in free-shear flows.

Menter (1994) noted two failings of the basic $k-\omega$ model, one minor and one major. The former is that it overpredicts the level of shear stress in adverse pressure-gradient boundary layers. The latter is its spurious sensitivity to free-stream conditions, discussed in Section 6.2.3. Just about any spreading rate can be obtained for free-shear flows, depending on the free-stream value of $\omega$. The model is not reliable in flows with detached shear layers.

To overcome the shortcomings of the basic $k-\omega$ model, Menter (1994) proposed the "shear-stress transport" (SST) model. This variant of $k-\omega$ has been found to be quite effective in predicting many aeronautical flows. The SST model is developed in two stages. The first is meant to improve predictions in adverse pressure-gradient boundary layers; the second to solve the problem of free-stream sensitivity.

The tendency to overestimate the shear stress is fixed by imposing a bound on the stress-intensity ratio, $|\overline{uv}|/k$. This ratio is often denoted $a_1$. Although in many flows $a_1 \approx 0.3$, lower values are observed in adverse pressure gradients (Cutler and Johnston, 1989). Invoking the formula $-\overline{uv} = k\,\partial_y U/\omega$ for parallel shear flow gives

$$\frac{\mathcal{P}}{\varepsilon} = \frac{\overline{uv}^2\omega}{k\varepsilon} = \frac{1}{C_\mu}\left|\frac{\overline{uv}}{k}\right|^2$$

or $a_1 = \sqrt{C_\mu}\sqrt{\mathcal{P}/\varepsilon} = 0.3\sqrt{\mathcal{P}/\varepsilon}$. In the outer part of an adverse pressure-gradient boundary layer, $\mathcal{P}/\varepsilon$ can be significantly greater than unity. Instead of decreasing, $a_1$ is predicted to increase. To prevent an increase of the stress-intensity ratio, Menter (1994) introduced the bound

$$\nu_T = \min\left[\frac{k}{\omega}, \frac{\sqrt{C_\mu}k}{|2\boldsymbol{\Omega}|}\right]. \tag{6.3.4}$$

In parallel shear flow, the magnitude of the mean flow rotation tensor is $|\boldsymbol{\Omega}| = \frac{1}{2}|\partial_y U|$; hence expression (6.3.4) gives

$$\frac{|\overline{uv}|}{k} = \min\left[\frac{|\partial_y U|}{\omega}, \sqrt{C_\mu}\right] = \min\left[\frac{|\partial_y U|}{\omega}, 0.3\right]. \tag{6.3.5}$$

The min function is sometimes called a limiter. The above limiter improves prediction of adverse pressure gradient and separated flow.

It is not unusual that a fix for one problem will introduce another. In this case, it is not so much that a problem is introduced by formula (6.3.4), but that it is an unwanted constraint in free-shear flow. To confine the limiter to the boundary layer, Menter (1994) introduced the blending function

$$F_2 = \tanh(arg_2^2),$$
$$arg_2 = \max\left[\frac{2\sqrt{k}}{C_\mu \omega y}, \frac{500\nu}{\omega y^2}\right]. \qquad (6.3.6)$$

The limiting behavior $\omega \to 6\nu/(C_{\omega 2} y^2)$, $k \to O(y^{3.23})$ as $y \to 0$ shows that $arg_2 \to 25/4$ near the wall. As $y \to \infty$, $arg_2 \to 0$. The blending function (6.3.6) is devised so that $F_2$ is nearly unity in most of the boundary layer, dropping to zero near the top and in the free stream. The second term in the min function of Eq. (6.3.4) is divided by $F_2$.

To rectify the spurious free-stream sensitivity of the original $k$–$\omega$ model, Menter (1994) developed a two-zone formulation that uses $k$–$\omega$ near the wall and $k$–$\varepsilon$ for the rest of the flow. The switch between these forms is by a smooth interpolation. Now the blending function is

$$F_1 = \tanh(arg_1^4),$$
$$arg_1 = \min\left[\max\left(\frac{\sqrt{k}}{C_\mu \omega y}, \frac{500\nu}{\omega y^2}\right), \frac{2k\omega}{y^2 \max(\nabla k \cdot \nabla \omega, 10^{-20})}\right]. \qquad (6.3.7)$$

This seemingly intricate function is simply an operational device to interpolate between the $k$–$\omega$ and $k$–$\varepsilon$ models. It is devised to be near unity in the inner half of the boundary layer and to decrease through the outer half, dropping to zero slightly inside its top edge. The extra term $S_\omega$ in (6.3.3) is faded out via $F_1$:

$$\partial_t \varepsilon + U_j\, \partial_j \varepsilon = \frac{C_{\varepsilon 1}\mathcal{P} - C_{\varepsilon 2}\varepsilon}{T} + \partial_j\left(\left(\nu + \frac{\nu_T}{\sigma_\varepsilon}\right)\partial_j \varepsilon\right) + F_1 S_\omega. \qquad (6.3.8)$$

Thereby, a transition between the $\varepsilon$ and $\omega$ equations typically is brought about across the middle of the boundary layer. To complete the interpolation, the model constants also are interpolated as

$$C_{\varepsilon 1} = 1 + (1 - F_1)0.44 + F_1 C_{\omega 1} \quad \text{and} \quad C_{\varepsilon 2} = 1 + (1 - F_1)0.92 + F_1 C_{\omega 2}/C_\mu.$$

These provide the $k$–$\varepsilon$ values when $F_1 = 0$ and the $k$–$\omega$ values when $F_1 = 1$. The coefficients $\sigma_k$ and $\sigma_\varepsilon$ are interpolated similarly.

## 6.4 Stagnation-point anomaly

The two equation models have been calibrated in boundary-free flows and in shear flows parallel to boundaries. Turn the flow around and let it impinge on the wall and an embarrassing phenomenon occurs – the stagnation-point anomaly (Durbin, 1996). The allusion is to a growth of $k$ and $\nu_T$ to excessive levels near the stagnation point. The anomaly can be seen in impinging jets and near the leading edge of airfoils, as in Figure 6.10(b). Three

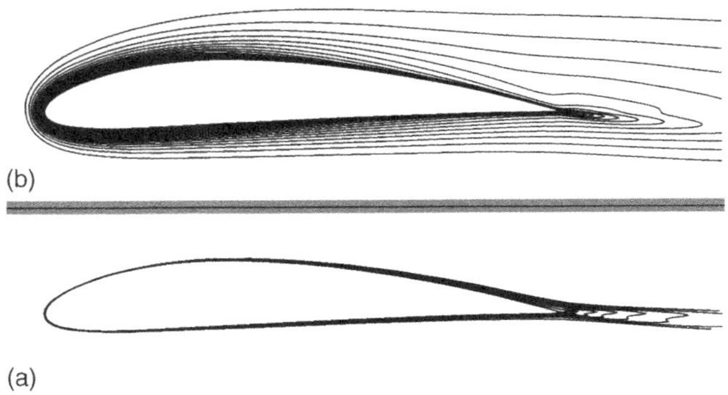


**Figure 6.10** The stagnation-point anomaly in flow round a leading edge. The figures show $k$ contours (a) with a bound and (b) without a bound on the turbulent time-scale.

explanations of this behavior can be offered. In all of them the eddy viscosity formula $\mathcal{P} = 2\nu_T|S|^2$ is seen to give rise to too large levels of production near the stagnation point.

The exact formula $\mathcal{P} = -\overline{u_i u_j} S_{ij}$ becomes $\mathcal{P} = A(\overline{u^2} - \overline{v^2})$ in the uniform straining flow $U = -Ax$, $V = Ay$. This represents the flow toward an obstacle at $x = 0$, as in Figure 6.10. The linear constitutive model $\overline{u_i u_j} = -2\nu_T S_{ij} + \frac{2}{3}k\delta_{ij}$ gives $\overline{u^2} - \overline{v^2} = 4\nu_T A$ and $\mathcal{P} = 4\nu_T A^2$ in this flow. Increasing $A$ always increases $\mathcal{P}$. Straining of turbulent vorticity is analyzed in Chapter 11: the plane strain presently under consideration is shown to amplify $\overline{u^2}$ relative to $\overline{v^2}$. If initially $\overline{u^2} < \overline{v^2}$, then at first $\mathcal{P}$ will be decreased by straining, while the eddy viscosity formula only permits it to increase. In this explanation of the stagnation-point anomaly, it is a deficient representation of normal stress anisotropy by the eddy viscosity formula that is the culprit.

In the airfoil flow, vortex stretching around the leading edge (illustrated by Figure 11.3, page 289), of incident isotropic turbulence, will amplify $\overline{u^2}$ and increase $\mathcal{P}$. Hence, the eddy viscosity formula is *qualitatively* correct. Its fault is a quantitative overestimate of the level of $\mathcal{P}$: at large rates of strain, production should grow linearly with rate of strain (Section 11.2.1) instead of quadratically, as in the formula $2\nu_T|S|^2$. A second perspective on the origin of the anomaly is that $\mathcal{P}$ should only grow as $|S|$ not as $|S|^2$.

A third perspective on the origin of the anomaly is that dissipation does not keep up with production. Consider the time evolution of $\varepsilon$, subject to a sudden increase of $\mathcal{P}$. Let $\mathcal{P}$ and $T$ be constant in

$$d_t\varepsilon = \frac{1}{T}(C_{\varepsilon 1}\mathcal{P} - C_{\varepsilon 2}\varepsilon).$$

Then

$$\varepsilon = \varepsilon(0)\,e^{-C_{\varepsilon 2}t/T} + C_{\varepsilon 1}\mathcal{P}/C_{\varepsilon 2}(1 - e^{-C_{\varepsilon 2}t/T}).$$

If $\varepsilon(0) \ll \mathcal{P}$, then $\varepsilon$ will be small compared to $\mathcal{P}$ until $t > T$. A large $T$ permits $k$ to grow large before dissipation becomes significant. Indeed, even in the limit $t \gg T$, the dissipation becomes $\varepsilon = C_{\varepsilon 1}\mathcal{P}/C_{\varepsilon 2}$, so the smaller is $C_{\varepsilon 1}/C_{\varepsilon 2}$, the larger is the imbalance between production and dissipation, and the more can $k$ grow in large strains.

Several proposals to alleviate the stagnation-point anomaly have evolved from the preceding considerations on its origin. If the problem is an excessive production of $k$ in irrotational strain, then replacing $2\nu_T|S|^2$ by $2\nu_T|S||\Omega|$ (Launder and Kato, 1993) will eliminate the problem; indeed, this equates production to zero in irrotational flow. Two obvious difficulties with this scheme are that production is not actually zero in pure straining flow and that a spurious production will occur in rotating frames of reference. Coordinate independence requires that $|\Omega|$ be understood as the absolute vorticity in an inertial frame. If $\Omega^F$ is the rate of frame rotation, then, for pure straining in a rotating frame, the production will be $2\nu_T|S||\Omega^F|$. The faster the rotation, the faster $k$ will grow. Frame rotation does not cause such an effect; in fact, rapid frame rotation can suppress turbulence.

A less obvious difficulty is that this approach violates energy conservation. As noted in Section 3.3.4, the equation for $\frac{1}{2}|\boldsymbol{U}|^2$ should contain $-\mathcal{P}$ (see Exercise 3.2). The eddy viscous term $\partial_j(2\nu_T S_{ij})$ in the equation for $U_i$, times $U_i$, contributes $-2\nu_T|S|^2$ to the equation for $\frac{1}{2}|\boldsymbol{U}|^2$. It is not consistent to use $2\nu_T|S||\Omega|$ for production in the $k$ equation.

Another proposal for eliminating the anomaly is to impose a "realizability" bound – see Durbin (1996) and Exercise 6.8. The eigenvalues of $\overline{u_i u_j}$ are bounded from below by zero and above by $2k$. When the lower bound is imposed on the formula $\overline{u_i u_j} = -2\nu_T S_{ij} + \frac{2}{3}\delta_{ij}k$, the constraint

$$2\nu_T \lambda^S_{\max} \leq \tfrac{2}{3}k \tag{6.4.1}$$

emerges. The upper bound is automatically satisfied if this lower bound is. In this inequality, $\lambda^S_{\max}$ is the maximum eigenvalue of the rate-of-strain tensor. It can be shown that $\lambda^S_{\max} < \sqrt{2|S|^2/3}$. Therefore, any method to impose the bound

$$\nu_T \leq \frac{k}{\sqrt{6}\,|\boldsymbol{S}|}$$

ensures realizability. A corollary to this bound is the limit $\mathcal{P} \leq k|\boldsymbol{S}|/\sqrt{6}$ on production. No more than linear growth of $\mathcal{P}$ is allowed at large rates of strain. A limiter like

$$\nu_T = \min(C_\mu k^2/\varepsilon, \alpha k/|\boldsymbol{S}|) \tag{6.4.2}$$

with $\alpha \leq 1/\sqrt{6}$ would affect the bound. This is the method used in part Figure 6.10(a). Introducing a formula like $\nu_T = C_\mu k T$ shows that (6.4.2) also can be written as an upper bound on the time-scale. When that is used in the $\varepsilon$ equation, it helps dissipation keep step with the rate of production.

## 6.5 The question of transition

The transition from laminar to turbulent flow in a boundary layer is preceded by velocity fluctuations within the laminar layer. Early ideas were built upon linear stability theory in which the fluctuations are instability waves. At sufficient Reynolds number, the shear layer becomes unstable: the instability takes the form of waves that grow exponentially with downstream distance. These growing waves are two-dimensional, possibly with their crests at an oblique angle to the flow. Two-dimensional waves cannot develop directly into turbulence. Turbulence requires a three-dimensional field of vorticity, which can develop complexity.

Secondary instabilities grow atop the primary instability wave. They develop into horseshoe vortices, much like those described in Chapter 5. Local patches of turbulence develop where these horseshoe vortices lift away from the wall. The patches are called turbulent *spots*. Spots occur sporadically in space and time. Initially they are very sparse; most of the boundary layer is laminar. Progressing downstream, they grow in size and increase in frequency until the shear flow becomes fully turbulent. The fraction of time that the flow at any point is turbulent is called the *intermittency*. In laminar flow, the intermittency is zero; in fully turbulent flow, it is unity.

Thus, transition develops in three stages: unsteady precursors develop in the laminar flow; spots form intermittently; and the spots grow and merge to form the fully turbulent boundary layer. Figure 6.11 illustrates how an instability wave breaks into turbulence. The amplitude of the oscillations grows slowly with $x$. Around $R_x \approx 4.7 \times 10^5$, a sharp spike appears, after which the skin friction rises rapidly to turbulent levels. The chain-dotted line is the time-averaged skin friction coefficient. The instability wave makes little change to the average, laminar skin friction. In the intermittent region, between $R_x \approx 4.7 \times 10^5$ and $R_x \approx 5.1 \times 10^5$, the average skin friction increases until it reaches the turbulent level. In fact, $C_f$ overshoots the curve for turbulent skin friction. The latter is a data correlation for high Reynolds number, turbulent boundary layers.

The first stage might develop as we have described: primary then secondary instability. That is called the *orderly* route. Or it can follow an alternative route, that is called *bypass* transition. Bypass occurs when a boundary layer is subjected to free-stream turbulence of greater than about 0.5% intensity; that is, $u'/U > 0.005$. In this case, the first stage consists of jet-like disturbances within the boundary layer. They are jet-like in the sense that the dominant velocity component is $u$ and the disturbance is long in the streamwise direction. They are reminiscent of the "streaks" discussed in Chapter 5. These disturbances lift away from the wall as they move downstream. Then secondary instabilities occur (Jacobs and Durbin, 2000). They break down locally to form intermittent turbulent spots.

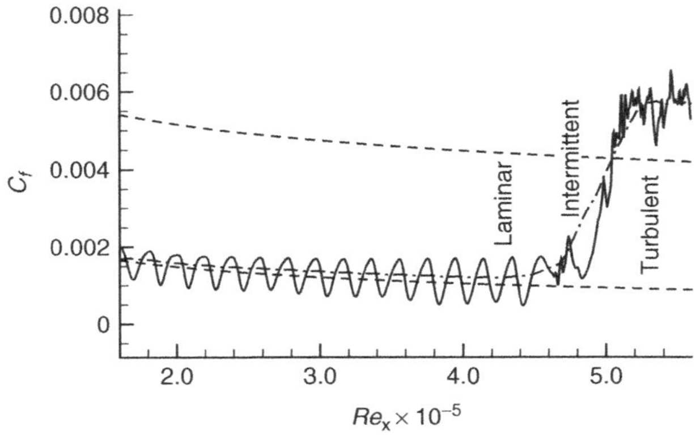


**Figure 6.11** Skin friction coefficient in orderly transition. The solid curve shows the evolution of the instability wave. The chain curve is its time average. The dashed curves are laminar and turbulent levels.

The top portion of Figure 6.12 is a plan view of the jets, observed in contours of the $u$ component of perturbation velocity. These streaky features, seen at the left-hand side of the $u$ contours, are often called Klebanoff modes. Spots are seen clearly in the lower portion of the figure, on the contours of $v$. A spot is located around $x = 290$. To either side of it, the flow remains laminar. The spots grow and merge to produce a fully turbulent boundary layer at the rightmost edge of the picture. The time-averaged skin friction coefficient, shown in Figure 6.13, starts at the laminar level and rises with downstream distance to the turbulent level. In between these two states, $C_f$ is an average of the turbulent level inside spots and the laminar level around them.

Bypass transition occurs at lower Reynolds number than orderly transition. It dominates when an attached boundary layer is subjected to free-stream disturbances. Orderly

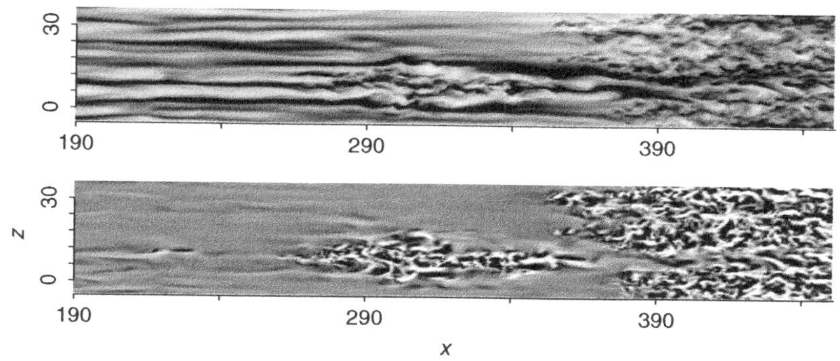


**Figure 6.12** Contours of $u$ (top) and $v$ (bottom) in a plane near the wall under conditions of bypass transition.

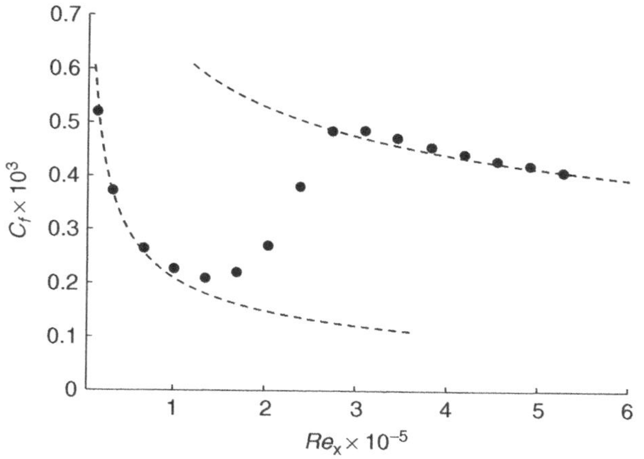


**Figure 6.13** Skin friction versus distance in bypass transition. Symbols are experimental data (Roach and Brierley, 1990).

transition is seen under a stream with low levels of perturbation. Instability waves also dominate in separated flow, and transition often occurs rapidly after separation.

Transition in attached flow has presented a great challenge to modelers. Three approaches have been devised: rely on the closure model to transition from laminar to turbulent solutions; use a data correlation to decide when to switch from laminar to turbulent flow; or devise additional model equations to represent transition. In this last category, two approaches have been pursued. The first is to develop an equation for the intermittency function, $\gamma(\boldsymbol{x}, t)$; the second is to develop an equation for the energy of fluctuations that occur in the laminar region upstream of transition.

### 6.5.1 Reliance on the turbulence model

Can turbulence models describe transition from laminar to turbulent flow? They are developed for fully turbulent conditions and calibrated with turbulence data; the answer would seem to be "no." However, most transport equation models do converge to a laminar solution at low Reynolds number and to a turbulent solution at sufficiently high Reynolds number; the model equations do evidence a transition between laminar and turbulent solution branches. This behavior is illustrated by the calculations of plane channel flow at various Reynolds numbers shown in Figure 6.14. Each point represents a separate computation. When the Reynolds number is low, the models converge to laminar flow; when it is high, they converge to turbulence. The figure contains one model that has not yet been encountered: the isotropization of production (IP) model is a second-moment closure that will be described in Section 7.1.4.

The two solid lines are theoretical and semi-empirical curves of $C_f$ versus $\mathrm{Re} = U_{\mathbb{C}}H/\nu$ for laminar and turbulent flow. The transition between them is set to be $\mathrm{Re} = 975$, which is a typical experimental value. No attempt has been made to smoothly characterize the transition region. All models, except the two-layer $k$–$\varepsilon$, display some form of transition from the laminar line to the turbulent line as Reynolds number increases. In all cases it occurs prior to the experimental critical Reynolds number.

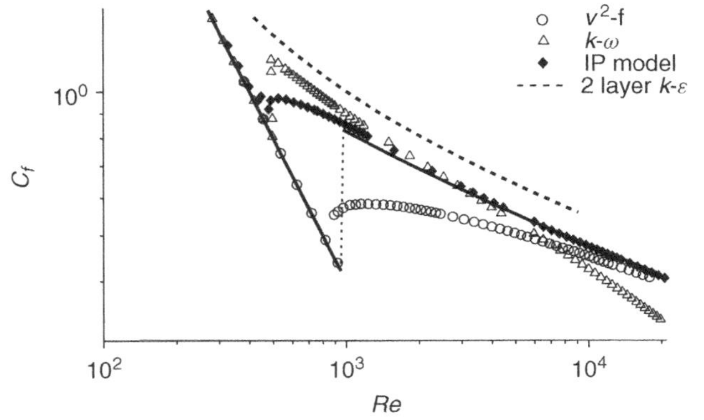


**Figure 6.14** Friction coefficient in a plane channel versus Reynolds number based on centerline velocity and channel half-width. Many turbulence models display a transition from laminar to turbulent solutions as Re increases.

A fully turbulent state is ingrained in the two-layer model through the Van Driest formula for $\ell$. This seems to preclude a plausible laminar solution. The other models obtain length scales from a transport equation. That feature influences their ability to emulate transition, although the precise mathematical origin of that capability is not well understood. The behavior evidenced in Figure 6.14 clearly is a property of the model equations, not of fluid dynamical mechanisms.

It is not uncommon to encounter regions of purely laminar, or buffeted laminar, flow in applications. For instance, turbine blades often operate at low enough Reynolds numbers to encounter significant portions of laminar flow on their surface–the blades are subjected to external turbulence, so their boundary layers are better described as buffeted laminar layers. In such instances, the bulk of the flow may be turbulent and the overall flow calculation must be done with a turbulence model. By what rationale, then, can turbulence models be applied when laminar regions exist?

Bypass transition is stochastic by nature. Turbulent spots are highly localized, irregular motions inside the boundary layer. So this lies within the province of statistical fluid dynamics. Turbulence models are not entirely irrelevant; but neither are they entirely justified. Very often the models are solved without revision, depending on behavior analogous to Figure 6.14 to accommodate the buffeted laminar regions. But when accurate predictions of the laminar and transitional regions are needed, the turbulence model must be supplemented explicitly by a method to predict transition.

One approach is to switch from a laminar to a turbulent computation at a prescribed transition point. For boundary layers under free-stream turbulence, the data correlation

$$R_{\theta \mathrm{tr}} = 163 + \mathrm{e}^{6.91-\mathrm{Tu}} \tag{6.5.1}$$

was proposed by Abu-Gannam and Shaw (1980) for zero pressure-gradient boundary layers. Tu is the turbulence intensity in percentage, $100\sqrt{\overline{u^2}}/U$, measured in the free stream. Transition is specified to occur where the local momentum thickness Reynolds number exceeds the above critical value.

Another approach is to modulate either the eddy viscosity or the production term in the $k$ equation to increase it from zero to its full value across a transition zone. The basic idea is to introduce an intermittency function, $\gamma$, that increases from zero to unity, and to replace the eddy viscosity by $\gamma \nu_{\mathrm{T}}$. It is not clear in general how to predict the transition location, aside from specifications such as (6.5.1). If the transition has been predicted to occur at $x_{\mathrm{tr}}$, formulas like

$$\gamma = 1 - \mathrm{e}^{-(x-x_{\mathrm{tr}})^2/\ell_{\mathrm{tr}}^2}, \qquad x \geq x_{\mathrm{tr}}, \tag{6.5.2}$$

have been used to ramp the eddy viscosity. Here $\ell_{\mathrm{tr}}$ is a transition length, which has been estimated to be about 126 times the momentum thickness ($\ell_{\mathrm{tr}} = 126\theta$) in zero pressure-gradient boundary layers.

### 6.5.2 Intermittency equation

The intermittency formula (6.5.2) can be developed into a transport equation. Note that

$$\frac{\mathrm{d}\gamma}{\mathrm{d}x} = 2\frac{x - x_{\mathrm{tr}}}{\ell_{\mathrm{tr}}^2}\,\mathrm{e}^{-(x-x_{\mathrm{tr}})^2/\ell_{\mathrm{tr}}^2} = \frac{2(\gamma - 1)}{\ell_{\mathrm{tr}}}\left[-\log(1-\gamma)\right]^{1/2}$$

for $x > x_{tr}$. If $x$ is regarded as the streamwise direction, this can be generalized to

$$\boldsymbol{u} \cdot \nabla\gamma = |\boldsymbol{u}|\frac{2(\gamma - 1)}{\ell_{tr}}\left[-\log(1-\gamma)\right]^{1/2}.$$

If $\gamma$ is small, $\sqrt{-\log(1-\gamma)} \approx \sqrt{\gamma}$. Adding a diffusion term provides a transport equation

$$D_t\gamma = 2(\gamma - 1)\sqrt{\gamma}\,\frac{|\boldsymbol{u}|}{\ell_{tr}} + \nabla \cdot [(\nu + \nu_T)\nabla\gamma]. \tag{6.5.3}$$

This is a starting point for more elaborate formulations. Steelant and Dick initiated this approach (see Menter *et al.*, 2004). The transport equation controls the rise of $\gamma$ from zero in laminar flow to unity in turbulent flow. The onset of transition is still determined by a data correlation like (6.5.1).

An alternative statement of the source term in (6.5.3) is (Menter *et al.*, 2004)

$$D_t\gamma = 2|\boldsymbol{S}|(\gamma - 1)\sqrt{\gamma f_{\text{onset}}}\, f_{\text{length}} + \nabla \cdot [(\nu + \nu_T)\nabla\gamma]. \tag{6.5.4}$$

This recognizes that the factor $|u|/\ell_{tr}$ in the previous transport equation determines the length of transition. Here $|\boldsymbol{S}|$ is the magnitude of the rate-of-strain tensor: $|\boldsymbol{S}|^2 = S_{ij}S_{ji}$. It is introduced for dimensional consistency and because turbulence is produced by mean straining. In (6.5.4), $f_{\text{onset}}$ implements a criterion to initiate transition.

The functions $f_{\text{onset}}$ and $f_{\text{length}}$ are empirical. Suluksna *et al.* (2009) have provided concrete formulas by fitting experimental data. They suggest

$$f_{\text{length}} = \min[0.1\,\mathrm{e}^{-0.022R_{\theta tr}+12} + 0.45, 300].$$

This decreases from 300 at low Reynolds number to 0.45 at high Reynolds number, and is unity at $R_{\theta tr} \approx 500$ – which by (6.5.1) occurs for a turbulent intensity of about 1%.

The onset function is rather more convoluted:

$$\begin{aligned} f_{\text{onset}} &= \max[f_{\text{onset2}} - f_{\text{onset3}}, 0], \\ f_{\text{onset2}} &= \min[\max[R_1, R_1^4], 2], \\ f_{\text{onset3}} &= \max[1 - (\nu_T/2.5\nu)^3, 0], \end{aligned} \tag{6.5.5}$$

where

$$R_1 = \frac{R_\nu}{R_{\theta c}}.$$

Suluksna *et al.* (2009), and others, replace the dependence of transition on momentum thickness with dependence on a local parameter

$$R_\nu = \frac{|S|d^2}{2.193\nu},$$

where $d$ is distance from the wall. In the Blasius boundary layer, $\max_y R_\nu = R_\theta$. In the laminar region, $\nu_T \ll \nu$ and $R_1 < 1$. Then $f_{\text{onset}} = R_1$. So transition begins where $R_\nu \sim R_{\theta c}$. In order to match data, the critical Reynolds number $R_{\theta c}$ is less than the transition Reynolds number $R_{\theta\text{tr}}$. A formula is provided in Suluksna *et al.* (2009). For low transition Reynolds numbers $R_{\theta c} \approx 0.8 R_{\theta\text{tr}}$.

Both the transition and critical Reynolds numbers depend on pressure gradient and ambient fluctuations. Correlations can be found in the cited literature. Adverse pressure gradient promotes transition and favorable pressure gradient delays it.

The intermittency function suppresses the eddy viscosity where $\gamma < 1$. For this approach to work, the eddy viscosity model must predict early transition; then $\nu_T$ will reach turbulent levels and $\gamma \nu_T$ can increase from 0 to $\nu_T$ under the control of $\gamma$. Fortunately, most eddy viscosity closure models predict early transition. The SST model (page 138) was used by Menter *et al.* (2004).

The critical Reynolds number is not predicted by the $\gamma$ equation. It is provided externally by a data correlation. Hence, the same model has been applied to orderly and to bypass transition. In each case, a suitable transition function is prescribed.

### 6.5.3 Laminar fluctuations

A second approach has closer connection to the phenomenology of transition. It postulates a transport equation for the energy of fluctuations in the laminar boundary layer – be they Klebanoff modes or instability waves.

These fluctuations grow and produce turbulent kinetic energy. The key elements of the equation for laminar fluctuations are production and transfer to turbulence. Walters and Cokaljat (2008) propose the form

$$\mathrm{D}_t k_\mathrm{L} = 2\nu_{\mathrm{T}\ell}|S|^2 - R - D_\mathrm{L} + \nabla \cdot (\nu \nabla k_\mathrm{L}), \tag{6.5.6}$$

in which $2\nu_{\mathrm{T}\ell}|S|^2$ is the rate of production of laminar fluctuations, $D_\mathrm{L}$ is a dissipative term, and $R$ will be described below. To accommodate both bypass and orderly transition, $\nu_{\mathrm{T}\ell}$ has two components,

$$\nu_{\mathrm{T}\ell} = \nu_{\mathrm{BP}} + \nu_{\mathrm{ord}},$$

associated with large-scale eddies and with instability.

Initially, the large-scale eddies are contained in free-stream turbulence. Klebanoff modes are spawned by these large-length-scale motions. The model is motivated by this phenomenology. Walters and Cokaljat (2008) write

$$\nu_{\mathrm{BP}} = 3.4 \times 10^{-6} f_{\tau\ell} \frac{\Omega \lambda_{\mathrm{eff}}^2}{\nu} \sqrt{k_{\mathrm{T}\ell}} \lambda_{\mathrm{eff}}, \tag{6.5.7}$$

with

$$\lambda_{\mathrm{eff}} = \min[2.495d, \sqrt{k}/\omega]$$

providing a length scale; and

$$k_{\mathrm{T}\ell} = k\left[1 - \left(\frac{\lambda_{\mathrm{eff}}}{L}\right)^{2/3}\right],$$

where $L = \sqrt{k}/\omega$, representing the large-scale component of the turbulent kinetic energy. The laminar fluctuation equation (6.5.6) is conjoined with the $k$–$\omega$ model.

In (6.5.7), $\Omega$ is the magnitude of the vorticity vector; and $f_{\tau\ell}$ is the damping function

$$f_{\tau\ell} = 1 - \exp\left[-4360\frac{k_{T\ell}}{2\lambda_{\text{eff}}^2|S|^2}\right].$$

The numerical coefficients were adjusted to fit data. They control the onset of transition similarly to functions like $f_{\text{onset}}$ in intermittency models. The component $\nu_{BP}$ becomes small where $d$ is large and where $d$ is small. This mimics the experimental observation that Klebanoff modes develop in the central part of the boundary layer.

Orderly transition is incorporated by

$$\nu_{\text{ord}} = 10^{-10}\beta_L\frac{\Omega d^2}{\nu}\Omega d^2,$$

with

$$\beta_L = \begin{cases} 0, & R_\Omega < 1000, \\ 1 - e^{-(0.005R_\Omega - 5)}, & R_\Omega > 1000, \end{cases} \tag{6.5.8}$$

where $R_\Omega = \Omega d^2/\nu$. This acts analogously to an instability criterion. In a Blasius boundary layer, $\max_y R_\Omega = 2.193R_\theta$. Thus the instability criterion is $R_\theta > 456$ (which is higher than the value of 200 from linear stability theory).

The term $R$ in Eq. (6.5.6) represents breakdown of laminar fluctuations into turbulence. The same term, with a positive sign, is added to the turbulent energy equation: $D_t k = P + R - \varepsilon \cdots$. Its form is

$$R = 0.21B_L\frac{K_L}{\tau_T},$$

where $\tau_T = \lambda_{\text{eff}}/\sqrt{k}$. As the turbulent energy grows, $\tau$ decreases, draining energy from the laminar fluctuations. The coefficient $B_L$ controls the onset of transition:

$$B_L = \begin{cases} 0, & R_k < 35, \\ 1 - e^{-(R_k - 35)/8}, & R_k > 35, \end{cases} \tag{6.5.9}$$

where $R_k = \sqrt{k}\,d/\nu$. The transition criterion is based on the Reynolds number $R_k$, which contains wall distance as well as turbulent energy. Thus breakdown initiates well above the wall, as occurs in experiments.

Walters and Cokaljat (2008) also modify $R$ for orderly transition, and introduce other limiting and interpolation functions to improve agreement with data.

## 6.6 Eddy viscosity transport models

Two equation models construct an eddy viscosity from velocity and time-scales. It might seem prudent to formulate a transport equation directly for the eddy viscosity. That idea has been proposed several times in the past. The most recent incarnation, initiated by Baldwin and Barth (1990), has proved quite effective. Flaws to the initial formulation were rectified by Spalart and Allmaras (1992), to produce the model described here. The

Spalart–Allmaras (SA) model has enjoyed great success in predicting aerodynamic flows (Bardina *et al.*, 1997).

Assume *a priori* that an effective viscosity, $\tilde{\nu}$, satisfies a prototype transport equation

$$\partial_t \tilde{\nu} + \boldsymbol{U} \cdot \nabla \tilde{\nu} = \mathcal{P}_\nu - \varepsilon_\nu + \frac{1}{\sigma_\nu}\left[\nabla((\nu + \tilde{\nu})\nabla \tilde{\nu}) + c_{b2}|\nabla \tilde{\nu}|^2\right]. \tag{6.6.1}$$

Aside from the term multiplying $c_{b2}$, this is analogous to the $k$, $\varepsilon$, or $\omega$ equations: the right-hand side consists of production, destruction, and transport. The $c_{b2}$ term is added to control the evolution of free-shear layers. This equation has a propagating front solution (Section 6.2.3), with the propagation speed depending on $c_{b2}$. Spalart and Allmaras (1992) argue that the front speed influences shear-layer development, and choose $c_{b2} = 0.622$, in conjunction with $\sigma_\nu = 2/3$, to obtain a good representation of the velocity profiles in wakes and mixing layers.

The cleverness in developing an equation for an effective viscosity, rather than the actual eddy viscosity, is that a numerical amenity can be added. Baldwin and Barth (1990) proposed to make $\tilde{\nu}$ vary linearly throughout the law-of-the-wall layer; in particular, to retain nearly the log-layer dependence $\tilde{\nu} = \kappa u_* y$ all the way to the wall. A nearly linear function can be discretized very accurately.

For production, choose the dimensionally consistent form

$$\mathcal{P}_\nu = c_{b1} S \tilde{\nu}.$$

Spalart and Allmaras (1992) selected the constant to be $c_{b1} = 0.1355$, which gave a good spreading rate of free-shear layers. They chose $S$ to be the magnitude of the mean vorticity. In a boundary layer, this is equivalent to $S = |\partial_y U|$, but the potential for spurious production to occur in applications with rotating surfaces exists – see the discussion in Section 6.4.

The wall distance $y$ is used for a length scale in the destruction term

$$\varepsilon_\nu = c_{w1} f_w \left(\frac{\tilde{\nu}}{y}\right)^2.$$

The function $f_w$ will be specified below. It is required to be unity in the region where $\tilde{\nu} \propto y$. Hence, in an equilibrium constant-stress layer, the model reduces to

$$0 = c_{b1} \tilde{S} \tilde{\nu} - c_{w1} \left(\frac{\tilde{\nu}}{y}\right)^2 + \frac{1}{\sigma_\nu}\left[\nabla((\nu + \tilde{\nu})\nabla \tilde{\nu}) + c_{b2}|\nabla \tilde{\nu}|^2\right]. \tag{6.6.2}$$

The need to introduce an effective vorticity, $\tilde{S}$, as done in this equation, will be described shortly. The log-layer solution should be $\tilde{\nu} = u_* \kappa y$ and $\tilde{S} = u_*/\kappa y$ (see Section 4.1.1). Substitution into Eq. (6.6.2) shows that these constraints require

$$c_{w1} = c_{b1} \kappa^{-2} + (1 + c_{b2})/\sigma_\nu. \tag{6.6.3}$$

Now the intriguing part: as already mentioned, numerical amenability is one of the guiding principles to this model – it is a good principle for all practical closure modeling. To that end, $\tilde{\nu}$ is contrived to vary nearly linearly all the way to the wall: in particular, the solution $\tilde{\nu} = u_* \kappa y$ is nearly retained. Substituting $\tilde{\nu} = u_* \kappa y$ into Eq. (6.6.2) shows that the effective vorticity must also follow its log-layer form $\tilde{S} = u_*/\kappa y$. Assume that the modified and unmodified mean vorticities are related by

$$\tilde{S} = S + F(\tilde{\nu}, y). \tag{6.6.4}$$

A form for $F$ is now sought.

In a constant-stress layer

$$(\nu_T + \nu)S = u_*^2.$$

The friction velocity, $u_*$, is eliminated by equating it to $\tilde{\nu}/\kappa y$; then the above becomes

$$(\nu_T + \nu)S = \tilde{\nu}^2/(\kappa y)^2. \tag{6.6.5}$$

The same substitution is made in the desired behavior, $\tilde{S} = u_*/\kappa y$: by (6.6.4)

$$S + F(\tilde{\nu}, y) = \tilde{\nu}/(\kappa y)^2.$$

After $S$ is eliminated via (6.6.5), the function becomes

$$F(\tilde{\nu}, y) = \frac{\tilde{\nu}}{(\kappa y)^2} - \frac{\tilde{\nu}^2}{(\nu_T + \nu)(\kappa y)^2}.$$

Combining this with (6.6.4) gives

$$\tilde{S} = S - \frac{\tilde{\nu}^2}{(\nu_T + \nu)(\kappa y)^2} + \frac{\tilde{\nu}}{(\kappa y)^2} \tag{6.6.6}$$

for the additive term in (6.6.4). Next to a wall, the second factor on the right-hand side cancels $S$ and the third provides the desired $\tilde{S}$.

A formulation with a nearly linear solution near the wall is attractive; but the real eddy viscosity is *not* linear near the wall. This is rectified by a nonlinear transformation: let

$$\nu_T = \tilde{\nu} f_\nu(\tilde{\nu}/\nu)$$

define the actual eddy viscosity. The argument of the function is a turbulent Reynolds number, so this is a bit like the damping function (6.2.20), but different in spirit. The transformation function

$$f_\nu(\tilde{\nu}/\nu) = \frac{(\tilde{\nu}/\nu)^3}{(\tilde{\nu}/\nu)^3 + 7.1^3} \tag{6.6.7}$$

was borrowed by Spalart and Allmaras (1992) from an earlier algebraic model of Mellor and Herring (1973).

The model is almost complete. It is

$$\partial_t \tilde{\nu} + \boldsymbol{U} \cdot \nabla \tilde{\nu} = \tilde{S}\tilde{\nu} - c_{w1} f_w \left(\frac{\tilde{\nu}}{d}\right)^2 + \frac{1}{\sigma_\nu}\left[\nabla((\nu + \tilde{\nu})\nabla\tilde{\nu}) + c_{b2}|\nabla\tilde{\nu}|^2\right], \tag{6.6.8}$$

with (6.6.6) to define the effective vorticity and (6.6.7) to transform to the true eddy viscosity. The coordinate $y$ has been replaced by the minimum distance to a wall, $d$.

The only remaining element is to specify the function $f_w$ in front of the destruction term. That function implements a constraint that the wall distance should drop out of the model far from the surface. Near to the wall, $f_w$ is required to be unity, so that the linear solution obtains; far from the wall $f \to 0$. Spalart and Allmaras (1992) make $f_w$ a function of $r \equiv \tilde{\nu}/\tilde{S}(\kappa d)^2$. The particular function has an arbitrary appearance. It is

$$f_w(r) = g\left[\frac{65}{g^6 + 64}\right]^{1/6}, \qquad \text{with} \qquad g = r + 0.3(r^6 - r). \tag{6.6.9}$$

In addition to the constraints $f_w(1) = 1$ and $f_w(0) = 0$, this function was selected to provide accurate agreement with the skin friction curve (Figure 6.8) beneath a flat-plate boundary layer.

An application of the SA model to the flow in a transonic turbine passage is illustrated in Figure 6.15. Flow is from bottom to top, angled 64° to the left at the inlet. The view is down the axis of a blade that is attached to a wall. The contours lie on the endwall. They show the distribution of non-dimensional heat-transfer coefficient. The flow accelerates as it runs between the blades, causing a heat-transfer maximum in the passage. The high heat transfer near the blunt leading edge of the blade is due to a vortex that wraps around it.

The performance of the model is quite good, given the complexity of the flow. It is especially gratifying to know that basic considerations about zero pressure-gradient boundary layers and simple free-shear layers can produce a model that functions well so far from its roots. Generally, turbulence models can be considered to be elaborate schemes to predict complex flows by incorporating data from much simpler flows. That is both the intellectual attraction and the practical value of this field.

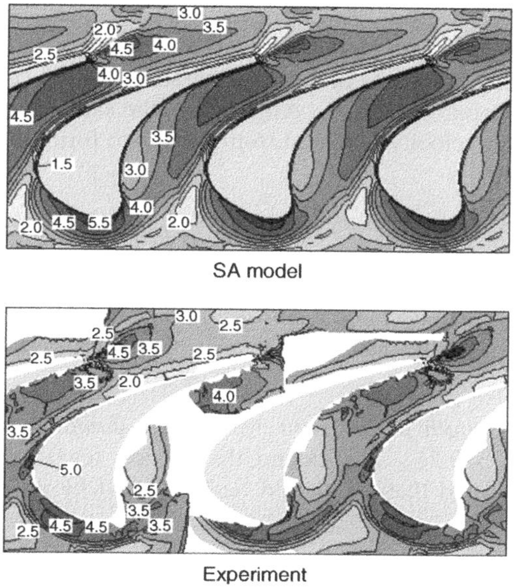


**Figure 6.15** Heat transfer in a transonic turbine passage: contours of Stanton number by SA model and experimental data courtesy of P. Giel (Giel *et al.*, 1998). Inflow is from bottom to top, angled 64° to the left. Calculation by G. Iaccarino and G. Kalitzin.

## Exercises

**Exercise 6.1.** *Integral equation closure.* Write a program to solve the model (6.1.1) through (6.1.3).

Plot solutions for $\theta(x/a)$, $H(x/a)$, and $C_f(x/a)$ with $U_\infty/U_0 = (1 - x/a)$, where $a$ is a characteristic length and $U_0$ a characteristic velocity. Start with $C_f(0) = 3 \times 10^{-3}$, $H(0) = 1.35$, and $R_a = 10^5$. (What is $\theta(0)/a$?) Stop the computation when the boundary layer separates. At what $x/a$ does separation occur?

Also plot these variables with $U_\infty/U_0 = (1 + x/a)$ in the range $0 \le x/a \le 0.5$. Discuss the effect of favorable and adverse pressure gradients on $C_f$ and $H$.

**Exercise 6.2.** *Golf ball revisited.* In Chapter 1, Exercise 1.3 asked why golf balls have dimples. An extension to Figure 6.2 provides an answer: the dimples promote transition. Provide a qualitative explanation of the figure given below.

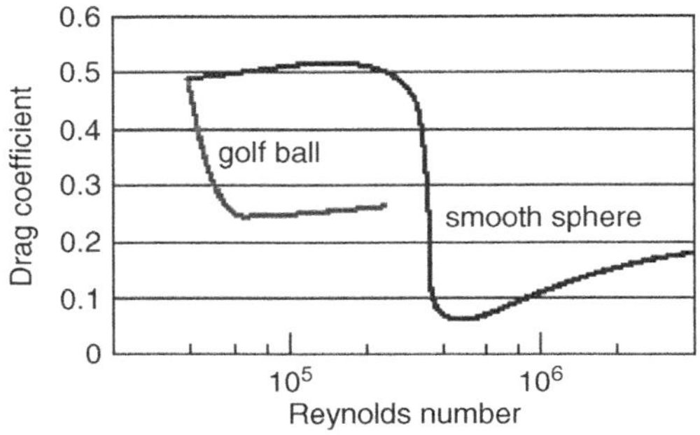


**Exercise 6.3.** *The mixing length model.* Show that, if $\ell_m = \kappa y$, then the mixing length formula (6.1.6) gives a log law for $U$ in the constant-stress layer ($-\overline{uv} = u_*^2$). Johnson and King (1985) suggested using $\nu_T = u_* \ell_m$, which gives the same result in a constant-stress layer. Compare the mean velocity obtained from these two formulations in a linear-stress layer, $-\overline{uv} = u_*^2 + \alpha y$.

**Exercise 6.4.** *The Stratford boundary layer.* Further to Exercise 6.3, Stratford proposed a clever idea for reducing skin friction. He suggested that a carefully designed adverse pressure gradient could reduce $u_*^2$ almost to zero along a substantial length of the boundary. The limiting case $u_*^2 = 0$ is called the *Stratford boundary layer*. The total stress then varies linearly, $\tau \propto y$, near the wall. What mean flow profile does the mixing length model predict in this case?

**Exercise 6.5.** *Mixing length model with Van Driest damping.* The log law applies in the region $40 \lesssim y_+ \lesssim 0.2\delta_{99}^+$. To extend the mixing length model all the way to $y = 0$, Van Driest suggested that $\kappa y$ should be multiplied by an exponential "damping function." Thus

$$\ell_m = \kappa y(1 - e^{-y_+/A_+}),$$

where $\kappa = 0.41$. Apply this to the constant total stress layer, $-\overline{uv} + \nu\, \partial_y U = u_*^2$ to find a formula for $\partial_y U$. Integrate this numerically and make a log–linear plot of $U_+$ versus $y_+$. What is the additive constant $B$ that you obtain for the log law if $A_+ = 26$? What value of $A_+$ gives $B = 5.5$?

**Exercise 6.6.** *A shortcoming to the $k-\varepsilon$ model.* Show that in incompressible flow the eddy viscosity constitutive formula (6.2.3) gives the rate of turbulent energy production as

$$\mathcal{P} = 2\nu_T S_{ij} S_{ji} \equiv 2\nu_T S_{ii}^2$$

irrespective of $\Omega_{ij}$ (see Eq. (3.3.8)). When turbulence is rotated, the centrifugal acceleration can affect the turbulent energy. Discuss whether the $k-\varepsilon$ model can predict such effects. There is an analogy between rotation and streamline curvature, so your conclusions apply to the limited ability of eddy viscosity to represent effects of curvature on the turbulence as well.

**Exercise 6.7.** *Realizability of $k-\varepsilon$.* Consider the $k-\varepsilon$ model (6.2.5) and (6.2.6) for the case of *homogeneous* turbulence. Prove that, if $k$ and $\varepsilon$ are greater than zero initially, they cannot subsequently become negative.

**Exercise 6.8.** *Bounds on production.* Show that

$$|\mathcal{P}| \leq 2k|\lambda_{\max}|,$$

where $|\lambda_{\max}|$ is the eigenvalue of the rate-of-strain tensor $\boldsymbol{S}$ with maximum absolute value. This suggests that eddy viscosity models should be constrained by

$$2\nu_T|\boldsymbol{S}|^2 \leq 2k|\lambda_{\max}|.$$

Let the eigenvalues of $\boldsymbol{S}$ be $\lambda_1$, $\lambda_2$, and $\lambda_3$. For incompressible flow, $\lambda_1 + \lambda_2 + \lambda_3 = 0$. If $\lambda_3 = 0$, show that

$$|\lambda_{\max}|^2 = |\boldsymbol{S}|^2/2$$

and generally that

$$|\boldsymbol{S}|^2 \geq 3|\lambda_{\max}|^2/2,$$

where $|\lambda_{\max}|$ is the maximum eigenvalue of $\boldsymbol{S}$. Hence conclude that the eddy viscosity ought to satisfy

$$\nu_T \leq \frac{2k}{3|\lambda_{\max}|}.$$

For the $k-\varepsilon$ model, $C_\mu \leq 2\varepsilon/(3k|\lambda_{\max}|)$ could be imposed.

**Exercise 6.9.** *Front solution.* Show that the mean shear at the propagating front solution corresponding to (6.2.33) is $\partial_y U \sim (ct - y)^{(\sigma_\varepsilon - 1)/(2-\sigma_\varepsilon)}$, for $y < ct$.

**Exercise 6.10.** *Calibrate an eddy viscosity transport model.* Eddy viscosity transport models, like SA, are an alternative to the $k-\varepsilon$ model for full-blown CFD analysis. They solve a single equation for the dependent variable $\nu_T$. Consider the model equation

$$D_t \nu_T = C_1|\boldsymbol{S}|\nu_T - (\partial_i \nu_T)(\partial_i \nu_T) + \partial_i\left[\left(\frac{\nu_T}{\sigma_\nu} + \nu\right)\partial_i \nu_T\right]$$

or, in vector notation,

$$D_t \nu_T = C_1|\boldsymbol{S}|\nu_T - |\nabla \nu_T|^2 + \nabla \cdot \left[\left(\frac{\nu_T}{\sigma_\nu} + \nu\right)\nabla \nu_T\right].$$

Note that $2|\boldsymbol{S}|^2 = |\partial_y U|^2$ in parallel shear flow.

Here $C_1$ and $\sigma_\nu$ are empirical constants. Use log-layer analysis to obtain a formula relating $C_1$ and $\sigma_\nu$ to the Von Karman constant, $\kappa$. What second experimental datum might be used to determine empirical values for $C_1$ and $\sigma_\nu$?

**Exercise 6.11.** *Channel flow by k–ε.* Use the velocity and length scales $u_*$ and $H$ to set up the non-dimensional $k$–$\varepsilon$ equations that must be solved in conjunction with (4.1.4) for fully developed channel flow. Non-dimensionalize the wall-function boundary conditions (6.2.22). [The $U$ boundary condition simplifies to $U_+(y_+ = 40) = \log(40)/\kappa + B$ in this non-dimensionalization.] Solve the resulting problem numerically for steady channel flow with $R_\tau = 10\,000$. Plot your solution for $U_+$ and $k_+$.

In the region $y_+ < 40$ use the profile computed in Exercise 6.5 and plot $U_+$ versus $y_+$ all the way to the wall.

Note that you are solving a pair of nonlinear diffusion equations; standard methods for that type of equation can be used. One simple approach is to retain the time derivative and use an Euler implicit time integration to steady state (Crank–Nicholson can be slow to converge). Central differencing in $y$ with tridiagonal, matrix inversion is quite effective. With this method, it helps if $\varepsilon$ is made implicit in the $k$ equation, either by using a block tridiagonal solver, or by solving the $\varepsilon$ equation before the $k$ equation.

**Exercise 6.12.** *Channel flow by the two-layer approach.* Repeat the previous exercise for the two-layer $k$–$\varepsilon$ model.

**Exercise 6.13.** *Turbulent dispersion.* Stochastic models provide a concrete physical representation of the ensemble-averaged smoothing effect of random convection. The following is simpler than Taylor's model. The position of a convected fluid element is given by

$$Y(t + \mathrm{d}t) = Y(t) + \sqrt{\alpha(t)\,\mathrm{d}t}\,\xi(t),$$

in which $\xi(t)$ is the random process defined below Eq. (2.2.10).

(i) Show that $\alpha(t) = 2K_{\mathrm{T}}$, where $K_{\mathrm{T}}$ is the eddy diffusivity.

(ii) Use the $k$–$\varepsilon$ formula $K_{\mathrm{T}} = C_\mu k^2/\varepsilon$ to show that, in decaying grid turbulence,

$$\overline{Y^2} \propto (t + t_0)^{(2C_{\varepsilon_2}-3)/(C_{\varepsilon_2}-1)}.$$

**Exercise 6.14.** *Non-existence of* 1/2 *power law.* In a linear-stress layer the total shear stress varies as $\tau = \alpha y$, with $\alpha$ being a constant that has dimensions $\ell/t^2$. By dimensional reasoning, $U \propto (\alpha y)^{1/2}$ and

$$k \propto \alpha y, \qquad \varepsilon \propto \alpha^{3/2} y^{1/2}, \qquad \omega \propto (\alpha/y)^{1/2}$$

(see Exercise 6.4). Show that the $k$–$\varepsilon$ and $k$–$\omega$ models do not admit a power-law solution of this form.

# 7

# Models with tensor variables

Try a new system or a different approach.

–Fortune Cookie

## 7.1 Second-moment transport

The limitations to the $k$–$\varepsilon$ model stem largely from the turbulence being represented by its kinetic energy, which is a scalar, and from the eddy viscosity assumption (6.2.3). The former does not correctly represent turbulence anisotropy. The latter assumes an instantaneous equilibrium between the Reynolds stress tensor and the mean rate of strain.

A shortcoming to representing the turbulent velocity fluctuations solely by the scalar magnitude $k$ is that sometimes an external force acts on one component more strongly than on others, producing very different component energies, $\overline{u^2} \neq \overline{v^2} \neq \overline{w^2}$. This is referred to as "normal stress anisotropy," alluding to the departure of the diagonal components of the Reynolds stress tensor from their isotropic form $\overline{u_i u_j} = \frac{2}{3}k\delta_{ij}$. For example, stable density stratification in the $y$ direction will suppress $\overline{v^2}$; similarly, stable streamline curvature will suppress the component directed toward the center of curvature.

Anisotropy exists in all real flows. In parallel shear flows, the dominant anisotropy is the shear stress. Eddy viscosity models are designed to represent shear stress; they are not designed to represent normal stress anisotropy. Second-moment closure (SMC)* incorporates many of these effects because it is based on the Reynolds stress transport equations (Chapter 3). For instance, curvature effects enter these equations through the production tensor $\mathcal{P}_{ij}$. The price paid for the increased physical content of the model is that more equations must be solved. The mathematics also become more intricate. Hanjalic (1994) discusses further pros and cons to developing models for the transport of Reynolds stresses.

* Also called second-order closure, Reynolds stress modeling (RSM), or Reynolds stress transport (RST).

*Statistical Theory and Modeling for Turbulent Flows, Second Edition* P. A. Durbin and B. A. Pettersson Reif

Another shortcoming to the eddy viscosity representation is that it causes the Reynolds stresses to change instantaneously when the mean rate of strain changes. Disequilibrium should come into play in rapidly changing flow conditions. Algebraic constitutive models, in general, assume an instantaneous equilibrium between turbulent stress and mean rate of strain–and mean rate of rotation in some cases, as formula (2.3.16). A frequently cited example of misalignment between stress and rate of strain is the three-dimensional boundary layer. As the boundary layer progresses downstream, the direction of the mean flow veers. This might be caused by an obstruction or by a pressure gradient transverse to the flow. The angle of the horizontal mean shear from the $x$ direction is $\tan^{-1}(\partial_y W/\partial_y U)$. The angle of the horizontal Reynolds shear stress, $\tan^{-1}(\overline{wv}/\overline{uv})$, also veers, but it lags behind the direction of the mean shear. Equation (6.2.3) predicts $-\overline{uv} = \nu_T\, \partial_y U$ and $-\overline{vw} = \nu_T\, \partial_y W$, so these two angles are assumed to be equal: the linear eddy viscosity formula does not account for the lag.

### 7.1.1 A simple illustration

We start by illustrating the ability of SMC to introduce disequilibrium between stress and mean rate of strain; for instance, to allow for the lag in three-dimensional boundary layers. Suppose that, instead of imposing the eddy viscosity constitutive relation (6.2.3), the Reynolds stresses were allowed to relax to that relation on the time-scale $T$. The linear relaxation of a variable $X$ to an equilibrium $X_{\text{eq}}$ is described by

$$\mathrm{d}_t X + \frac{C_1 X}{T} = \frac{C_1 X_{\text{eq}}}{T}.$$

Writing the eddy viscosity as $\nu_T = C_\mu k T$ and introducing relaxation into (6.2.3) suggests the evolution equation

$$\mathrm{d}_t[\overline{u_i u_j} - \tfrac{2}{3}\delta_{ij}k] + C_1 \frac{[\overline{u_i u_j} - \frac{2}{3}\delta_{ij}k]}{T} = -2C_\mu C_1 k S_{ij}. \tag{7.1.1}$$

The bracketed expression is the dependent variable; it is just the departure of $\overline{u_i u_j}$ from its isotropic value. $C_1$ is an empirical constant that was introduced so that the relaxation time-scale is proportional to $T$ instead of exactly equaling it. The constant $C_1$ is usually referred to as the "Rotta constant" after a pioneering paper by J. C. Rotta. It is readily verified that, in the steady state, where $\mathrm{d}_t[\overline{u_i u_j} - \frac{2}{3}\delta_{ij}k] = 0$, Eq. (7.1.1) reduces to (6.2.3). Contracting the subscripts and using the incompressibility constraint, $S_{ii} = 0$, verifies that (7.1.1) preserves the identity $\overline{u_i u_i} = 2k$.

If isotropic turbulence with initial state $(\overline{u_i u_j})_0 = \frac{2}{3}\delta_{ij}k_0$ were subjected to a sudden mean rate of strain, (6.2.3) predicts that the Reynolds stresses would change instantly to $\overline{u_i u_j} = \frac{2}{3}\delta_{ij}k - 2\nu_T S_{ij}$. The non-equilibrium model (7.1.1) allows them to adjust with time. The first two terms of a Taylor series expansion in time give the solution

$$\overline{u_i u_j} = \tfrac{2}{3}\delta_{ij}k_0 - 2C_\mu C_1 k_0 S_{ij} t.$$

The off-diagonal stresses grow linearly with time; for example, $-\overline{u_1 u_2} = 2C_\mu C_1 k_0 S_{12} t$. This is of the eddy viscosity form, but the time-scale is $C_1 t$ instead of $T$. Thus if the direction of the mean rate of strain veers, as in a three-dimensional boundary layer, the direction of the Reynolds stress will follow, but now with a time lag.

The simplistic evolution equation (7.1.1) introduces the basic idea of incorporating temporal relaxation effects into the constitutive relation between stress and rate of strain. However, there are other physics that are as desirable. For instance, the model should respond to streamline curvature and system rotation, as well as to skewing of the flow direction. A systematic approach is to develop the model from the exact, but unclosed, Reynolds stress transport equations. That is the topic of this chapter.

### 7.1.2 Closing the Reynolds stress transport equation

The essence of second-moment closure modeling can be described by reference to homogeneous turbulence. Our discussion begins there. Under the condition of homogeneity, the exact transport equation is (3.3.10), page 54:

$$\partial_t \overline{u_i u_j} = \mathcal{P}_{ij} + \underbrace{\left(\tfrac{2}{3}\varepsilon\delta_{ij} - \phi_{ij} - \varepsilon_{ij}\right)} - \tfrac{2}{3}\varepsilon\delta_{ij}. \tag{7.1.2}$$

Note that $\varepsilon$ is the dissipation rate of $k$ so that $\varepsilon \equiv \frac{1}{2}\varepsilon_{ii}$. Indeed, the trace of (7.1.2) is two times the $k$ equation. In that sense, SMC modeling can be looked on as unfolding the $k$–$\varepsilon$ model to recover a better representation of stress anisotropy.

The dependent variable in (7.1.2) is $\overline{u_i u_j}(t)$. The underbrace indicates the only new unclosed term. It is assumed that the $\varepsilon$ equation is retained as (6.2.6), or something similar, so that the last term of (7.1.2) has already been closed. The mean flow gradients must be spatially constant in homogeneous turbulence, and therefore can be considered as given; they prescribe the type of flow, such as homogeneous shear, strain, and so on. The explicit form for the production tensor is stated in (3.3.5) to be $\mathcal{P}_{ij} = -\overline{u_j u_k}\,\partial_k U_i - \overline{u_i u_k}\,\partial_k U_j$, which involves the dependent variable $\overline{u_i u_j}$ and the given flow gradients; hence, it is a closed term.

Modeling involves developing formulas and equations to relate the unclosed term to the mean flow gradients and to the dependent variable, $\overline{u_i u_j}$. Denote the unclosed term by

$$\wp_{ij} = -(\phi_{ij} + \varepsilon_{ij} - \tfrac{2}{3}\varepsilon\delta_{ij}). \tag{7.1.3}$$

This will be referred to as the redistribution tensor. In Section 3.3, $\phi_{ij}$ alone was called redistribution. For the purpose of modeling, dissipation anisotropy has been lumped in as well. Then (7.1.2) can be written compactly as

$$\partial_t \overline{u_i u_j} = \mathcal{P}_{ij} + \wp_{ij} - \tfrac{2}{3}\varepsilon\delta_{ij}. \tag{7.1.4}$$

In order to close this evolution equation, the unknown term $\wp_{ij}$ must be modeled in terms of already known quantities. Standard practice is to develop the closure as a sum of a "slow" contribution and a "rapid" contribution. The basic slow model is given by (7.1.8), and the linear, rapid model is either (7.1.32), or its alternative statement as (7.1.33). Operationally, those formulas replace $\wp_{ij}$ in (7.1.4) to provide a closed equation. At this point the reader might want to write out the closed equation for $\overline{u_i u_j}$ as motivation for what follows. The theory behind these models will now be developed.

What is needed is a function of the form

$$\wp_{ij} = F_{ij}(\overline{u_i u_j}, \partial_j U_i; k, \varepsilon, \delta_{ij}).$$

This ought to be written in a dimensionally consistent manner. The redistribution tensor $\wp$ has the same dimensions as the rate of dissipation, $\varepsilon$, which can be used to scale $F_{ij}$. The mean velocity gradient $\partial_j U_i$ has dimensions of $1/t$, so it can be non-dimensionalized as $k\partial_j U_i/\varepsilon$. It is common to non-dimensionalize $\overline{u_i u_j}$ by $k$ and subtract $\frac{2}{3}\delta_{ij}$ to form a trace-free tensor called the anisotropy tensor

$$b_{ij} \equiv \frac{\overline{u_i u_j}}{k} - \frac{2}{3}\delta_{ij} \tag{7.1.5}$$

(see Exercise 3.4). It can be verified that $b_{ii} = 0$. With these non-dimensionalizations, the functional dependence of $\wp_{ij}$ is written more appropriately as

$$\wp_{ij} = \varepsilon \mathcal{F}_{ij}[b_{ij},\ k/\varepsilon\partial_j U_i,\ \delta_{ij}]. \tag{7.1.6}$$

As an example, the relaxation term in (7.1.1) can be written $C_1 b_{ij}\varepsilon$, which corresponds to $\mathcal{F}_{ij} = -C_1 b_{ij}$. A high Reynolds number assumption is implicit in (7.1.6): dependence on molecular viscosity could be added through the turbulent Reynolds number $k^2/\varepsilon\nu$. Again, that parameter is important in near-wall modeling, but it is of secondary interest in homogeneous flows.

There is an implicit assumption of locality in (7.1.6): in the present case of homogeneous flow, it is locality in time. In particular, $\mathcal{F}$ could be a functional of $b_{ij}(t')$, $t' \le t$. However, all SMC models currently in use invoke a temporally local redistribution model; all variables in (7.1.6) are evaluated at the same time $t$. History effects are present, but only through the evolution equation (7.1.4).

Common practice is to separate $\wp$ into slow and rapid contributions, $\wp^{\text{slow}} + \wp^{\text{rapid}}$, and to model their functional dependence, $\mathcal{F}^{\text{slow}} + \mathcal{F}^{\text{rapid}}$, separately. Terms that do not depend on $\partial_j U_i$ are referred to as the *slow terms* of the redistribution model. The rapid terms depend on velocity gradients; they are usually tensorally linear in $\partial_j U_i$. The terminology "rapid" and "slow" originated in the idea that only the former terms alter instantaneously with mean flow changes. The division into slow and rapid parts is addressed in the next two sections.

It is sometimes desirable to let the anisotropy tensor, $b_{ij}$, be the dependent variable rather than $\overline{u_i u_j}$. Equation (7.1.4) can be rearranged as the evolution equation

$$\partial_t b_{ij} = -b_{ik}\,\partial_k U_j - b_{jk}\,\partial_k U_i - \frac{4}{3}S_{ij} - \left(b_{ij} + \frac{2}{3}\delta_{ij}\right)\frac{\mathcal{P}}{k} + b_{ij}\frac{\varepsilon}{k} + \frac{\mathcal{F}_{ij}}{T} \tag{7.1.7}$$

for $b_{ij}$. As previously, $S_{ij}$ is the rate-of-strain tensor. This form invokes the production tensor as a function of $b_{ij}$:

$$\frac{\mathcal{P}_{ij}}{k} = -b_{ik}\,\partial_k U_j - b_{jk}\,\partial_k U_i - \frac{4}{3}S_{ij}.$$

The evolution equation (7.1.7) must preserve the condition that $b_{ii} = 0$: to ensure this, the constraint $\mathcal{F}_{ii} = 0$ must be imposed. Equation (7.1.7) must also preserve the symmetry condition $b_{ij} = b_{ji}$, so $\mathcal{F}_{ij} = \mathcal{F}_{ji}$ is a further constraint. Various other constraints on the form of closure can be identified. These will be discussed in due course.

### 7.1.3 Models for the slow part

The slow term is associated with the problem of *return to isotropy*. To isolate slow terms, consider the case $\partial_j U_i = 0$. Then there is no directional preference imposed on the turbulence and hence no driving force toward anisotropy. Removing the mean shear also means that there is no turbulence production. In that case, (7.1.6) becomes $\wp_{ij} = \varepsilon \mathcal{F}_{ij}[\boldsymbol{b}, \boldsymbol{\delta}]$. The most commonly used form for the slow redistribution is the *Rotta model:*

$$\wp_{ij}^{\text{slow}} = -C_1 \varepsilon b_{ij}. \tag{7.1.8}$$

This is just a linear relaxation of the anisotropy tensor $b_{ij}$ toward 0, or similarly of $\overline{u_i u_j}$ toward $\frac{2}{3}k\delta_{ij}$. For that reason it is called a return-to-isotropy model. Typical empirical values of $C_1$ are in the range 1.5–2.0. As a word of caution, sometimes $q^2 = 2k$ is used to scale the Reynolds stress and $T$ is defined as $q^2/\varepsilon$. Then the Rotta constant will be doubled to 3.0–4.0. This warning should be taken broadly: if there is confusion over a factor of 2 in the constants, check whether $k$ or $q^2$ is used as the intensity scale.

The Rotta model is usually quite effective. However, the representation theorem (2.3.8) derived in Section 2.3.2 shows that the most general functional dependence of the slow redistribution model is

$$\wp_{ij}^{\text{slow}} = -\varepsilon C_1 b_{ij} + \varepsilon C_1^n (b_{ij}^2 - \tfrac{1}{3} b_{kk}^2 \delta_{ij}). \tag{7.1.9}$$

Recall that $b_{ij}^2$ is defined as the matrix times itself: $b_{ij}^2 = b_{ik}b_{kj}$ and that $b_{kk}^2 = b_{km}b_{km}$ is its trace. The term containing $\delta_{ij}$ makes formula (7.1.9) consistent with the trace-free constraint $\wp_{kk} = 0$. The coefficients $C_1$ and $C_1^n$ can be functions of the invariants $II_b = -\frac{1}{2}b_{kk}^2$ and $III_b = \frac{1}{3}b_{kk}^3$, defined below Eq. (2.3.10). The Rotta model is $C_1 =$ constant and $C_1^n = 0$.

While there are grounds for including the $C_1^n$ term, as a practical matter it should be noted that the nonlinearity can adversely affect numerical convergence when SMC models are used in complex geometries. For instance, the SSG (Speziale–Sarkar–Gatski) model (Speziale *et al.*, 1991) suffers from stiffness unless $C_1^n$ is set to zero; doing so has a negligible effect on predictions in wall-bounded flow. Given that caveat, it is instructive to consider the more general form (7.1.9).

In the absence of mean flow gradients, the evolution equation (7.1.7) with (7.1.9) becomes

$$d_t b_{ij} = (1 - C_1)\frac{b_{ij}}{T} + C_1^n \frac{(b_{ij}^2 - \frac{1}{3}b_{kk}^2\delta_{ij})}{T}. \tag{7.1.10}$$

The isotropic state $b_{ij} = 0$ is a solution. The condition that this should be a stable equilibrium is $C_1 > 1$ (see Exercise 7.3). It is an intuitive notion that turbulent scrambling should drive the flow toward isotropy in the absence of external forcing. Experiments on homogeneous turbulence have been performed, in which initially anisotropic turbulence was observed to relax toward isotropy (Townsend, 1976).

A coordinate system can be chosen in which $b_{ij}$ is diagonal because it is a symmetric tensor. This is called the system of principal axes. With mean flow gradients, the direction of the principal axes would rotate in time. However, for (7.1.10) that does not happen because when $i \neq j$ that equation gives $\mathrm{d}_t b_{ij} = 0$ if $b_{ij} = 0$ initially: no off-diagonal components are generated and $\boldsymbol{b}$ remains diagonal. The trace-free condition $b_{kk} = 0$ shows the anisotropy tensor to be of the form

$$\boldsymbol{b} = \begin{vmatrix} b_{11} & 0 & 0 \\ 0 & b_{22} & 0 \\ 0 & 0 & -(b_{11} + b_{22}) \end{vmatrix}. \tag{7.1.11}$$

Note that $b_{kk}^2 = 2(b_{11}^2 + b_{22}^2 + b_{11}b_{22})$ and let $\mathrm{d}\tau = \mathrm{d}t/T$. Then the evolution equations become

$$\begin{aligned} \mathrm{d}_\tau b_{11} &= (1 - C_1)b_{11} + C_1^n[\tfrac{1}{3}b_{11}^2 - \tfrac{2}{3}(b_{22}^2 + b_{11}b_{22})], \\ \mathrm{d}_\tau b_{22} &= (1 - C_1)b_{22} + C_1^n[\tfrac{1}{3}b_{22}^2 - \tfrac{2}{3}(b_{11}^2 + b_{11}b_{22})]. \end{aligned} \tag{7.1.12}$$

If $C_1^n = 0$ and $C_1 > 1$ this represents exponential decay toward isotropy. Before continuing, some properties of the anisotropy tensor warrant discussion. They are embodied in the curvilinear triangle plotted in Figure 7.1, which is explained in the following.

The one-component, two-component, and axisymmetric states are signposts for characterizing anisotropy. From $b_{11} = \overline{u_1^2}/k - \frac{2}{3}$ and $0 \le \overline{u_1^2}/k \le 2$ it follows that $b_{11}$ falls between $-\frac{2}{3}$ and $\frac{4}{3}$. When $b_{11} = -\frac{2}{3}$ the turbulence has two non-zero components, $\overline{u_2^2}$ and $\overline{u_3^2}$; when $b_{11} = \frac{4}{3}$ it has only one, $\overline{u_3^2} = \overline{u_2^2} = 0$ and $\overline{u_1^2} = 2k$. These define the one- and two-component states.

The other special state is the axisymmetric condition $b_{11} = b_{22}$; then $\boldsymbol{b}$ has the form

$$\boldsymbol{b} = \begin{vmatrix} b_{11} & 0 & 0 \\ 0 & b_{11} & 0 \\ 0 & 0 & -2b_{11} \end{vmatrix}.$$

If $b_{11} < 0$ then $\overline{u_1^2}$ and $\overline{u_2^2}$ are less than $\frac{2}{3}k$ and $\overline{u_3^2}$ is greater. This is called the case of axisymmetric *expansion*. Such anisotropy could be produced by expansion in the plane normal to the $x_3$ axis and compression along that axis. The turbulence can be thought of

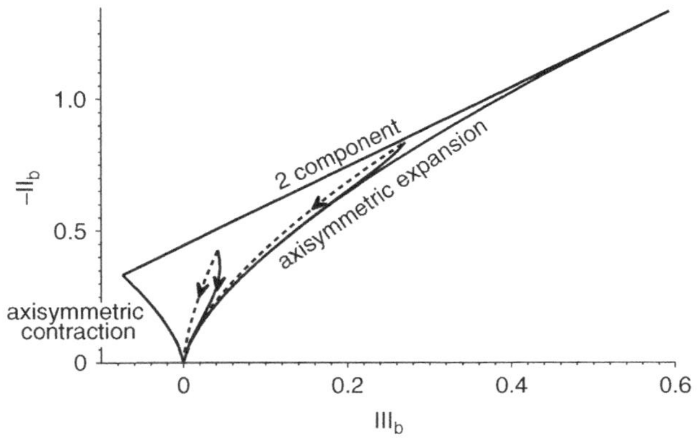


**Figure 7.1** Anisotropy-invariant triangle: $C_1 = 1.7$, $C_1^n = 1.05$ (——), and $II_b \propto (III_b)^{2/3}$ (- - - -). All the trajectories flow toward the origin.

as having been squashed along $x_3$ and stretched along $x_1$ and $x_2$. Stretching the vorticity along $x_1$ and $x_2$ amplifies the $u_3^2$ component of turbulence. The opposite case $b_{11} > 0$ is of axisymmetric *contraction* along $x_1$ and $x_2$, with stretching along $x_3$.

Now consider the boundaries to the triangle in Figure 7.1 (sometimes called the Lumley triangle). The state of turbulence can be characterized by the invariants (Section 2.3.2) $II_b = -\frac{1}{2}b_{kk}^2$ and $III_b = \frac{1}{3}b_{kk}^3$. From (7.1.11)

$$II_b = -(b_{11}^2 + b_{22}^2 + b_{11}b_{22}),$$
$$III_b = -(b_{11}^2 b_{22} + b_{11}b_{22}^2) = b_{11}II_b + b_{11}^3. \tag{7.1.13}$$

In the axisymmetric state, $b_{11} = b_{22}$, so $II_b = -3b_{11}^2$ and $III_b = -2b_{11}^3$. Eliminating $b_{11}$ between these gives

$$III_b = \pm 2(|II_b|/3)^{3/2}.$$

The "+" sign corresponds to axisymmetric expansion, the "−" sign to axisymmetric contraction. In the two-component state, $b_{11} = -\frac{2}{3}$, and the last equality of Eqs. (7.1.13) becomes

$$III_b = -\tfrac{2}{3}II_b - \tfrac{8}{27}.$$

The boundaries of the triangle have been found.

The minimum of the second invariant is at $b_{22} = -\frac{1}{2}b_{11}$ and equals $-\frac{3}{4}b_{11}^2$. This is seen by minimizing the expression (7.1.13) with respect to $b_{22}$ for fixed $b_{11}$. Since $b_{11} \le \frac{4}{3}$, the range of $II_b$ is $0 \le -II_b \le \frac{4}{3}$. Rearranging the last of (7.1.13) as

$$-II_b = |II_b| = b_{11}^2 - III_b/b_{11}$$

enables the range of $III_b$ to be identified. For a given $III_b$, the minimum of $|II_b|$ occurs at $b_{11} = -(III_b/2)^{1/3}$ and equals $3(III_b/2)^{2/3}$. Since $b_{11} > -\frac{2}{3}$, the maximum of $|II_b|$ is $\frac{4}{9} + \frac{3}{2}III_b$. The upshot is that the range of realizable values of $III_b$ all fall inside the curvilinear triangle

$$\max\left[-2(|II_b|/3)^{3/2}, \frac{2}{3}|II_b| - \frac{8}{27}\right] \le III_b \le 2(|II_b|/3)^{3/2}, \qquad 0 \le -II_b \le \tfrac{4}{3}$$

displayed in Figure 7.1 (Lumley, 1978). The max on the left becomes the straight line $\frac{2}{3}|II_b| - \frac{8}{27}$ when $|II_b| > \frac{1}{3}$. This straight line is the two-component state. The curvilinear sides of the triangle are the axisymmetric states.

We now return from the diversion, to consider the evolution (7.1.12) of anisotropy. Return to isotropy in the $II_b$–$III_b$ plane of Figure 7.1 consists of a trajectory starting inside the triangle and moving toward the origin. Four such trajectories are shown in the figure. The dashed lines are for the linear, Rotta model. It is readily shown (Exercise 7.4) that the linear model produces the trajectory $III_b = III_0(II_b/II_0)^{3/2}$ originating at $(III_0, II_0)$ and flowing to (0, 0). The solid lines are for a quadratic nonlinear model, with the constants

$$C_1 = 1.7 \qquad \text{and} \qquad C_1^n = 1.05$$

used by Speziale *et al.* (1991). Speziale *et al.* also revised the Rotta coefficient to

$$C_1 = 1.7 + 0.9\mathcal{P}/\varepsilon$$

in shear flows.

The tendency of the nonlinear model to swerve toward the axisymmetric expansion agrees qualitatively with data of Choi and Lumley (1984), but it has been questioned whether return to isotropy can be described by a function of $\boldsymbol{b}$ alone–for instance, the anisotropy of dissipation $(\varepsilon_{ij} - \frac{2}{3}\varepsilon\delta_{ij})$ might need to be included.

If an eigenvalue of $\overline{u_i u_j}$ becomes zero, the turbulence reaches a two-component state. It would then lie on the top line of the triangle in Figure 7.1. Any further decrease will produce the unphysical condition that one of the component energies, say $\overline{u_1^2}$, becomes negative; the trajectory would exit the top of the triangle. Negative $\overline{u_1^2}$ is called an unrealizable state because no real statistic can have a negative variance. Hence, any violation of realizability will occur by a trajectory crossing through the two-component state; that is, exiting through the upper boundary of the triangle in Figure 7.1. This particular feature of Lumley's triangle can be used as a consistency check for numerical, or experimental, data.

If $\overline{u_1^2} < 0$ the corresponding component of the anisotropy tensor is less than $-\frac{2}{3}$. In general, a turbulence model will not ensure that $b_{11} \geq -\frac{2}{3}$ unconditionally. A constraint that does guarantee this is called a realizability condition. Clearly, $b_{11}$ will not drop below $-\frac{2}{3}$ if $d_t b_{11} > 0$ when $b_{11} = -\frac{2}{3}$. In the case of (7.1.12), when $b_{11} = -\frac{2}{3}$,

$$d_\tau b_{11} = -\tfrac{2}{3}[(1 - C_1) + C_1^n(b_{22}^2 - \tfrac{2}{3}b_{22} - \tfrac{2}{9})].$$

The right-hand side has a minimum value of $\frac{2}{3}[C_1 - 1 - \frac{2}{3}C_1^n]$ when $b_{22} = \frac{4}{3}$. Requiring this to be positive gives the realizability constraint

$$C_1^n < \tfrac{3}{2}(C_1 - 1). \tag{7.1.14}$$

The marginal value $C_1^n = \frac{3}{2}(C_1 - 1)$ was used in Figure 7.1. When $C_1^n = 0$ the condition $C_1 > 1$ is recovered.

### 7.1.4 Models for the rapid part

The rapid part of the $\wp_{ij}$ model is defined as the portion of the model that explicitly involves the tensor components $\partial_j U_i$. A good deal of research on SMC has focused on the rapid pressure–strain model for homogeneous turbulence. The relevant theory will be developed in the present section. Outside the vicinity of walls, quasi-homogeneity may be a reasonable approximation even when the flow is not strictly homogeneous; so modeling concepts developed by reference to homogeneous flow are of broad interest. However, in some regions, such as the immediate vicinity of walls, the assumptions are strongly violated and most models fail. The subject of wall effects will be covered in Section 7.3; the present section is restricted to homogeneity.

The pressure contribution to the redistribution term is

$$\phi_{ij} = (\overline{u_j\,\partial_i p} + \overline{u_i\,\partial_j p})/\rho.$$

In homogeneous turbulence, this equals minus the pressure–strain correlation,

$$\phi_{ij} = -\overline{p(\partial_i u_j + \partial_j u_i)}/\rho$$

(see Section 3.3). A closure model relating this to the Reynolds stress tensor is needed. A first step is to relate pressure to velocity.

In incompressible flow, the pressure satisfies the Poisson equation

$$\nabla^2 \tilde{p} = -\rho\, \partial_k \tilde{u}_l\, \partial_l \tilde{u}_k,$$

as is found by taking the divergence of the Navier–Stokes equations, and invoking continuity. Recall that $\tilde{\mathbf{u}} = \mathbf{U} + \mathbf{u}$ is the total velocity. The fluctuating pressure equation is found by subtracting its average from the right-hand side:

$$\nabla^2 p = -\rho(\partial_k u_l\, \partial_l u_k - \overline{\partial_k u_l\, \partial_l u_k}) - 2\rho\, \partial_l U_k\, \partial_k u_l. \tag{7.1.15}$$

The form of the right-hand side is the motivation for decomposing redistribution into a sum of rapid and slow contributions: the first term does not depend explicitly on the mean velocity and so is associated with the "slow" part of the redistribution; the last term is the "rapid part." A better terminology might be "linear" and "nonlinear" parts, alluding to the nature of the dependence on $u$. The rapid, or linear, part is also linear in the mean velocity gradient. That term is the subject of the following.

The equation

$$\nabla^2 p = -2\rho\, \partial_l U_k\, \partial_k u_l \tag{7.1.16}$$

has the formal solution

$$p(\boldsymbol{x}) = \frac{1}{4\pi} \iiint_{-\infty}^{\infty} \frac{2\rho\, \partial_l U_k\, \partial'_k u_l(\boldsymbol{x}')}{|\boldsymbol{x} - \boldsymbol{x}'|}\, \mathrm{d}^3 x' \tag{7.1.17}$$

in unbounded space. This solution is obtained with the free-space Green function $1/4\pi|\boldsymbol{x} - \boldsymbol{x}'|$ for Laplace's equation (Copson, 1975).

Differentiating (7.1.17) with respect to $x_i$ and integrating by parts give

$$\begin{aligned}
\partial_i p(\boldsymbol{x}) &= \frac{1}{2\pi} \iiint_{-\infty}^{\infty} \rho\, \partial_l U_k\, \partial'_k u_l(\boldsymbol{x}') \frac{\partial}{\partial x_i} \frac{1}{|\boldsymbol{x} - \boldsymbol{x}'|}\, \mathrm{d}^3 \boldsymbol{x}' \\
&= -\frac{1}{2\pi} \iiint_{-\infty}^{\infty} \rho\, \partial_l U_k\, \partial'_k u_l(\boldsymbol{x}') \frac{\partial}{\partial x'_i} \frac{1}{|\boldsymbol{x} - \boldsymbol{x}'|}\, \mathrm{d}^3 \boldsymbol{x}' \\
&= \rho\, \partial_l U_k \frac{1}{2\pi} \iiint_{-\infty}^{\infty} \partial'_i \partial'_k u_l(\boldsymbol{x}') \frac{1}{|\boldsymbol{x} - \boldsymbol{x}'|}\, \mathrm{d}^3 \boldsymbol{x}',
\end{aligned}$$

where the fact that $\rho\, \partial_l U_k$ is constant in homogeneous flow has been used to move it outside the integral. Note that a sign change occurred on switching from an $x$ derivative to an $x'$ derivative.

Next form the velocity–pressure gradient correlation

$$
\begin{aligned}
-\rho\phi_{ij} &= -\overline{u_j\,\partial_i p} - \overline{u_i\,\partial_j p} \\
&= -\frac{\rho\,\partial_l U_k}{2\pi}\iiint\limits_{-\infty}^{\infty}\left[\overline{u_j(\boldsymbol{x})\,\partial'_i\partial'_k u_l(\boldsymbol{x}')} + \overline{u_i(\boldsymbol{x})\,\partial'_j\partial'_k u_l(\boldsymbol{x}')}\right]\frac{1}{|\boldsymbol{x}-\boldsymbol{x}'|}\,\mathrm{d}^3\boldsymbol{x}'.
\end{aligned}
\tag{7.1.18}
$$

Although the quantity on the left-hand side is a single-point correlation, the integrand on the right-hand side contains two-point correlations. This substantiates the observation that the Reynolds stress transport equations are unclosed because they depend on nonlocal effects: single-point equations contain two-point correlations. This nonlocality is due to pressure forces acting at a distance. The Green function diminishes the distant pressure fluctuations, but the fall-off is slow, like $1/r$.

In homogeneous turbulence, two-point correlations are a function only of the difference between the points, cf. also Chapter 9. A correlation of the form $\overline{a(x)b(x')}$ is a function $\overline{ab}(x - x')$. The derivative with respect to $x'$ is distinguished from that with respect to $x$, so $a(x)\,\partial_{x'}b(x') = \partial_{x'}(a(x)b(x'))$; the $x'$ derivative does not operate on $a(x)$. Owing to the latter, the bracketed term of (7.1.18), $\overline{u_j(\boldsymbol{x})\,\partial'_i\partial'_k u_l(\boldsymbol{x}')}$ is equivalent to $\partial'_i\partial'_k\left[\overline{u_j(\boldsymbol{x})u_l(\boldsymbol{x}')}\right]$; and owing to the former, this is equivalent to $\partial_i\partial_k\overline{u_j u_l}(\boldsymbol{\xi})$ with $\boldsymbol{\xi} = \boldsymbol{x} - \boldsymbol{x}'$. The derivative is with respect to the relative position vector $\boldsymbol{\xi}$. With these substitutions the formula (7.1.18) becomes

$$
\begin{aligned}
-\phi_{ij} &= -\frac{\partial_l U_k}{2\pi}\iiint\limits_{-\infty}^{\infty}\left[\partial_i\partial_k\overline{u_j u_l} + \partial_j\partial_k\overline{u_i u_l}\right](\boldsymbol{\xi})\frac{1}{|\boldsymbol{\xi}|}\,\mathrm{d}^3\boldsymbol{\xi} \\
&= 2\partial_l U_k\left[(\nabla^2_\xi)^{-1}\partial_i\partial_k\overline{u_j u_l}(\boldsymbol{\xi}) + (\nabla^2_\xi)^{-1}\partial_j\partial_k\overline{u_i u_l}(\boldsymbol{\xi})\right] \\
&= M_{ijkl}\,\partial_l U_k,
\end{aligned}
\tag{7.1.19}
$$

where $(\nabla^2_\xi)^{-1}$ represents the convolution integral symbolically: that is, it is simply a shorthand for integration over the function $-1/4\pi|\boldsymbol{x} - \boldsymbol{x}'|$. The significance of the representation (7.1.19) is that the right-hand side of the first line is a definite integral, so the components of $\boldsymbol{M}$ are constants.

The fourth-order tensor $M_{ijkl}$ is defined as

$$
M_{ijkl} = 2\left[(\nabla^2_\xi)^{-1}\partial_i\partial_k\overline{u_j u_l}(\boldsymbol{\xi}) + (\nabla^2_\xi)^{-1}\partial_j\partial_k\overline{u_i u_l}(\boldsymbol{\xi})\right]. \tag{7.1.20}
$$

A normalization of $\boldsymbol{M}$ follows from contracting on $j$ and $k$ in (7.1.20):

$$
\begin{aligned}
M_{ijjl} &= 2\left[(\nabla^2_\xi)^{-1}\partial_i\partial_j\overline{u_j u_l}(\boldsymbol{\xi}) + (\nabla^2_\xi)^{-1}\nabla^2\overline{u_i u_l}(\boldsymbol{\xi})\right] \\
&= 2\left[(\nabla^2_\xi)^{-1}\partial_i\overline{\partial_j u_j u_l}(\boldsymbol{\xi}) + \overline{u_i u_l}\right] = 2\,\overline{u_i u_l},
\end{aligned}
\tag{7.1.21}
$$

upon noting that the inverse Laplacian of the Laplacian is the identity,

$$
(\nabla^2)^{-1}\nabla^2 = 1
$$

and invoking continuity to set $\partial_i \overline{\partial_j u_j u_l} = 0$. This last step is based on previous reasoning about homogeneous correlation functions, applied in reverse:

$$\partial_i \partial_j \overline{u_j u_l}(\boldsymbol{\xi}) = \partial'_i \partial'_j \overline{u_j u_l}(\boldsymbol{x}' - \boldsymbol{x}) = \partial'_i \partial'_j \overline{u_j(\boldsymbol{x}')u_l(\boldsymbol{x})}$$
$$= \partial'_i \overline{\partial'_j u_j(\boldsymbol{x}')u_l(\boldsymbol{x})} = \partial'_i \overline{\nabla \cdot \mathbf{u}(\boldsymbol{x}')u_l(\boldsymbol{x})} = 0.$$

In addition to the normalization, $\boldsymbol{M}$ must preserve the symmetry in $i$, $j$ and the trace-free property of contracting on $i = j$ that are built into the anisotropy tensor, $\mathbf{b}$. In short, the following constraints are imposed upon $\boldsymbol{M}$:

$$M_{ijkl} = M_{jikl}, \qquad M_{iikl} = 0, \qquad M_{ijjl} = 2\,\overline{u_i u_l}. \tag{7.1.22}$$

The first implements the symmetry in $i$ and $j$; the second preserves $b_{ii} = 0$; and the third normalizes $\boldsymbol{M}$. It is sometimes argued that the further condition $M_{ijkk} = 0$ should be imposed on the basis of continuity. While this seems attractive, and follows from (7.1.20), it actually is not compelling. It would be incorrect to characterize $M_{ijkk} = 0$ as an incompressibility condition because the single-point Reynolds stress tensor gives no direct information on compressibility of the turbulence. Only the two-point correlation can satisfy a divergence-free condition, $\partial'_j \overline{u_j(\boldsymbol{x}')u_l(\boldsymbol{x})} = 0$. In the rapid redistribution term $M_{ijkl}\,\partial_l U_k$, the last two indices of $\boldsymbol{M}$ are contracted. The only constraints apparent to the Reynolds stress evolution are those on the first two indices. To impose $M_{ijkk} = 0$ would overly constrain the model, without cause.

Based on (7.1.19), the rapid contribution to (7.1.6) is represented as

$$\wp_{ij}^{\mathrm{rapid}} = \varepsilon \mathcal{F}_{ij}^{\mathrm{rapid}} = M_{ijkl}\,\partial_l U_k, \tag{7.1.23}$$

where $M_{ijkl}$ is an unspecified tensor that is constrained by (7.1.22). There is still considerable freedom in how to select it. The following section develops one possible form. The method is a systematic expansion in powers of anisotropy.

#### 7.1.4.1 Expansion of $M_{ijkl}$ in powers of $b_{ij}$

The dependence of (7.1.6) on mean gradients is expressed by the form (7.1.23). The remaining arguments of $\mathcal{F}_{ij}$ show that $M_{ijkl}$ is a function of $b_{ij}$ and $\delta_{ij}$. A technique that has been employed to develop closure formulas is expansion in powers of $b_{ij}$. The first term in this expansion is the zeroth power, which means that it involves $\delta_{ij}$ alone; the next involves $b_{ij}$ to the first power. According to the Cayley–Hamilton theorem (Eq. (2.3.7)), the tensoral expansion would stop at order $b_{ij}^2$. However, we will stop at the first power, thereby deriving the general linear model. The coefficients in the tensoral expansion could be functions of the invariants. In that case the model derived at first order would be called the general quasi-linear model. In principle, the expansion in powers of anisotropy should include expansion of the coefficients. For instance, this was done by Speziale *et al.* (1991).

Since four subscripts are needed, the first term in the power series consists of products of two $\delta$. The most general form is obtained by a linear combination of all possible, distinct products:

$$M^0_{ijkl} = A\delta_{ij}\delta_{kl} + B\delta_{ik}\delta_{jl} + C\delta_{il}\delta_{jk}. \tag{7.1.24}$$

Symmetry in the $i$, $j$ indices, the first of (7.1.22), requires that $B = C$. The second of (7.1.22) then requires that

$$A\delta_{ii}\delta_{kl} + B(\delta_{ik}\delta_{il} + \delta_{il}\delta_{ik}) = 0. \tag{7.1.25}$$

But $\boldsymbol{\delta}$ is just the identity matrix, so $\delta_{ii} = 3$ and $\delta_{ik}\delta_{il} = \delta_{kl}$. Thus the above equation is $(3A + 2B)\delta_{kl} = 0$, or $3A = -2B$.

Finally, the last of (7.1.22) can be written

$$M_{ijjl} = 2k(b_{il} + \tfrac{2}{3}\delta_{il}). \tag{7.1.26}$$

Expression (7.1.24) is the first term in the expansion in powers of $b_{il}$. Only the $\delta_{il}$ term of Eq. (7.1.26) contributes at this order; so this constraint on (7.1.25) with $A = -\frac{2}{3}B$ reduces to

$$M^0_{ijjl} = (-\tfrac{2}{3}B + 4B)\delta_{il} = 2k(\tfrac{2}{3}\delta_{il}). \tag{7.1.27}$$

Thus $B = \frac{2}{5}k$, $A = -\frac{4}{15}k$, and

$$M^0_{ijkl} = \tfrac{1}{15}(6\delta_{ik}\delta_{jl} + 6\delta_{il}\delta_{jk} - 4\delta_{ij}\delta_{kl})k, \tag{7.1.28}$$

which has no empirical constants! Substituting this into (7.1.23) and using incompressibility, $\partial_i U_i = 0$, gives

$$\wp^{\mathrm{rapid}\,0}_{ij} = \tfrac{2}{5}k(\partial_j U_i + \partial_i U_j). \tag{7.1.29}$$

This leading-order perturbation of isotropy is sometimes referred to as the Crow constraint (Crow, 1968).

The general linear model (GLM) is obtained by carrying the expansion in powers of $b_{ij}$ to the linear term. By analogy with (7.1.24), the next term is of the form

$$\begin{aligned} M^1_{ijkl} = {} & Ab_{ij}\delta_{kl} + C_3(b_{ik}\delta_{jl} + b_{jk}\delta_{il} - \tfrac{2}{3}\delta_{ij}b_{kl}) \\ & + C_2(b_{il}\delta_{jk} + b_{jl}\delta_{ik} - \tfrac{2}{3}\delta_{ij}b_{kl}), \end{aligned} \tag{7.1.30}$$

where the first two of (7.1.22) have already been imposed. The last constraint is $M^1_{ijjl} = 2kb_{il}$, so

$$A + \tfrac{10}{3}C_2 + \tfrac{1}{3}C_3 = 2k.$$

However, substituting $M^1_{ijkl}$ into (7.1.23) gives

$$\begin{aligned} \wp^{\mathrm{rapid}\,1}_{ij} = {} & kC_2(b_{ik}\,\partial_k U_j + b_{jk}\,\partial_k U_i - \tfrac{2}{3}\delta_{ij}b_{kl}\,\partial_k U_l) \\ & + kC_3(b_{ik}\,\partial_j U_k + b_{jk}\,\partial_i U_k - \tfrac{2}{3}\delta_{ij}b_{kl}\,\partial_k U_l), \end{aligned}$$

so the value of $A$, and hence the normalization constraint, is irrelevant. The two free constants $C_3$ and $C_2$ can be chosen empirically.

Adding $\wp^{\text{rapid 1}}$ to $\wp^{\text{rapid 0}}$ gives the GLM:

$$\wp_{ij}^{\text{rapid}} = \tfrac{2}{5}k(\partial_j U_i + \partial_i U_j) + kC_2(b_{ik}\,\partial_k U_j + b_{jk}\,\partial_k U_i - \tfrac{2}{3}\delta_{ij} b_{kl}\,\partial_k U_l)$$
$$+ kC_3(b_{ik}\,\partial_j U_k + b_{jk}\,\partial_i U_k - \tfrac{2}{3}\delta_{ij} b_{kl}\,\partial_k U_l). \qquad (7.1.31)$$

Special cases of (7.1.31) are the LRR (Launder–Reece–Rodi) model (Launder *et al.*, 1975) and the IP (isotropization of production) model (Launder, 1989). An example of the general quasi-linear model is the SSG (Speziale–Sarkar–Gatski) model (Speziale *et al.*, 1991).

The formula (7.1.31) can be rearranged into other forms that appear in the literature. These express it in terms of the production tensor as

$$\wp_{ij}^{\text{rapid}} = [\tfrac{2}{5} - \tfrac{2}{3}(C_2 + C_3)]k(\partial_j U_i + \partial_i U_j)$$
$$- C_2(\mathcal{P}_{ij} - \tfrac{2}{3}\delta_{ij}\mathcal{P}) - C_3(D_{ij} - \tfrac{2}{3}\delta_{ij}\mathcal{P}), \qquad (7.1.32)$$

which is obtained by substituting $b_{ij} = \overline{u_i u_j}/k - \frac{2}{3}\delta_{ij}$, $\mathcal{P}_{ij} = -\overline{u_i u_k}\,\partial_k U_j - \overline{u_j u_k}\,\partial_k U_i$, and $\mathcal{P} = \frac{1}{2}\mathcal{P}_{kk}$, and defining

$$D_{ij} = -\overline{u_i u_k}\,\partial_j U_k - \overline{u_j u_k}\,\partial_i U_k.$$

Alternatively, it can be expressed in terms of rate of strain and rate of rotation as

$$\wp_{ij}^{\text{rapid}} = \tfrac{4}{5}kS_{ij} + k(C_2 + C_3)(b_{ik}S_{kj} + b_{jk}S_{ki} - \tfrac{2}{3}\delta_{ij}b_{kl}S_{lk})$$
$$+ k(C_2 - C_3)(b_{ik}\Omega_{kj} + b_{jk}\Omega_{ki}), \qquad (7.1.33)$$

which is obtained by substituting $\partial_i U_j = S_{ij} + \Omega_{ij}$, $S_{ij} = S_{ji}$, and $\Omega_{ij} = -\Omega_{ji}$ (see Eq. (3.3.8)).

At this point the reader might want to pause and take stock of what has been accomplished. The evolution equation (7.1.4) for the Reynolds stresses

$$\partial_t \overline{u_i u_j} = \mathcal{P}_{ij} + \wp_{ij}^{\text{rapid}} + \wp_{ij}^{\text{slow}} - \tfrac{2}{3}\varepsilon\delta_{ij}$$

has now been closed. The explicit closure is simply to replace $\wp^{\text{rapid}}$ with (7.1.32) and $\wp^{\text{slow}}$ with (7.1.8), and invoke the $\varepsilon$ equation (6.2.6).

The form (7.1.32) was introduced by Launder *et al.* (1975), who imposed the additional constraint $M_{ijkk} = 0$ to eliminate one of the empirical constants. As discussed above, this is not necessary, and was not used in the other models. In the LRR model,

$$C_2 = (c + 8)/11 \quad \text{and} \quad C_3 = (8c - 2)/11,$$

where $c$ is an empirical constant, for which the value 0.4 was selected. The IP model uses

$$C_2 = 3/5 \quad \text{and} \quad C_3 = 0,$$

so that

$$\wp_{ij}^{\text{rapid}} = -\tfrac{3}{5}(\mathcal{P}_{ij} - \tfrac{2}{3}\delta_{ij}\mathcal{P}). \qquad (7.1.34)$$

This provides a simple, commonly used SMC model.

The form (7.1.33) was invoked by Speziale *et al.* (1991) with $C_2 = 0.4125$ and $C_3 = 0.2125$. They also added the term

$$-C_s^* \sqrt{II_b}\, k S_{ij} \tag{7.1.35}$$

with $C_s^* = 0.65$ to Eq. (7.1.33). Adding such a term is consistent with the expansion to first order in anisotropy. (Note that Speziale *et al.* (1991) used $2k$ to scale turbulent intensity, so the anisotropy tensor in their paper is defined as $\overline{u_i u_j}/2k - \frac{1}{3}\delta_{ij}$, which equals our $b_{ij}/2$.)

The relative importance of the rapid and slow redistribution models depends somewhat on flow conditions. The split between them for the IP and SSG models is shown in Figure 7.2 for plane channel flow. The specifics vary with the model, but both rapid and slow components make a significant contribution throughout the flow. Both contributions

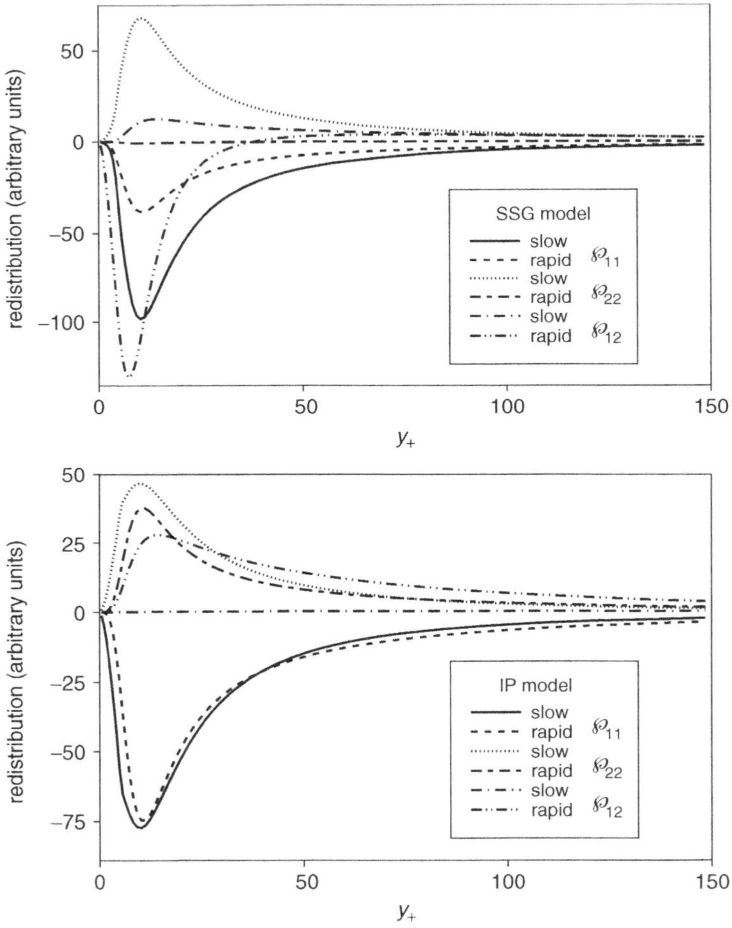


**Figure 7.2** Contributions of the rapid and slow redistribution model in plane channel flow, for SSG and IP models.

to the streamwise component, $\wp_{11}$, are negative, and both contributions to the wall normal component, $\wp_{22}$, are positive. These correspond to redistribution from $\overline{u^2}$ into $\overline{v^2}$.

One should be warned that the predictions of these models are not correct in the region $y_+ \lesssim 80$. The large negative peak in the SSG rapid model for $\wp_{12}$ is a notable instance of this failure: $\wp_{12}$ should be positive near the wall. DNS data for this term can be found later in Figure 7.9. The erroneous behavior comes from $D_{12}$ in Eq. (7.1.32), which contains $\overline{u^2}\partial_y U$. Near the wall both $\partial_y U$ and $\overline{u^2}$ become large (Figure 6.6, page 127). The behavior of $D_{ij}$ near the wall can produce quite anomalous predictions by the redistribution model. Methods to correct the near-wall behavior of quasi-homogeneous models are discussed in Section 7.3.2.

Note that $\mathcal{P}_{12}$ contains $\partial_y U$ in the form $-\overline{v^2}\,\partial_y U$. Near the wall $\overline{v^2}$ is small. The IP model (7.1.34) is better behaved than the SSG model near the wall because it contains only $\mathcal{P}_{12}$, not $D_{12}$. This model is quite intuitive: the isotropization of production rapid model counteracts the production tensor. Near the wall, production of $\overline{uv}$ is suppressed and the IP model follows that behavior.

## 7.2 Analytic solutions to SMC models

This section describes some analytical solutions. They have several motivations: the solution in homogeneous shear flow is used to calibrate the empirical constants, and it illustrates how SMC predicts normal stress anisotropy; and the analysis of curvature illustrates how the influence of external forces is captured by the production tensor. Further solutions can be found in Section 8.3.

To start, the complete general linear model (GLM) will be assembled. Substituting the rapid (7.1.31) and slow (7.1.8) models into the evolution equation (7.1.7), in which $\mathcal{F}_{ij} = \wp_{ij}/\varepsilon$, gives the closed equation

$$\begin{aligned} \mathrm{d}_t b_{ij} = {} & (1 - C_1) b_{ij}\frac{\varepsilon}{k} - b_{ik}\,\partial_k U_j - b_{jk}\,\partial_k U_i - (b_{ij} + \tfrac{2}{3}\delta_{ij})\frac{\mathcal{P}}{k} - \tfrac{8}{15}S_{ij} \\ & + C_2(b_{ik}\,\partial_k U_j + b_{jk}\,\partial_k U_i - \tfrac{2}{3}\delta_{ij} b_{kl}\,\partial_k U_l) \\ & + C_3(b_{ik}\,\partial_j U_k + b_{jk}\,\partial_i U_k - \tfrac{2}{3}\delta_{ij} b_{kl}\,\partial_k U_l) \end{aligned} \qquad (7.2.1)$$

for the anisotropy tensor. It is assumed that $k$ and $\varepsilon$ are found from their own evolution equations. For the present case of homogeneous turbulence, the complete model is a set of ordinary differential equations. They could be solved numerically by standard Runge–Kutta methods, but analytical solutions are instructive and give valuable insight into the model.

### 7.2.1 Homogeneous shear flow

Parallel and self-similar shear layers are discernible elements of many complex engineering flow fields. It is a long-standing tradition to calibrate model constants with data from simple shear flow. A closed-form solution to the GLM that relates its constants to readily measurable quantities is developed in this section.

Homogeneous, parallel shear flow has the mean velocity $U_j = \mathcal{S}x_i\delta_{i2}\delta_{j1}$, or $U(y) = \mathcal{S}y$. The velocity gradient and rate-of-strain tensors are $\partial_i U_j = \mathcal{S}\delta_{i2}\delta_{j1}$ and

$S_{ij} = \frac{1}{2}\mathcal{S}(\delta_{i1}\delta_{j2} + \delta_{i2}\delta_{j1})$. Explicitly, $\partial_2 U_1 = \mathcal{S}$ and $S_{12} = S_{21} = \frac{1}{2}\mathcal{S}$, with all other components equal to zero.

The equilibrium solution to the $k$–$\varepsilon$ model was given in Section 6.2.1: it was $k$ and $\varepsilon \propto e^{\lambda t}$. Define the ratio of production to dissipation by

$$\mathcal{P}/\varepsilon \equiv \mathcal{P}_{\mathcal{R}}.$$

The equilibrium value of $\mathcal{P}_{\mathcal{R}}$ is given by the second of Eqs. (6.2.11) on page 124:

$$\mathcal{P}_{\mathcal{R}} = \frac{C_{\varepsilon 2} - 1}{C_{\varepsilon 1} - 1}.$$

This remains valid for the present solution; the new element is that the anisotropy tensor is determined by the evolution equation (7.2.1) instead of by the constitutive equation (6.2.3). The latter gives

$$b_{12} = -C_\mu \mathcal{S}\frac{k}{\varepsilon}, \qquad b_{11} = b_{22} = b_{33} = 0, \tag{7.2.2}$$

in parallel shear flow. The first of these is reasonable; the second is certainly erroneous. This illustrates that the linear, eddy viscosity constitutive relation is likely to predict erroneous normal stresses ($\overline{u_i^2}$). Second-moment closures should be able to do better.

The concept of moving equilibrium is used to good effect when analyzing turbulent flows. Moving equilibrium means that $k$ and $\overline{u_i u_j}$ evolve in time such that $b_{ij}$ remains constant. In particular, the equilibrium solutions for $k$ and $\overline{u_i u_j}$ both grow with the same exponential rate, so that their ratio is time-independent. Mathematically, $\mathrm{d}_t(\overline{u_i u_j}/k) = \mathrm{d}_t b_{ij} = 0$ is the condition of moving equilibrium. Another way to state this is (Rodi, 1976)

$$\mathrm{d}_t \overline{u_i u_j} = \frac{\overline{u_i u_j}}{k}\,\mathrm{d}_t k.$$

Thus the rate of change of all the components of the Reynolds stress tensor are proportional to the rate of change of $k$.

This moving equilibrium solution will be sought. After setting $\mathrm{d}_t b_{ij} = 0$ in (7.2.1) and substituting $\partial_j U_i = \mathcal{S}\delta_{i1}\delta_{j2}$ and $S_{ij} = \frac{1}{2}\mathcal{S}(\delta_{i1}\delta_{j2} + \delta_{i2}\delta_{j1})$, it becomes

$$\begin{aligned} 0 = {} & (1 - C_1)b_{ij} - b_{i2}\frac{\mathcal{S}k}{\varepsilon}\delta_{j1} - b_{j2}\frac{\mathcal{S}k}{\varepsilon}\delta_{i1} - \left(b_{ij} + \frac{2}{3}\delta_{ij}\right)\mathcal{P}_{\mathcal{R}} \\ & - \frac{4}{15}\frac{\mathcal{S}k}{\varepsilon}(\delta_{i1}\delta_{j2} + \delta_{i2}\delta_{j1}) \\ & + C_2\left(b_{i2}\frac{\mathcal{S}k}{\varepsilon}\delta_{j1} + b_{j2}\frac{\mathcal{S}k}{\varepsilon}\delta_{i1} + \frac{2}{3}\delta_{ij}\mathcal{P}_{\mathcal{R}}\right) \\ & + C_3\left(b_{i1}\frac{\mathcal{S}k}{\varepsilon}\delta_{j2} + b_{j1}\frac{\mathcal{S}k}{\varepsilon}\delta_{i2} + \frac{2}{3}\delta_{ij}\mathcal{P}_{\mathcal{R}}\right). \end{aligned} \tag{7.2.3}$$

In component form, Eqs. (7.2.3) can be solved as

$$
\begin{aligned}
Gb_{11} &= \tfrac{2}{3}(2 - 2C_2 + C_3), \\
Gb_{22} &= \tfrac{2}{3}(C_2 - 2C_3 - 1), \\
Gb_{33} &= \tfrac{2}{3}(C_2 + C_3 - 1), \\
(Gb_{12})^2 &= \tfrac{4}{15}G + 2C_3^2 - \tfrac{2}{3}(1 + 2C_3 - C_2)^2
\end{aligned} \tag{7.2.4}
$$

where $G \equiv (C_1 - 1)/\mathcal{P}_\mathcal{R} + 1$. The exact formula for the ratio of production to dissipation $-b_{12}\mathcal{S}k/\varepsilon = \mathcal{P}_\mathcal{R}$ was used to eliminate $\mathcal{S}k/\varepsilon$ in the process of finding this solution. Formulas (7.2.4) can be employed either to determine the constants of a model from experimental measurements of $b_{ij}$, or to assess the accuracy of a given set of constants. Usually the constants are a compromise between these and other data.

Tavoularis and Karnik (1989) summarize experimental measurements of $b_{ij}$ and $\mathcal{P}_\mathcal{R}$. Early experiments on homogeneous shear flow did not reach equilibrium. The data cited here are from experiments in which $\mathcal{S}t$ was large enough for $b_{ij}$ to equilibrate. An average of data in table 2 of Tavoularis and Karnik (1989) gives $\mathcal{P}_\mathcal{R} = 1.6 \pm 0.2$. They cite

$$
\{b_{ij}\} = \begin{Bmatrix} 0.36 \pm 0.08 & -0.32 \pm 0.02 & 0 \\ -0.32 \pm 0.02 & -0.22 \pm 0.05 & 0 \\ 0 & 0 & -0.14 \pm 0.06 \end{Bmatrix} \tag{7.2.5}
$$

for the anisotropy tensor.

Energy is fed from the mean shear to $\overline{u_1^2}$ and then redistributed to $\overline{u_2^2}$ and $\overline{u_3^2}$, so one expects $b_{11}$ to be greater than zero and $b_{22}$ to be less than zero. This is found experimentally to be true. From the first of (7.2.4), $b_{11} > 0$ only if $2 - 2C_2 + C_3 > 0$. This is satisfied by both the IP and LRR models (the latter uses $C_2 = 8.4/11$ and $C_3 = 1.2/11$). The IP model has $C_2 = 3/5$ and $C_3 = 0$. With $C_3 = 0$, expression (7.2.4) shows that $b_{22} = b_{33} = -2b_{11}$. This is a fault of the IP model; the experimental data show that $b_{22} < b_{33}$, although $b_{22} \approx b_{33}$ is not too bad an approximation. In order to calibrate a model to fit (7.2.5), three model constants must be available; IP has only one.

Substituting model constants, the IP model, with $\mathcal{P}_\mathcal{R} = 1.6$ and $C_1 = 1.8$, predicts

$$
\{b_{ij}\} = \begin{Bmatrix} 0.356 & -0.361 & 0 \\ -0.361 & -0.178 & 0 \\ 0 & 0 & -0.178 \end{Bmatrix}.
$$

The value $C_1 = 1.7$ was selected by the developers of the SSG model. With $\mathcal{P}_\mathcal{R} = 1.6$ that model gives

$$
\{b_{ij}\} = \begin{Bmatrix} 0.433 & -0.328 & 0 \\ -0.328 & -0.282 & 0 \\ 0 & 0 & -0.151 \end{Bmatrix}.
$$

It should be noted that the SSG model was calibrated using older data than (7.2.5), with which it agreed more closely.

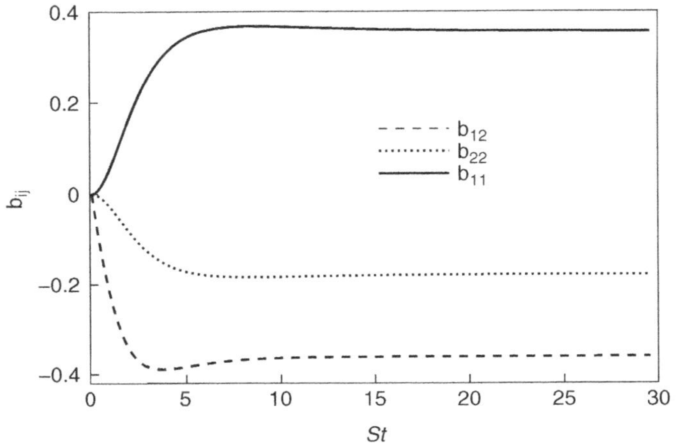


**Figure 7.3** Evolution of the anisotropy tensor in homogeneous shear flow. The IP model is used for illustration.

The approach of the model to equilibrium can be obtained by numerically integrating (7.2.1). Figure 7.3 illustrates the evolution of anisotropy for the IP model. The initial condition is isotropy, $b_{ij} = 0$ at $t = 0$. Equilibrium is reached at $\mathcal{S}t \sim 6$. Using the experimental values in $\mathcal{S}k/\varepsilon = -\mathcal{P}_{\mathcal{R}}/b_{12}$ gives $\mathcal{S}k/\varepsilon = 5$ in equilibrium. Hence, equilibrium is attained when $t \sim k/\varepsilon = T$. This is consistent with intuitive interpretations of scaling analysis. The integral correlation time-scale is regarded as that on which the turbulence equilibrates and is of order $k/\varepsilon$.

A final remark on solving more complex models is needed. The constants, $C_i$, in the general quasi-linear model can depend on invariants of $\boldsymbol{b}$. In that case, closed-form solutions can still be found; further equations are simply added. For instance, the SSG model contains a dependence on $II_b$ as in (7.1.35): in that case (7.2.1) contains the additional term $-C_s^*\sqrt{II_b}S_{ij}$ and the last of (7.2.4) has an extra term $-\frac{1}{2}GC_s^*\sqrt{II_b}$. Substituting (7.2.4) into

$$II_b = b_{11}^2 + b_{22}^2 + b_{22}b_{11} + b_{12}^2$$

provides an additional equation that determines $II_b$ implicitly (see Exercise 7.5).

### 7.2.2 Curved shear flow

The topic in this section is not so much a solution as it is an analysis of the physical content of such models. One of the motivations for second-moment closure is their greater adherence to fluid dynamical physics than scalar models, such as $k$–$\varepsilon$. The intent of this example is to illustrate that.

Surface curvature can suppress or amplify turbulence, depending on whether it is convex or concave curvature. The turbulence in a boundary layer entering a convex curve is diminished by the centrifugal acceleration; the turbulence in one entering a concave curve is amplified. Figure 7.4 illustrates this configuration. The shear is

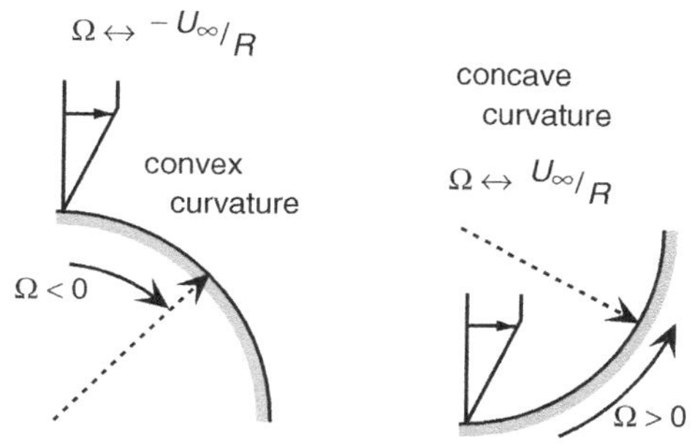


**Figure 7.4** Schematic of boundary layers on curved surfaces and the analogy between rotation and curvature. Convex curvature is stabilizing and concave is destabilizing.

toward the center of curvature in the destabilizing case and outward from it in the stabilizing case.

Curvature is analogous to rotation. A parallel shear layer placed into rotation will be destabilized if the rotation is contrary in sign to the mean vorticity. In Figure 7.4 this correspondence between concave curvature and rotation is noted. A lengthy analysis of rotation can be found in Sections 8.3.2 and 8.3.3.

The origin of these effects of curvature and rotation can also be understood by an examination of the Reynolds stress production tensor. They are most clear-cut in cylindrical coordinates, with $x_1 = R\theta$ and $x_2 = r$. Then the shear flow in Figure 7.4 is in the $x_1$ direction, $\boldsymbol{U} = U(r)\boldsymbol{e}_1$ where $\boldsymbol{e}_1$ is a unit vector in the circumferential direction, $\boldsymbol{e}_1 = (-\sin\theta, \cos\theta)$. The unit vector in the radial direction is $\boldsymbol{e}_2 = (\cos\theta, \sin\theta)$. Note that $\partial_{x_1} e_1 = 1/R\, \partial_\theta(-\sin\theta, \cos\theta) = -e_2/R$. Then the velocity gradient tensor has components

$$e_1\, \partial_1(U(r)e_1) = e_1 U(r)\, \partial_1 e_1 = -e_1 e_2 \frac{U(r)}{R},$$
$$e_2\, \partial_2(U(r)e_1) = e_2 e_1\, \partial_r U(r)$$

or

$$\{\partial_i U_j\} = \begin{Bmatrix} 0 & -U/R & 0 \\ \partial_r U & 0 & 0 \\ 0 & 0 & 0 \end{Bmatrix}.$$

With the above velocity gradient, the non-zero components of the production tensor (3.3.5) become

$$\begin{aligned}
\mathcal{P}_{11} &= -2\overline{u_1 u_2}\partial_2 U_1 &&= -2\overline{uv}\partial_r U, \\
\mathcal{P}_{22} &= -2\overline{u_2 u_1}\partial_1 U_2 &&= 2\overline{uv}\, U/R, \\
\mathcal{P}_{12} &= -\overline{u_2 u_2}\partial_2 U_1 - \overline{u_1 u_1}\, \partial_1 U_2 &&= -\overline{v^2}\partial_r U + \overline{u^2}\, U/R, \\
\mathcal{P} &= \tfrac{1}{2}(\mathcal{P}_{11} + \mathcal{P}_{22}) &&= -\overline{uv}(\partial_r U - U/R).
\end{aligned} \tag{7.2.6}$$

On a convex wall the velocity increases in the radial direction; hence $\partial_r U > 0$. The two terms of $\mathcal{P}$ are opposite in sign and the curvature acts to diminish the production of turbulent kinetic energy by the mean shear.

Curvature effects increase as the radius of curvature, $R$, decreases. The relative magnitude of curvature and shear can be estimated as

$$\frac{U/R}{\partial_r U} = \frac{\nu_T U/R}{\nu_T\, \partial_r U} \sim \frac{\nu_T U/R}{u_*^2} \sim \frac{\delta_* U}{R u_*}$$

in a turbulent boundary layer. The formula $u_* \delta_*$ (see expressions (6.1.12)) was used as an order-of-magnitude estimate for $\nu_T$, and the turbulent Reynolds stress, $\nu_T \partial_r U$, was estimated to be on the order of the surface kinematic stress. It follows that curvature effects become important when

$$\frac{\delta_*}{R} \gtrsim \frac{u_*}{U}. \tag{7.2.7}$$

The right-hand side is typically of order 1/20. This leads to the characterization of "strong curvature" as that for which $\delta_*/R \gtrsim 0.05$ (Bradshaw, 1976); that is, if the radius of curvature is less than 20 times the displacement thickness, curvature exerts a profound influence. Without the above order-of-magnitude analysis, this association of such a small ratio with a strong effect would be puzzling.

The qualitative effect of curvature is also apparent in the rate of shear stress production $\mathcal{P}_{12}$. In practice, the dominant influence of curvature on a turbulent flow is through this term. The estimates that gave (7.2.7) now give

$$\frac{\delta_*}{R} \gtrsim \frac{u_* \overline{v^2}}{U \overline{u^2}}.$$

The ratio $\overline{v^2}/\overline{u^2}$ is less than unity in the vicinity of a wall, so curvature plays a correspondingly larger role in the $\overline{uv}$ equation than in the $k$ equation.

On a concave wall, $\partial_r U < 0$. The two terms on the right-hand side of equations (7.2.6) that contribute to $\mathcal{P}_{12}$ and $\mathcal{P}$ then have the same sign. In this case, curvature supplements the production by mean shear. Concave curvature is destabilizing and amplifies the turbulence. Its physical effects are embodied in the production tensor.

If the eddy viscosity formula $-\overline{uv} = \nu_T\, \partial_r U$ is adopted, then $\mathcal{P}_{11} \propto (\partial_r U)^2 > 0$ in Eq. (7.2.6). Similarly $\mathcal{P}_{22} \propto -U\, \partial_r U/R$. On a concave wall, $U/R$ and $\partial_r U$ are of opposite sign (Figure 7.4) and $\mathcal{P}_{22} > 0$: the wall normal component of intensity is amplified. Conversely, the wall normal component is suppressed on a convex wall. The normal stress is responsible for the shear production of $-\overline{uv}$ in (7.2.6). Increased $\overline{v^2}$ on the concave wall will increase the magnitude of $-\overline{uv}$ because $\overline{v^2}$ enters that formula with a negative sign.

The effects of convex and concave curvature are illustrated in Figures 7.5 and 7.6. These shows Reynolds shear stress and skin friction in a boundary layer that starts on a flat wall, then enters either a convex or a concave bend. The symbols are experimental data; the lines were computed with a Reynolds stress model.

The $-\overline{uv}$ data on the convex wall illustrate how centrifugal acceleration suppresses the turbulence. The turbulent kinetic energy, not shown in the figures, is similarly suppressed.

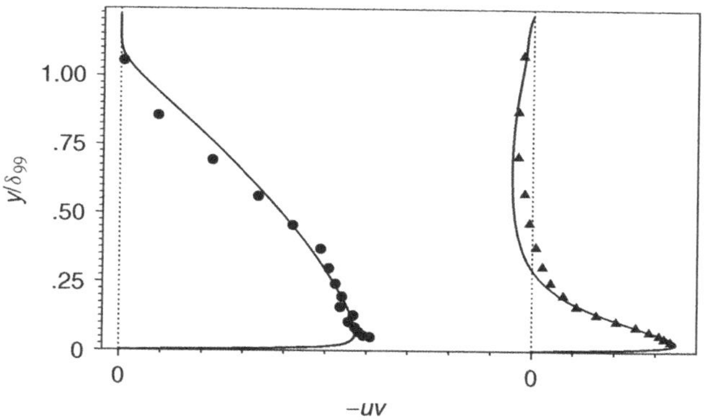


Reynolds shear stress in a boundary layer that starts on a flat plate, then enters a convex curve. The leftmost profile is on the flat section; the right is 21° around the bend.

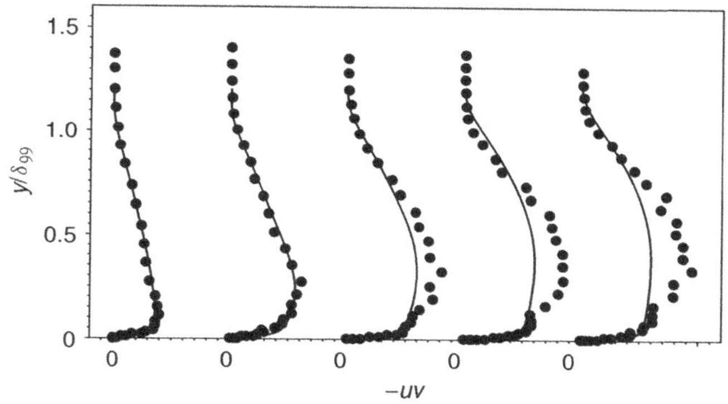


The same, but for a concave curve. The leftmost profile is on the flat section; the others are at 15°, 30°, 45°, 60° around the bend.

**Figure 7.5** Reynolds shear stress in curved wall boundary layers. Experimental data are from Johnson and Johnston (1989), Barlow and Johnston (1988) and Gillis *et al.* (1980); lines are SMC computations.

A curious feature illustrated by the upper pane of Figure 7.5 is that the shear stress actually becomes positive in the outer region of the boundary layer at the location 21° around the bend. At that location, $\partial_r U > 0$, so the eddy viscosity model $-\overline{uv} = \nu_T \, \partial_r U$ gives the incorrect sign for $\overline{uv}$. This is referred to as a region of counter-gradient transport. Counter-gradient transport can occur under the influence of strong stabilizing curvature (or stratification). Whether this has a practical impact is uncertain. When solving a Reynolds stress transport model, it is not an issue: the computations in Figure 7.5 demonstrate that the turbulent shear stress can have the same sign as the mean shear in computations that use Reynolds stress transport models.

The $-\overline{uv}$ data on the concave wall, in the lower pane of Figure 7.5, develop a bulge around $0.25 < y/\delta_{99} < 0.75$. This is not predicted by the model. The bulge is probably associated with the "Görtler vortices" that form on concave walls. These are long,

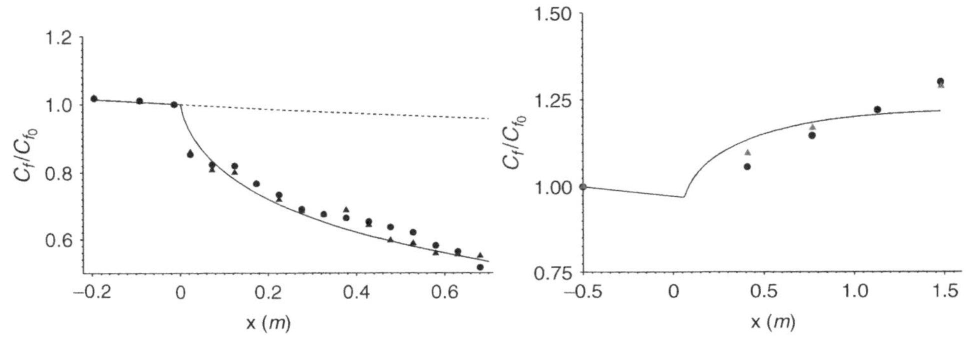


**Figure 7.6** Skin friction coefficient on curved walls. The wall is flat for $x < 0$, then enters a convex section (left) in which the skin friction is reduced, or a concave section (right) in which it increases. Symbols are experimental data; the lines are SMC computations, the dashed line being a solution for a flat wall.

quasi-steady vortices aligned in the $x_1$ direction. They are a manifestation of centrifugal instability. They occur in laminar flows as well; in that case they are truly steady and rather more coherent. When deterministic eddies, like Görtler vortices, occur in a flow, one cannot expect the statistical turbulence model to represent them. If those vortices contribute significantly to transport phenomena, then they must be computed as part of the mean flow field. The issue of computing deterministic flow structures will be discussed in Section 7.4.2.3.

The Reynolds shear stress $-\overline{uv}$ is responsible for turbulent transport of mean momentum (see Eq. (4.2.6)). Hence, when $-\overline{uv}$ is increased or reduced by curvature, the mean momentum transport from the free stream, through the boundary layer, to the underlying surface will also be enhanced or suppressed. At the surface, this momentum flux is balanced by viscous skin friction. Curvature effects on the turbulent shear stress will be manifested in the surface skin friction. It follows that convex curvature will diminish the skin friction coefficient and concave curvature will enhance it. This is verified by the experimental data and SMC computations in Figure 7.6.

### 7.2.3 Algebraic stress approximation and nonlinear eddy viscosity

In the early days of Reynolds stress transport modeling, the computational expense of solving five additional transport equations was daunting. Even now these models can be computationally stiff and present many other challenges. Such considerations lead Rodi (1976) to propose an approximation to reduce them from partial differential equations to a set of algebraic equations. Thus a foundation was laid for nonlinear constitutive modeling.

If the transport terms in (3.2.4) are added to (7.1.4), the result can be written symbolically as

$$D_t\overline{u_iu_j} - \mathcal{T}_{ij} - \nu\nabla^2\overline{u_iu_j} = \mathcal{P}_{ij} + \wp_{ij} - \tfrac{2}{3}\varepsilon\delta_{ij}, \tag{7.2.8}$$

where $\mathcal{T}_{ij}$ is the term marked "turbulent transport" in (3.2.4). In the algebraic approximation, the transport terms are assumed to be proportional to the Reynolds stress tensor times a scalar. That scalar is the transport of kinetic energy. Specifically, the approximation is

$$\mathrm{D}_t\overline{u_iu_j} - \mathcal{T}_{ij} - \nu\nabla^2\overline{u_iu_j} \approx \frac{\overline{u_iu_j}}{k}(\mathrm{D}_t k - \mathcal{T} - \nu\nabla^2 k), \qquad (7.2.9)$$

where $\mathcal{T} = \frac{1}{2}\mathcal{T}_{kk}$. The kinetic energy transport equation is

$$\mathrm{D}_t k - \mathcal{T} - \nu\nabla^2 k = \mathcal{P} - \varepsilon.$$

This and the approximation (7.2.9) reduce (7.2.8) to the algebraic equation

$$\frac{\overline{u_iu_j}}{k}(\mathcal{P} - \varepsilon) = \mathcal{P}_{ij} + \wp_{ij} - \frac{2}{3}\varepsilon\delta_{ij}. \qquad (7.2.10)$$

If closure models for $\wp_{ij}$ and the definition of the production tensor $\mathcal{P}_{ij}$ are substituted, this becomes a set of algebraic equations for the unknown $\overline{u_iu_j}$. Transport equations are solved for $k$ and $\varepsilon$; in particular, $k$ and $\varepsilon$ are treated as known quantities. That is the essence of the algebraic stress model.

For the general linear model, the algebraic approximation is identical to the evolution equation (7.2.1) with the left-hand side set to zero. Hence, it is equivalent to an equilibrium assumption for the anisotropy tensor.

A broader view is that algebraic stress models (ASM) are a type of constitutive relation. Formula (7.1.33) expresses redistribution as a function of rate of strain $\boldsymbol{S}$ and rate of rotation $\boldsymbol{\Omega}$; thus (7.2.10) is an implicit relation between stress, rate of strain, and rate of rotation. A formula relating stress to rates of strain and rotation is called a constitutive relation; so (7.2.10) is the implicit statement of a constitutive relation. An explicit solution – an explicit algebraic stress model (EASM)–can be derived for the linear and the quasi-linear redistribution models, as will be seen on page 229. We will leave those mathematics to the next chapter and proceed more informally with the idea of a nonlinear constitutive model.

The general functional form of a nonlinear constitutive model is

$$\overline{u_iu_j} = F_{ij}(\boldsymbol{S}, \boldsymbol{\Omega}; k, \varepsilon).$$

Note that the functional dependence can include the non-dimensional magnitudes $|\boldsymbol{S}|k/\varepsilon$ and $|\boldsymbol{\Omega}|k/\varepsilon$, where $|\boldsymbol{\Omega}|^2 = \Omega_{ij}\Omega_{ij}$ and similarly for $|\boldsymbol{S}|$. For the $k$–$\omega$ model, the ratio $k/\varepsilon$ is replaced by $1/\omega$.

The linear eddy-viscosity model is the special case

$$\overline{u_iu_j} = -c_1 S_{ij} + \tfrac{2}{3}\delta_{ij}k,$$

with $c_1 = 2\nu_{\mathrm{T}} = 2C_\mu k^2/\varepsilon$. The quasi-linear form $C_\mu = f(|\boldsymbol{S}|k/\varepsilon,\ |\boldsymbol{\Omega}|k/\varepsilon)$ can be introduced as an *ad hoc* extension. The term *quasi-linear* refers to the tensoral dependence being linear in the rate of strain, but the coefficient introduces nonlinear dependence on the magnitude of rate of strain and rotation. As an example, Shih *et al.* (1995) invoked a realizability constraint and postulated that

$$c_1 \sim \frac{\alpha_0}{\alpha_1 + k/\varepsilon(|\boldsymbol{S}|^2 + |\boldsymbol{\Omega}|^2)^{1/2}}. \qquad (7.2.11)$$

This is often called the "realizable" $k-\varepsilon$ model. One of the failures of the strictly linear form is that it is insensitive to rotation. This formula introduces a dependence on rotation. Strong rotation can stabilize the turbulence; this formula does not correct that flaw, but other quasi-linear formulations have that motive.

The starting point for a fully nonlinear constitutive relation is the material in Section 2.3.2 on representation theorems for tensor functions of two tensors. There are a large number of nonlinear eddy viscosity models in the literature; the majority retain only quadratic terms

$$\begin{aligned}\overline{u_i u_j} - \tfrac{2}{3}\delta_{ij}k = &-c_1 S_{ij} + c_2(S_{ik}S_{kj} - \tfrac{1}{3}|\boldsymbol{S}|^2\delta_{ij}) + c_3(\Omega_{ik}S_{kj} + \Omega_{jk}S_{ki}) \\ &+ c_4(\Omega_{ik}\Omega_{jk} - \tfrac{1}{3}|\boldsymbol{\Omega}|^2\delta_{ij}).\end{aligned} \tag{7.2.12}$$

Craft *et al.* (1996) provide a brief overview of various quadratic models. Let us consider how the coefficients might be prescribed.

Often models retain $c_4 \neq 0$. However, they behave erroneously in rotating isotropic flow; more importantly, $c_4 \neq 0$ causes the model to violate realizability quite dramatically. It is desirable to take $c_4 = 0$. Note that this term does not appear in the strictly two-dimensional representation (2.3.16).

For parallel shear flow, $U(y) = \mathcal{S}y$. We have seen (Eq. (7.2.2)) that the linear model incorrectly predicts

$$\overline{u^2} = \overline{v^2} = \overline{w^2} = \tfrac{2}{3}k.$$

With the definitions (3.3.8) of rate of strain and rotation, the nonlinear model (7.2.12) predicts

$$\begin{aligned}&\overline{u^2} - \tfrac{2}{3}k = \tfrac{1}{12}c_2\mathcal{S}^2 + \tfrac{1}{2}c_3\mathcal{S}^2, \qquad &&\overline{v^2} - \tfrac{2}{3}k = \tfrac{1}{12}c_2\mathcal{S}^2 - \tfrac{1}{2}c_3\mathcal{S}^2, \\ &\overline{w^2} - \tfrac{2}{3}k = -\tfrac{1}{6}c_2\mathcal{S}^2, &&\overline{uv} = -\tfrac{1}{2}c_1\mathcal{S}.\end{aligned} \tag{7.2.13}$$

The coefficient $c_2$ creates anisotropy. The shear stress retains the linear eddy viscosity form.

For a general 2D incompressible flow, the components of the mean rate of strain and mean vorticity tensors are given by

$$\boldsymbol{S} = \begin{pmatrix} \lambda & 0 & 0 \\ 0 & -\lambda & 0 \\ 0 & 0 & 0 \end{pmatrix}, \quad \boldsymbol{\Omega} = \begin{pmatrix} 0 & \omega & 0 \\ -\omega & 0 & 0 \\ 0 & 0 & 0 \end{pmatrix}$$

in principal axes of $\boldsymbol{S}$. We have $|\boldsymbol{S}|^2 = 2\lambda^2$ and $|\boldsymbol{\Omega}|^2 = 2\omega^2$. Using (7.2.12), with $c_4 = 0$, the non-zero components of the Reynolds stress tensor are

$$\begin{aligned}&\overline{u^2} - \tfrac{2}{3}k = -c_1\lambda + \tfrac{1}{3}c_2\lambda^2, \qquad &&\overline{v^2} - \tfrac{2}{3}k = c_1\lambda + \tfrac{1}{3}c_2\lambda^2, \\ &\overline{w^2} - \tfrac{2}{3}k = -\tfrac{2}{3}c_2\lambda^2, &&\overline{u_1u_2} = 2c_3\lambda\omega.\end{aligned} \tag{7.2.14}$$

The coefficients $c_i$ can be functions of the mean deformation rate. However, the functional dependence is *a priori* unknown, so it would be beneficial to impose constraints. One such constraint is realizability (discussed in Section 8.1.2). The realizability constraints are equivalent to the Schwartz inequality $(\overline{u_i u_j})^2 \le \overline{u_i^2}\,\overline{u_j^2}$:

$$\tfrac{4}{3} \ge b_{ij} \ge -\tfrac{2}{3}, \qquad b_{12} \le b_{11}b_{22} - \tfrac{2}{3}b_{33} + \tfrac{4}{9}. \tag{7.2.15}$$

The first insists on the variances being positive. Imposing these constraints on (7.2.14) gives

$$c_1 \le \frac{c_2\lambda^2 - 4}{3\lambda}, \qquad c_2 \le \frac{1}{\lambda^2}, \qquad c_3 \le \frac{9(c_1\lambda)^2 - (c_2\lambda^2)^2 - 4c_2\lambda^2 - 4}{9\lambda\omega}.$$

These inequalities sometimes serve as constraints to derive an allowable mathematical form of the model coefficients.

Craft *et al.* (1996) noted that, while modest improvements could be made using the simple quadratic expansion, a cubic constitutive equation was needed to account for streamline curvature. They proposed

$$b_{ij} = b_{ij}^{2D} + c_5 S_{kl}(\Omega_{lj}S_{ki} + \Omega_{li}S_{kj}) + c_7 S_{ij}(|\boldsymbol{S}|^2 - |\boldsymbol{\Omega}|^2), \tag{7.2.16}$$

where $b_{ij}^{2D}$ denotes the quadratic expansion (7.2.12). Optimization over a wide range of flows determined the linear coefficient:

$$c_1 \sim \left[1 + \max(|\boldsymbol{S}|^2,\ |\boldsymbol{\Omega}|^2)^{3/2}\right]^{-1}. \tag{7.2.17}$$

## 7.3 Non-homogeneity

All real flows are non-homogeneous. This does not mean that the material in Section 7.1 is irrelevant. It should be understood as development of the quasi-homogeneous portion of SMC models. Quasi-homogeneous means that the equations were developed for homogeneous conditions, but they can be applied to situations in which all variables are functions of position but do not vary rapidly. Truly non-homogeneous effects must be added explicitly to the model. These additional terms may control the spreading rate of shear layers, the skin friction in boundary layers, or the strength of mean flow vortices.

There are two critical effects of non-homogeneity on the mathematical modeling: the turbulent transport terms in the Reynolds stress budget do not vanish; and the steps to derive the velocity–pressure gradient correlation (7.1.19) from Eq. (7.1.17) are not correct. In the former case, turbulent transport terms play a role in all inhomogeneous flows of engineering relevance. The latter failing is critical in the vicinity of walls, although starting well above the viscous sublayer.

A turbulence model must also be capable of satisfying suitable boundary conditions at the wall. The non-homogeneous parts of the model permit this. In addition to these modifications of the turbulence closure by non-homogeneity, molecular transport becomes dominant within viscous wall layers. However, the molecular terms are exact and need simply be added to the equations.

Transport terms will be discussed first, then the more difficult topic of wall effects on the redistribution model is covered.

### 7.3.1 Turbulent transport

The Reynolds stress budget (3.2.4), page 49, contains both turbulent and molecular transport on its right-hand side. The relevant terms are

$$-\partial_k \overline{u_k u_i u_j} + \nu \nabla^2 \overline{u_i u_j}.$$

The second term, molecular transport, is closed because it involves only the dependent variable $\overline{u_i u_j}$. Often, modeling turbulent transport is characterized as representing $\overline{u_k u_i u_j}$ as a tensor function of $\overline{u_i u_j}$. That philosophy leads to rather complex formulas because the symmetry in $i, j, k$ should be respected. However, the term being modeled, $\partial_k \overline{u_k u_i u_j}$, is only symmetric in $i, j$. The three-fold symmetry is not apparent in the Reynolds stress transport equation. Hence, there is little motive to constrain the model to satisfy the hidden symmetry–especially when it causes a great deal of added complexity. The inviolable constraints are that the model must preserve the conservation form and be symmetric in $i$ and $j$.

The notion that the third velocity moment represents random convection by turbulence, cited below (6.2.4), is again invoked. The Markovian assumption, that this can be modeled by gradient diffusion, is made in most engineering models. An exception occurs in buoyantly driven flows, where third-moment transport models occasionally have been found necessary. The most common closure is the Daly and Harlow (1970) formula. This is a gradient transport model with a tensor eddy viscosity:

$$-\partial_k \overline{u_k u_i u_j} = \partial_k (C_s T\, \overline{u_k u_l}\, \partial_l \overline{u_i u_j}). \tag{7.3.1}$$

Here the eddy viscosity tensor is $C_s T\, \overline{u_k u_l} \equiv \nu_{T_{kl}}$. A typical value for $C_s$ is 0.22 (Launder, 1989). Near to a wall, the dominant component of the gradient is in the wall normal direction, $y$. Then (7.3.1) is approximately $\partial_y (C_s T\, \overline{v^2}\, \partial_y \overline{u_i u_j})$. The dominant eddy viscosity is $\nu_T = C_s T\, \overline{v^2}$. One influence of a wall is to suppress $\overline{v^2}$ relative to the other intensities. This is due, in part, to the impenetrability of the boundary; but also, a strong shear, $\partial_y U$, will reduce $\overline{v^2}/k$, even without a direct boundary influence. Both of these phenomena will reduce the effective eddy viscosity near a wall beneath a boundary layer. If the closure model is able to capture the correct near-wall behavior of $\overline{v^2}$, then the Daly–Harlow transport model is able to represent the suppression of wall normal turbulent diffusion. With the caveat that the model must be able to predict $\overline{v^2}$ correctly, (7.3.1) proves to be more accurate than a simpler model proposed by Mellor and Herring (1973):

$$-\partial_k \overline{u_k u_i u_j} = \partial_k (C_s T k\, \partial_k \overline{u_i u_j}).$$

The eddy viscosity $C_s T k$ does not capture wall damping, as was seen in Figure 6.7 on page 130.

A technicality should be mentioned. Equation (3.3.7) defined the trace-free redistribution tensor

$$\rho \Pi_{ij} = \overline{u_j\, \partial_i p} + \overline{u_i\, \partial_j p} - \tfrac{2}{3} \partial_k \overline{u_k p}\, \delta_{ij}.$$

When this is used in the Reynolds stress transport equation (3.2.4), "pressure-diffusion" is usually absorbed into gradient transport so that the model (7.3.1) is presumed to be

$$-\partial_k (\overline{u_k u_i u_j} - \tfrac{2}{3} \overline{u_k p}\, \delta_{ij}/\rho) = \partial_k (C_s T\, \overline{u_k u_l}\, \partial_l \overline{u_i u_j}).$$

Evidence is that pressure-diffusion has a similar spatial distribution to turbulent self-transport, but has an opposite sign. Hence its inclusion is reflected in the value of the empirical coefficient $C_s$. For example, the model constant obtained from data without including pressure-diffusion is about 20% greater than 0.22, but it is reduced when the pressure contribution is included.

The closed Reynolds stress transport equation with (7.3.1) is

$$\partial_t \overline{u_i u_j} + U_k\, \partial_k \overline{u_i u_j} = \wp_{ij} - \tfrac{2}{3}\delta_{ij}\varepsilon + \partial_k (C_s T\, \overline{u_k u_l}\, \partial_l \overline{u_i u_j}) \\ - \overline{u_j u_k}\, \partial_k U_i - \overline{u_i u_k}\, \partial_k U_j + \nu \nabla^2 \overline{u_i u_j}. \qquad (7.3.2)$$

Only the turbulent stresses $\overline{u_i u_j}$ and dissipation rate $\varepsilon$ appear as dependent variables if one of the previous algebraic formulas in Section 7.1.3 and subsequent pages is used for $\wp_{ij}$: closure has been achieved. This closure can be used to predict free-shear flows. But near to walls the algebraic formulas for $\wp_{ij}$ can be rather erroneous, as will be seen in connection with Figure 7.8. Either walls must be avoided, or the subject of near-wall second-moment modeling must be broached.

To skirt the problem, wall function boundary conditions can be applied in the log layer; *in lieu* of a suitable model, practical computations often resort to this. The approach is quite similar to the method discussed in connection with the $k$–$\varepsilon$ model. The wall function consists of adding $\overline{u_i u_j}$ to Eqs. (6.2.22) by prescribing anisotropy ratios $b_{ij}$, such as (7.2.5), to give a $\overline{u_i u_j}$ boundary condition. The procedure is self-evident (Exercise 7.10); the intent is to circumvent the SMC model in the region where it fails. Although there is no need to discuss wall functions further, one should be aware that their use in computations often is tacit.

Wall function boundary conditions for $\overline{u_i u_j}$ are harder to justify than is the logarithmic specification for $U$. The data in Figure 6.6, page 127, show that the constant-stress assumption can be quite inaccurate for $\overline{u_i u_j}$. Pressure gradients can make a constant-stress layer even harder to locate. The subject of near-wall second-moment modeling therefore warrants scrutiny.

### 7.3.2 Near-wall modeling

In an equilibrium boundary layer, the near-wall region refers to the zone between the log layer and the wall. It includes the viscous dominated region next to the surface and the strongly inhomogeneous region above it. Generally it is a region in which non-homogeneity and viscosity play dominant roles. The near-wall region is one of "high impedance" to turbulent transport, in the sense that the wall suppresses the normal component of turbulence–as discussed in the previous subsection and in Section 4.4. This means that the layer adjacent to the wall controls skin friction and heat transfer disproportionately, making it critical to engineering applications. It is also of great interest to turbulence theory because it is the region of high shear and large rates of turbulence production.

The primary mathematical issues in near-wall modeling are boundary conditions and nonlocal wall effects on redistribution. The issue of nonlocal influences of the wall upon the redistribution model presents rather a challenge to an analytical model. These wall influences can have pronounced effects. Various methods have been developed to represent nonlocal wall influences. The two discussed in Section 7.3.4 are wall-echo and elliptic relaxation.

Asymptotic ordering of Reynolds stress components is the subject of the next subsection. Often a model cannot satisfy the exact asymptotic scaling near a wall. That may or may not be important: the exact behaviors described in the next section might not be as significant as obtaining correct orders of magnitude, and relative strengths, of the Reynolds stress components.

### 7.3.3 No-slip condition

It might seem that the boundary condition at a no-slip wall for the Reynolds stress tensor is simply $\overline{u_i u_j} = 0$. While that is correct, the power of $y$ with which the zero value is approached is often of importance. Models should be designed such that their near-wall asymptotic behavior is reasonable. We will examine the consequences of the no-slip boundary condition on the asymptotic behavior of turbulence statistics near a wall, determining the power of $y$ with which various quantities vary as $y \to 0$.

Let the no-slip wall be the plane $y = 0$. The no-slip condition is that all components of velocity vanish: $\boldsymbol{u} = 0$. Even if the wall is moving, all components of the turbulent velocity vanish, provided that the wall motion is not random. If the velocity is a smooth function of $y$, it can be expanded in a Taylor series,

$$u_i = a_i + b_i y + c_i y^2 + \cdots,$$

where $a_i$, $b_i$, and $c_i$ are functions of $x$ and $z$. The no-slip condition requires that $a_i = 0$. Thus the tangential components satisfy $u = O(y)$ and $w = O(y)$ as $y \to 0$. However, the continuity equation, $\partial_y v = -\partial_x u - \partial_z w$, shows that $b_2 = 0$ and thus $v = O(y^2)$. From these limits of the fluctuating velocity, the Reynolds stresses are found to behave like

$$\begin{aligned} &\overline{u^2} = O(y^2), \qquad &&\overline{v^2} = O(y^4), \qquad &&\overline{w^2} = O(y^2),\\ &\overline{uv} = O(y^3), &&\overline{uw} = O(y^2), &&\overline{vw} = O(y^3), \end{aligned} \tag{7.3.3}$$

as $y \to 0$. The solution to a Reynolds stress model should in principle be consistent with these. However, in practice, it may be sufficient to ensure that $-\overline{uv}$ and $\overline{v^2}$ are small compared to $\overline{u^2}$ when $y_+ \ll 1$ (Figure 6.6, page 127). This implements the suppression of normal transport in the immediate vicinity of the wall. The formality $y_+ \ll 1$ can be taken with a grain of salt; the powers (7.3.3) are satisfied by experimental data when $y_+ \lesssim 5$.

Derivatives of velocity in the plane of the wall are $O(y)$. For instance,

$$\partial_x u = \lim_{\Delta x \to 0} (u(x + \Delta x) - u(x))/\Delta x = O(y)$$

because $u(x + \Delta x)$ and $u(x)$ are both $O(y)$. Derivatives of the tangential velocity in the normal direction are seen to be $O(1)$. Thus the limiting behavior of the dissipation rate is

$$\varepsilon = \nu[\,\overline{\partial_i u_j\, \partial_i u_j}\,] \to \nu[\overline{(\partial_y u)^2} + \overline{(\partial_y w)^2}] + O(y^2), \qquad y \to 0. \tag{7.3.4}$$

This is $O(1)$ at the surface: see Figure 3.4 on page 52. The components of the dissipation rate tensor, $\varepsilon_{ij} = 2\nu(\,\overline{\partial_i u_j\, \partial_i u_k})$, behave as

$$\begin{aligned} &\varepsilon_{11} = O(1), \qquad &&\varepsilon_{22} = O(y^2), \qquad &&\varepsilon_{33} = O(1),\\ &\varepsilon_{12} = O(y), &&\varepsilon_{13} = O(1), &&\varepsilon_{23} = O(y). \end{aligned} \tag{7.3.5}$$

These are derived by considerations such as $\varepsilon_{12} \to 2\nu(\overline{\partial_y u\, \partial_y v}) = O(y)$, using the near-wall behavior of the fluctuation velocity cited above. Note that $\varepsilon_{ij} = O(\overline{u_i u_j}/k)$ as $y \to 0$. This proves to be a useful observation about near-wall scaling.

A consideration of the various contributions to the Reynolds stress budget (3.2.4) shows that the dominant balance near a surface is between dissipation, molecular diffusion, and the pressure term. The budget reduces simply to

$$\nu\, \partial_y^2 \overline{u_i u_j} = \varepsilon_{ij} \tag{7.3.6}$$

if $i$ and $j$ do not equal 2, and to

$$\nu\, \partial_y^2 \overline{u_i u_j} = \overline{u_i\, \partial_j p} + \overline{u_j\, \partial_i p} + \varepsilon_{ij}$$

if $i$ or $j$ equal 2. The first recovers the limit found previously,

$$\varepsilon_{ij} \to 2\nu\, \overline{u_i u_j}/y^2 = \overline{u_i u_j}\, \varepsilon/k,$$

if $i, j \neq 2$–here, use was made of the fact that $k \to \varepsilon y^2/2\nu$ as $y \to 0$. The second shows that $\varepsilon_{ij}$ is of the same order in $y$ as $\overline{u_i u_j}/k$, although not exactly equal to it.

Asymptotic considerations have motivated some researchers to define the redistribution tensor as

$$\wp_{ij} = -(\Pi_{ij} + \varepsilon_{ij} - \overline{u_i u_j}\, \varepsilon/k), \tag{7.3.7}$$

instead of $\wp_{ij} = -(\Pi_{ij} + \varepsilon_{ij} - \frac{2}{3}\delta_{ij}\varepsilon)$. The previous observation that $\varepsilon_{ij} \to \overline{u_i u_j}\, \varepsilon/k$ shows that all components of this $\wp_{ij}$ vanish at the wall. The Reynolds stress transport equation now becomes

$$\begin{aligned} \partial_t \overline{u_i u_j} + U_k\, \partial_k \overline{u_i u_j} &= \wp_{ij} - \overline{u_i u_j}\, \varepsilon/k + \partial_k (C_s T\, \overline{u_k u_l}\, \partial_l \overline{u_i u_j}) \\ &\quad - \overline{u_j u_k}\, \partial_k U_i - \overline{u_i u_k}\, \partial_k U_j + \nu \nabla^2 \overline{u_i u_j}. \end{aligned} \tag{7.3.8}$$

If the homogeneous redistribution models developed in Section 7.1 are denoted $\wp_{ij}^{\mathrm{h}}$, then the revised definition (7.3.7) for non-homogeneous flow is

$$\wp_{ij} = \wp_{ij}^{\mathrm{h}} + \varepsilon b_{ij}. \tag{7.3.9}$$

This transformation is accomplished quite simply by subtracting 1 from the Rotta constant, $C_1$. While that would seem to have no effect on (7.3.8), the redefinition can become nontrivial when the redistribution model is revised for nonlocal wall effects. Revisions are made to $\wp^{\mathrm{h}}$ in the wall-echo method and to $\wp$ in elliptic relaxation. These methods are our next topic.

### 7.3.4 Nonlocal wall effects

The elliptic nature of wall effects was recognized early in the literature on turbulence modeling (Chou, 1945) and has continued to influence thoughts about how to incorporate nonlocal influences of boundaries (Launder *et al.*, 1975; Durbin, 1993; Manceau *et al.*, 1999). In the literature on closure modeling, the nonlocal effect is often referred to as "pressure reflection" or "pressure echo" because it originates with the surface boundary condition imposed on the Poisson equation (7.1.15) for the perturbation pressure, $p$. The boundary condition at a wall, $y = 0$ say, is found from the Navier–Stokes equations to be $\partial_y \tilde{p} = \mu\, \partial_y^2 \tilde{v}$. However, this is usually taken to be $\partial_y \tilde{p} = 0$, ignoring the small viscous contribution. The boundary condition influences the pressure interior to the fluid through the solution to (7.1.15). Mathematically this is quite simple: the solution to the linear equation (7.1.15) consists of a particular part, forced by the right-hand side, and a homogeneous part, forced by the boundary condition. The fact that the boundary condition adds to the solution interior to the fluid can be described as a nonlocal, kinematic effect.

For the case of a plane boundary with $\partial_y p = 0$ on the surface, (7.1.15) can be solved by the method of images (Figure 7.7). A source equal to that in the fluid is imagined to lie in the wall at the negative $y$ coordinate, such that its pressure field is superposed on that in the fluid. By symmetry, the normal derivative is then zero at the wall. Mathematically, the solution (7.1.17) is modified to

$$p(\boldsymbol{x}) = \frac{1}{2\pi} \iint_{-\infty}^{\infty} \int_0^{\infty} \frac{\rho\, \partial_l U_k\, \partial'_k u_l(\boldsymbol{x}')}{[(x-x')^2 + (z-z')^2 + (y-y')^2]^{1/2}}\, \mathrm{d}y'\, \mathrm{d}x'\, \mathrm{d}z'$$

$$+ \frac{1}{2\pi} \iint_{-\infty}^{\infty} \int_0^{\infty} \frac{\rho\, \partial_l U_k\, \partial'_k u_l(\boldsymbol{x}')}{[(x-x')^2 + (z-z')^2 + (y+y')^2]^{1/2}}\, \mathrm{d}y'\, \mathrm{d}x'\, \mathrm{d}z'. \qquad (7.3.10a)$$

The second integral is the image term–note that $y + y'$ appears in its denominator. It can be verified that $\partial_y p = 0$ at $y = 0$. This solution can be combined into a single integral by extending the source symmetrically into the wall:

$$p(\boldsymbol{x}) = \frac{1}{2\pi} \iiint_{-\infty}^{\infty} \frac{\rho\, \partial_l U_k\, \partial'_k u_l(x', z', |y'|)}{|\boldsymbol{x} - \boldsymbol{x}'|} \mathrm{d}^3 \boldsymbol{x}'. \qquad (7.3.10b)$$

The two forms (7.3.10a) and (7.3.10b) represent two ways to think about wall influences on the redistribution term. The first is an additive correction. The second is more difficult to summarize, but, essentially, it is a modified operator on the source of pressure fluctuations.

Figure 7.7 schematizes nonlocality in the Poisson equation as a reflected pressure wave; however, for incompressible turbulent fluctuations, the wall effect is instantaneous, though nonlocal. The two integrals in (7.3.10a) are represented by the direct and reflected waves in the figure. Either the equation, or the figure, illustrate that wall reflection enhances pressure fluctuations: the "reflected" pressure has the same sign as the "incident" pressure and enhances it. As a corollary, Manceau *et al.* (1999) showed that pressure reflection can increase the redistribution of Reynolds stress anisotropy. The notion that the

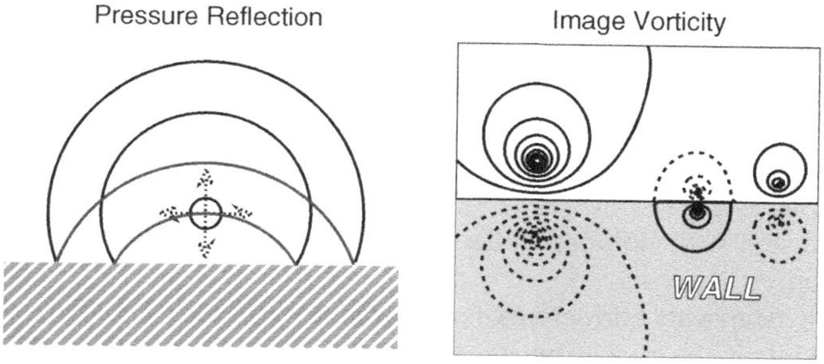


**Figure 7.7** Schematic representations of nonlocal wall influences. Pressure reflection is characterized by a reflected pressure wave at the left. At the right, image vorticity characterizes the blocking effect of a solid wall.

redistribution is *increased* by wall reflection is contrary to many second-moment closure (SMC) models: they represent pressure echo as a *reduction* of the redistribution term.

The association of inviscid wall effects with pressure reflection is natural because the pressure enters the Reynolds stress transport equation through the velocity–pressure gradient correlation. Suppression of the normal component of pressure gradient by the wall should have an effect on the rate of redistribution of variance between components of the Reynolds stress tensor that contain the normal velocity component. This effect should enter the evolution equation for the Reynolds stress (3.2.4). However, there is another notion about how the anisotropy of the Reynolds stress tensor is altered nonlocally by the presence of a wall – one that is not so easily identified in the transport equations. The inviscid boundary condition on the normal component of velocity is the no-flux condition $\boldsymbol{u} \cdot \hat{\boldsymbol{n}} = 0$. The normal component is suppressed within an extended neighborhood of the wall. As in the case of pressure reflection, the wall boundary condition alters the flow interior to the fluid through nonlocal kinematics. This perspective on ellipticity is often referred to as "kinematic blocking" (Hunt and Graham 1978). It explains why, intuitively, one feels that wall effects should suppress transfer of energy into the normal component of intensity. At a plane boundary, this can be termed the "image vorticity" effect (Figure 7.7).

Kinematic blocking is not an alteration of the redistribution tensor. It cannot be directly identified in the Reynolds stress transport equations. Blocking is a continuity effect. Suppose that a field of homogeneous turbulence $\boldsymbol{u}_\infty$ exists, and instantaneously a wall is inserted at $y = 0$. Then the velocity will instantaneously be altered to $\boldsymbol{u} = \boldsymbol{u}_\infty - \nabla\phi$, where $\nabla\phi$ is a potential flow caused by the wall. The no-penetration boundary condition, $\boldsymbol{u} \cdot \hat{\boldsymbol{n}} = 0$ requires that $\hat{\boldsymbol{n}} \cdot \nabla\phi = \boldsymbol{u}_\infty \cdot \hat{\boldsymbol{n}}$ on $y = 0$. Incompressibility applied to the entire velocity $\nabla \cdot \boldsymbol{u} = 0$ and to the turbulent velocity far above the wall $\nabla \cdot \boldsymbol{u}_\infty = 0$ implies that $\nabla^2\phi = 0$. This problem will be solved in Section 11.2.2. The analysis is purely kinematic: solve Laplace's equation with $\hat{\boldsymbol{n}} \cdot \nabla\phi = \boldsymbol{u}_\infty \cdot \hat{\boldsymbol{n}}$ on the wall; there is no role for Reynolds stress dynamics. Indeed, without invoking a continuity equation, it is difficult to locate kinematic blocking. The Reynolds stress transport equations are moments of the momentum equations alone; continuity does not add extra single-point moment equations. Hence, continuity is not invoked and the blocking effect has no direct representation in single-point models. However, the physical concept of blocking does guide the development of near-wall modifications.

The concept is that nonlocal, elliptic wall effects originate in exact kinematics; but the practical question of how to incorporate nonlocality into single-point moment closures is somewhat elusive. Exact expressions derived from kinematics–Eqs. (7.3.10) are an example–are unclosed because single-point statistics, such as $\overline{u_i \, \partial_j p}$, are found to be functions of two-point correlations. It is not possible to include the exact elliptic formulations in a Reynolds stress transport model. Research in this area has sought circuitous methods to represent wall influences. Two approaches will be described: wall-echo and elliptic relaxation.

The need for near-wall corrections is illustrated by Figure 7.8. This shows the wall normal component of the redistribution tensor in plane channel flow. The symbols are DNS data and the lines are models. The IP and SSG models both grossly overpredict $\wp_{22}$ near the surface. The definition (7.3.7) was used in Figure 7.8. This definition causes $\wp_{ij}$ to be zero at the wall. Without the redefinition of $\boldsymbol{\wp}$, the quasi-homogeneous model would be even more at odds with the data.

The IP and SSG models include the slow term in the form of Eq. (7.1.10). In that form, Eq. (6.2.18), page 129, was used for the time-scale so that the Rotta slow model is

$$\wp_{22}^{\text{slow}} = -(C_1 - 1)\frac{\overline{v^2} - \frac{2}{3}k}{T}$$

(recall that the definition (7.3.9) reduces $C_1$ by 1). The time-scale (6.2.18) goes to the finite limit $6\sqrt{\nu/\varepsilon}$ at the wall, while $k$ and $\overline{v^2}$ go to zero, so that the slow term vanishes as it should when (7.3.9) is used. Note that, if only $T = k/\varepsilon$ were used, the time-scale would vanish and $\wp_{22}$ would not correctly go to zero. Even with this, Figure 7.8 shows that when $y_+ < 100$ the models are sorely in need of fixing. It was seen in Figure 7.2 that the rapid and slow contributions to $\wp_{22}$ are comparable for the IP model while the slow term dominates in SSG. In general, it is the whole redistribution formula that is in error. Let us consider this problem from the perspective that the quasi-homogeneous approximation for redistribution has failed.

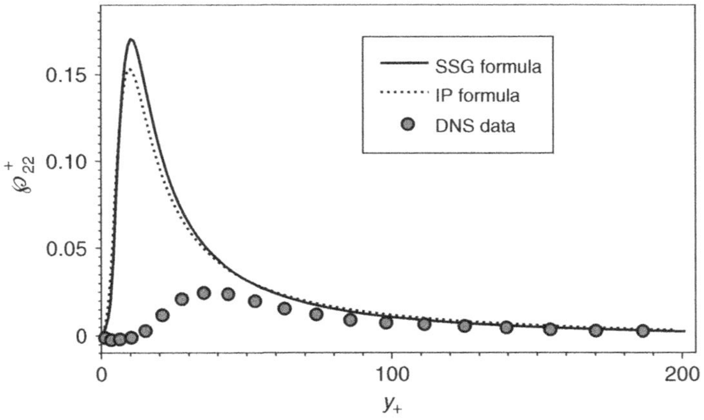


**Figure 7.8** The SSG and IP formulas for the homogeneous redistribution model $\wp_{22}^{\text{h}}$ in plane channel flow with $R_\tau = 395$. The models are erroneous near the wall; DNS data show the correct profile.

#### 7.3.4.1 Wall echo

The velocity–pressure gradient correlation corresponding to (7.1.19) can be rederived for the pressure (7.3.10a). The derivation is a matter of formal manipulation. It suffices to state the result symbolically as

$$-\phi_{ij} = (\nabla_\xi^2)_+^{-1} S_{ij}(\boldsymbol{x}') + (\nabla_\xi^2)_-^{-1} S_{ij}(\boldsymbol{x}'),$$

where $S_{ij}$ is the source term $2\,\partial_i \partial_k \overline{u_j u_l}(\boldsymbol{x})\,\partial_l U_k(\boldsymbol{x})$ plus its transpose, as in (7.1.19). The notation $(\nabla_\xi^2)_\pm^{-1}$ is used to represent the two integrations in (7.3.10a). The "+" subscript is the integral with $y - y'$ in the denominator and the "−" has $y + y'$ in the denominator. For the purposes of modeling, this form of $\phi_{ij}$ suggests a decomposition of the redistribution formula into a homogeneous model, plus an additive wall correction

$$\wp_{ij} = \wp_{ij}^{\mathrm{h}} + \wp_{ij}^{\mathrm{w}}, \tag{7.3.11}$$

corresponding to the direct contribution $(\nabla_\xi^2)_+^{-1}$ and the image contribution $(\nabla_\xi^2)_-^{-1}$, respectively. The $\wp_{ij}^{\mathrm{h}}$ term represents one of the models developed in Section 7.1, such as the IP formula (7.4.2.6) (see Exercise 7.7) plus the Rotta return to isotropy (7.1.8).

The additive wall correction $\wp_{ij}^{\mathrm{w}}$ is often referred to as the wall-echo contribution. It is modeled as a function of the unit wall normal $\hat{\boldsymbol{n}}$ and of the shortest distance to the wall, $d$. The wall normal is used interior to the fluid, where it is ill-defined: it is usually chosen as the wall normal at the nearest point on the wall. With this choice, $\hat{\boldsymbol{n}}$ is discontinuous at a sharp corner; it changes abruptly at a surface in the fluid, across which the nearest point moves from one side of the corner to the other.

A simple example of a wall-echo term is the formula

$$\begin{aligned}\wp_{ij}^{\mathrm{w}} &= -C_1^{\mathrm{w}}\frac{\varepsilon}{k}[\overline{u_i u_m}\,\hat{n}_{\mathrm{m}}\hat{n}_j + \overline{u_j u_m}\,\hat{n}_{\mathrm{m}}\hat{n}_i - \tfrac{2}{3}\overline{u_m u_l}\,\hat{n}_{\mathrm{m}}\hat{n}_l\delta_{ij}]\frac{L}{d} \\ &\quad - C_2^{\mathrm{w}}[\wp_{im}^{\mathrm{rapid}}\hat{n}_{\mathrm{m}}\hat{n}_j + \wp_{jm}^{\mathrm{rapid}}\hat{n}_{\mathrm{m}}\hat{n}_i - \tfrac{2}{3}\wp_{ml}^{\mathrm{rapid}}\hat{n}_{\mathrm{m}}\hat{n}_l\delta_{ij}]\frac{L}{d}\end{aligned} \tag{7.3.12}$$

proposed by Gibson and Launder (1978). Here $L = k^{3/2}/\varepsilon$ and the $\hat{n}_i$ are components of the unit wall normal vector. The factor of $L/d$ causes this correction to vanish far from the surface. The idea is that wall effects decay at a distance on the order of the correlation scale of the turbulent eddies. Gibson and Launder (1978) used (7.3.12) in conjunction with the IP model for $\wp^{\mathrm{rapid}}$. The model constants $C_1^{\mathrm{w}} = 0.3$ and $C_2^{\mathrm{w}} = 0.18$ were suggested.

The dependence on wall distance in (7.3.12) is somewhat arbitrary. The log-layer solution $k \sim$ constant and $\varepsilon \sim 1/d$ lead to $L/d \sim$ constant. This means that the wall-echo term does not vanish in the log region. It has an influence on the Reynolds stresses up to some height in the outer portion of the boundary layer. Whether or not it is physically correct to include wall-echo contributions in the log layer has been a matter of debate. However, since the wall correction does play a role in the log layer, its empirical constants can be selected to improve the agreement between the model and measurements of $\overline{u_i u_j}$. This has been done in the case of (7.3.12) and in Launder and Shima (1989).

If the wall normal is the $x_2$ direction, then the rapid contribution to (7.3.12) is

$$\wp^{w}_{22} = -\frac{4}{3} C^{w}_{2} \wp^{\text{rapid}}_{22} \frac{L}{x_2}$$

for the wall normal intensity. In shear flow parallel to the wall, energy is redistributed from $\overline{u^2}$ into $\overline{v^2}$ so $\wp^{\text{rapid}}_{22} < 0$. With $\wp^{\text{rapid}}_{22} < 0$, the wall correction is negative. This is consistent with the idea that blocking suppresses the wall normal component of intensity.

However, in a flow toward the wall, the velocity has a component $V(y)$. On the stagnation streamline, the mean rate of strain $\partial_y V$ will produce $\overline{v^2}$ and energy will be redistributed out of this component: $\wp^{\text{rapid}}_{22} < 0$. The above wall correction is then positive. It therefore has the erroneous effect of enhancing the normal component of intensity (Craft *et al.*, 1993). Clearly (7.3.12) is not a generally suitable model of wall-echo and blocking.

The formula (7.3.12) illustrates that wall corrections are tensoral operators that act on the Reynolds stress tensor. The $\hat{n}_i$ dependence of these operators has to be adjusted to properly damp each component of $\overline{u_i u_j}$. Equation (7.3.12) is designed to suppress the component of intensity normal to the wall if the flow is parallel to the wall. However, when the flow impinges perpendicularly to the wall, it erroneously amplifies the normal component of intensity. It is no easy matter to enforce blocking correctly. The formula for the correction function, $\phi^{w}_{ij}$, has to be readjusted in a suitable manner for each homogeneous redistribution model to which it is applied. For instance, we have seen that the relative magnitudes of the rapid and slow contributions to $\wp_{ij}$ differ between the IP and SSG models. This demands that wall-echo be adapted differently in each instance (Lai and So 1990). Such complications have led to the additive wall-echo methodology being largely abandoned by the research community.

#### 7.3.4.2 Elliptic relaxation

Elliptic relaxation (Durbin, 1993) is a rather different approach to wall effects. It is based on the second perspective on the pressure boundary condition, Eq. (7.3.10b). Effectively, instead of adding a wall-echo term, the integral operator is changed. In practice, the integrated form (7.3.10b) is not used; rather, a non-homogeneous, elliptic equation is derived. Contact with homogeneous redistribution models, such as those described in Section 7.1, is made via the source term in the elliptic equation. One is truly building on the foundation developed earlier in this chapter; but the outcome can be a significant revision to the redistribution tensor, as will be seen.

The elliptic relaxation method can be developed by a modification to the usual rationale for pressure–strain modeling. The analysis is rather informal; it is in the vein of formulating a template for elliptic relaxation, rather than being a derivation *per se*.

For simplicity, write (7.3.10b) as

$$p(\boldsymbol{x}) = \frac{1}{4\pi} \iiint_{-\infty}^{\infty} \frac{S(\boldsymbol{x}')}{|\boldsymbol{x} - \boldsymbol{x}'|} \, \mathrm{d}^3 \boldsymbol{x}'. \tag{7.3.13}$$

The redistribution term in the exact, unclosed second-moment closure equations includes the velocity–pressure gradient correlation,

$$\overline{u_i\,\partial_j p(\boldsymbol{x})} = \frac{1}{4\pi}\iiint_{-\infty}^{\infty} \frac{\overline{u_i(\boldsymbol{x})\,\partial_j S(\boldsymbol{x}')}}{|\boldsymbol{x}-\boldsymbol{x}'|}\,\mathrm{d}^3\boldsymbol{x}'. \tag{7.3.14}$$

When the turbulence is homogeneous, the nonlocal closure problem is masked by translational invariance of two-point statistics. The analyses for homogeneous turbulence in Section 7.1.4 noted at this point in the derivation that $\overline{u_i(\boldsymbol{x})\,\partial_j S(\boldsymbol{x}')}$ is a function of $\boldsymbol{x}-\boldsymbol{x}'$ alone, so that the integral in (7.3.14) became a product of a constant fourth-order tensor, $M_{ijkl}$, and the velocity gradient, as in Eq. (7.1.19).

On the other hand, if the turbulence is not homogeneous, then (7.1.19) is not applicable, and the role of non-homogeneity in (7.3.14) must be examined. This requires a representation of the spatial correlation function in the integrand. To this end, an exponential function will be used as a device to introduce the correlation length of the turbulence into the formulation (Manceau *et al.*, 1999). Assuming the representation $\overline{u_i(\boldsymbol{x})\,\partial_j S(\boldsymbol{x}')} = Q_{ij}(\boldsymbol{x}')\,\mathrm{e}^{-|\boldsymbol{x}-\boldsymbol{x}'|/L}$ and substituting it into (7.3.14) gives

$$\overline{u_i\,\partial_j p(\boldsymbol{x})} = \iiint_{-\infty}^{\infty} Q_{ij}(\boldsymbol{x}')\frac{\mathrm{e}^{-|\boldsymbol{x}-\boldsymbol{x}'|/L}}{4\pi|\boldsymbol{x}-\boldsymbol{x}'|}\,\mathrm{d}^3\boldsymbol{x}'. \tag{7.3.15}$$

The function $Q$ is simply a placemarker for a source term; interest is in the kernel of the integral. The representation of nonlocality in this formula can be described as both geometrical and statistical. The exponential term in the integrand is caused by statistical decorrelation between distant eddies. The denominator is the deterministic fall-off with distance of the pressure field of a point source:

$$\iiint Q_{ij}(\boldsymbol{x}')\;\underbrace{\mathrm{e}^{-|\boldsymbol{x}-\boldsymbol{x}'|/L}}_{\text{statistical decorrelation}}\;\Big/\;\underbrace{4\pi|\boldsymbol{x}-\boldsymbol{x}'|}_{\text{geometrical spreading}}\;\mathrm{d}^3\boldsymbol{x}'.$$

The kernel in (7.3.15) is the Green function for the modified Helmholtz equation. In other words, if $L$ is constant, then (7.3.15) is the solution to

$$\nabla^2\overline{u_i\,\partial_j p} - \frac{\overline{u_i\,\partial_j p}}{L^2} = -Q_{ij}. \tag{7.3.16}$$

Ultimately, the motivation for the elliptic relaxation method is to enable boundary conditions and anisotropic wall effects to be introduced into the second-moment closure model in a flexible and geometry-independent manner. In applications, $L$ is not constant and the boundaries are not plane. Formulation (7.3.16) is used, not (7.3.15).

The derivation of (7.3.16) is a justification for using the modified† Helmholtz form of the equation

$$L^2\nabla^2 f_{ij} - f_{ij} = -\frac{\wp_{ij}^{\mathrm{h}} + \varepsilon b_{ij}}{k} \equiv -\frac{\wp_{ij}^{\mathrm{qh}}}{k} \tag{7.3.17}$$

† The modified Helmholtz equation is $\nabla^2\phi - k^2\phi = 0$; the unmodified equation has a + sign.

to represent nonlocal wall effects in general geometries. On the right-hand side, the form (7.3.9) is adopted for the source term. The superscript "qh" acknowledges that this is the quasi-homogeneous model. The solution to (7.3.17) provides the non-homogeneous model. In the above equation, $f_{ij}$ is an intermediate variable, related to the redistribution tensor by $\wp_{ij} = kf_{ij}$. The turbulent kinetic energy, $k$, is used as a factor in order to enforce the correct behavior, $\wp_{ij} \to 0$, at a no-slip boundary (Section 7.3.3). The anisotropic influence of the wall on Reynolds stresses interior to the fluid arises by imposing suitable boundary conditions on the components of the $\overline{u_i u_j} - f_{ij}$ system. Boundary conditions are described in the next subsection. The wall normal now enters only into the wall boundary condition.

The length scale in (7.3.17) is prescribed by analogy to (6.2.18) as

$$L = \max\left\{c_L \frac{k^{3/2}}{\varepsilon},\ c_\eta \left(\frac{\nu^3}{\varepsilon}\right)^{1/4}\right\}. \tag{7.3.18}$$

In fully turbulent flow, it has been found that Kolmogoroff scaling collapses near-wall data quite effectively. Hence the Kolmogoroff scale is used for the lower bound in (7.3.18). Although all implementations of elliptic relaxation to date have used these simple formulas for $L$ and $T$, they are not crucial to the approach. The only important feature is that $L$ and $T$ do not vanish at no-slip surfaces. If they vanished, then the equations would become singular.

The precise form of elliptic relaxation models can be found in Durbin (1993), Wizman *et al.* (1996) or Pettersson and Andersson (1997). The elliptic relaxation procedure (7.3.17) accepts a homogeneous redistribution model on its right-hand side and operates on it with a Helmholtz type of Green function, imposing suitable wall conditions. The net result can be a substantial alteration of the near-wall behavior from that of the original redistribution model. Figure 7.9 illustrates how elliptic relaxation modifies the SSG homogeneous redistribution model. The flow is plane channel flow, with a friction velocity Reynolds number of $R_\tau = 395$. The dashed lines in this figure are the $\wp_{ij}^{\mathrm{qh}}$, computed from the original model. When the quasi-homogeneous model is used as the source term in (7.3.17), the black solid line is obtained as solution. The circles are DNS data for the exact redistribution term (Mansour *et al.*, 1988). Solutions are shown for the $\overline{uv}$, $\overline{v^2}$, and $\overline{u^2}$ components.

The elliptic relaxation solution alters the redistribution term rather dramatically when $y_+ \lesssim 80$. The magnitude and sign of $\wp_{12}^{\mathrm{qh}}$ are quite wrong; however, $\wp_{12}$ predicted by elliptic relaxation agrees quite well with the data. How can such a large alteration occur? Equation (7.3.17) is linear, so its general solution can be written as a particular part, forced by the source, plus a homogeneous part, that satisfies the boundary condition. The particular part would tend to have the same sign as the source and could not cause the sign reversal shown by the solution for $\wp_{12}$. It is the homogeneous part of the general solution that causes the sign of $\wp_{12}$ to be opposite to $\wp_{12}^{\mathrm{qh}}$ near the wall, and brings $\wp_{12}$ into agreement with the DNS data.

A similar reduction in magnitude, and drastic improvement in agreement with the data, is seen also for $\wp_{11}$ and $\wp_{22}$. Near the wall, the homogeneous model shows too large transfer out of $\overline{u^2}$ and into $\overline{v^2}$ (and $\overline{w^2}$). The elliptic relaxation procedure greatly improves agreement with the data. This includes creation of a very small negative lobe

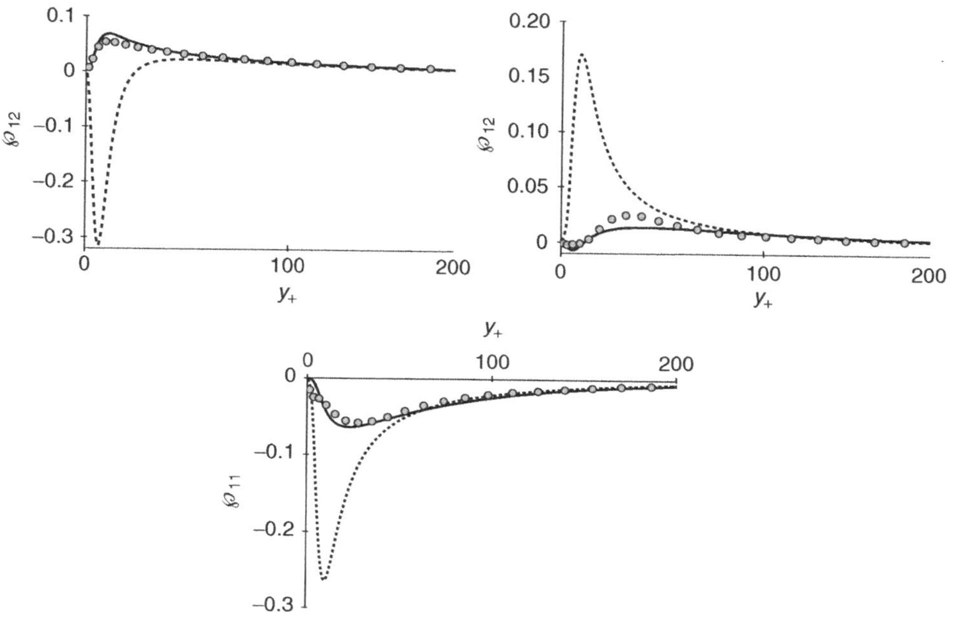


**Figure 7.9** The effect of elliptic relaxation on the SSG formula for $\wp_{ij}^{\text{qh}}$. Computation of plane channel flow with $R_\tau = 395$: solution for $\wp_{ij}$ by elliptic relaxation (———); formula for $\wp_{ij}^{\text{qh}}$ (········); and DNS data (○ ○ ○).

near the wall in the $\wp_{22}$ profile, corresponding with the data. The negative lobe is required if $\overline{v^2}$ is to be non-negative (see (7.3.20)).

Elliptic relaxation is not a panacea, but it has intriguing properties. Other avenues to geometry-independent near-wall modeling have been proposed. Tensorally nonlinear representations is one such approach (Launder and Li 1994). We will not delve into that topic.

#### 7.3.4.3 Elliptic relaxation with Reynolds stress transport

The near-wall behavior of turbulence models must be analyzed by considering the limit as $y \to 0$ of the solution. The combination of elliptic relaxation with the Reynolds stress transport equations serves to illustrate the general manner in which turbulence models can be analyzed. Such considerations play a role in the development of closure schemes for the near-wall region.

The complete set of closed Reynolds stress transport and redistribution equations that have been used in most elliptic relaxation models to date are

$$\mathrm{D}_t\overline{u_iu_j} + \varepsilon\frac{\overline{u_iu_j}}{k} = \mathcal{P}_{ij} + \wp_{ij} + \partial_k[\nu_{T_{kl}}\,\partial_l\overline{u_iu_j}] + \nu\nabla^2\overline{u_iu_j},$$

$$L^2\nabla^2 f_{ij} - f_{ij} = -\frac{\wp_{ij}^{\text{h}}}{k} - \frac{b_{ij}}{T}, \tag{7.3.19}$$

where $\wp_{ij} = kf_{ij}$. A variety of closures, additional to IP and SSG, have been inserted for $\boldsymbol{\wp}^{\mathrm{h}}$ (Pettersson and Andersson 1997).

If the turbulence is homogeneous, then $f_{ij}$ is constant and $\nabla^2 f_{ij} = 0$. Then the solution to the second of Eqs. (7.3.19) is simply

$$\wp_{ij} = \wp_{ij}^{\mathrm{h}} + \varepsilon \left( \frac{\overline{u_i u_j}}{k} - \tfrac{2}{3}\delta_{ij} \right).$$

When this is inserted into the first of Eqs. (7.3.19), the aggregate effect is to replace $\wp_{ij}$ by $\wp_{ij}^{\mathrm{h}}$ on the right-hand side and $\varepsilon\overline{u_i u_j}/k$ by $\frac{2}{3}\varepsilon\delta_{ij}$ on the left. This recovers the quasi-homogeneous model far from the wall.

The mathematical motive for inserting $\varepsilon\overline{u_i u_j}/k$ on the left-hand side of (7.3.19) is to ensure that all components of $\overline{u_i u_j}$ go to zero at least as fast as $k$ as the wall is approached. In particular, the tangential components of $\overline{u_i u_j}$ are $O(y^2)$ as $y \to 0$. The normal component becomes $O(y^4)$ if the boundary condition

$$f_{nn} = -5 \lim_{y \to 0} \left[ \frac{\varepsilon \overline{u_n u_n}}{k^2} \right] \tag{7.3.20}$$

is imposed on $y = 0$ ($n$ indicates normal direction: there is no summation over $n$). This condition is derived by noting that, as $y \to 0$, the dominant balance for the normal stress in (7.3.19) is

$$\varepsilon \frac{\overline{u_n u_n}}{k} = \wp_{nn} + \nu\, \partial_y^2 \overline{u_n u_n}.$$

Assuming $\overline{u_n u_n} \sim y^4$ gives

$$\varepsilon \frac{\overline{u_n u_n}}{k} - 12\nu \frac{\overline{u_n u_n}}{y^2} = \wp_{nn} = kf_{nn}.$$

But $k \to \varepsilon y^2/2\nu$, so the left-hand side is $-5\,\varepsilon\overline{u_n u_n}/k$, giving (7.3.20). (Recall that the asymptote $k \to \varepsilon y^2/2\nu$ follows from the limiting behavior of the $k$ equation $\varepsilon = \nu\, \partial_y^2 k$ upon integrating twice with the no-slip condition $k = \partial_y k = 0$ on $y = 0$.) It can be shown that all components containing the normal, $f_{ni}$, have the above behavior.

The dominant balance for the tangential intensity is

$$\varepsilon \frac{\overline{u_t u_t}}{k} = \nu\, \partial_y^2 \overline{u_t u_t} + O(y^2)$$

if $f_{tt} = O(1)$, so that $\wp_{tt} = O(y^2)$; the viscous and dissipation terms balance to lowest order. After substituting $k/\varepsilon = y^2/2\nu$, the above equation is found to have the solution $\overline{u_t u_t} = \frac{1}{2} A y^2$. Here $A$ is an arbitrary constant; it equals $\varepsilon_{tt}(0)/\nu$. Hence the correct asymptote $\overline{u_t u_t} \sim O(y^2)$, given in (7.3.3), is satisfied. Indeed, the term $\varepsilon\overline{u_i u_i}/k$ was used to ensure this behavior.

The tangential components $f_{t_i t_j}$ are only required to be $O(1)$ as $y \to 0$. Demuren and Wilson (1995) use the condition $f_{t_1 t_1} = f_{t_2 t_2} = -1/2 f_{nn}$ on the diagonalized tensor to ensure that $\wp_{ij}$ is trace-free. The correct tangential boundary conditions are somewhat of an open issue, although these usually suffice.

#### 7.3.4.4 The $v^2-f$ model

Reynolds stress transport equations can add to the complexity of computing fluid flow. It is not simply that the number of equations to be solved is increased, but coupling between the individual equations through production and redistribution can impede numerical convergence. Simpler models are used in most practical computations.

Elliptic relaxation adds further differential equations to be solved. A simplified variant of the approach is desirable, and this is provided by the $v^2-f$ model (Durbin, 1995; Parneix, *et al.*, 1998). The system of equations for the Reynolds stress tensor is replaced by a transport equation for a velocity scalar $\overline{v^2}$ and an elliptic equation is introduced for a function $f$. This function is analogous to a redistribution term. The objective of this model is to retain a characterization of near-wall stress anisotropy, but to embed it in a more computationally tractable formulation. Accurate prediction of heat transfer, skin friction, and boundary-layer separation in turbulent flows depends crucially on anisotropy; suppression of the normal component of turbulence relative to the tangential components impedes heat, mass, and momentum transport. The $v^2-f$ formulation attempts to capture these important processes.

The equations of the model are

$$\partial_t \overline{v^2} + U_j\, \partial_j \overline{v^2} + \varepsilon \frac{\overline{v^2}}{k} = kf + \partial_k[\nu_T\, \partial_k \overline{v^2}] + \nu\nabla^2\overline{v^2}$$

$$L^2\nabla^2 f - f = -c_2\frac{\mathcal{P}}{k} + \frac{c_1}{T}\left(\frac{\overline{v^2}}{k} - \frac{2}{3}\right). \qquad (7.3.21)$$

The source on the right-hand side of the $f$ equation is analogous to the IP form of closure. The constants are

$$c_2 = 0.3 \qquad \text{and} \qquad c_1 = 0.4,$$

which differ from the IP values for homogeneous turbulence. (Actually, the corresponding constants in the IP model would be $C_2 = \frac{3}{2}c_2$ and $C_1 = c_1 + 1$.)

The boundary condition to (7.3.21) is that of Eq. (7.3.20) for the normal component of Reynolds stress:

$$f = -5 \lim_{y\to 0}\left[\frac{\varepsilon\overline{v^2}}{k^2}\right] = -20\nu^2 \lim_{y\to 0}\left[\frac{\overline{v^2}}{\varepsilon y^4}\right] \qquad (7.3.22)$$

on a wall, $y = 0$. This ensures that $\overline{v^2} \to O(y^4)$ as $y \to 0$. The mean flow is computed with the eddy viscosity

$$\nu_T = c_\mu \overline{v^2} T. \qquad (7.3.23)$$

This formula is found to agree well with the DNS curve in Figure 7.10. Comparison to Figure 6.7 shows the advantage of working with $\overline{v^2}$ as the velocity scale.

The essential desire of this model is to represent the tendency of the wall to suppress transport in the normal direction without resort to a full second-moment closure. Representing this physical effect eliminates the need to patch the model, or to damp the eddy

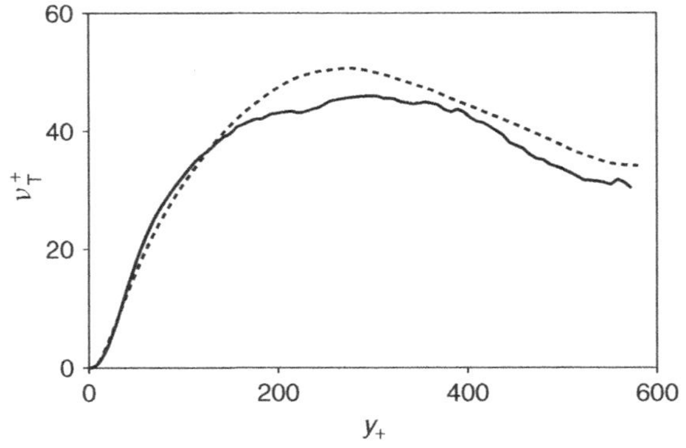


**Figure 7.10** Exact eddy viscosity (———) compared to the $\overline{v^2}-f$ formula (·········). Both curves were computed from the DNS channel flow data of Moser *et al.* (1999).

viscosity. It should be realized, however, that $\overline{v^2}$ is a scalar, not the normal component of a tensor. The boundary condition on $f$ just makes $\overline{v^2}$ behave like $\overline{u_n^2}$ near to solid walls. This is more explicit in a variant of $\overline{v^2}-f$ by Hanjalic *et al.* (2005) called the $\zeta-f$ model. It replaces $\overline{v^2}$ by $\zeta \equiv \overline{v^2}/k$, where $\zeta$ is a scalar measure of anisotropy. The operational motive for this variation is to change the wall boundary condition to $f = -2\nu\zeta/y^2$ in the interest of numerical robustness.

Equations (7.3.21) require values of $k$ and $\varepsilon$. They are determined by Eqs. (6.2.5) and (6.2.6), the only revision being to replace $C_{\varepsilon 1}$ either by $1.4[1 + 0.045(k/\overline{v^2})^{1/2}]$ or by $1.3 + 0.25/[1 + (0.15d/L)^2]^4$ (Parneix *et al.*, 1998). Either of these increases dissipation near the wall and improves the predictions of $k$. The first formula avoids reference to the wall distance, $d$. The second is a simple blending function that interpolates between $C_{\varepsilon 1} = 1.55$ as $d \to 0$ and $C_{\varepsilon 1} = 1.3$ as $d \to \infty$. This embodies one of the fundamental dilemmas in turbulence modeling: model constants are not physical constants. The constant $C_{\varepsilon 1}$ controls the growth rate of shear layers. The value of 1.55 is obtained from boundary-layer data; the value 1.3 agrees with mixing-layer data. Interpolation seems the only method to produce a flexible model that can predict both attached and free-shear layers.

## 7.4 Reynolds averaged computation

Closure models are ultimately meant to be used in computational fluid dynamics (CFD) codes for the purpose of predicting mean flow fields, and possibly the Reynolds stress components. The only information that they can provide is these low-order statistics. Any deterministic structures in the flow, such as a Von Karman vortex street, or Görtler vortices, must be computed as part of the flow; it is erroneous to assume that they are represented by the turbulence closure. On the other hand, it must be assumed that any stochastic feature of the flow is represented on a statistical level by the closure.

Random eddying motion cannot be computed correctly if a full Reynolds averaged closure is being used. When the complete randomness is computed, no model is used: this is called direct numerical simulation (DNS). A properly resolved DNS is an exact, albeit numerical, solution to the Navier–Stokes equations; it is quite analogous to a laboratory experiment. Turbulence simulation is the topic of Chapter IV. We mention it in advance to avoid confusion between Reynolds averaged CFD (referred to as RANS) and turbulence simulation.

Even though flows that are not statistically stationary may require a time-dependent RANS computation, that is not synonymous with large eddy simulation (LES). The output from a RANS computation is the mean flow, with no averaging needed. The output from DNS or LES is a random field that must be ensemble-averaged. The average flow calculated by DNS can be used as data to test RANS models. Examples of this have already been seen; further examples occur below in Section 7.4.2.

Standard discretization methods (finite-difference, finite-volume, or finite-element) can be applied to turbulent transport equations, such as those for $k$, $\varepsilon$, or $\overline{u_i u_j}$. Techniques of computational fluid dynamics are addressed in texts on the subject (Fletcher, 1991; Tannehill *et al.*, 1997). Those references should be consulted on matters of numerical analysis. The following is a brief mention of a few peculiarities that might lie outside the scope of many CFD books.

### 7.4.1 Numerical issues

Virtually all practical engineering computations are done with some variety of eddy viscosity formulation (Chapter 6). Second-moment closures (Chapter 7) promise greater fidelity to turbulence physics, provided that the closure retains the virtues of the Reynolds stress transport equations. However, the computational difficulties they present are manifold. The absence of numerically stabilizing eddy viscous terms in the mean flow equations, the strong coupling between Reynolds stress components via production and redistribution, the increased number of equations, and other computationally adverse properties lead to slow, tenuous convergence. Special methods to overcome this have been explored; the work of Lien and Leschziner (1994, 1996a) provides many suggestions. Productive research in this area continues, and increasingly complex flows are being broached.

By contrast, eddy viscosities, as a rule, assist convergence. How else could steady solutions to the Navier–Stokes equations be obtained at Reynolds numbers in the millions? The tendency for the flow equations to develop chaotic solutions is overcome by the enhanced viscous dissipation; or actually, by the transfer of mean flow energy to turbulent velocity variance. The eddy viscosity provides an explicit statement of the enhanced dissipation. It can be incorporated in the numerics so as to improve convergence; for instance, it might enhance diagonal dominance of a spatial discretization scheme.

The precise method of implementing transport equations for turbulence variables into a RANS computer code is a function of the numerical algorithm. Some general observations can be made that are relevant in many cases. However, it is especially true of second-moment closures that no hard-and-fast rules exist. Sometimes the numerical techniques are recipes that often, but not always, assist convergence.

It is common practice to decouple the turbulence model solver from the mean flow solver. The only communication from the model might be an eddy viscosity that is passed to the mean flow. The mean flow solver would then compute the Navier–Stokes equations with variable viscosity. Most applied CFD codes incorporate a selection of more than one eddy viscosity scheme; isolating the model solution from the mean flow solution simplifies the implementation of the various models.

It is not just the plethora of models that motivates a segregated solution. As a rule, turbulence models are solved more readily with implicit numerical schemes, while explicit schemes are sometimes preferred for the mean flow (Turner and Jenions, 1993). A case in point is provided by the Spalart–Allmaras (SA) model (Spalart and Allmaras, 1992): explicit methods (Tannehill *et al.*, 1997) are popular for compressible aerodynamic flows; the SA eddy viscosity transport model (Section 6.6) is also popular for compressible aerodynamics; unfortunately, there has been little success solving SA with explicit schemes. On the other hand, the SA equations are readily integrated with alternating direction implicit (ADI) or other implicit methods.

It has been argued that first-order, upwind discretization of the convective derivative is acceptable for the turbulence variables. This is not acceptable for the mean flow because of the excessive numerical diffusion. But, it is argued, the turbulence equations are dominated by source and sink terms, so inaccuracies in the convection term are quantitatively small. In most cases, that line of reasoning has been verified: solutions with first-order and with higher-order convection are quite close to one another. However, where production is small, or changes are rapid, local inaccuracies exist. A standard strategy is first to compute a preliminary solution with first-order convection, then reconverge it with higher-order accuracy. This is often effective because many turbulence models are computationally stiff. The low-order computation generates a solution close to the higher-order solution. Starting from the preliminary solution makes convergence to the more accurate solution easier. In the same vein, an eddy viscosity solution is sometimes used to initiate a second-moment computation.

A general rule of thumb for discretization is to make dissipation implicit and production explicit. To see why, consider the simple equation

$$d_t k = \mp \frac{k}{T}.$$

The "$-$" sign corresponds to dissipation and the "$+$" sign to production. A finite-difference approximation is $(k^{n+1} - k^n)/\Delta t = \mp k^{n[+1]}/T$. The notation $[+1]$ in the exponent represents an optional choice of whether to evaluate at $n+1$ or at $n$. In the former case, the right-hand side is "implicit." In the latter case, it is "explicit." With the implicit choice,

$$k^{n+1} = \frac{k^n}{(1 \pm \Delta t/T)}.$$

In the dissipative case, the factor in the denominator is $1 + \Delta t/T$, which cannot be 0; the time step can be large without worry of a singularity. In an implicit spatial discretization, this treatment of dissipation adds to diagonal dominance.

But, for the case of production, the denominator is $1 - \Delta t/T$, which can run into difficulty. The explicit case uses $(k^{n+1} - k^n)/\Delta t = \mp k^n/T$, so that

$$k^{n+1} = k^n(1 \mp \Delta t/T).$$

For production, the factor on the right is $1 + \Delta t/T$, which cannot become negative; the finite-difference approximation cannot inadvertently change production into dissipation.

With both production and dissipation,

$$d_t k = \frac{k}{T_p} - \frac{k}{T_d},$$

the rule that production is explicit and dissipation implicit gives the finite-difference formula $(k^{n+1} - k^n)/\Delta t = k^n/T_p - k^{n+1}/T_d$. Solving for $k^{n+1}$:

$$k^{n+1} = k^n \frac{1 + \Delta t/T_p}{1 + \Delta t/T_d}.$$

This rule of thumb generalizes to the evolution equation for any variable $\phi$, where $\phi$ would be $k$ in the $k$ equation, or $\overline{u_i u_j}$ in a Reynolds stress equation. In order to distinguish the implicit and explicit parts, the source term is arranged into the form $A - B\phi$, where $A, B > 0$; that is, the evolution equation is put into the form

$$D_t \phi = A - B\phi + \cdots,$$

where $A$ and $B$ can be functions of the dependent variables. The rule of thumb for treating dissipation and production is implemented by updating the source term as $A^n - B^n \phi^{n+1}$. However, the splitting between the explicit contribution $A$ and the implicit contribution $B\phi$ is not unique. As an example of the leeway, consider the right-hand side of the $k$ equation: $\mathcal{P} - \varepsilon$ might be rewritten $\mathcal{P} - (\varepsilon/k)k$, for which $A = \mathcal{P}$ and $B = \varepsilon/k$. The update is then $\mathcal{P}^n - |\varepsilon^n/k^n| k^{n+1}$; the absolute value is a good idea in case $k$ or $\varepsilon$ becomes negative in the course of iterations. Lien and Leschziner (1994) discuss the treatment of source terms in second-moment closures.

Numerical stiffness is often encountered when solving transport models for $\overline{u_i u_j}$. The root cause is the functional coupling between the Reynolds stress components introduced by the production and redistribution tensors. For example, the evolution equation for $\overline{uv}$ contains $\overline{v^2}\, \partial_y U$ in the production term and the equation for $\overline{v^2}$ contains $\overline{uv}\, \partial_y U$ in the redistribution term. When $\overline{uv}$ and $\overline{v^2}$ are solved separately – often called a segregated numerical solution – justice is not done to this intimate coupling. On occasion it has been proposed to solve the full set of components simultaneously, as a coupled system. However, no general scheme has been offered for robustly solving Reynolds stress transport equations.

The Reynolds stress appears as a body force if it is treated explicitly in the mean flow equation (3.2.2). But the Reynolds stress is inherently a diffusive term that should provide numerical stability. To recover the diffusive property, eddy diffusion can be subtracted from both sides of the mean flow equation, treating it implicitly on one side and explicitly on the other, as in

$$\begin{aligned} &\partial_t U_i + U_j\, \partial_j U_i - \underline{\partial_j [\nu_T (\partial_j U_i + \partial_i U_j)]^{(n+1)}} \\ &\quad = -\frac{1}{\rho} \partial_i P + \nu \nabla^2 U_i - \partial_j \overline{u_j u_i}^{(n)} - \underline{\partial_j [\nu_T (\partial_j U_i + \partial_i U_j)]^{(n)}}. \end{aligned} \tag{7.4.1}$$

The superscripts on the underlined terms denote the level of iteration. Lien and Leschziner (1996a) argue that the eddy viscosity should be chosen by formal approximation to the

Reynolds stress transport equations. This requires that it be a matrix, $\boldsymbol{\nu}_T$. The rationale is that the procedure (7.4.1) will be the most effective if

$$\nabla \cdot \overline{\boldsymbol{uu}} + \nabla \cdot (2\boldsymbol{\nu}_T \cdot \boldsymbol{S} - \tfrac{2}{3}k\boldsymbol{I})$$

is made small. In practice, simply using the $k$–$\varepsilon$ formula (6.2.2) for $\nu_T$ in Eq. (7.4.1) adds greatly to the ease of convergence.

Boundary conditions can be problematic, especially with elliptic relaxation methods. The appearance of $y^4$ in the boundary condition (7.3.22) is a source of computational stiffness. It is ameliorated if the $\overline{v^2}$ and $f$ equations are solved as a coupled system so that the boundary condition can be treated implicitly. In an iterative scheme, this means that

$$f^{(n)}(0) = -20\nu^2 \lim_{y\to 0}\left[\overline{v^2}^{(n)}(y)/\varepsilon y^4\right]$$

is used on the $n$th iteration to update $f$ and $\overline{v^2}$ simultaneously. The same can be done when the $k$–$\varepsilon$ model is integrated to a no-slip wall: then

$$\varepsilon^{(n)}(0) = 2\nu \lim_{y\to 0}\left[k^{(n)}(y)/y^2\right]$$

is used on the $n$th iteration. A ratio evaluated at the first computational point from the wall is substituted for the limit on the left-hand sides of these equations: $\lim[k(y)/y^2]$ is replaced by $k(y_1)/y_1^2$.

Another approach is to modify the model. The modified $\varepsilon$ equation (6.2.21), page 131, satisfies $\tilde{\varepsilon}(0) = 0$. A modified $v^2$–$f$ model (Lien *et al.*, 1997) has also been developed to convert the boundary condition to $f(0) = 0$, and the $\zeta$–$f$ model is similarly motivated by the boundary condition.

The singular boundary condition to the $k$–$\omega$ model (Section 6.1.3) is

$$\omega(0) = \lim_{y\to 0}[6\nu/(C_{\omega 2}y^2)].$$

Wilcox (1993) suggests specifying this as the solution for the first few grid points. Menter (1994) recommends using $\omega = 10[6\nu/(C_{\omega 2}y_1^2)]$ as the boundary value. The factor of 10 is arbitrary, but he states that the results are not sensitive to the precise value.

Frequently, boundary conditions are a source of numerical stiffness. They require attention, but are a surmountable obstacle.

### 7.4.2 Examples of Reynolds averaged computation

The examples in this section are relatively simple cases of RANS computation. The experimental and DNS data cited have been used in the literature to test models. The criteria for a good test case is that it consists of reproducible, accurately measured data, in a well-defined geometrical configuration, with well-defined inlet and boundary conditions. Generally the configuration is selected to highlight a phenomenon that has bearing on a wide range of practical applications. Examples are effects of wall curvature, favorable and adverse pressure gradients, and separation.

The criterion of well-defined inlet conditions is usually met by establishing a fully developed flow upstream of the test section. This might be a fully developed boundary-layer or channel flow. Variables like $\varepsilon$ or $\omega$ that arise in turbulence models are difficult, if not impossible, to measure accurately. A fully developed inlet obviates the need for such data. One simply solves the model itself to provide inflow profiles for use in the computation. An example of this has already been seen in Figure 7.6. The flat-plate boundary layer upstream of the convex wall was fully developed, with $R_\theta = 4200$. A separate computation of this boundary layer preceded the calculation of flow round the convex bend.

The following examples are presented to illustrate the process of assessing models. No attempt is made to review the performance of the wide variety of models available in the literature. The reader can consult that literature for information on the merits of any particular model. We will, however, make a few general observations on the predictions of different types of turbulence models.

### 7.4.2.1 Plane diffuser

The flow in an asymmetric plane diffuser provides an example of separation from a smooth wall. The data of Obi *et al.* (1993) are an attractive test case. These data were reproduced and extended by Buice and Eaton (1997). Therefore, they meet the criterion of reproducibility. The inlet is a fully developed channel flow with a Reynolds number of $10^4$, based on centerline velocity and channel half-height. Figure 7.11 shows the geometry, along with computed streamlines. The dashed curve delimits a region of reversed flow. The streamlines were computed with the $v^2-f$ model, which predicts the same extent of separation as seen in the experiments.

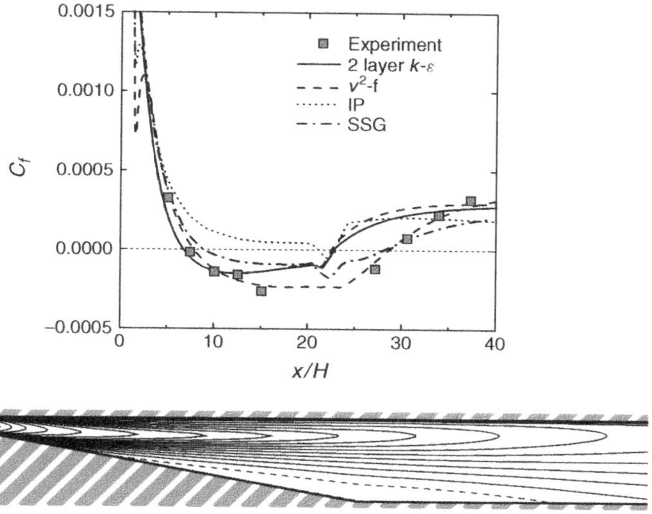


**Figure 7.11** Flow in a plane asymmetric diffuser. The inlet duct height Reynolds number is $2 \times 10^4$ and the expansion ratio is 4.7 : 1. Skin friction is plotted above, and $U$ is contoured below. The dotted line demarcates a region of negative $U$.

The upper portion of Figure 7.11 contains skin friction predictions by various models, along with the experimental data of Buice and Eaton (1997). The critical feature is the region in which $C_f < 0$. The start of this separated zone is near $x/H = 7$ and it terminates near $x/H = 30$. The predictions illustrate a general property of second-moment closures: they underpredict backflow in separated regions and sometimes fail to predict separation when it should occur. The IP curve has only a tiny extent of negative $C_f$ at the bottom of the ramp. The SSG model performs better than IP, as is generally the case. However, the IP model is less stiff computationally–although omitting the nonlinear slow term ($C_1^n$ in Eq. (7.1.9)) from the SSG model improves its computability. Both the IP and SSG quasi-homogeneous models were solved with elliptic relaxation. Similar predictions would be obtained with wall functions, although the tendency to separate would be reduced even further.

It was found by Obi *et al.* (1993) that the $k$–$\varepsilon$ model failed to predict separation when solved with wall functions. It is generally observed that wall functions tend to underpredict separation. The two-layer formulation (Section 6.2.2.2) has been found to be more accurate than wall functions in many tests (Rodi, 1991), so only it is included in Figure 7.11. Separation is indeed predicted correctly, although reattachment is premature–that is, $C_f$ crosses zero from above at the right location, but rises and crosses from below too early.

The $\overline{v^2}$–$f$ model (Eqs. (7.3.21)) shows good agreement with the data. This stems from using $\overline{v^2}$ to represent transport normal to the surface. It should be recalled that $\overline{v^2}$ in this model is not the $y$ component of intensity; it is a velocity scale that is meant to behave like the normal component near the wall, whatever Cartesian direction that may be.

#### 7.4.2.2 Backward-facing step

A standard example of massively separated flow is provided by the backward-facing step. It combines boundary-layer flow with a fully detached mixing layer in the lee of the step. Many experiments have been done on this flow. We choose the Driver and Seegmiller (1985) dataset as representative. It has a fully developed boundary layer at the inlet, atop the step. The inlet momentum thickness Reynolds number is $R_\theta = 5000$. The step-height Reynolds number is $R_H = 37\,500$ and the expansion ratio is 9 : 8. The expansion ratio is the height of the flow domain downstream of the step divided by that upstream of it. Figure 7.12 illustrates the flow via computed streamlines. The dotted lines show the separated zone, which extends six step heights downstream of the base of the step. This is a typical extent for the separated zone behind a two-dimensional high Reynolds number backstep (10 step heights would be a typical reattachment length for a step inside an axisymmetric duct).

The predictions of skin friction downstream of the step show how second-moment closures tend not to fare well in grossly separated flow. The IP and SSG models both underpredict the magnitude of $C_f$ in the region where $C_f < 0$. The skin friction also does not rise sufficiently steeply after its minimum.

The low Reynolds number $k$–$\varepsilon$ computation is representative of that class of models (see Eq. (6.2.20) and following text); the particular computation was made with a variant of the Launder and Sharma (1974) formulation. While the reattachment point, where $C_f$ crosses zero from below, is approximately at the correct location of $x/H = 6$, the

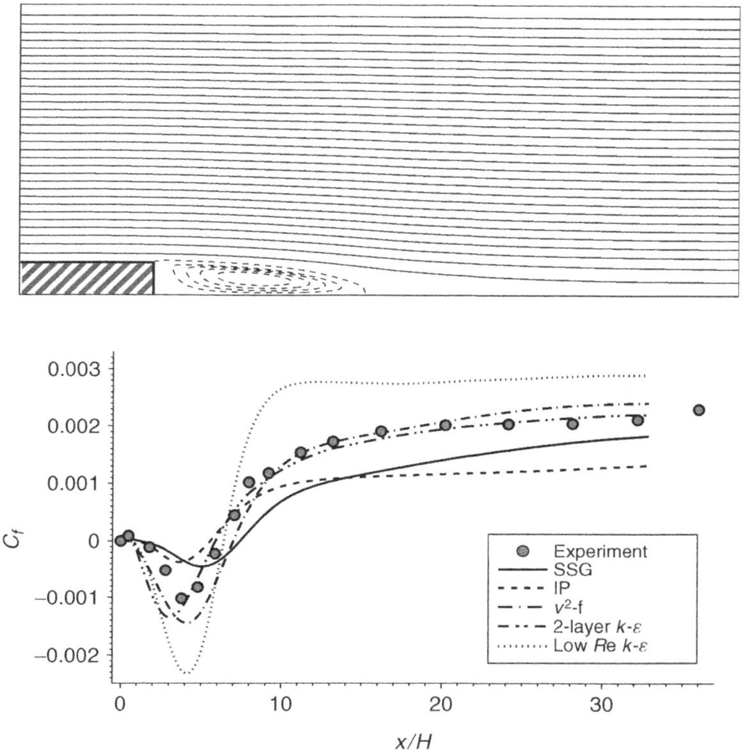


**Figure 7.12** Flow over a backward-facing step, $R_H = 35\,700$: (top) streamlines and (bottom) skin friction on the wall downstream of the step for SSG, IP, low Reynolds number $k-\varepsilon$, and $v^2-f$ models.

magnitude of the minimum $C_f$ is overpredicted, and the recovery when $x/H > 6$ is too rapid. As noted previously, the two-layer formulation is a more effective near-wall patch on the standard $k-\varepsilon$ model. It is in rather good accord with the data. The $v^2-f$ prediction is similar to the two-layer $k-\varepsilon$ result.

The velocity profiles at the top of Figure 7.13 were computed with the SSG model, but they evidence a failure common to all current closures. Downstream of reattachment, the experimental velocity profile is observed to fill out more rapidly than predicted. Even models that predict the correct rate of recovery of the skin friction predict a too slow recovery of the velocity profile. Recovery prediction is an unresolved problem in turbulence modeling. (There is one case in which models tend to predict velocity profiles correctly, that is, a low Reynolds number backstep with $R_H = 5000$. We have intentionally chosen a more representative, high Reynolds number case, even though a DNS database is available for the low Reynolds number.)

Predictions of $\overline{u^2}$ in the lower part of Figure 7.13 are in reasonable agreement with experimental data, except for the profile in the backflow region, at $x/H = 4$. Even the last profile, at $x/H = 15$, has not recovered to the fully developed flat-plate form of Figure 6.5, page 126. The latter form has a sharp peak near the wall, with no maximum

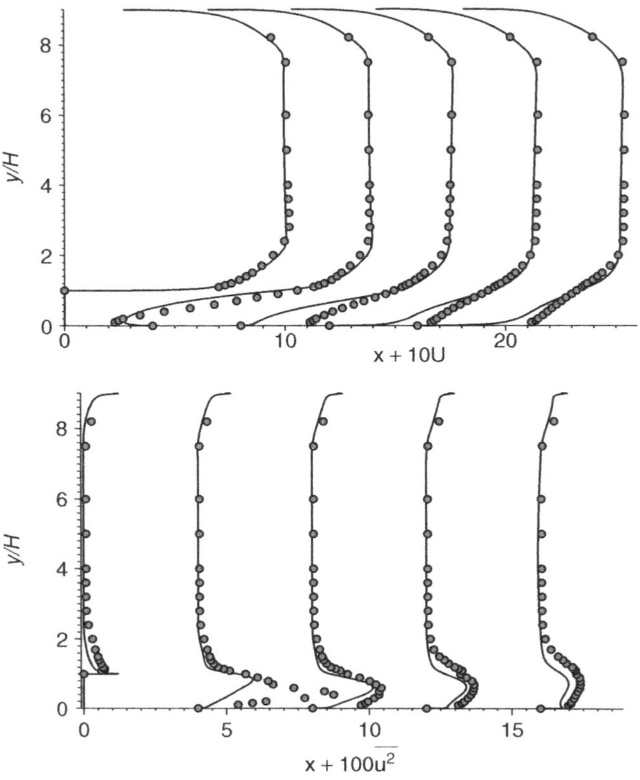


**Figure 7.13** Flow over a backward-facing step, continued: (top) $U$ component of velocity and (bottom) $\overline{u^2}$ component of turbulence for the SSG model. Data of Driver and Seegmiller (1985).

away from the surface. The maximum away from the wall, that is seen in Figure 7.13, is a residue of the mixing layer that detaches from the upstream step. Even though the mean flow has reattached at $x/H = 6$, the turbulent intensity profiles have the appearance of a detached mixing layer well beyond that point. This is a nice example of disequilibrium.

#### 7.4.2.3 Vortex shedding by unsteady RANS

Unsteady flows occur in turbomachinery, in stirring tanks, and behind bluff bodies, to name a few examples. A proper treatment of such flows requires a time-dependent computation. The terminology "unsteady RANS" has evolved. This is not a new topic in modeling. The model equations were developed with a convective derivative; an unsteady RANS computation simply requires including the time derivative in the computation. Of course, this is not to say that non-equilibrium effects are not an issue in modeling; but, to date, no special features of temporal unsteadiness have arisen.

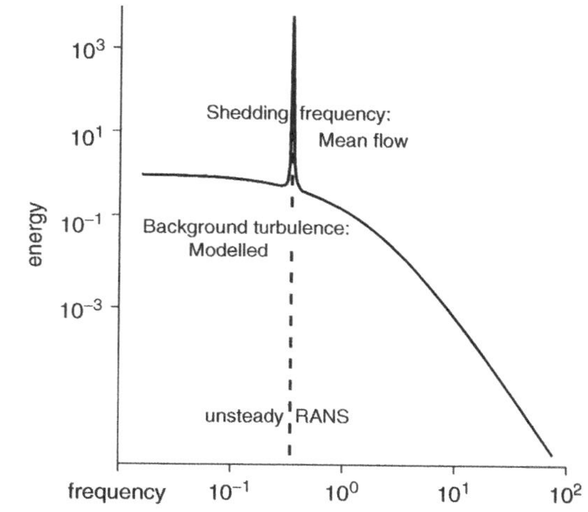


**Figure 7.14** Schematic spectrum of turbulence with coherent, periodic unsteadiness.

One can understand the role of the turbulence model by considering the energy spectrum in Figure 7.14. The sharp peak is produced by coherent, periodic vortex shedding. The broadband spectrum is caused by background turbulence. The turbulence closure model represents the broadband component; it does not model the sharp peak. An unsteady RANS computation contains periodic energy at this frequency, as suggested by the dashed line labelled "unsteady RANS" in the figure. This contribution to the unsteady energy is computed explicitly as part of the mean flow.

Vorticity contours in the flow around a triangular cylinder inside a channel were portrayed in Figure 5.12, page 102. These are instantaneous mean flow vortices created by periodic shedding. Time-averaged velocity contours for that same case are shown in Figure 7.15(a). This is a composite view, showing the time average of an unsteady, periodic solution in the upper half, versus a steady computation of the same flow in the lower half. (The steady solution can be computed by imposing top–bottom symmetry in order to suppress vortex shedding.) Clearly the time average of the solution with vortex shedding is quite different from a steady solution. To capture the correct flow field, it is necessary to include deterministic shedding in the computation and only use the turbulence model to represent statistics of the irregular component of motion. In an experiment, the averaging can be along the span, or over measurements at a fixed phase of oscillation; time averaging is not equivalent to statistical averaging in this case because the flow is not statistically stationary.

Figure 7.15(c) shows experimental data measured on the centerline behind the triangle. The data are time-averaged, mean velocities. The time-averaged, computed velocity is displayed by the solid line. Its agreement with the data is far better than the steady computation, shown by a dashed line. Figure 7.15(a) shows that in the unsteady RANS computation vortex shedding causes the wake to fill in quite rapidly–an effect not captured by the steady RANS computation. The wake profiles in Figure 7.15(b) also are time averages of the unsteady computation.

Looking back at Figure 5.12 on page 102, it can be seen that the vortex street decays shortly downstream of the cylinder, in marked contrast to laminar vortex shedding, where

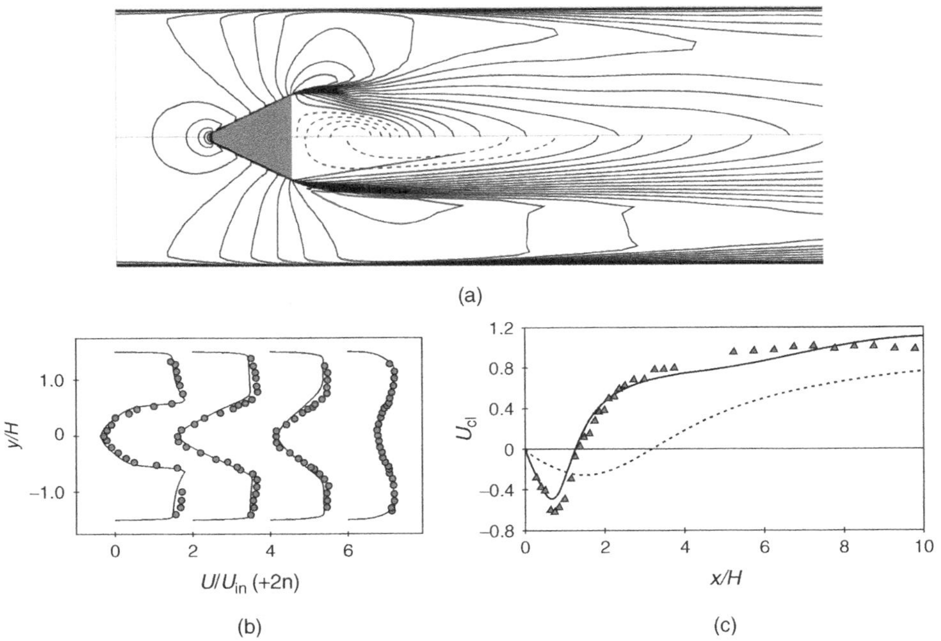


**Figure 7.15** Vortex shedding from a triangular cylinder. (a) Composite showing time-average contours of $U$ in the upper half, versus a steady solution in the lower half. The dashed lines indicate negative velocity. (b) Time-averaged velocity profiles in the wake. (c) Velocity along centerline: time average (——) and steady computation (- - - -) Reprinted from *International Journal of Heat and Fluid Flow*, Vol 24, G. Iaccarino, A. Ooi, P. A. Durbin, M. Behnia, 'Reynolds averaged simulation of unsteady separated flow', 147–156. Copyright 2003, with permission from Elsevier.

the vortices persist for quite a distance. Turbulence increases the rate at which the vortices dissipate. To predict this flow correctly, both the unsteady mean flow and the turbulence model are required.

Similar unsteadiness of the ensemble-averaged flow can occur in three-dimensional geometries. For example, a cube mounted on a flat wall sheds arch vortices. The legs of the arch oscillate periodically so that streamlines in a plane near the wall have the appearance of vortex shedding. The view in Figure 7.16 is looking down on the cube. A periodic, oscillatory pattern is seen on the downstream side. The two foci are created by the legs of vortices that arch over the height of the cube. Vortex legs shed from alternative sides in the course of one period. The vortices dissipate more rapidly than in a two-dimensional vortex street, so that they are not noticeable beyond a few cube diameters downstream.

Again, the time average of the unsteady RANS computation differs from the steady RANS computation. The two are compared in Figure 7.17. The unsteady case has a shorter recirculation region behind the cube. The oscillatory arch vortices enhance mixing, causing the wake to fill more rapidly. Experimental data are closely similar to the unsteady case.

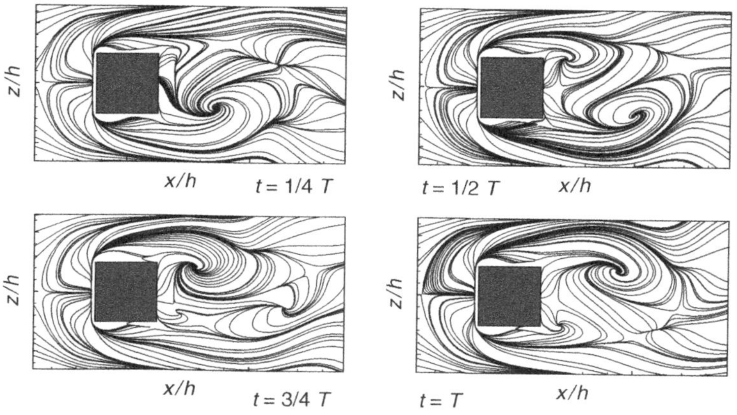


**Figure 7.16** Surface stress lines at four instants in flow over a surface-mounted cube. Reprinted from *International Journal of Heat and Fluid Flow*, Vol 24, G. Iaccarino, A. Ooi, P. A. Durbin, M. Behnia, 'Reynolds averaged simulation of unsteady separated flow', 147–156. Copyright 2003, with permission from Elsevier.

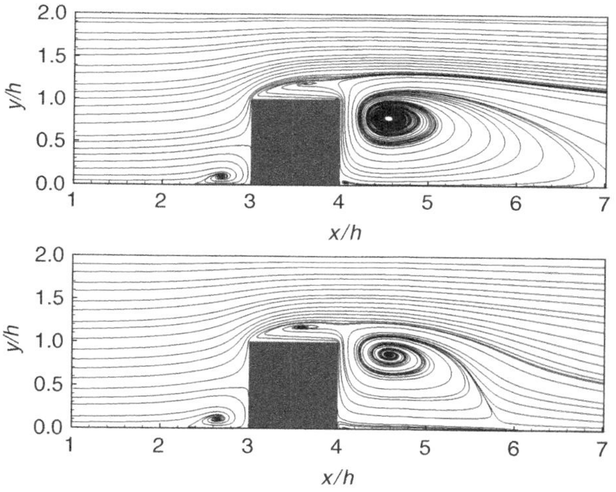


**Figure 7.17** Streamlines in the central plane of flow over a surface-mounted cube: (top) steady computation, (bottom) unsteady computation. From Iaccarino *et al.* (2003).

A necklace vortex wraps around the front of the cube. Its influence is visible in both of Figures 7.16 and 7.17. In the latter, it appears as a small recirculating region at the base of the cube.

#### 7.4.2.4 Jet impingement

Jet impingement heating and cooling is used in many engineering and industrial applications. It also serves as a prototype for stagnating flows as a class. In most applications, a

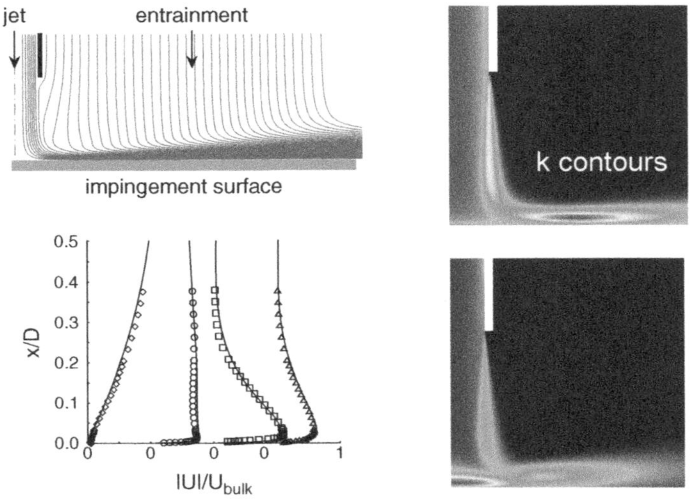


**Figure 7.18** Left: (top) streamlines of flow in impinging jet and (bottom) profiles of the velocity magnitude at various radial locations: $r/D = 0$ (◇), 0.5 (○), 1 (□), and 2.5 (△). Right: contours of $k$ as obtained from $v^2\!-\!f$ (top) and low Reynolds number $k\!-\!\varepsilon$ (bottom) models.

jet of gas or liquid is directed toward a target area, as at the upper right of Figure 7.18. This is a radial section through an axisymmetric jet. The full flow field can be visualized by rotating this view 360° about the centerline. As the jet exits the pipe, it begins to spread and to entrain ambient fluid (Section 4.3). This draws a flow through the upper boundary, as labelled on the figure. In the computation, this boundary was maintained at constant pressure and the entrainment velocity was allowed to adapt.

Impinging jets have several features that make them a good vehicle for evaluation of turbulence models. In the impingement region, the mean flow is nearly perpendicular to the surface; it then turns and follows the surface to form a wall jet. Adjacent to the wall, there are thin stagnation point and wall jet boundary layers on the target plate. Large total strains occur near the stagnation streamline.

In the experiment of Cooper *et al.* (1993), the jet issues from a fully developed pipe flow at Reynolds number 23 000 based on diameter. It is situated two diameters above the plate. Mean flow profiles at several radii are plotted at the lower right of Figure 7.18. The data are the magnitude of the mean velocity; the curves are the same, computed with the $v^2\!-\!f$ model (Behnia *et al.*, 1998). On the stagnation line, $r = 0$, the mean velocity is in the $-y$ direction. Therefore, the leftmost profile shows how the component normal to the wall is blocked by the surface. The other profiles show how a wall jet develops and spreads along the surface. The turbulence model is quite able to predict the flow evolution; but some qualification is in order. The entire flow field, including the wall jet, is axisymmetric. When the wall jet issues from an orifice tangential to the surface, the jet will be three-dimensional, spreading at different rates in directions normal and tangential to the wall. The ratio of tangential to normal spreading is measured to be about 5 (Launder and Rodi, 1989). Turbulence models, as a whole, significantly underpredict this ratio. Wall jets and impingement continue to challenge models, despite the impressive predictive ability often demonstrated.

We have already mentioned the difficulties experienced by otherwise acceptable closure models when they are applied to impingement flow. Sections 6.4 and 7.3.4 cite the stagnation-point anomaly of two-equation models and of wall-echo corrections. The present example of an impinging round jet was studied by Craft *et al.* (1993), who showed these faults in $k-\varepsilon$ and SMC models. The tendency is toward spuriously high turbulent kinetic energy and excessive eddy viscosity in the vicinity of the stagnation point. Without correcting this anomaly, the predicted profiles at the lower left of Figure 7.18 would spread more quickly into the free stream than the data. The lower contour plot at the right of Figure 7.18 was produced with a low Reynolds number $k-\varepsilon$ model. The large maximum of $k$ at the stagnation point is erroneous. The potential core of the incident jet carries a low level of turbulence; it should not become a maximum near the wall. Instead, the turbulence in the shear layers should be amplified by streamline convergence at the side of the jet, as in the upper right contour plot–this is the turbulence field of the computation shown in the left portion of the figures. Erroneous predictions caused by the stagnation-point anomaly are displayed in Behnia *et al.*, (1998). Evidence is that the anomaly can be corrected, but the reader is warned of its potential harm.

#### 7.4.2.5 Square duct

Secondary currents in the plane perpendicular to the primary flow direction are frequently encountered in geophysical and engineering flows. This secondary mean fluid motion is created and maintained by two fundamentally different mechanisms: (i) transverse pressure gradients, or inertial forces such as the Coriolis acceleration, can cause quasi-inviscid deflection of the flow; and (ii) turbulence anisotropy can generate a new component of the mean motion. Prandtl formally separated instances of these mechanisms into two categories that are known as secondary motions of Prandtl's first and second kind, respectively. The latter, which can only exist in turbulent flows, are in general associated with a substantially weaker secondary motion than the former, which can exist in laminar or turbulent flow. Secondary flow of the second kind may, nevertheless, substantially alter characteristics of the flow field; an example is provided by turbulent flows inside non-circular ducts. From a turbulence modeling perspective, secondary flow prediction constitutes a demanding test. A delicate imbalance between gradients of the Reynolds stress components is responsible for generating secondary flow. This imbalance has to be accurately predicted.

The velocity vectors depicted in Figure 7.19 are an example of secondary flow of the second kind. The primary flow is along the $x$ axis. Flow in the $y-z$ plane would not be present under laminar conditions. It can be attributed to streamwise vorticity generated by turbulent stresses. To see its origin, consider the transport equation of streamwise mean vorticity:

$$V\,\partial_y\Omega_x + W\,\partial_z\Omega_x = \nu(\partial_y^2\Omega_x + \partial_z^2\Omega_x) + \partial_y\partial_z(\overline{v^2} - \overline{w^2}) - \partial_y^2\overline{vw} + \partial_z^2\overline{vw}, \qquad (7.4.2)$$

where $\Omega_x = \partial_y W - \partial_z V$. In laminar flow, the Reynolds stresses are absent and the solution is $\Omega_x = 0$. In turbulent flow, the three last terms become sources of mean streamwise vorticity: the first involves normal stress anisotropy; the second and third involve the secondary Reynolds shear stress–the component $\overline{vw}$ is secondary in the sense that the

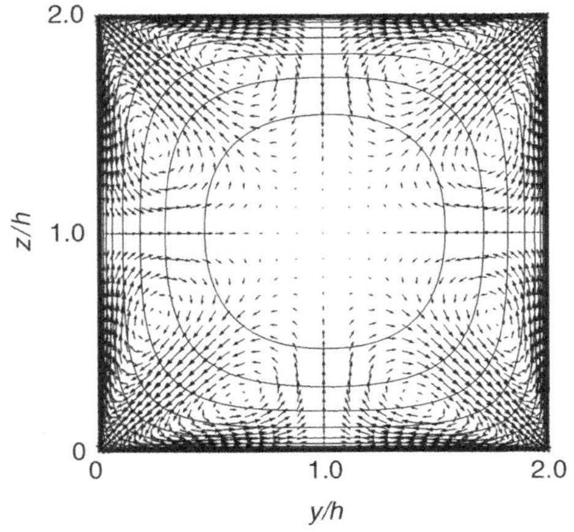


**Figure 7.19** Fully developed flow in a straight square duct at $R_\tau \equiv 2hu_*/\nu = 600$. The vectors show the secondary mean flow field.

primary shear stress is $\overline{uv}$. Normal stress anisotropy, $\overline{v^2} \neq \overline{w^2}$, has traditionally been recognized as driving the streamwise vorticity. However, evidence from numerical simulations is that the shear stress contributions may dominate the net source term. This highlights the complexity of the flow field: the $\overline{vw}$ shear stress component would be zero if the mean secondary flow field were not present, yet it dominates and maintains the secondary motion once it is created.

The flow pattern of Figure 7.19 displays the characteristic eight-fold symmetry about the two wall mid-planes and the corner bisectors, seen in fully developed flow inside a square duct. The computation was performed with the SSG model with the elliptic relaxation treatment (Section 7.3.4) of the near-wall region. Elliptic relaxation enables a full SMC to be integrated all the way to the wall in this complex flow. Figure 7.20 compares model predictions to DNS data. The agreement of the primary mean velocity and Reynolds stresses to data is generally good. The discrepancy most noted by Pettersson Reif and Andersson (1999) is that the strength of the weak secondary flow was underpredicted, a fault traced back to deficient prediction of the secondary shear stress component, $\overline{vw}$.

#### 7.4.2.6 Rotating shear flow

Flows in rotating frames of reference are encountered in a variety of applications, including turbomachinery and geophysics. When the momentum equations are transformed to a rotating frame of reference, a Coriolis acceleration, $2\mathbf{\Omega}^{\mathrm{F}} \wedge \boldsymbol{U}$, is added (see Eq. (8.1.2)). Here $\mathbf{\Omega}^{\mathrm{F}}$ is the frame rotation vector. One half of the Coriolis acceleration comes from transforming the time derivative from the inertial frame to the rotating system; the other half comes from rotation of the velocity components relative to the inertial frame. When the Reynolds stress transport equations are similarly transformed, as in Eq. (8.3.4) on page 228, the rotation adds the term $-\Omega_l^{\mathrm{F}}(\epsilon_{ikl}\overline{u_k u_j} + \epsilon_{jkl}\overline{u_k u_i})$ to the time derivative, and subtracts the same term from the production tensor, by the same reasoning. It is important to distinguish these two contributions, because the production tensor often appears in

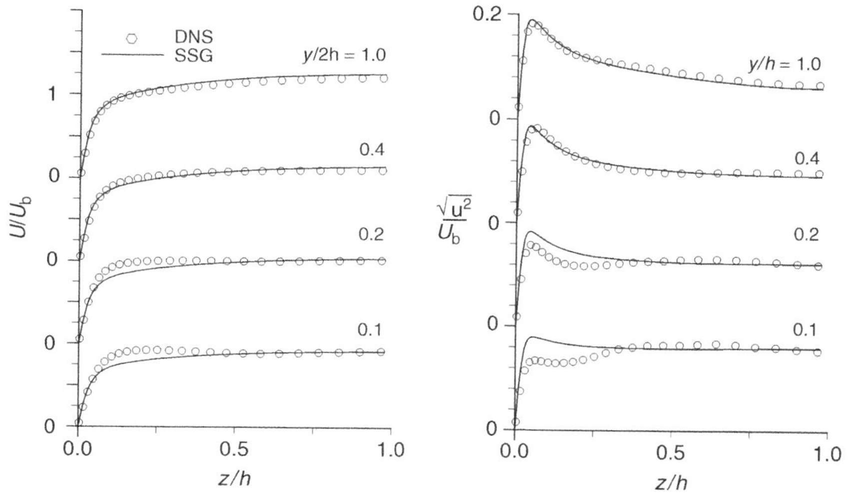


**Figure 7.20** Fully developed flow in a straight square duct at $R_\tau \equiv 2hu_*/\nu = 600$. Lines are SSG model with elliptic relaxation; symbols are DNS (Huser and Biringen 1993).

closure models: only the contribution of rotation to the production tensor should be added to the closure formula. If this is not done correctly, the equations will not be coordinate frame-independent. In the case of Eq. (8.3.4), the correct accounting is accomplished by introducing the absolute vorticity, $\boldsymbol{\Omega}^A$, into the production term and into the redistribution model, while adding one half of the Coriolis acceleration explicitly to the time derivative.

The Coriolis acceleration can profoundly affect turbulent flows. These effects are traditionally investigated by studying parallel shear flows subjected to orthogonal mode rotation; in other words, flows where the axis of rotation is either parallel or antiparallel to the mean flow vorticity. Depending on the magnitude and orientation of the rotation vector ($\boldsymbol{\Omega}^F$) relative to the mean flow vorticity ($\boldsymbol{\omega} = \nabla \wedge \boldsymbol{U}$), turbulence can be augmented or reduced: the turbulence is usually suppressed if the rotation vector and the vorticity vector are parallel, and enhanced if they are antiparallel. However, a large rate of antiparallel rotation can suppress turbulence.

To elucidate the predominant dynamic processes associated with imposed rotation, consider the exact Reynolds stress generation terms due to mean shear ($\mathcal{P}_{ij}$) and due to system rotation ($R_{ij}$),

$$\mathcal{P}_{ij} + R_{ij} = -\overline{u_i u_k}\,\partial_k U_j - \overline{u_j u_k}\,\partial_k U_i + 2\Omega^F(\varepsilon_{ik3}\,\overline{u_k u_j} + \varepsilon_{jk3}\,\overline{u_k u_i}).$$

In parallel shear flow, rotating about the spanwise, $z$, axis,

$$\mathcal{P}_{12} + R_{12} = -\overline{v^2}\,\mathrm{d}U/\mathrm{d}y - 2\Omega^F(\overline{u^2} - \overline{v^2}) \tag{7.4.3}$$

and

$$\mathcal{P} = -\overline{uv}\,\mathrm{d}U/\mathrm{d}y$$

represent the local generation of turbulent shear stress and of kinetic energy, respectively. Note that the latter does not directly depend on the imposed frame rotation. We will consider the case where $\mathrm{d}U/\mathrm{d}y > 0$; recall that the corresponding vorticity, $\omega_z = -\partial_y U$, is negative.

Consider a positive imposed angular velocity, $\Omega^{\mathrm{F}} > 0$. Usually $\overline{uv}$ is negative in parallel shear flow. According to Eq. (7.4.3), frame rotation will further decrease $\overline{uv}$ if $\overline{u^2} > \overline{v^2}$. Hence, positive rotation increases the magnitude of the shear stress. Negative rotation has the opposite effect. These are consistent with the conception that rotation in the same direction as the mean vorticity will suppress turbulence.

However, sufficiently strong positive rotation can drive $\overline{u^2}$ toward $\overline{v^2}$. The difference between the production of $\overline{u^2}$ and the production of $\overline{v^2}$ is

$$\mathcal{P}_{11} + R_{11} - R_{22} = -2\overline{uv}(\mathrm{d}U/\mathrm{d}y - 4\Omega^{\mathrm{F}}). \tag{7.4.4}$$

A reduction of the normal stress anisotropy can be expected if this becomes negative. If $\overline{uv} < 0$, then, when $\Omega^{\mathrm{F}} < \frac{1}{4}\,\mathrm{d}U/\mathrm{d}y$, normal stress anisotropy will begin to decrease, insofar as Eq. (7.4.4) is concerned. The rotation number‡ is defined by $\mathrm{Ro} = -2\Omega^{\mathrm{F}}/(\mathrm{d}U/\mathrm{d}y)$. Often $\mathrm{Ro} < 0$ is referred to as the anticyclonic, and $\mathrm{Ro} > 0$ as the cyclonic, direction of rotation. The previous criterion becomes $\mathrm{Ro} < -\frac{1}{2}$. Suppression of anisotropy by strong anticyclonic rotation, in turn, reduces the rate of production of $-\overline{uv}$, and can ultimately lead to reduction of kinetic energy. In fact, linear theory suggests that the turbulence intensity will be suppressed by rotation if $\mathrm{Ro} < -1$. Hence, even when the rotation vector is antiparallel to the background mean vorticity ($\mathrm{Ro} < 0$), the turbulence is not expected always to be enhanced.

**Plane Poiseuille flow** Highly simplified flows are attractive because the predictive capability of a closure model can be accurately assessed. In trivial geometries, the mean flow might respond solely via alterations to the turbulence field. Thereby, the pure turbulence effects are isolated.

Fully developed turbulent flow between two infinite parallel planes in orthogonal mode rotation constitutes one example of a trivial geometry. DNS databases are available to assist the model development. These data have contributed to making this particular flow a standard benchmark test case. The most frequently adopted configuration is pressure-driven, Poiseuille flow, illustrated by Figure 7.21(a). Note that the $y$ axis points downward; $y/h = 1$ in Figure 7.22 is the lower side of the channel.

Figure 7.22 displays model predictions of unidirectional, fully developed, rotating Poiseuille flow: $\boldsymbol{U} = [U(y), 0, 0]$ and $\boldsymbol{\Omega}^{\mathrm{F}} = [0, 0, \Omega]$. In the figure, the rotation number is defined as $\mathrm{Ro} \equiv 2\Omega h/U_{\mathrm{b}}$, where the bulk velocity is

$$U_{\mathrm{b}} = \frac{1}{2h}\int_{-h}^{h} U(y)\,\mathrm{d}y.$$

The transport of mean momentum is governed by

$$\begin{aligned} 0 &= -\partial_x P^* + \partial_y(\nu\,\partial_y U - \overline{uv}), \\ 0 &= -\partial_y P^* - 2\Omega U - \partial_y \overline{v^2}, \end{aligned} \tag{7.4.5}$$

‡ The Rossby number, defined as 1/Ro, is frequently used in geophysical flows; the rotation number is mostly used in engineering flows.

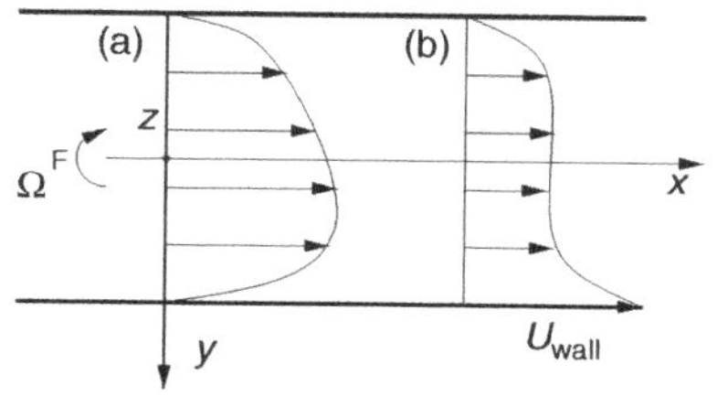


**Figure 7.21** Schematic of (a) plane Poiseuille and (b) plane Couette flow in orthogonal mode rotation. The $y$ axis points down to form a right-handed coordinate system with $z$ into the page.

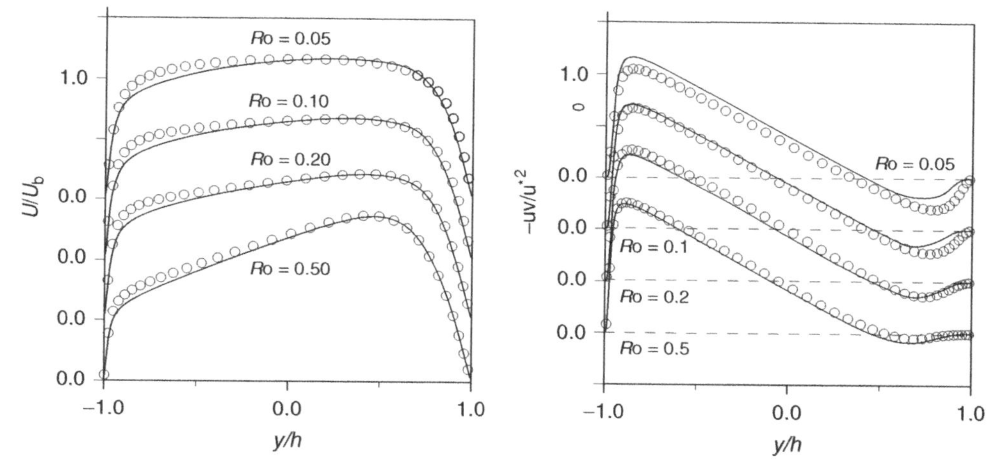


**Figure 7.22** Spanwise rotating Poiseuille flow at $R_\tau \equiv u_* h/\nu = 194$: curves, second-moment closure; symbols, DNS (Kristoffersen and Andersson, 1993).

where $P^* = P - \frac{1}{2}\Omega^2(x^2 + y^2)$ is the mean reduced pressure. The Coriolis force does not directly affect the mean velocity; it is balanced by the gradients of the pressure and the wall normal turbulent stress through the second of Eqs. (7.4.5). Nevertheless, the imposed rotation breaks the symmetry of the mean flow field, as seen in Figure 7.22. How does this occur? The asymmetry would not be seen in laminar flow; it is an example of Prandtl's "secondary flow of the second kind." Rotation alters the profile of $\overline{uv}$, as seen at the right of Figure 7.22. It thereby induces asymmetry into the $U(y)$ profile. Because the mean flow vorticity changes sign across the channel, the flow field is simultaneously subjected to both cyclonic and anticyclonic rotation; it is cyclonic on the lower side and anticyclonic on the upper. Cyclonic rotation suppresses the turbulence, causing the asymmetry.

The model predictions exhibit many of the effects of Coriolis acceleration upon the mean flow field. An almost irrotational core region, where $dU/dy \approx 2\Omega$, is seen at $\mathrm{Ro} = 0.5$; and the steeper mean flow profile near $y/h = -1$ than near $y/h = 1$ is correctly predicted. The mean flow is asymmetric because turbulence mixes high-speed fluid from the central region toward the wall, maintaining a steep gradient on the turbulent side. The stable side profile is less steep, but the velocity is higher. The same pressure gradient is imposed on both sides of the channel, but the laminar side has lower resistance, and this permits the higher flow rate. The stable side has essentially laminarized when Ro $= 0.5$, as seen in the Reynolds shear stress plots of Figure 7.22.

From a practical standpoint, Poiseuille flow may overemphasize the impact of rotationally modified Reynolds stresses. In non-trivial geometries, imposed rotation can directly alter the mean flow field; for instance, if the walls of the channel are not parallel, Coriolis accelerations alter the mean velocity. It may be possible to predict this first-order response with a turbulence model that is not correctly sensitive to rotation. However, the Coriolis acceleration does change the turbulence field, as just described, and this contributes to the mean flow prediction as well.

**Plane Couette flow** Plane turbulent Couette flow subjected to spanwise rotation is rather interesting. In contrast to pressure-driven channel flow, the flow field in Figure 7.21(b) is exposed entirely to cyclonic or anticyclonic rotation, depending on the sign of $\Omega$, because of the antisymmetric mean velocity profile. Figure 7.23 shows the predicted secondary flow in a cross-plane of the channel. The intriguing feature of this case is that rotationally induced streamwise vortices span the channel in this flow. The counter-rotating cells shown by the velocity vectors in Figure 7.23 are repeated across the span of the flow. Figure 7.23 was computed with left–right periodic boundary conditions. A wider domain would contain further cells, their width remaining approximately equal to the channel height. The predicted magnitude of the secondary flow at Ro $= 0.1$ is approximately 20% of the mean streamwise bulk velocity.

Turbulence models account only for the random component of the flow. Accurate predictions require that the deterministic roll cells be calculated explicitly. Therefore, the secondary flow was solved as an integral part of a two-dimensional, three-component

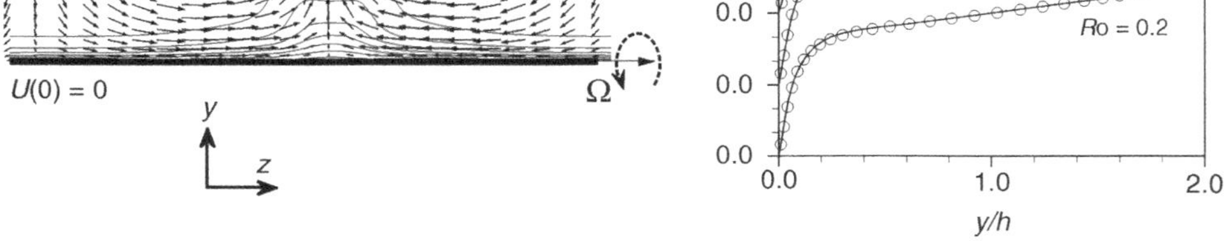

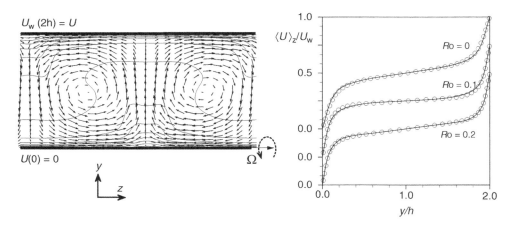

**Figure 7.23** Spanwise rotating plane Couette flow at $\mathrm{Re} \equiv U_w h/\nu = 2600$, $\mathrm{Ro} \equiv 2\Omega h/U_w = 0.1$: arrow vectors, secondary flow field in a plane perpendicular to the streamwise direction; contours, streamwise mean velocity; right, mean streamwise component.

flow field, $\boldsymbol{U} = [U(y,z), V(y,z), W(y,z)]$. The turbulence model is then left to represent nothing but real turbulence. The mean flow predicted by this approach agrees quite well with the data (Figure 7.23). Without resolving the vortices, the predictions would be less accurate.

# Exercises

**Exercise 7.1.** *Three-dimensional boundary layers.* Let $\alpha_{\mathrm{RS}} = \overline{u_2 u_3}/\overline{u_1 u_2}$ and $\alpha_{\mathrm{U}} = S_{23}/S_{12}$ denote the tangents of the direction of Reynolds stress and mean rate of strain. Show that (7.1.1) implies the evolution equation

$$\mathrm{d}_t \alpha_{\mathrm{RS}} = \frac{-2kC_\mu C_1 S_{12}}{\overline{u_1 u_2}}(\alpha_{\mathrm{U}} - \alpha_{\mathrm{RS}}).$$

Let the initial state be an equilibrium shear flow, $U(y)$, at $t = 0$.

Suppose that initially $\alpha_{\mathrm{U}} = 0 = \alpha_{\mathrm{RS}}$ but the flow subsequently veers direction. It then develops a component $W(y)$ and $\alpha_{\mathrm{U}} > 0$. Show that, if $\alpha_{\mathrm{U}}$ increases monotonically with time, then $\alpha_{\mathrm{RS}} \le \alpha_{\mathrm{U}}$. Thus the angle of the Reynolds stress lags that of the mean rate of strain.

**Exercise 7.2.** *The anisotropy tensor.* Why must the diagonal components $b_{11}$, $b_{22}$, and $b_{33}$ of the anisotropy tensor (7.1.5) lie between $-\frac{2}{3}$ and $\frac{4}{3}$? Show that, generally, the eigenvalues of $b_{ij}$ also lie between $-\frac{2}{3}$ and $\frac{4}{3}$.

**Exercise 7.3.** *Return to isotropy.* Use the Rotta model for $\mathcal{F}_{ij}$ to solve (7.1.7) for the relaxation of $b_{ij}$ in homogeneous turbulence with no mean flow, $\partial_i U_j = 0$, and initially $b_{ij} = b^0_{ij}$. Obtain the turbulence time-scale, $T = k/\varepsilon$, from the $k$–$\varepsilon$ solution for decaying isotropic turbulence (Section 6.2.1). Why is it satisfactory to use the isotropic solution, even though $b_{ij} \neq 0$? Rewrite your solution for $b_{ij}$ as a solution for $\overline{u_i u_j}$. Also explain why $C_1 > 1$ is necessary.

**Exercise 7.4.** *Invariants of the anisotropy tensor.* The second ($II$) and third ($III$) invariants of $b_{ij}$ are defined in (2.3.10) as $II = -\frac{1}{2}b_{ij}b_{ji} = -\frac{1}{2}b^2_{ii}$ and $III = \frac{1}{3}b_{ij}b_{jk}b_{ki} = \frac{1}{3}b^3_{ii}$. (The first invariant $b_{ii}$ is identically 0.) For the same case of relaxation toward isotropy via Rotta's model as in the previous exercise, write the evolution equations for $II$ and $III$.

The two-dimensional plane with coordinates $II$ and $III$ is called the "invariant map." Find the equation for $III$ as a function of $II$ for the Rotta model by forming the ratio

$$\frac{\mathrm{d}III}{\mathrm{d}II} = \frac{\mathrm{d}_t III}{\mathrm{d}_t II}.$$

Then show that this model predicts $III \propto II^{3/2}$.

**Exercise 7.5.** *Solution to SSG for homogeneous shear.* Derive (7.2.4) with (7.1.35) included. Substitute the SSG values for the coefficients of the rapid model. Use either the empirical value $\mathcal{P}_{\mathcal{R}} = 1.6$ or the $k$–$\varepsilon$ value (6.2.11). Set the Rotta constant to $C_1 = 1.7$. Compare numerical values to the experimental results. (The values cited in the text include a nonlinear term in the slow model.)

**Exercise 7.6.** *Skew symmetric tensor* $\varepsilon_{ijk}$. If you have not done so in Exercise 5.1), show that

$$\varepsilon_{ijk}\varepsilon_{ilm} = \delta_{jl}\delta_{km} - \delta_{jl}\delta_{mk}.$$

From this show that, if $\Omega_{ij} = \varepsilon_{ijk}\Omega_k$, then $\omega_i = \varepsilon_{ijk}\Omega_{jk} = 2\Omega_i$.

**Exercise 7.7.** *General linear model (GLM).* Show that (7.1.31) can be written in terms of the production tensor as

$$\wp_{ij}^{\text{rapid}} = [\tfrac{4}{5} - \tfrac{4}{3}(C_2 + C_3)]kS_{ij} - C_2(\mathcal{P}_{ij} - \tfrac{2}{3}\delta_{ij}\mathcal{P}) - C_3(D_{ij} - \tfrac{2}{3}\delta_{ij}\mathcal{P}),$$

where $D_{ij} \equiv -\overline{u_i u_k}\,\partial_k U_j - \overline{u_j u_k}\,\partial_i U_k$. From this, show that the IP model corresponds to canceling the terms involving $D_{ij}$ and $S_{ij}$ to leave

$$\wp_{ij}^{\text{rapid}} = -\tfrac{3}{5}(\mathcal{P}_{ij} - \tfrac{2}{3}\delta_{ij}\mathcal{P}).$$

This is a simple, popular model.

**Exercise 7.8.** *Solution for axisymmetric, homogeneous rate of strain.* Find the equilibrium solution to (7.2.1) for the incompressible, homogeneous straining flow, $U = \mathcal{S}x$, $V = -\frac{1}{2}\mathcal{S}y$, and $W = -\frac{1}{2}\mathcal{S}z$.

**Exercise 7.9.** *Equations in a rotating frame.* If Eqs. (7.2.1) are referred to a reference frame rotating about the $x_3$ axis, the $i, j$ component of the mean velocity gradient becomes

$$\partial_j U_i - \varepsilon_{ij3}\Omega_3^{\text{F}}$$

in the absolute frame. Here $\Omega_3^{\text{F}}$ is the rate of frame rotation and $\partial_j U_i$ is the velocity gradient in the rotating frame. Evaluate the components of the production tensor for rotating, homogeneous shear flow, $U_i(x_j) = \delta_{i1}\delta_{j2}\mathcal{S}x_2$, and compare to (7.2.6).

The time derivative of $b_{ij}$ in the stationary frame is replaced by

$$\partial_t b_{ij} + \Omega_3^{\text{F}}(b_{ik}\varepsilon_{kj3} + b_{jk}\varepsilon_{ki3})$$

relative to the rotating frame. Substitute this and the transformed velocity gradient tensor into (7.2.1) to derive the evolution equation (8.3.4) for $b_{ij}$ in the rotating frame. There are two choices for the last part of this exercise.

(i) Either solve this equation numerically for rotating, homogeneous shear flow. The initial condition is isotropy, $b_{ij} = 0$. Non-dimensionalize time by $\mathcal{S}$ so that the equations contain the ratio $\Omega^{\text{F}}/\mathcal{S}$. Investigate how the solution depends on this parameter–experiment with its value. The $k$–$\varepsilon$ equations

$$\mathrm{d}_t k = -k\mathcal{S}b_{12} - \varepsilon, \qquad \mathrm{d}_t\varepsilon = -(C_{\varepsilon 1}k\mathcal{S}b_{12} + C_{\varepsilon 2}\varepsilon)\frac{\varepsilon}{k}$$

will be needed as well. Include a plot of $b_{12}$ versus time.

(ii) Or derive the closed-form equilibrium solution (8.3.12) for $b_{11}$, $b_{22}$, $b_{12}$, and $Sk/\varepsilon$ as functions of $\Omega^F/S$ in homogeneous shear flow. You can use (6.2.11) for the equilibrium value of $\mathcal{P}/\varepsilon$.

In either case, discuss the stabilizing or destabilizing effects of rotation.

**Exercise 7.10.** *Wall functions for SMC.* State the wall function boundary conditions for $k$, $\varepsilon$, $U$, and $\overline{u_i u_j}$ that can be applied somewhere inside the log layer. Write a program to solve plane channel flow with one of the SMC formulations and wall functions. Compare the solution to channel flow data at $R_\tau = 590$ presented in Figure 4.3.

**Exercise 7.11.** *Limiting budget.* Consider the Reynolds stress budget (3.2.4). By examining the scaling of each term with $y$ as $y \to 0$ at a no-slip wall, derive the limiting behaviors given in Eq. (7.3.6) and following pages.

**Exercise 7.12.** *Near-wall analysis.* Verify that the first of Eqs. (7.3.19) has the behavior

$$\varepsilon \frac{\overline{u_i u_j}}{k} - \nu \frac{\partial^2 \overline{u_i u_j}}{\partial y^2} = \wp_{ij} + O(y^3) = k f_{ij} + O(y^3)$$

as $y \to 0$. Show that, if $f_{ij}(0) = O(1)$ and $\overline{u_i u_j}$ is not singular, then the only possible behaviors are $\overline{u_i u_j} = O(y^4)$ or $\overline{u_i u_j} = O(y^2)$.

# 8

# Advanced topics

Though this be madness, yet there is method in't.

– William Shakespeare

Attempts to formulate principles of modeling often lead to intricate analysis. That, in turn, leads to a deepened understanding of the models. Many of these developments are not essential to practical applications. Some represent directions of contemporary research; others may provide a basis for future improvements to predictive capabilities. They have been left to this final chapter of Part II.

Equilibrium analysis provides an understanding of how solutions respond to imposed forcing. Closed-form solutions provide a systematic development for nonlinear constitutive modeling. Material in this vein also has been assigned to the current chapter in order to present it without encumbering the more elementary development in the forerunning chapters.

In addition, we have relegated a discussion of closure modeling for scalar transport to this chapter on advanced topics. Some practical considerations on heat and mass transfer have been covered in Chapter 4. What is described here is a more elaborate level of closure. The three main topics of this chapter – principles (Section 8.1), equilibria (Section 8.3), and scalars (Section 8.4) – are largely independent and can be read separately.

## 8.1 Further modeling principles

From time to time, attempts have been made to formulate systematic guidelines for developing second-moment closures. The idea of expansion in powers of anisotropy was used in Section 7.1.4 to derive second-moment closure (SMC) models. A more formal discussion consists in identifying overarching principles. Some of the principles

*Statistical Theory and Modeling for Turbulent Flows, Second Edition* P. A. Durbin and B. A. Pettersson Reif

are inviolable, others are optional (a less compelling name might be preferred, but we will stay with "principles"). Those discussed in this section are:

- Dimensional consistency
- Coordinate system independence
- Galilean invariance
- Realizability.

*Physical coherence* could be included as well. This is not so much a concrete guideline as a conceptual principle. It means that closure models should be physically plausible substitutions for the real process. To mention one example, an inviscid effect should not be modeled in terms of viscous parameters.

The first of the above list is obvious: all terms in any equation must have the same dimensions. The model equation for the dissipation rate, $\varepsilon$, was described as a dimensionally consistent analog to the turbulent kinetic energy equation. Dimensional consistency demanded that production of $k$ be divided by the turbulence time-scale to serve as a model for production of $\varepsilon$. This principle is trivial, but still a powerful force in turbulence modeling.

The second item in the list, coordinate system independence, demands that the model must be independent of its expression by a particular set of components. Physically, it implies that a closure model should exhibit the same response to translation, acceleration, and reflections of the coordinate system as the real process. Chapter 2 contains mathematical techniques that are used in coordinate-independent modeling. Consistent use of tensor algebra and exact coordinate transformation are its essence. In Section 2.3.2, on page 35, tensors were defined as coordinate-independent mathematical objects.

At an operational level, consistent use of tensor algebra is invoked by ensuring that all terms in the equations have the same free subscripts and that matrix multiplication is done correctly. For instance, the flux of a scalar, $\theta$, having mean concentration $\Theta$, could be modeled by

$$\overline{\theta u_i} = K_{ij}\, \partial_j \Theta$$

because both sides have free index $i$ and the right-hand side is the product of a matrix and a vector. It should not be modeled by $\overline{\theta u_i} = T\, \overline{u_i u_i}\, \partial_i \Theta$. With summation over the repeated index $i$, the right-hand side has no free indices, while the left-hand side is a vector with $i$ being an unsummed index. Suspending the summation convention, to make $i$ free on the right, amounts to suspending correct matrix multiplication. The result would depend on the coordinate system used in solving the model. In numerical computations, this could mean that the solution depends on the grid if $i$ is the direction of a grid line. Indeed, a corollary to the principle of coordinate system independence is the principle of grid system independence. This means that no functions in the model should change if the computational mesh is altered, assuming that the mesh provides good numerical resolution.

The remainder of the list of principles leads to new mathematical developments.

### 8.1.1 Galilean invariance and frame rotation

Galilean invariance means that the equations should be the same in any two frames of reference that move with a *constant* relative velocity, $U_{\rm rel}$. This is satisfied in SMC modeling by using the convective derivative, $\mathrm{D}f/\mathrm{D}t = \partial_t f + U_i\,\partial_i f$, where $f(x,t)$ is any function, and by allowing the model to depend on velocity derivatives but not velocity itself.

If one frame is "stationary" and the other moves with velocity $U_{\rm rel}$, then the fluid velocity $U$, relative to the stationary frame, becomes $U - U_{\rm rel}$ in the moving frame. The position of a fixed point $x$ moves as $x = x' - U_{\rm rel}t$ relative to the translating frame. Hence the convective derivative $\mathrm{D}f(x,t)/\mathrm{D}t$ becomes

$$
\begin{aligned}
\partial_t f(\boldsymbol{x}) + U_i \frac{\partial f(\boldsymbol{x})}{\partial x_i} &= \partial_t f(\boldsymbol{x}' - U_{\rm rel}t, t) + U_i \frac{\partial f(\boldsymbol{x}' - U_{\rm rel}t, t)}{\partial x_i'} \\
&= \partial_t f' + (U - U_{\rm rel})_i \frac{\partial f'}{\partial x_i'} \\
&= \partial_t f' + U_i' \frac{\partial f'}{\partial x_i'},
\end{aligned}
$$

where the primes denote functions evaluated in the moving frame. Therefore, the convective derivative retains its form in the new inertial frame. Similarly,

$$
\partial_i U_j = \partial_i'(U - U_{\rm rel})_j = \partial_i' U_j'.
$$

The velocity gradient is unchanged by uniform translation; it is Galilean invariant. It follows that the way the models in Chapter 7 were stated satisfies Galilean invariance, as it should. Galilean invariance is best checked in Cartesian coordinates. A term like $U_i\overline{u_i u_j}$ is not invariant, but a term somewhat similar to this arises correctly upon transformation to cylindrical coordinates.

If the transformation is to a non-inertial frame, then additional accelerations arise. A case that has been studied at length is transformation to a rotating frame. If a fluid is in solid-body rotation, then relative to an inertial frame it has the rotation tensor $\varepsilon_{ijk}\Omega_k^{\rm F}$, where $\Omega_k^{\rm F}$ is the component of the rate of frame rotation along the $x_k$ axis, and $\varepsilon_{ijk}$ is the cyclic permutation tensor. If $\Omega_{ij}$ is the fluid rotation tensor relative to the rotating frame, then

$$
\Omega_{ij}^{\rm A} = \Omega_{ij} + \varepsilon_{ijk}\Omega_k^{\rm F} \tag{8.1.1}
$$

is called the absolute rotation, measured relative to an inertial frame.

It stands to reason that models like (7.1.33) should be made to depend on $\Omega_{ij}^{\rm A}$ if they are to be frame-invariant. To demonstrate this requires that Coriolis acceleration be added to the convective derivative. In a rotating frame, the fluctuating acceleration of a fluid element is

$$
\partial_t u_i + U_k\,\partial_k u_i + u_k\,\partial_k U_i + 2\varepsilon_{ijk}\Omega_j^{\rm F} u_k + \partial_k u_k u_i - \partial_k \overline{u_k u_i}. \tag{8.1.2}
$$

The new term, containing $\Omega^{\rm F}$, is the Coriolis acceleration. It is an exact consequence of expressing the acceleration relative to a non-inertial frame.

Taking minus the divergence of Eq. (8.1.2), equating it to the Laplacian of pressure, and keeping only the linear terms, we obtain

$$\begin{aligned}\frac{1}{\rho}\nabla^2 p &= -2\,\partial_i U_k\,\partial_k u_i - 2\varepsilon_{kji}\Omega_j^{\rm F}\,\partial_k u_i = -2(\partial_i U_k + \varepsilon_{kji}\Omega_j^{\rm F})\,\partial_k u_i \\ &= -2(S_{ik} + \Omega_{ik} + \varepsilon_{ikj}\Omega_j^{\rm F})\,\partial_i u_k = -2(S_{ik} + \Omega_{ik}^{\rm A})\,\partial_i u_k\end{aligned}$$

for the rapid pressure. As in (7.1.33), $\partial_i U_j = S_{ij} + \Omega_{ij}$ was used. It follows that the rapid redistribution model should depend either on the absolute rate of rotation or on the absolute vorticity. The vorticity and rotation tensor are related by $\omega_i = \varepsilon_{ijk}\Omega_{jk} = \varepsilon_{ijk}\,\partial_j U_k$ (see Exercise 7.6) and the absolute vorticity is

$$\omega_i^{\rm A} = \varepsilon_{ijk}\Omega_{jk}^{\rm A} = \omega_i + 2\Omega_i^{\rm F}. \tag{8.1.3}$$

It might at first seem odd that frame rotation *should* enter into the turbulence model. Frame rotation *should not* enter constitutive models for material properties. Viscous stresses vanish in a fluid under solid-body rotation, so the viscous constitutive model should not depend on frame rotation or on absolute vorticity. However, turbulent motions are most certainly affected by rotation; for instance, it can enhance or suppress turbulent energy. The Coriolis acceleration therefore *should* appear in the closure model.

But a warning is in order. The replacement of mean rate of rotation, $\Omega_{ij}$, by absolute mean rate of rotation, $\Omega_{ij}^{\rm A}$, makes only *algebraic* closure formulas frame-independent (as explained in the next section). In a *stress transport* model, frame rotation appears additionally to the absolute fluid rotation tensor. This is seen by forming the convective derivative of Reynolds stresses. From (8.1.2) the convective derivative is

$$\begin{aligned}&\partial_t\overline{u_iu_j} + U_k\,\partial_k\overline{u_iu_j} + \overline{u_iu_k}\,\partial_k U_j + 2\varepsilon_{jlk}\Omega_l^{\rm F}\overline{u_iu_k} + \cdots \\ &\quad= {\rm D}_t\overline{u_iu_j} + \overline{u_iu_k}(\partial_k U_j + 2\varepsilon_{kjl}\Omega_l^{\rm F}) + \cdots \\ &\quad= {\rm D}_t\overline{u_iu_j} + \overline{u_iu_k}(S_{kj} + \Omega_{kj} + 2\varepsilon_{kjl}\Omega_l^{\rm F}) + \cdots \\ &\quad= {\rm D}_t\overline{u_iu_j} + \overline{u_iu_k}(S_{kj} + \Omega_{kj}^{\rm A} + \varepsilon_{kjl}\Omega_l^{\rm F}) + \cdots.\end{aligned} \tag{8.1.4}$$

The "$+\cdots$" signify that a transpose with respect to $i$ and $j$ is to be added. Both $\boldsymbol{\Omega}^{\rm A}$ and $\boldsymbol{\Omega}^{\rm F}$ enter. The algebraic formulas and scalar transport models of Chapter 6 accommodate frame rotation only through the absolute vorticity. The convective derivative contains another effect that will be missed in such models: as the frame rotates, the direction of the unit direction vectors rotate as well. A vector that is constant in an inertial frame rotates in time relative to the non-inertial frame. This gives the explicit contribution of frame rotation, $\boldsymbol{\Omega}^{\rm F}$, to Eq. (8.1.4).

#### 8.1.1.1 Invariance and algebraic models

Properly formulated algebraic models should be both Galilean invariant, and invariant under frame rotation. For example, in solving a rotor–stator flow, it should be irrelevant whether the rotor or the stator frame of reference is adopted.

The simple eddy viscosity model takes turbulent stresses to be proportional to spatial gradients of the velocity field, in particular to the rate-of-strain tensor $S_{ij} = \frac{1}{2}(\partial V_i/\partial x_j + \partial V_j/\partial x_i)$. This particular combination of velocity gradients is not arbitrary: it is due to a type of kinematic frame invariance. Consider a general, time-dependent change of reference frame that can be expressed algebraically as

$$\tilde{x}_k = b_k(t) + E_{kl}(t)x_l. \tag{8.1.5}$$

The coordinates $\tilde{x}_k$ and $x_k$ denote position in two different reference frames. These are related by the transformation matrix $\boldsymbol{E}$, as described in Eq. (2.3.2). The origins of the two coordinate systems can be in relative translation, as allowed by the $\boldsymbol{b}(t)$. The time dependence of $\boldsymbol{E}$ allows the coordinate systems to be in relative rotation. The velocity in the $\tilde{\boldsymbol{x}}$ frame, $\tilde{V}_k \equiv \dot{\tilde{x}}_k$, is given by

$$\tilde{V}_k = \dot{b}_k + \dot{E}_{kl}x_l + E_{kl}V_l. \tag{8.1.6}$$

Frame invariance requires that $\tilde{V}_k = E_{kl}V_l$ – see Section 2.3, page 34. Hence the velocity vector obviously is not frame-invariant under the transformation (8.1.5). From (8.1.6), the gradient of the velocity becomes

$$\frac{\partial \tilde{V}_k}{\partial \tilde{x}_l} = \dot{E}_{km}\frac{\partial x_{\mathrm{m}}}{\partial \tilde{x}_l} + E_{km}\frac{\partial V_m}{\partial \tilde{x}_l}$$

or

$$\frac{\partial \tilde{V}_k}{\partial \tilde{x}_l} = \dot{E}_{km}E_{lm} + E_{km}\frac{\partial V_m}{\partial x_n}E_{ln}, \tag{8.1.7}$$

which differs from the transformation $E_{km}(\partial V_m/\partial x_n)E_{ln}$ described in Section 2.3. Consequently, velocity gradients are not invariant either. However, it is readily seen that the mean rate-of-strain tensor $S_{kl}$ *is* invariant;

$$\begin{aligned}\tilde{S}_{kl} &= \frac{1}{2}\left(E_{km}\frac{\partial V_m}{\partial \tilde{x}_n}E_{ln} + E_{lm}\frac{\partial V_m}{\partial \tilde{x}_n}E_{kn} + \dot{E}_{km}E_{lm} + \dot{E}_{lm}E_{km}\right)\\ &= E_{km}S_{mn}E_{nl}.\end{aligned} \tag{8.1.8}$$

The last equality is obtained upon noting that $\dot{E}_{km}E_{lm} + \dot{E}_{lm}E_{km} = \partial_t(E_{km}E_{lm}) = \partial_t(\delta_{kl}) = 0$. The local vorticity tensor is not invariant under the arbitrary change of frame, but it can be shown that the absolute vorticity $\boldsymbol{\Omega}^{\mathrm{A}}$ (defined by Eq. (8.1.3)) transforms properly (Exercise 8.1) between the stationary and rotating frames.

Hence, it is concluded that algebraic stress formulations should be functions of $\boldsymbol{S}$ and $\boldsymbol{\Omega}^{\mathrm{A}}$, but not of $\boldsymbol{\Omega}$ or $\boldsymbol{\Omega}^{\mathrm{F}}$.

### 8.1.2 Realizability

The last of the principles stated at the outset of this section is realizability. Realizability means that the model should not violate the Schwartz inequality

$$(\overline{u_i u_j})^2 \le \overline{u_i^2}\;\overline{u_j^2},$$

and that component energies $\overline{u_i^2}$ should be non-negative (Schumann, 1977). These two conditions can alternatively be stated as one: the eigenvalues of $\overline{u_i u_j}$ should be non-negative. Generally, the differential equations of the turbulence model will not ensure these inequalities in all circumstances. Indeed, most models can be made to predict the equivalent of $\overline{u^2} < 0$ in extreme conditions. One can debate whether ensuring realizability in all circumstances is necessary. Realizability is a property of each solution; if a model predicts realizable solutions in virtually all cases of interest, is there any need for concern? As long as violations are not catastrophic, the answer is probably "no." However, there are circumstances in which potentially negative eigenvalues could be a matter for concern. For instance, $C_\mu \overline{u_i u_j} T$ is commonly used as an eddy viscosity tensor. A negative eigenvalue leads to negative viscosity, which is numerically unstable. In such a context, realizability offers a useful mathematical tool for nonlinear eddy viscosity closure development.

Much of the literature on realizability is rather misleading. It has been implied that tensoral nonlinearity in the $M_{ijkl}$ expansion developed in Section 7.1.4 is required in order to guarantee realizability. In fact, certain inequalities need only be satisfied, and this is readily done by constraints on the model coefficients. It is possible to revise any model to ensure realizability in all circumstances with only a minor alteration of its coefficients, as will be explained at the conclusion of the present section (also see Section 8.2).

A realizability inequality (7.1.14), page 162, has already been encountered in the discussion of the slow redistribution model. Corresponding results can be derived for the rapid redistribution model. First adopt a coordinate system in which the Reynolds stress tensor is diagonal; in other words, $\tilde{\tau}_{ij} = 0$, $i \neq j$. The issue at hand can be stated as being to ensure unconditionally that $\tilde{\tau}_{11} \geq 0$, where $\tilde{\tau}_{11}$ is defined as the smallest component of $\tilde{\boldsymbol{\tau}}$. In the notation $\tau_{ij} \equiv \overline{u_i u_j}$, Eq. (7.1.4), page 157, can be written as

$$\mathrm{d}_t \tau_{ij} = H_{ij}, \tag{8.1.9}$$

where

$$H_{ij} = \mathcal{P}_{ij} + \wp_{ij} - \tfrac{2}{3}\varepsilon\delta_{ij}.$$

Let $\boldsymbol{\mathcal{U}}$ represent the matrix of orthonormal eigenvectors of $\boldsymbol{\tau}$ and let ${}^t\boldsymbol{\mathcal{U}}$ denote its transpose. These are matrix inverses: $\boldsymbol{\mathcal{U}} \cdot {}^t\boldsymbol{\mathcal{U}} = \boldsymbol{\delta}$. In terms of these matrices, $\tilde{\boldsymbol{\tau}} = \boldsymbol{\mathcal{U}} \cdot \boldsymbol{\tau} \cdot {}^t\boldsymbol{\mathcal{U}}$ is the diagonalized Reynolds stress. $\boldsymbol{\mathcal{U}}$ can also be thought of as a product of rotations about each of the coordinate axes, that diagonalizes $\boldsymbol{\tau}$. Then the present transformation is analogous to those discussed in Section 2.3.

As the turbulence evolves in time, $\boldsymbol{\mathcal{U}}$ will evolve too. Thus the evolution equation for the diagonalized Reynolds stress tensor is

$$\mathrm{d}_t \tilde{\boldsymbol{\tau}} = \tilde{\boldsymbol{H}} + (\mathrm{d}_t\boldsymbol{\mathcal{U}} \cdot {}^t\boldsymbol{\mathcal{U}}) \cdot \tilde{\boldsymbol{\tau}} + \tilde{\boldsymbol{\tau}} \cdot (\boldsymbol{\mathcal{U}} \cdot \mathrm{d}_t {}^t\boldsymbol{\mathcal{U}}). \tag{8.1.10}$$

The matrices in parentheses are transposes of one another. They are easily shown to be antisymmetric matrices: differentiating $\boldsymbol{\mathcal{U}} \cdot {}^t\boldsymbol{\mathcal{U}} = \boldsymbol{\delta}$ shows that

$$\boldsymbol{W} \equiv (\mathrm{d}_t\boldsymbol{\mathcal{U}} \cdot {}^t\boldsymbol{\mathcal{U}}) = -(\boldsymbol{\mathcal{U}} \cdot \mathrm{d}_t {}^t\boldsymbol{\mathcal{U}}) = -{}^t\boldsymbol{W}.$$

$\boldsymbol{W}$ is defined by this equation as the rotation matrix of the principal axes of $\boldsymbol{\tau}$. The only property of $\boldsymbol{W}$ needed here is that it is antisymmetric, so that $W_{ij} = 0$ if $i = j$. The evolution equation (8.1.10) is then seen to be

$$\mathrm{d}_t \tilde{\boldsymbol{\tau}} = \tilde{\boldsymbol{H}} + \boldsymbol{W} \cdot \tilde{\boldsymbol{\tau}} + \tilde{\boldsymbol{\tau}} \cdot {}^t\boldsymbol{W} = \tilde{\boldsymbol{H}} + \boldsymbol{W} \cdot \tilde{\boldsymbol{\tau}} - \tilde{\boldsymbol{\tau}} \cdot \boldsymbol{W}.$$

Because $\tilde{\tau}$ is a diagonalized matrix, by definition $\tilde{\tau}_{ij}$ vanishes at all times if $i \neq j$. In component form, the evolution equation is therefore

$$\begin{aligned} \mathrm{d}_t \tilde{\tau}_{ij} &= \tilde{H}_{ij}, & i &= j, \\ 0 &= \tilde{H}_{ij} + W_{ik}\tilde{\tau}_{kj} - \tilde{\tau}_{ik}W_{kj}, & i &\neq j. \end{aligned}$$

The second equation determines the rate of rotation of the principal axes; for instance the 1, 2 component gives

$$W_{12} = \tilde{H}_{12}/(\tilde{\tau}_{11} - \tilde{\tau}_{22}).$$

However, $\boldsymbol{W}$ does not factor into present considerations.

When $\tilde{\tau}_{11} = 0$, realizability demands that $\tilde{\tau}_{11}$ decrease no further. This is ensured by requiring that $\mathrm{d}_t \tilde{\tau}_{11} > 0$, or that $\tilde{H}_{11} > 0$, when $\tilde{\tau}_{11} = 0$. For the GLM (Eq. (7.1.31), page 167) with Rotta return to isotropy (Eq. (7.1.8)) for the slow model,

$$\begin{aligned} \tilde{H}_{ij} = \tilde{\mathcal{P}}_{ij} - \tfrac{2}{3}\varepsilon\delta_{ij} - \frac{\varepsilon}{k}C_1(\tilde{\tau}_{ij} - \tfrac{2}{3}k\delta_{ij}) + [\tfrac{4}{5} - \tfrac{4}{3}(C_2 + C_3)]k\tilde{S}_{ij} \\ - C_2(\tilde{\mathcal{P}}_{ij} - \tfrac{2}{3}\delta_{ij}\mathcal{P}) - C_3(\tilde{D}_{ij} - \tfrac{2}{3}\delta_{ij}\mathcal{P}). \end{aligned} \tag{8.1.11}$$

If $\tau_{11} = 0$, both $\tilde{\mathcal{P}}_{11}$ and $\tilde{D}_{11}$ vanish. This follows from $\tilde{\mathcal{P}}_{11} = -2\tilde{\tau}_{11}\,\partial_1\tilde{U}_1$, which is a consequence of the diagonalization

$$-\boldsymbol{\mathcal{U}} \cdot (\boldsymbol{\tau} \cdot \nabla \boldsymbol{U}) \cdot {}^{t}\boldsymbol{\mathcal{U}} = -(\boldsymbol{\mathcal{U}} \cdot \boldsymbol{\tau} \cdot {}^{t}\boldsymbol{\mathcal{U}}) \cdot (\boldsymbol{\mathcal{U}} \cdot \nabla \boldsymbol{U} \cdot {}^{t}\boldsymbol{\mathcal{U}}) = -\tilde{\boldsymbol{\tau}} \cdot \widetilde{\nabla \boldsymbol{U}}$$

and the fact that $\tilde{\boldsymbol{\tau}}$ is diagonal. The production tensor $\tilde{\boldsymbol{\mathcal{P}}}$ equals the above plus its matrix transpose. Note that only $\tilde{\boldsymbol{\tau}}$ is diagonal; the velocity gradient matrix is not.

The 1, 1 component of (8.1.11) with $\tilde{\tau}_{11} = 0$ is

$$\tilde{H}_{11} = \tfrac{2}{3}(C_1 - 1)\varepsilon + [\tfrac{4}{5} - \tfrac{4}{3}(C_2 + C_3)]k\tilde{S}_{11} + \tfrac{2}{3}(C_2 + C_3)\mathcal{P}. \tag{8.1.12}$$

This is positive if

$$C_1 - 1 + (C_2 + C_3)\frac{\mathcal{P}}{\varepsilon} + [\tfrac{6}{5} - 2(C_2 + C_3)]\tilde{S}_{11}k/\varepsilon > 0. \tag{8.1.13}$$

Only this inequality need be ensured to guarantee realizability. However, the rate of strain, $\tilde{S}_{11}$, is a projection onto the principal axes of the Reynolds stress tensor. That makes the inequality (8.1.13) rather tenuous. A more readily imposed bound is needed.

It will be assumed that $\frac{6}{5} - 2(C_2 + C_3) < 0$ in Eq. (8.1.13), as is usually the case: for example, the LRR model equates it to $(6 - 90 \times 0.4)/55 = -6/11$. To formulate a general constraint, $\tilde{S}_{11}$ in Eq. (8.1.13) can be replaced by the largest eigenvalue of $S_{ij}$. This will suffice to prevent the smallest eigenvalue, $\tilde{\tau}_{11}$, from becoming negative. Eigenvalues are invariant under coordinate rotation, so it is the largest eigenvalue of $S_{ij}$ that is desired. Denote this by $\lambda^S_{\max}$. Then the realizability constraint (8.1.13) will be met if

$$C_1 > 1 - (C_2 + C_3)\frac{\mathcal{P}}{\varepsilon} + [2(C_2 + C_3) - \tfrac{6}{5}]\frac{\lambda^S_{\max}k}{\varepsilon} \tag{8.1.14}$$

(Durbin and Speziale, 1994). For the LRR model, $C_2 = (c + 8)/11$, $C_3 = (8c - 2)/11$, and $c = 0.4$, and condition (8.1.14) becomes

$$C_1 > 1 - \frac{9.6}{11}\frac{\mathcal{P}}{\varepsilon} + \frac{6}{11}\frac{\lambda^S_{\max}k}{\varepsilon}.$$

For the IP model, $C_2 = 3/5$ and $C_3 = 0$, and constraint (8.1.14) is simply

$$C_1 > 1 - \frac{3}{5}\frac{\mathcal{P}}{\varepsilon}. \tag{8.1.15}$$

At this point it should be clear that realizability is not a daunting matter. It has been reduced to a straightforward inequality.

A typical value of $C_1$ is 1.8. So the IP model is realizable if the rate of turbulent energy production is not negative, which is usually the case. Non-realizable solutions can be prevented by specifying $C_1 = \max[1.8, 1 - \frac{3}{5}\mathcal{P}/\varepsilon]$. If sufficiently negative production were to occur in any computation, this limiter would come into effect; but in most cases it would be irrelevant.

The same approach can be applied to (8.1.14). If $C_1 = 1.8$ is usually a satisfactory value, then

$$C_1 = \max[1.8,\ 1 - (C_2 + C_3)\mathcal{P}/\varepsilon + (2C_2 + 2C_3 - \tfrac{6}{5})\lambda^S_{\max}k/\varepsilon] \tag{8.1.16}$$

will ensure realizability in extreme cases without altering the model in typical cases. Certainly, more elaborate schemes can be devised to satisfy the inequality. Such schemes would be warranted if they were designed to improve agreement with data; otherwise, it is hard to justify the added complexity. What is revealed by the analysis is that realizability can be implemented readily. It is not a cumbersome constraint, and it might prove valuable in some applications.

## 8.2 Second-moment closure and Langevin equations

Probability density models are widely used in reacting flows (Fox, 2003). In practice, they are rarely formulated as evolution equations for the PDF; invariably they are formulated as stochastic models. A Langevin equation plus a stochastic mixing model is commonly the concrete realization of the PDF method.

In Section 2.2.2.1 the Langevin equation introduced concepts of Lagrangian dispersion theory. Indeed, its origin in the study of turbulence was as a Lagrangian model. However, the application to PDF methods led to a recognition that it also forms a basis of an Eulerian model – and, particularly, that it can be linked to second-moment closure (SMC). Pope (1994) noted that the second moment of a Langevin equation is an SMC, and Durbin and Speziale (1994) described how a Langevin equation can be derived for any particular SMC.

We consider a stochastic differential equation of the general form

$$\mathrm{d}u_i = -\frac{1}{T}A_{ij}u_j\,\mathrm{d}t + B_{ij}\,\mathrm{d}W_j(t) \tag{8.2.1}$$

(see Eq. (2.2.23)). This is a particular version of an Ito type of stochastic differential equation (Section 2.2.2.1); or, rather, it is a differential equation in which randomness is introduced by white noise. The function $\mathrm{d}W_j(t)$ can be thought of as a set of independent random variables, selected at a time interval of $\mathrm{d}t$ and having the properties

$$\begin{aligned} \overline{\mathrm{d}W_i(t)} &= 0, \\ \overline{\mathrm{d}W_i(t)\,\mathrm{d}W_j(t)} &= \mathrm{d}t\,\delta_{ij}, \\ \overline{u_j(t)\,\mathrm{d}W_i(t)} &= 0. \end{aligned} \tag{8.2.2}$$

Its average vanishes, its components are uncorrelated, and the increment $\mathrm{d}W(t)$ is uncorrelated with the present value of the dependent variable $u(t)$.

The second of properties (8.2.2) states that $\mathrm{d}W_i$ is an isotropic random process with magnitude of order $\sqrt{\mathrm{d}t}$. When deriving moment equations, therefore, it is necessary to retain terms to $O(\mathrm{d}W_i)^2$. The first step to evaluate $\mathrm{d}\,\overline{u_i u_j}/\mathrm{d}t$ is to expand the differential

$$\begin{aligned} \mathrm{d}(u_i u_j) &= (u_i + \mathrm{d}u_i)(u_j + \mathrm{d}u_j) - u_i u_j \\ &= u_i\,\mathrm{d}u_j + u_j\,\mathrm{d}u_i + \mathrm{d}u_i\,\mathrm{d}u_j. \end{aligned} \tag{8.2.3}$$

Then substitute (8.2.1) for $\mathrm{d}u$, using the rules (8.2.2) and keeping terms up to order $\mathrm{d}t$:

$$\frac{\mathrm{d}\,\overline{u_i u_j}}{\mathrm{d}t} = -\frac{1}{T}(\overline{u_i u_k} A_{jk} + \overline{u_j u_k} A_{ik}) + B_{ik} B_{jk}. \tag{8.2.4}$$

This is in the format of an SMC.

As an example, consider the Reynolds stress evolution equation

$$\frac{\mathrm{d}\,\overline{u_i u_j}}{\mathrm{d}t} = \mathcal{P}_{ij} - \tfrac{2}{3}\delta_{ij}\varepsilon - \frac{C_1}{T}(\overline{u_i u_j} - \tfrac{2}{3}k\delta_{ij}) - C_2(\mathcal{P}_{ij} - \tfrac{2}{3}\delta_{ij}\mathcal{P}).$$

With $T = k/\varepsilon$ and $C_2 = 3/5$, this is the Rotta slow model and the IP rapid model. Its right-hand side can be rearranged to

$$(1 - C_2)\mathcal{P}_{ij} - \frac{C_1}{T}\overline{u_i u_j} + \frac{2}{3}\left(\frac{C_1 k}{T} - \varepsilon + C_2\mathcal{P}\right)\delta_{ij}.$$

Writing out the production tensor puts this into the form

$$-\left[(1 - C_2)\overline{u_i u_k}\,\partial_k U_j + \frac{C_1}{2T}\overline{u_i u_j}\right] - \left[(1 - C_2)\overline{u_j u_k}\,\partial_k U_i + \frac{C_1}{2T}\overline{u_j u_i}\right] + \frac{2}{3}\left(\frac{C_1 k}{T} - \varepsilon + C_2\mathcal{P}\right)\delta_{ij}.$$

Now, by comparison to (8.2.4), the coefficients of the Langevin equation (8.2.1) are found to be

$$\begin{aligned} A_{ik} &= (1 - C_2)T\,\partial_k U_i + \frac{1}{2}C_1\delta_{ik}, \\ B_{ik} &= \delta_{ik}\sqrt{\frac{2}{3}\left(\frac{C_1 k}{T} - \varepsilon + C_2\mathcal{P}\right)}. \end{aligned}$$

Hence a Langevin equation has been found for the IP model. The argument of the square root in the definition of $B_{ij}$ must be positive, so

$$C_1 > \frac{\varepsilon T}{k} - C_2 \frac{\mathcal{P}T}{k}.$$

This is just the realizability condition (8.1.15) of the previous section. The solution to a Langevin equation is a realization of a random process. Hence, any model that can be derived from a Langevin equation is by definition realizable. The criterion that $B_{ij}$ be real-valued is a first-principles derivation of a realizability constraint.

The general linear model (7.1.32) is obtained from (8.2.4) via the prescriptions

$$\begin{aligned} A_{ij} &= \tfrac{1}{2} C_1 \delta_{ij} + (1 - C_2) T\, \partial_j U_i - C_3 T\, \partial_i U_j, \\ B_{ij} &= \delta_{ij} \sqrt{\tfrac{2}{3}[C_1 \varepsilon - \varepsilon + (C_2 + C_3) P] - \tfrac{1}{3} c_s \varepsilon M_{kl} M_{lk}} + \sqrt{c_s \varepsilon} M_{ij} \end{aligned} \qquad (8.2.5)$$

(Durbin and Speziale, 1994), where $\boldsymbol{M}$ is a symmetric matrix related to the rate of strain by

$$M_{ij}^2 - \tfrac{1}{3} M^2 \delta_{ij} = T S_{ij},$$

and $c_s$ is the coefficient of the rate-of-strain tensor, $c_s = \frac{4}{5} - \frac{4}{3}(C_2 + C_3)$. The argument of the square root must be non-negative for this to be a properly posed Langevin equation, which recovers the realizability constraint (8.1.14) of the previous section. As this example illustrates, $\boldsymbol{B}$ is not, in general, isotropic.

To go from the Langevin equation (8.2.1) to the SMC equation (8.2.4) is a straightforward exercise. Going in the other direction is less obvious. A given Reynolds stress closure must be arranged into a form that is like (8.2.4), then suitable matrices $\boldsymbol{A}$ and $\boldsymbol{B}$ can be devised.

## 8.3 Moving equilibrium solutions of SMC

We next explore equilibrium analysis in a widened context. Equilibrium analysis provides insights into properties of closure schemes, for example showing how the model responds to imposed forcing. It is also the basis for a systematic derivation of nonlinear, algebraic constitutive formulas.

Equation (7.2.4) is a special case of a broader equilibrium solution that can be found to the general linear model, or, for that matter, to the general quasi-linear model. The more comprehensive solution described below is valid for any two-dimensional, steady, homogeneous mean flow. That solution is a tensoral formula that relates $\overline{u_i u_j}$ to $S_{ij}$ and $\Omega_{ij}$. In other words, it is a *constitutive relation* between turbulent stress and mean velocity gradient.

A constitutive relation is essentially a broadening of the concept of an eddy viscosity. The terminology "constitutive relation" is inherited from the theory of continuum mechanics. The objective of a constitutive equation is to establish a mathematical relationship between kinematical variables and stresses – in the simplest case, between rate of strain and stress. However, there is an important distinction between constitutive relations in continuum mechanics and in turbulence modeling: the former describes a material property, whereas the latter approximates statistical properties of the flow field.

The simplest turbulent constitutive relations are the linear eddy viscosity models. An extension of this class is the so-called nonlinear eddy viscosity closure. That terminology alludes to *tensoral* nonlinearity, as described in Section 2.3.2. For instance, $S^2_{ij}$ might appear in addition to $S_{ij}$; and $k$ and $\varepsilon$ enter as dimensionalizing factors. When used in conjunction with the $k$–$\varepsilon$ transport equations, the constitutive relation is called a nonlinear $k$–$\varepsilon$ model.

In a sense this section is an introduction to nonlinear constitutive modeling. However, the approach has a very attractive element: if the full Reynolds stress transport equations are considered to encompass the correct physical processes, then the formula derived from equilibrium approximation to those equations should be imbued with some of their physics. That is the method explained in this section. Formulations derived by systematic equilibrium approximation to the transport equations are called *explicit algebraic stress models* (EASM).

But the value of this section goes beyond introducing nonlinear algebraic stress relations. By allowing for coordinate system rotation, we will set the stage to explore its stabilizing and destabilizing effects. That forms the topic of Section 8.3.3 on *bifurcations*.

The matter immediately at hand is to find the moving equilibria of (7.2.1) for three-dimensional turbulence in two-dimensional mean flows. The method of solution invokes integrity basis expansions (Pope, 1975). A mathematical background can be found in Section 2.3.2, especially the material leading to (2.3.16). As explained there, the two-dimensional integrity basis is far more manageable than the three-dimensional basis (Eq. (2.3.17)). That is one reason why we consider the solution for two-dimensional mean flow. Equilibrium solutions for the three-dimensional case were derived by Gatski and Speziale (1993). Those solutions are not true equilibria because they do not correspond to steady mean flow; however, they are a nice example of the power of tensor analysis.

## 8.3.1 Criterion for steady mean flow

A basic assumption in the analysis to follow is that the mean flow is constant in time. As a preliminary to the main analysis, it is instructive to consider the criterion for the mean velocity gradient to be homogeneous and steady. Homogeneity requires that a mean flow be of the form $U_i = (S_{ki} + \Omega_{ki})x_k$ (if the origin of the $x$ coordinate system is suitably chosen). The steady mean momentum equation is $U_j\,\partial_j U_i = -(1/\rho)\,\partial_i P$. Substituting for $U_i$ and differentiating with respect to $x_k$ gives

$$(S_{kj} + \Omega_{kj})(S_{ji} + \Omega_{ji}) = S^2_{ik} + \Omega^2_{ik} + S_{kj}\Omega_{ji} + \Omega_{kj}S_{ji} = -\frac{1}{\rho}\,\partial_i\partial_k P.$$

The symmetric part of this equation determines the mean pressure

$$P = P_0 - \tfrac{1}{2}\rho(S^2_{ik} + \Omega^2_{ik})x_k x_i.$$

The antisymmetric part,

$$S_{kj}\Omega_{ji} + \Omega_{kj}S_{ji} = 0, \qquad (8.3.1)$$

is equivalent to the steady vorticity equation $\omega_i(S_{ij} + \Omega_{ij}) = 0$. That can be shown by contracting Eq. (8.3.1) with $\varepsilon_{lik}$ and invoking the relation $\Omega_{ji} = \frac{1}{2}\varepsilon_{jim}\omega_m$. Upon noting that $2\omega_i\Omega_{ij} = \omega_i\varepsilon_{ijk}\omega_k = 0$, the condition for steady flow becomes the statement

$$\omega_i S_{ij} = 0$$

that the mean vorticity be perpendicular to the principal axes of strain. Physically, to attain steady state, the strain cannot be stretching the vorticity.

If the $\boldsymbol{\omega}$ axis is $x_3$, then the axes of the rate-of-strain tensor must lie in the $x_1$–$x_2$ plane. For instance, homogeneous shear $U_1 = Ax_2$ corresponds to $\boldsymbol{\omega} = (0, 0, -A)$ and

$$\boldsymbol{S} = \begin{pmatrix} 0 & A/2 & 0 \\ A/2 & 0 & 0 \\ 0 & 0 & 0 \end{pmatrix}.$$

Thus $\boldsymbol{\omega} \cdot \boldsymbol{S} = 0$ is satisfied.

Generally, if $\boldsymbol{\omega} = (0, 0, 2\Omega)$, then steady flow requires the velocity gradient matrix to be

$$[\partial_i U_j] = \begin{pmatrix} S & \Omega & 0 \\ -\Omega & -S & 0 \\ 0 & 0 & 0 \end{pmatrix} \tag{8.3.2}$$

in the coordinate system of the principal axes of the rate of strain. (In homogeneous shear flow, $S = \frac{1}{2}\,\mathrm{d}U/\mathrm{d}y$ and $\Omega = -\frac{1}{2}\,\mathrm{d}U/\mathrm{d}y$.) The rate-of-strain and rate-of-rotation tensors are

$$\boldsymbol{S} = \begin{pmatrix} S & 0 & 0 \\ 0 & -S & 0 \\ 0 & 0 & 0 \end{pmatrix}, \qquad \boldsymbol{\Omega} = \begin{pmatrix} 0 & \Omega & 0 \\ -\Omega & 0 & 0 \\ 0 & 0 & 0 \end{pmatrix}. \tag{8.3.3}$$

The streamlines of this flow are proportional to $\frac{1}{2}\Omega(x^2 + y^2) - Sxy$. Cases where $|\Omega| > |S|$ are referred to as elliptic flow because, when rotation is larger than strain, the streamlines are ellipses.

### 8.3.2 Solution in two-dimensional mean flow

We now seek the solution to the general linear model for steady, two-dimensional, incompressible mean flow. First, a slight extension will be made: the possibility of coordinate system rotation will be included. In a rotating frame, Coriolis acceleration must be added to the evolution equation (7.2.1) of the Reynolds stress anisotropy, on page 169. From (8.1.4) the revised evolution equation is seen to be

$$\begin{aligned} \mathrm{d}_t b_{ij} = {} & (1 - C_1) b_{ij}\frac{\varepsilon}{k} - b_{ik}\varepsilon_{kjl}\Omega_l^{\mathrm{F}} - b_{jk}\varepsilon_{kil}\Omega_l^{\mathrm{F}} - b_{ij}\frac{\mathcal{P}}{k} - \tfrac{8}{15}S_{ij} \\ & + (C_2 + C_3 - 1)(b_{ik}S_{kj} + b_{jk}S_{ki} - \tfrac{2}{3}\delta_{ij}b_{kl}S_{lk}) \\ & + (C_2 - C_3 - 1)(b_{ik}\Omega_{kj}^{\mathrm{A}} + b_{jk}\Omega_{ki}^{\mathrm{A}}). \end{aligned} \tag{8.3.4}$$

In addition to adding the non-inertial acceleration, involving $\boldsymbol{\Omega}^{\mathrm{F}}$, the rate of rotation of the fluid has been replaced by the absolute rotation, $\boldsymbol{\Omega}^{\mathrm{A}}$, in the last term (see page 219). The latter accommodates the contribution of frame rotation to Reynolds stress production and to the redistribution model.

Following Gatski and Speziale (1993), this is further rearranged to

$$\begin{aligned}\frac{k}{\varepsilon}\,\mathrm{d}_t a_{ij} &= (1 - C_1)a_{ij} - a_{ij}\mathcal{P}_{\mathcal{R}} - \tfrac{8}{15}\mathcal{S}_{ij}\\ &\quad - a_{ik}\mathcal{S}_{kj} - a_{jk}\mathcal{S}_{ki} + \tfrac{2}{3}\delta_{ij}a_{kl}\mathcal{S}_{lk} - a_{ik}W_{kj} - a_{jk}W_{ki}\end{aligned} \tag{8.3.5}$$

(recall that $\mathcal{P}_{\mathcal{R}} \equiv \mathcal{P}/\varepsilon$), where

$$\begin{aligned}W_{ij} &= [(1 - C_2 + C_3)\Omega^{\mathrm{A}}_{ij} + \varepsilon_{ijl}\Omega^{\mathrm{F}}_l]\frac{k}{\varepsilon},\\ \mathcal{S}_{ij} &= [(1 - C_2 - C_3)S_{ij}]\frac{k}{\varepsilon},\\ a_{ij} &= (1 - C_2 - C_3)b_{ij}.\end{aligned} \tag{8.3.6}$$

This form has a non-dimensional right-hand side, and constants have been absorbed into the definitions of $\boldsymbol{W}$, $\boldsymbol{\mathcal{S}}$, and $\boldsymbol{a}$ to simplify subsequent algebra. In equilibrium, $\mathrm{d}_t a_{ij} = 0$; then in matrix notation, Eq. (8.3.5) becomes

$$\begin{aligned}0 &= (1 - C_1 - \mathcal{P}_{\mathcal{R}})\boldsymbol{a} - \tfrac{8}{15}\boldsymbol{\mathcal{S}} - \boldsymbol{a}\cdot\boldsymbol{\mathcal{S}} - \boldsymbol{\mathcal{S}}\cdot\boldsymbol{a} + \tfrac{2}{3}\boldsymbol{\delta}\,\mathrm{trace}(\boldsymbol{a}\cdot\boldsymbol{\mathcal{S}})\\ &\quad - \boldsymbol{a}\cdot\boldsymbol{W} + \boldsymbol{W}\cdot\boldsymbol{a}.\end{aligned} \tag{8.3.7}$$

Note that the ratio of production to dissipation is

$$\mathcal{P}_{\mathcal{R}} = -\mathrm{trace}(\boldsymbol{b}\cdot\mathbf{S})k/\varepsilon = -\mathrm{trace}(\boldsymbol{a}\cdot\boldsymbol{\mathcal{S}})/(1 - C_2 - C_3)^2$$

in the scaled variables. If for now $\mathcal{P}_{\mathcal{R}}$ is considered to be known, then Eq. (8.3.7) is analogous to Exercise 2.11: it is a tensor equation with a tensor unknown, $a$. Such equations can be solved by integrity basis expansion, as explained in Section 2.3.2.

As explained in Section 2.3.2, it is a mathematical theorem that any solution to Eq. (8.3.7) must be of the form

$$\boldsymbol{a} = C_\mu\boldsymbol{\mathcal{S}} + b(\boldsymbol{W}\cdot\boldsymbol{\mathcal{S}} - \boldsymbol{\mathcal{S}}\cdot\boldsymbol{W}) + c(\boldsymbol{\mathcal{S}}^2 - \tfrac{1}{3}|\boldsymbol{\mathcal{S}}|^2\boldsymbol{\delta}), \tag{8.3.8}$$

where $|\boldsymbol{\mathcal{S}}|^2 \equiv |\mathrm{trace}(\boldsymbol{\mathcal{S}}^2)| = \mathcal{S}_{ij}\mathcal{S}_{ji}$. This is substituted into (8.3.7) to find the unknown coefficients $C_\mu$, $b$, and $c$. Substituting (8.3.8) into (8.3.7), simplifying using (8.3.11), and equating separately the coefficients of $\boldsymbol{\mathcal{S}}$, $\boldsymbol{\mathcal{S}}^2$, and $\boldsymbol{\mathcal{S}}\cdot\boldsymbol{W}$ gives

$$\begin{aligned}C_\mu &= \frac{-\tfrac{8}{15}g}{1 - \tfrac{2}{3}g^2|\boldsymbol{\mathcal{S}}|^2 + 2g^2|\boldsymbol{W}|^2},\\ b &= gC_\mu,\\ c &= -2gC_\mu,\end{aligned} \tag{8.3.9}$$

where

$$g = 1/(C_1 - 1 + \mathcal{P}_\mathcal{R}).$$

Finally, equating the coefficients of $\boldsymbol{\delta}$ and inserting the definition $\mathcal{P}_\mathcal{R} \equiv \mathcal{P}/\varepsilon$ gives

$$(1 - C_2 - C_3)^2 \frac{\mathcal{P}}{\varepsilon} = \frac{\frac{8}{15} g |\boldsymbol{\mathcal{S}}|^2}{1 - \frac{2}{3} g^2 |\boldsymbol{\mathcal{S}}|^2 + 2g^2 |\boldsymbol{W}|^2} \tag{8.3.10}$$

to complete the solution.

For those who wish to verify this solution, the identities

$$\boldsymbol{W}^2 = -\tfrac{1}{2} |\boldsymbol{W}^2| \boldsymbol{\delta}_2, \qquad \boldsymbol{\mathcal{S}}^2 = \tfrac{1}{2} |\boldsymbol{\mathcal{S}}^2| \boldsymbol{\delta}_2, \qquad \boldsymbol{\mathcal{S}} \cdot \boldsymbol{W} + \boldsymbol{W} \cdot \boldsymbol{\mathcal{S}} = 0 \tag{8.3.11}$$

are useful. They follow from the Cayley–Hamilton theorem (2.3.12) for trace-free two-dimensional tensors. The two-dimensional identity $\boldsymbol{\delta}_2$ is defined as the diagonal matrix

$$\boldsymbol{\delta}_2 = \begin{pmatrix} 1 & 0 & 0 \\ 0 & 1 & 0 \\ 0 & 0 & 0 \end{pmatrix}.$$

In the first of (8.3.11), $\mathrm{trace}(\boldsymbol{W}^2) \le 0$ was used. This follows from

$$4\,\mathrm{trace}(\boldsymbol{W}^2) \equiv 4 W_{ij} W_{ji} = \varepsilon_{ijk}\omega_k \varepsilon_{jil}\omega_l = (\delta_{kj}\delta_{jl} - \delta_{jj}\delta_{kl})\omega_k\omega_l = -2|\omega|^2 \le 0.$$

Equation (8.3.11) shows that $\boldsymbol{\mathcal{S}}^2$ is interchangeable with the two-dimensional identity matrix. But, by expressing the solution in terms of $\boldsymbol{\mathcal{S}}^2$, a formula is obtained that could be adopted in three-dimensional flow.

The solution (8.3.8) with coefficients (8.3.9) can be written as the extended eddy viscosity formula

$$\begin{aligned} \overline{u_i u_j} - \tfrac{2}{3} k \delta_{ij} = & - C_\mu \frac{k^2}{\varepsilon} S_{ij} + C_\mu \frac{g k^3}{\varepsilon^2} [(1 - C_2 + C_3)(\Omega^{\mathrm{A}}_{ik} S_{kj} - S_{ik}\Omega^{\mathrm{A}}_{kj}) \\ & + \varepsilon_{ikl}\Omega^{\mathrm{F}}_l S_{kj} - S_{ik}\varepsilon_{kjl}\Omega^{\mathrm{F}}_l - 2(1 - C_2 - C_3)(S^2_{ij} - \tfrac{1}{3}|S|^2\delta_{ij})]. \end{aligned} \tag{8.3.12}$$

Formula (8.3.12) has reverted to the original variables used in (8.3.6) to define $\boldsymbol{W}$, $\boldsymbol{\mathcal{S}}$, and $\boldsymbol{a}$. The solution (7.2.4) is a special case of the more general formula (8.3.12). This can be verified by substituting the $\boldsymbol{S}$ and $\boldsymbol{\Omega}$ corresponding to homogeneous shear.

The leading term on the right-hand side of (8.3.12) is the linear, eddy viscosity formula (although we have used $C_\mu$ rather than $2C_\mu$ as the coefficient). The bracketed term is a tensorally nonlinear extension.

The closed-form solution (8.3.12) is commonly known as an explicit algebraic stress model (EASM). The algebraic stress approximation (ASM) was introduced by Rodi (1976) (and earlier in his PhD thesis); the explicit solution was first obtained by Pope (1975), who noted that it serves as a constitutive model. It is a particular case of the concept of a nonlinear $k$–$\varepsilon$ model. In this case, it was derived by solving a second-moment closure. More generally, a formula like (8.3.8) with empirical coefficients

$C_\mu$, $b$, and $c$ could be postulated as an *ad hoc* constitutive relation. There would then be complete freedom in choosing these coefficients as functions of $|\boldsymbol{S}|$ and $|\boldsymbol{\Omega}|$. The EASM is attractive precisely because it eliminates that freedom in favor of a systematic solution of the second-moment equations.

When Eq. (8.3.12) is used to predict the Reynolds stress tensor, the $k$–$\varepsilon$ model is first solved, then $\mathcal{P}$ is found from (8.3.10), which is entirely equivalent to $\mathcal{P} = -\overline{u_i u_j} S_{ij}$, with $\overline{u_i u_j}$ substituted by expression (8.3.12). The mean flow is obtained by a RANS computation in which (8.3.12) closes (3.2.2). Examples of RANS computation with nonlinear constitutive models can be found in Lien and Leschziner (1996b). It presently is not clear when, or whether, nonlinear models are actually needed for predictive flow computation. Here we are more concerned to gain further understanding of Reynolds stress models than to expound on nonlinear $k$–$\varepsilon$ methodology.

What improvements can be expected with the added complexity contained in a nonlinear constitutive equation? Linear eddy viscosity models are not properly sensitive to rotation. The nonlinear formulation corrects it. This would seem an obvious advantage, but it must be qualified. It will be shown in Section 8.3.3 that the ability to predict rotational stabilization is contained solely in the *linear* term, $C_\mu \boldsymbol{S}$; the new dependence on $|\boldsymbol{W}|$ is the critical feature. As an example, consider the case of pipe flow rotating about its axis. The most prominent feature of axially rotating pipe flow is a stabilization of turbulence that can be caused by the superimposed axial rotation. As the turbulence intensity is suppressed, the axial mean velocity distribution becomes more laminar-like; given a pressure gradient, the mass flow through the pipe is thereby increased. This feature can be predicted by the explicit algebraic stress model (8.3.12).

As another example, a nonlinear constitutive formula is necessary in order to account for normal stress anisotropy in parallel shear flows; *any* linear model erroneously predicts $\overline{u^2} = \overline{v^2} = \overline{w^2} = \frac{2}{3}k$. Normal stress anisotropy plays a crucial role in the generation of mean streamwise vorticity – also known as Prandtl's second kind of secondary flow. A three-dimensional turbulent jet exiting from a rectangular nozzle constitutes an example: the mixing of the jet with the surrounding ambient fluid is significantly increased due to secondary motion generated by normal stress anisotropy. A phenomenon referred to as axis switching occurs, which means that the major and minor axes of the jet rotate by 90° about the streamwise axis as the jet proceeds downstream. Turbulent flow in a non-circular duct constitutes another example in which normal stress anisotropy alters the mean flow field. The particular case of the flow in a square duct is illustrated in Figure 7.19 on page 208.

### 8.3.3 Bifurcations

Although the first term on the right of (8.3.12) is just the linear eddy viscosity formula, the coefficient $C_\mu$ is no longer a constant; it is now the function (8.3.9) of $\boldsymbol{S}$ and $\boldsymbol{W}$. That dependence produces an interesting mathematical behavior: the solution to the $k$–$\varepsilon$ equations can bifurcate between exponential and algebraic time dependence.

The origin of the bifurcation can be understood by attributing it to an external stabilizing force. If the stabilizing force is weak, then the equilibrium state is fully turbulent and $k$ grows exponentially. But what if the force is sufficient to suppress the turbulence? A structural equilibrium still exists, in the sense that the anisotropy tensor, $b_{ij}$, is constant, but $k$ and $\overline{u_i u_j}$ decay with time in this case. This is a different solution branch to that

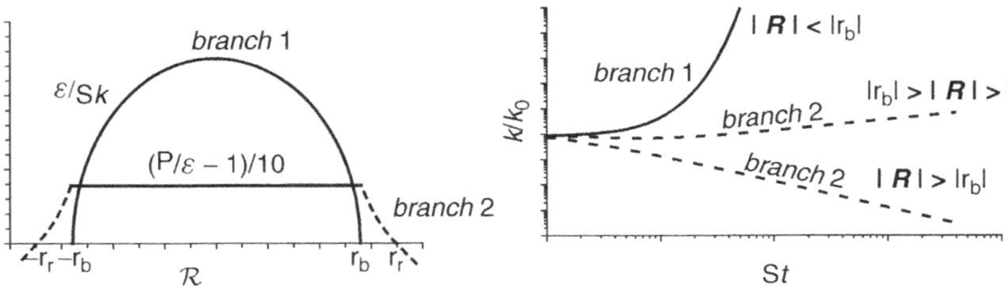


**Figure 8.1** Left: bifurcation diagram showing parametric dependence of turbulent time-scale and of production. Right: illustration of $k$ as a function of time on the two solution branches.

of healthy turbulence where $k$ and $\overline{u_i u_j}$ both grow exponentially. Mathematically, the model has undergone a bifurcation between solution branches.

The bifurcation occurs at a critical value, $r_b$, of a parameter characterizing the stabilizing force. The parameter in the present analysis is the ratio of the rate of rotation to the rate of strain, represented by $\mathcal{R}$ in Figure 8.1. The region $|\mathcal{R}| < r_b$ is the region of healthy turbulence. Non-rotating shear flow lies in that region. If the rate of rotation is small in comparison to the mean rate of strain, rotation has little effect. But if the rate of rotation is sufficiently large, $\mathcal{R}$ moves out of the unstable region and the turbulence decays. This is where $\mathcal{R} > r_r$ in the figure. More correctly, the time dependence of the solution shifts from exponential to algebraic when $\mathcal{R}$ crosses $r_b$. There is a short range $r_b < |\mathcal{R}| < r_r$ in which the solution actually grows algebraically as $t^\lambda$, $\lambda > 0$. This short range is followed by the stable region $|\mathcal{R}| > r_r$ in which the solution decays as $t^\lambda$, $\lambda < 0$ (Durbin and Pettersson Reif, 1999). These temporal behaviors are illustrated to the right of Figure 8.1. The formal analysis corresponding to these diagrams follows from the solution (8.3.10), as explained in the following.

The $k$ and standard $\varepsilon$ equations (6.2.5) and (6.2.6) on page 122 can be combined to

$$\mathrm{d}_t(\varepsilon/k) = (\varepsilon/k)^2[(C_{\varepsilon 1} - 1)\mathcal{P}_\mathcal{R} - (C_{\varepsilon 2} - 1)].$$

This admits two equilibria, obtained by setting $\mathrm{d}_t(\varepsilon/k) = 0$ on the left-hand side (Speziale and Mac Giolla Mhuiris, 1989). The two solutions to

$$(\varepsilon/k)^2[(C_{\varepsilon 1} - 1)\mathcal{P}_\mathcal{R} - (C_{\varepsilon 2} - 1)] = 0 \tag{8.3.13}$$

are $\mathcal{P}_\mathcal{R} = (C_{\varepsilon 2} - 1)/(C_{\varepsilon 1} - 1)$ and $\varepsilon/k = 0$. These are the two solution branches of the bifurcation analysis.

On the first branch, $\mathcal{P}_\mathcal{R}$ attains the fixed value $(C_{\varepsilon 2} - 1)/(C_{\varepsilon 1} - 1)$, as occurs in homogeneous shear or straining flow, for example. The exponentially growing solution for $k$ and $\varepsilon$, found in Section 6.2.1, occurs on this solution branch. It has the form

$$k \propto \mathrm{e}^{\lambda t}, \qquad \varepsilon \propto \mathrm{e}^{\lambda t},$$

with

$$\lambda = \frac{C_{\varepsilon 2} - C_{\varepsilon 1}}{C_{\varepsilon 1} - 1}\left(\frac{\varepsilon}{k}\right)_\infty.$$

The subscript $\infty$ denotes the equilibrium value. This solution is readily verified by substituting it into $d_t\varepsilon = \varepsilon(C_{\varepsilon 1}\mathcal{P}_\mathcal{R} - C_{\varepsilon 2})(\varepsilon/k)$ and $d_t k = \varepsilon(\mathcal{P}_\mathcal{R} - 1)$.

However, the exponential solution is only valid if (8.3.10) can be satisfied. Indeed, that equation determines the value of $(\varepsilon/k)_\infty$ which is needed to complete the solution for $\lambda$. Equation (8.3.10), with the definitions (8.3.6), can be rearranged to read

$$\frac{(\varepsilon/Sk)^2_\infty}{g^2(1 - C_2 - C_3)^2} = \frac{2}{3} + \frac{8}{15(1 - C_2 - C_3)^2 g\mathcal{P}_\mathcal{R}} - 2\frac{(1 - C_2 + C_3)^2}{(1 - C_2 - C_3)^2}\mathcal{R}^2, \qquad (8.3.14)$$

where $\mathcal{R}$ is defined as

$$\mathcal{R}^2 \equiv \left(\frac{\Omega}{S} + \frac{(2 - C_2 + C_3)\Omega^\mathrm{F}}{(1 - C_2 + C_3)S}\right)^2$$

for the general 2D flow represented by (8.3.3). The parameter $\mathcal{R}$ is a function of the mean flow vorticity and rate of strain, and of the coordinate system rotation. For instance, with the constants of the SSG model, it is simply $\mathcal{R}^2 = (2.25\Omega^\mathrm{F}/S + \Omega/S)^2$; and for the LRR constants, it is $\mathcal{R}^2 = (6.0\Omega^\mathrm{F}/S + \Omega/S)^2$.

Given a set of model constants, (8.3.14) is an equation for $(\varepsilon/Sk)_\infty$ as a function of $\mathcal{R}$ of the form $A(\varepsilon/Sk)^2_\infty = B - C\mathcal{R}^2$. This is the equation of an ellipse in the $\mathcal{R}$–$\varepsilon/Sk$ plane. Half of the ellipse is portrayed as *branch 1* on the left-hand side of Figure 8.1. That semi-ellipse is called the *bifurcation curve*.

The left-hand side of (8.3.14) is non-negative, but the right becomes negative when $\mathcal{R}$ is large. It follows that the exponentially growing solution exists only if $\mathcal{R}$ is sufficiently small. Indeed, the curve marked *branch 1* in Figure 8.1 is the solution to (8.3.14) for $(\varepsilon/Sk)_\infty$ as a function of $\mathcal{R}$. If $\mathcal{R}$ is larger than a critical value, marked $r_\mathrm{b}$ in the figure, this solution no longer exists. The equilibrium then bifurcates to *branch 2*. On that branch, $\varepsilon/Sk = 0$, as explained below Eq. (8.3.13).

The occurrence of a bifurcation is solely a consequence of the functional dependence of $C_\mu$; tensoral nonlinearity plays no role. To see this, notice that, when (8.3.8) is contracted with $\mathcal{S}$ to obtain the formula (8.3.10) for production,

$$a_{ij}\mathcal{S}_{ji} = C_\mu \mathcal{S}_{ij}\mathcal{S}_{ji} + b(W_{ij}\mathcal{S}^2_{ji} - \mathcal{S}^2_{ij}W_{ji}) + c(\mathcal{S}^3_{ii} - \tfrac{1}{3}|\boldsymbol{\mathcal{S}}|^2\mathcal{S}_{ii}) = C_\mu|\boldsymbol{\mathcal{S}}|^2$$

is obtained. The last equality follows because $\mathcal{S}^3_{ii} = 0 = \mathcal{S}_{ii}$ for incompressible two-dimensional straining, and $W_{ij}\mathcal{S}^2_{ji} = 0$ because $\boldsymbol{W}$ is antisymmetric and $\boldsymbol{\mathcal{S}}^2$ is symmetric. The important message for turbulence modeling is that the response to rotation can be obtained by an eddy viscosity model with a variable $C_\mu$. Instead of using the present analytical solution for $C_\mu$, an *ad hoc* functional dependence could be designed to capture the bifurcation. Constant $C_\mu$ precludes bifurcations; indeed, Exercise 6.6 shows that with constant $C_\mu$ the solution is independent of system rotation.

The nature of the second branch follows from setting $\varepsilon/Sk = 0$. In this case $\mathcal{P}/\varepsilon = \mathcal{P}_\mathcal{R}$ is no longer constant; it varies as a function of $\mathcal{R}$ as illustrated by the dashed line in Figure 8.1. When $|\mathcal{R}|$ just crosses $r_\mathrm{b}$, $\mathcal{P}_\mathcal{R}$ is still greater than unity, although it is decreasing. As long as $\mathcal{P} > \varepsilon$, the turbulence will grow with time; so there is a region between $r_\mathrm{b}$, the bifurcation point, and $r_\mathrm{r}$, the restabilization point, in which $k$ grows with time, albeit the solution branch has bifurcated.

The solution for $k$ on the second branch is found to be (Durbin and Pettersson Reif, 1999)

$$k \sim t^{\lambda}, \qquad \varepsilon \sim t^{\lambda-1},$$

with

$$\lambda = \frac{\mathcal{P}_{\mathcal{R}} - 1}{C_{\varepsilon 2} - 1 - \mathcal{P}_{\mathcal{R}}(C_{\varepsilon 1} - 1)}. \tag{8.3.15}$$

This is verified by substitution into the $k$ and $\varepsilon$ equations. Now (8.3.14) plays the role of an equation for $\mathcal{P}_{\mathcal{R}}$:

$$\frac{C_1 - 1 + \mathcal{P}_{\mathcal{R}}}{\mathcal{P}_{\mathcal{R}}} = \frac{15(1 - C_2 + C_3)^2 \mathcal{R}^2}{4} - \frac{5(1 - C_2 - C_3)^2}{4} \tag{8.3.16}$$

after using $g = 1/(C_1 - 1 + \mathcal{P}_{\mathcal{R}})$ and recognizing that $\varepsilon/Sk = 0$. This formula provided the dashed curve plotted in the left half of Figure 8.1.

The bifurcation point, $\mathcal{R} = r_b$, is found by rearranging Eq. 8.3.16) to

$$\mathcal{R} = \pm \left( \frac{1}{3} + \frac{4(C_1 - 1 + \mathcal{P}_{\mathcal{R}})}{15\mathcal{P}_{\mathcal{R}}(1 - C_2 - C_3)^2} \right)^{1/2} \frac{(1 - C_2 - C_3)}{(1 - C_2 + C_3)}. \tag{8.3.17}$$

This pertains to *branch 2*, while $\mathcal{P}_{\mathcal{R}} = (C_{\varepsilon 2} - 1)/(C_{\varepsilon 1} - 1)$ obtains on *branch 1*. The bifurcation point is common to both branches. Hence, the value of $\mathcal{R}$ at bifurcation is deduced by substituting $\mathcal{P}_{\mathcal{R}} = (C_{\varepsilon 2} - 1)/(C_{\varepsilon 1} - 1)$ into (8.3.17). These points are evaluated on the two top lines of Table 8.1 for two turbulence models. The table cites the frame rotation rates for bifurcation. They are not symmetric with respect to $\Omega^F = 0$.

With the standard constants $C_{\varepsilon 1} = 1.44$ and $C_{\varepsilon 2} = 1.92$ the ratio of production to dissipation is $\mathcal{P}_{\mathcal{R}} \approx 2.1$ on branch 1. As $\mathcal{R}$ increases beyond the bifurcation point $\mathcal{P}_{\mathcal{R}}$ decreases along branch 2. When $\mathcal{P}_{\mathcal{R}}$ decreases below unity $\lambda$ becomes negative in (8.3.15) and $k$ decays algebraically with time. The value where $\mathcal{P}_{\mathcal{R}} = 1$ is $\mathcal{R} = r_r$ in Figure 8.1. Its numerical value is given by (8.3.17) upon substitution of $\mathcal{P}_{\mathcal{R}} = 1$. At that point the solution changes from algebraic growth to algebraic decay. Restabilization points are evaluated on the two lower lines of Table 8.1. The physical source of this stabilization is rotation of the fluid. However, the same concepts apply to stabilization by other external forces, such as stratification.

Linear stability theory gives the values $\Omega^F/S = 1$ and 0 as the stability limits for laminar parallel shear flow (Craik, 1989). Speziale *et al.* (1991) proposed that an approximate correspondence to linear theory is a requirement for any model that intends to

**Table 8.1** Bifurcation and restabilization points in homogeneous shear.

| Model | $\mathcal{P}/\varepsilon$ | $r_b$ | $\Omega_+^F/S$ | $\Omega_-^F/S$ |
|---|---|---|---|---|
| SSG | 2.09 | ±1.358 | 1.048 | −0.159 |
| LRR | 2.09 | ±3.136 | 0.689 | −0.356 |
| **Model** | $\mathcal{P}/\varepsilon$ | $r_r$ | $\Omega_+^F/S$ | $\Omega_-^F/S$ |
| SSG | 1.00 | ±1.427 | 1.078 | −0.190 |
| LRR | 1.00 | ±3.552 | 0.759 | −0.425 |

represent rotational effects on turbulence. Formula (8.3.17) gives the bifurcation and restabilization points for linear models such as LRR. Analogous formulas can be derived for quasi-linear models such as SSG (Pettersson Reif *et al.*, 1999). Table 8.1 gives the bifurcation and restabilization points, $r_b$ and $r_r$, for the SSG and LRR models. The fourth and fifth columns, stating values of frame rotation, were obtained by setting $\mathcal{R} = \pm r_b$ or $\pm r_r$. For parallel shear flow, $\Omega/S = -1$, and the definition of $\mathcal{R}$ becomes

$$\mathcal{R} = -1 + \frac{(2 - C_2 + C_3)}{(1 - C_2 + C_3)} \frac{\Omega^F}{S}.$$

By design, the SSG model is close to the stability values $\Omega^F_+/S = 1$ and $\Omega^F_-/S = 0$.

## 8.4 Passive scalar flux modeling

Research into turbulence closure modeling has focused primarily on momentum transport; less attention has been paid to transport of scalars. There is a good reason for this: the mean velocity and turbulent stresses constitute very important inputs to the scalar flux equations. The opposite is not true: the passive scalar field is mathematically decoupled from the dynamical equations governing the flow field. A solution of the flow field is thus a prerequisite to solution of the scalar field. From a practical point of view, a 15% error in the prediction of skin friction might be unacceptable in aerospace applications, but the same level of accuracy of the heat-transfer coefficient would probably suffice in most applications. In view of this, elaborate closure schemes seem more justified for the turbulent stresses than for the turbulent fluxes. But there is no compulsion to dismiss the topic of scalar flux transport in a text on turbulence modeling, and we broach the subject in the ensuing sections.

### 8.4.1 Scalar diffusivity models

The notion that turbulence transports passive contaminants in much the same way as momentum has led to closure formulations of the same ilk as eddy viscosity stress models. The simplest scalar flux model adopts the gradient diffusion hypothesis

$$\overline{u_i c} = -\alpha_T \frac{\partial C}{\partial x_i}, \tag{8.4.1}$$

where the scalar diffusivity is $\alpha_T = \nu_T/\mathrm{Pr}_T$, as discussed in Section 4.4, page 77. The eddy viscosity $\nu_T$ is known from the solution to one of the models from Chapter 6, and a value of the turbulent Prandtl number $\mathrm{Pr}_T$ is simply prescribed. The value chosen for $\mathrm{Pr}_T$ depends on the flow configuration. Experimental measurements suggest that a value of $\mathrm{Pr}_T \approx 0.9$ can be used in turbulent boundary layers, while $\mathrm{Pr}_T \approx 0.7$ is often more suitable in free-shear flows. The turbulent Prandtl number is not a material property. Inevitably, it can depend on many factors that influence the flow field, and a constant value is not generally acceptable.

In many practical calculations the dominant concern is with transport across boundary layers, and the scalar diffusivity assumption, with constant $\mathrm{Pr}_T = 0.9$, is reasonably effective; then nothing more needs to be said. However, some researchers have explored scalar flux transport models, with more complex applications in mind.

## 8.4.2 Tensor diffusivity models

For many years, development and refinement of scalar flux closures suffered tremendously from a shortage of available experimental data. For every scalar field measurement, there were many, many velocity field measurements. However, as more data have been compiled over the years, it has become obvious that the gradient diffusion model (8.4.1) has serious limitations. The use of a scalar diffusivity $\alpha_T$ implies alignment between the scalar flux and the mean scalar gradient. This is not always true.

To illustrate this point, consider the transport of a passive scalar in fully developed plane channel flow with a temperature differential between the upper and lower walls. The only mean temperature gradient is in the $y$ direction. Figure 8.2 displays profiles of wall normal and streamwise scalar flux obtained in a direct numerical simulation. The maximum streamwise flux is almost an order of magnitude larger than the wall normal flux component; but the model (8.4.1) predicts $\overline{uc}/\overline{vc} = \partial_x C/\partial_y C = 0$. This failure to predict the streamwise turbulent flux is not as severe as it first seems: $\overline{uc}$ does not influence the mean scalar concentration profile across the channel because the derivative $\partial_x \overline{uc}$ is zero. Only the wall normal flux $\partial_y \overline{vc}$ contributes to the mean flux divergence, and only this enters the mean concentration equation. The wall normal flux can be reasonably well predicted by a simple formula like (8.4.1). However, this example highlights a shortcoming of scalar diffusivity models that has the potential to adversely affect predictions in complex flows of engineering interest.

The inability to predict the scalar flux vector with a gradient diffusion model could be corrected by replacing the scalar diffusivity $\alpha_T$ by a tensor diffusivity, also referred to as a *dispersion tensor*. The simplest form is the *generalized* gradient diffusion hypothesis

$$\overline{u_i c} = -\alpha_{T_{ij}} \frac{\partial C}{\partial x_j}, \tag{8.4.2}$$

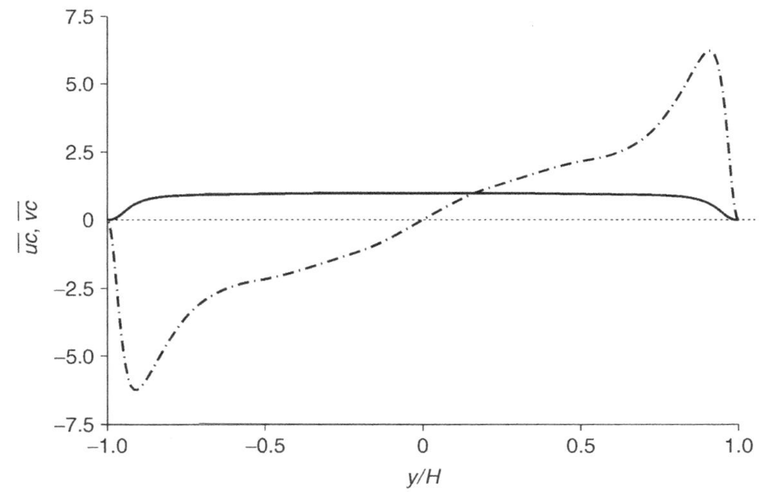


**Figure 8.2** Streamwise and cross-stream scalar fluxes in plane channel flow at $R_\tau = 180$ (Kim and Moin, 1989): $\overline{u\theta}$ (— · —); and $\overline{v\theta}$ (———). A temperature difference is imposed between the top and bottom walls.

as in Exercise 3.3. This expression is termed "generalized" because, in contrast to (8.4.1), scalar fluxes arise in directions other than that of the mean scalar gradient. In the presence of mean velocity gradients, $\alpha_{T_{ij}}$ will be an asymmetric tensor. For instance, Dr. M. Rogers informs us that he found $\alpha_{T_{12}}/\alpha_{T_{22}} \approx -2$ and $\alpha_{T_{21}}/\alpha_{T_{22}} \approx 0$ in a direct numerical simulation of homogeneously sheared turbulence.

A formula that does not respect the asymmetry is that of Daly and Harlow (1970):

$$\alpha_{T_{ij}} = C_c \overline{u_i u_j} \frac{k}{\varepsilon}.$$

Near a wall with a concentration gradient in the $y$ direction, this gives $\overline{vc}/\overline{uc} = \alpha_{T_{22}}/\alpha_{T_{12}} = \overline{v^2}/\overline{uv}$. Although a non-zero streamwise flux is produced, the predicted value is only half of that observed in numerical simulations. Including gradients of the mean velocity field

$$\overline{u_i c} = -C_c \frac{k}{\varepsilon} \left( \overline{u_i u_j} \frac{\partial C}{\partial x_j} + \overline{c u_j} \frac{\partial U_i}{\partial x_j} \right) \tag{8.4.3}$$

further improves the predictions. The introduction of the mean velocity field into (8.4.3) might seem arbitrary. However, the term inside the bracket in (8.4.3) is nothing but the *exact* rate of production of turbulent fluxes, as appears in Eq. (8.4.5), which depends on both the mean scalar and mean velocity gradients. On dimensional grounds, it has been multiplied by a suitable time-scale, $k/\varepsilon$.

Substituting the representation (8.4.2) into formula (8.4.3) shows that the corresponding dispersion tensor satisfies

$$\alpha_{T_{ij}} = C_c \frac{k}{\varepsilon} \left( \overline{u_i u_j} - \alpha_{T_{kj}} \frac{\partial U_i}{\partial x_k} \right). \tag{8.4.4}$$

This algebraic formula for $\boldsymbol{\alpha}_{\mathrm{T}}$ can be solved by the methods of Section 2.3.2, or, in this case, by direct matrix inversion. It can been seen from that solution, or by inspection, that $\boldsymbol{\alpha}_{\mathrm{T}}$ is not symmetric in the presence of velocity gradients.

This formula embodies an important concept in passive scalar transport: the dispersion tensor is a function of the turbulence, not of the contaminant. Concentration does not appear in Eq. (8.4.4); only statistics of the turbulent velocity do. From another perspective, this embodies the superposition principle: any representation of the flux must be linear in the concentration. Some researchers have proposed to normalize $\overline{\boldsymbol{u}c}$ by the nonlinear term $\sqrt{k\overline{c^2}}$. That normalized flux vector produces no simplification because superposition demands that it cannot be independent of the scalar field – as is readily apparent from substitution into Eq. (8.4.3).

A more refined algebraic model than (8.4.4) would be needed to accurately predict scalar flux anisotropy in complex flows. This is especially the case for flows affected by body forces, such as buoyancy. Rather than postulating an algebraic formula for the scalar flux vector $\overline{u_i c}$, a more physically appealing approach is to solve a closed version of the scalar flux *transport* equation (3.4.3). Starting from the exact equation is physically appealing because the generation rate of fluxes is retained in its exact form. This particular feature makes second-moment closures for Reynolds stress transport seem attractive in many complex flow computations. The case that it is necessary to solve a full set of transport equations for the scalar flux distribution is probably less strong; if the velocity field is properly modeled, it might suffice to adopt an algebraic model, or perhaps just simple gradient diffusion, for the turbulent fluxes. But, even if an algebraic model for the dispersion tensor is the route adopted, the scalar flux transport equation provides an attractive starting point. Let us consider closure of the scalar flux equation.

### 8.4.3 Scalar flux transport

The exact transport equation (3.4.3), page 55, for $\overline{u_i c}$ can be written symbolically as

$$\mathrm{D}_t \overline{u_i c} = \mathcal{P}_{ic} + \wp_{ic} + D_{ic} - \varepsilon_{ic}, \tag{8.4.5}$$

where

$$\begin{aligned}
\wp_{ic} &= -\frac{1}{\rho}\overline{c\,\partial_i p},\\
D_{ic} &= -\partial_j\left(\overline{u_i u_j c} - \alpha\,\overline{u_i\,\partial_j c} - \nu\,\overline{c\,\partial_j u_i}\right),\\
\varepsilon_{ic} &= (\alpha+\nu)\,\overline{\partial_j c\,\partial_j u_i},\\
\mathcal{P}_{ic} &= -\overline{u_i u_j}\,\partial_j C - \overline{c u_j}\,\partial_j U_i.
\end{aligned}$$

The first three terms represent the pressure scalar correlation, molecular and turbulent diffusion, and dissipation rate. They are all unclosed and need to be modeled. The last term is the rate of production of scalar fluxes by mean scalar and velocity gradients – this is the term that was used in Eq. (8.4.3) to model the turbulent flux.

The problem consists of how to close Eq. (8.4.5). It will suffice to consider the limit of high Reynolds and Peclet numbers. Molecular diffusion and turbulent scrambling rapidly remove small-scale directional preferences, so at high Reynolds numbers the small-scale motion is nearly isotropic; this is the well-known assumption of local isotropy of the small scales. The dissipation rate $\varepsilon_{ic}$ is associated with small scales. Since a first-order isotropic tensor does not exist, the isotropic value of the scalar flux dissipation rate is zero:

$$\varepsilon_{ic} = 0.$$

Dispersion by random convection dominates over molecular diffusion provided the Peclet number is large. The reason for this is provided in Section 4.4: if the flow field has characteristic turbulent velocity, $u$, and length, $L$, the ratio of the turbulent diffusivity ($\alpha_{\mathrm{T}} = Lu$) to the molecular diffusivity is $\alpha_{\mathrm{T}}/\alpha = \mathrm{Pe}$. So if the Peclet number is large, turbulent dispersion controls the rate of mixing, and molecular diffusion can be neglected.

Different models for the turbulent transport term have been put forward by different researchers. The Daly–Harlow form (Daly and Harlow, 1970) is widely used:

$$-\partial_j \overline{u_i u_j c} = C_c \frac{k}{\varepsilon}\,\partial_j\left[\overline{u_j u_k}\,\partial_k \overline{u_i c}\right]. \tag{8.4.6}$$

The value of $C_c$ varies in the range $0.11 \le C_c \le 0.20$; a common value is 0.15.

In the absence of dissipative terms, only the pressure scalar correlation, $\wp_{ic}$, is responsible for counteracting production of $\overline{u_i c}$. The problem of closing (8.4.5) is now reduced to providing a model for $\wp_{ic}$. If buoyancy is neglected, this term can be decomposed into slow and rapid parts:

$$\wp_{ic} = \wp_{ic1} + \wp_{ic2}.$$

As will be shown later, a third term appears if the flow is affected by buoyancy.

The slow part $\wp_{ic1}$ is associated with the problem of return to isotropy; if there are no mean scalar or velocity gradients, an initially anisotropic scalar flux field would be driven towards isotropy. The simplest way to achieve this is to adopt a formula in the spirit of Rotta's linear return model (7.1.8) for the turbulent stresses:

$$\wp_{ic1} = -C_{1c}\frac{\varepsilon}{k}\overline{u_i c}. \tag{8.4.7}$$

The coefficient $C_{1c}$ is usually termed the Monin constant. It is not as constant as the corresponding coefficient used in the Rotta model. A wide range of values has been proposed in the literature; perhaps the most commonly used is $C_{1c} = 3.0$. Modifications to (8.4.7) have been proposed to accommodate effects of turbulence anisotropy. In practice, this has meant that nonlinear terms like $\overline{u_i u_k}\,\overline{u_k c}$ were added. At present there is little motive for such elaboration.

In order to gain insight into modeling of the rapid term, $\wp_{ic2}$, the approach described in Section 7.1.4 can be adopted. The fluctuating pressure gradient is eliminated by first solving the Poisson equation, which, for the rapid contribution, has the formal solution given by (7.1.17). Then, by differentiating this solution with respect to $x_i$ and substituting the result into $\wp_{ic}$,

$$\begin{aligned}\wp_{ic2} \equiv -\frac{1}{\rho}\overline{c\,\partial_i p}^{\text{rapid}} &= -\frac{\partial_l U_k}{2\pi}\iiint_{-\infty}^{\infty} [\partial_i\partial_k\overline{u_l c}\,]\,(\boldsymbol{\xi})\frac{1}{|\boldsymbol{\xi}|}\,\mathrm{d}^3\boldsymbol{\xi}\\ &= M_{ikl}\,\partial_l U_k\end{aligned} \tag{8.4.8}$$

is obtained. The right-hand side of the first line in (8.4.8) is a definite integral, so the components of $\boldsymbol{M}$ are constants. This is exactly the same procedure as was used in Section 7.1.4 to evaluate the rapid pressure–strain correlation term that appears in the Reynolds stress transport equations. Obviously, the third-order tensor $M_{ikl}$ is analogous to the fourth-order tensor $M_{ijkl}$ in Eq. (7.1.20). A normalization is obtained from the contraction $M_{kkl}$, which, for the integral in (8.4.8), is readily evaluated as $M_{kkl} = 2\overline{u_l c}$. The reasoning is the same as used to derive (7.1.21).

In Section 7.1.4 rapid pressure–strain models were developed by expansion in anisotropy. In isotropic turbulence, $\overline{u_i c} = 0$. If $M_{ikl}$ is assumed to depend only on turbulent fluxes, in accord with (8.4.8), then

$$M_{ikl} = C_{2c}\,\overline{u_l c}\,\delta_{ik} + C_{3c}\,(\overline{u_k c}\,\delta_{li} + \overline{u_i c}\,\delta_{lk}) \tag{8.4.9}$$

is the most general form that is linear in anisotropy and symmetric in $i$ and $k$. The rapid term is correspondingly modeled as

$$\wp_{ic2} = C_{2c}\,\overline{cu_k}\,\partial_k U_i + C_{3c}\,\overline{cu_k}\,\partial_i U_k.$$

The normalization constraint is $3C_{2c} + 2C_{3c} = -2$; however, this is not usually imposed in practice.

A further term is often added to counteract the production term, based partly on dimensional reasoning, and partly on pragmatic grounds. The general linear model,

$$\wp_{ic2} = C_{2c}\,\overline{cu_k}\,\partial_k U_i + C_{3c}\,\overline{cu_k}\,\partial_i U_k + C_{4c}\,\overline{u_i u_j}\,\partial_j C, \tag{8.4.10}$$

is then obtained. There is no consensus in the literature on values for the constants: Launder (1989) gives $C_{2c} = 0.4$, $C_{3c} = 0$, $C_{4c} = 0$; Craft *et al.* (1993) give $C_{2c} = 0.5$, $C_{3c} = 0$, $C_{4c} = 0$; and Durbin (1993) gives $C_{2c} = 0$, $C_{3c} = 0$, $C_{4c} = 0.45$ with $C_{1c} = 2.5$. Based on these, we will adopt $C_{2c} = 0$ and $C_{3c} = 0$, which greatly simplifies this model. On substituting the closure (8.4.10) into (8.4.5), it becomes

$$\mathrm{d}_t\overline{u_i c} = -C_{1c}\frac{\varepsilon}{k}\,\overline{u_i c} + (C_{4c} - 1)\,\overline{u_i u_j}\,\partial_j C - \overline{cu_j}\,\partial_j U_i \tag{8.4.11}$$

in homogeneous turbulence.

Closure has now been devised for the set of equations governing the transport of passive scalar fluxes. The advantages of solving the additional partial differential equations in practical computations are debatable; it may suffice to simplify them to obtain a set of algebraic equations. In general, this is a difficult exercise; but results are readily obtained in the important case of homogeneously sheared turbulence.

#### 8.4.3.1 Equilibrium solution for homogeneous shear

The form of Eq. (8.4.11) suggests that a solution be sought in the form

$$\overline{u_i c} = -\alpha_{T_{ij}}\,\partial_j C.$$

The mean scalar gradient evolves under homogeneous distortion according to $\mathrm{d}(\partial_j C)/\mathrm{d}t = -\partial_j U_k\,\partial_k C$. Hence, the evolution equation for $\boldsymbol{\alpha}_{\mathrm{T}}$, obtained from (8.4.11) by inserting the above representation of $\overline{u_i c}$, is

$$\frac{\mathrm{d}\alpha_{T_{ij}}}{\mathrm{d}t} = (1 - C_{4c})\,\overline{u_i u_j} - \frac{C_{1c}\varepsilon}{k}\,\alpha_{T_{ij}} + \alpha_{T_{ik}}\partial_k U_j - \alpha_{T_{kj}}\partial_k U_i. \tag{8.4.12}$$

Under conditions of equilibrium, $\boldsymbol{\alpha}_{\mathrm{T}}$ will not be constant, but the ratio $\boldsymbol{\alpha}_{\mathrm{T}}/\nu_{\mathrm{T}}$ will. This implies that a solution to Eq. (8.4.12) should be of the form

$$\alpha_{T_{ij}} = K_{ij}k^2/\varepsilon,$$

with $K_{ij}$ being constant in equilibrium. Substituting this form, setting $\mathrm{d}_t K_{ij} = 0$, and invoking the $k$–$\varepsilon$ equations for $\mathrm{d}_t k$ and $\mathrm{d}_t\varepsilon$, provides a set of algebraic equations for $K_{ij}$ (Shabany and Durbin, 1997).

For a uniform shear flow, $U = Sy$, the equilibrium solution to (8.4.12) obtained by the above procedure is

$$\begin{aligned}
K_{11} &= (1 - C_{4c})g_k[\tau_{11} - 2(g_k\mathcal{S})^2\tau_{22}],\\
K_{22} &= (1 - C_{4c})g_k\tau_{22},\\
K_{12} &= (1 - C_{4c})g_k[\tau_{12} - g_k\mathcal{S}\tau_{22}],\\
K_{21} &= (1 - C_{4c})g_k[\tau_{12} + g_k\mathcal{S}\tau_{22}],\\
K_{33} &= (1 - C_{4c})g_k\tau_{33},
\end{aligned} \tag{8.4.13}$$

where

$$\frac{1}{g_k} = C_{1c} + (2 - C_{\varepsilon 1})\mathcal{P}/\varepsilon + C_{\varepsilon 2} - 2,$$

$\mathcal{S} = Sk/\varepsilon$, and $\tau_{ij} = \overline{u_i u_j}/k$. A log-layer analysis shows that this solution applies there, as well, except that

$$g_k^{-1} = C_{1c}.$$

Solution (8.4.13) illustrates the ability of the full transport closure to capture the asymmetry of the dispersion tensor. Because $\tau_{12} < 0$, the qualitative effect is that $K_{21}$ is decreased in magnitude by shear and the magnitude of $K_{12}$ is increased. Also $K_{12} < 0$. The solution for $K_{11}$ raises the concern that $K_{11}$ might become negative at large $\mathcal{S}$, which is unphysical.

If the temperature gradient is in the $y$ direction, then a turbulent Prandtl number can be defined as

$$\mathrm{Pr_T} = \frac{\nu_\mathrm{T}}{\alpha_{T_{22}}} = \frac{C_{1c} C_\mu}{1 - C_{4c}}, \tag{8.4.14}$$

where the $v^2\!-\!f$ formula $\nu_\mathrm{T} = C_\mu \tau_{22} k^2/\varepsilon$ was used and the log-layer value for $g_k$ was substituted. The model constants should be chosen to produce a plausible value of $\mathrm{Pr_T}$.

### 8.4.4 Scalar variance

A section on scalar variance is interposed here, in preparation for the ensuing material on buoyancy. Analysis of scalar variance is of interest in its own right, but it also appears explicitly in the scalar flux equation for turbulent buoyant flow.

The equation governing scalar variance is obtained by multiplying Eq. (3.4.1), for the instantaneous value of the scalar $(C + c)$, by $c$ and averaging. The result is

$$\mathrm{D}_t \overline{c^2} = \alpha \nabla^2 \overline{c^2} - \partial_j \overline{u_j c^2} - 2\,\overline{u_j c}\,\partial_j C - 2\alpha\,\overline{\partial_j c\,\partial_j c}. \tag{8.4.15}$$

The first two terms on the right-hand side of (8.4.15) represent molecular and turbulent transport, whereas the third is the rate of variance production due to mean scalar gradients. The last term is the scalar variance dissipation rate, which will be denoted $\varepsilon_c$. This equation resembles the turbulent energy transport equation, except that pressure-diffusion is absent. The majority of proposals to close the turbulent transport term in (8.4.15) have adopted gradient transport:

$$-\partial_i \overline{u_i c^2} = \partial_i \left( C_c\, \overline{u_i u_j}\, T\, \partial_j \overline{c^2} \right). \tag{8.4.16}$$

The more important term to close in (8.4.15) is the rate of dissipation, $\varepsilon_c$. It could be obtained from its own modeled transport equation, in the vein of two-equation stress models. We will describe the simpler approach of invoking a *time-scale ratio*. The ratio of mechanical to scalar time-scales is defined by

$$\mathcal{R} \equiv \frac{k/\varepsilon}{\overline{c^2}/\varepsilon_c}. \tag{8.4.17}$$

Assuming $\mathcal{R}$ to be a universal constant gives the unostentatious closure

$$\varepsilon_c = \mathcal{R}\frac{\overline{c^2}}{k}\varepsilon. \tag{8.4.18}$$

The rate of scalar dissipation is made proportionate to the rate of energy dissipation

Any experiment can be used to measure $\mathcal{R}$ if it is fundamentally a constant. Decaying isotropic turbulence is a natural. In homogeneous isotropic turbulence, the scalar variance equation reduces to

$$\mathrm{d}_t\overline{c^2} = -\mathcal{R}\frac{\varepsilon}{k}\overline{c^2} = -n\mathcal{R}\frac{\overline{c^2}}{t},$$

where $n$ is the decay exponent of $k$. The solution is $\overline{c^2} \propto t^{-m}$, where $m = n\mathcal{R}$. One need only measure the decay exponents, $n$ and $m$, of $k$ and $\overline{c^2}$ to obtain the time-scale ratio.

While experiments on the decay rate of turbulent kinetic energy show that $n$ is reasonably constant, and equal to about 1.2, the scalar decay exponent, $m$, has been found to vary over the range 1.0–3.0 (Warhaft, 2000). Hence $\mathcal{R}$ is not a constant in grid turbulence. The range of measured values does not reflect lack of reproducibility. The evidence is that the variation of $\mathcal{R}$ is systematic and that it depends on the initial length scale of the contaminant relative to that of the velocity.

That tendency for scalar variance to be sensitive to source conditions is found for concentrated sources too (Borgas and Sawford, 1996). Such dependence on initial conditions can be explained theoretically by relating scalar variance to relative dispersion of particle pairs (Durbin, 1980) (see Exercise 2.3). When the integral length scale of the contaminant concentration is comparable to, or less than, that of the velocity, scalar variance is predicted by the theory to be a function of the shape of the two-point velocity correlation. Unfortunately, practical prediction methods have not benefited from this theory.

In the presence of a uniform mean scalar gradient, $\mathcal{R}$ is found to be more nearly constant and equal to about 1.5 (Warhaft, 2000). This gives some hope that a constant time-scale ratio model (8.4.18) will be a reasonable estimate in some circumstances.

## 8.5 Active scalar flux modeling: effects of buoyancy

Buoyancy effects arise in a group of problems that include the dynamics of atmospheric and oceanic boundary layers, and convective cooling. Predictions of heat transfer with buoyancy are needed in many engineering applications. Apart from generating turbulence anisotropy, buoyancy gives rise to changes of turbulence structure and intensity in much the same way as streamline curvature does: it may either enhance or suppress turbulence intensity. When density increases in the direction of gravity, the stratification is stable and tends to suppress turbulent energy.

Buoyant forces are due to variations in the density of the fluid caused by differences in the scalar concentration. They give rise to a fluctuating body force in the vertical momentum equation. This body force affects both the mean flow field and the turbulence. We will focus our attention on incompressible fluids and will invoke the Boussinesq approximation. The latter approximation is that density variations can be neglected in inertial terms, but not when they are multiplied by gravity.

When the contaminant is dynamically active, the equations governing its concentration are coupled to the momentum equations. The Reynolds stress equations depend on the scalar flux and vice versa. A fluctuating concentration field causes density fluctuations $\rho' = -\rho\beta c$, where $\beta$ is defined as the expansion coefficient at a fixed mean state,

$$\beta \equiv [-\partial \log(\rho)/\partial c]_C.$$

Note that, if $c$ is temperature, $\beta$ is positive because increase of temperature causes decrease of density. It is assumed that density variations are not large, consistent with the Boussinesq assumption.

In the presence of gravity, the fluctuating density enters the Navier–Stokes equations through a fluctuating body force $g_i\rho'/\rho$, which can be rewritten as $-g_i\beta c$. This extra term is added to the right-hand side of the fluctuation momentum equation (3.2.3). Note that $\boldsymbol{g}$ is a vector in the direction of gravity, which is often the $-y$ direction. In some applications, such as to rotating machinery, the effective gravity is due to centrifugal acceleration and is directed toward the center of curvature.

Let us start by writing down the set of governing equations under the condition of homogeneity. This is sufficiently general since the terms containing gravity are the same in homogeneous and non-homogeneous flow. If buoyancy effects are added, the exact transport equation for the turbulent flux is

$$\mathrm{d}_t\overline{u_i c} = \mathcal{P}_{ic} + \wp_{ic} - \varepsilon_{ic} + G_{ic}, \tag{8.5.1}$$

where only the last term is new to (8.4.5). Its explicit statement is

$$G_{ic} = -g_i\beta\overline{c^2}. \tag{8.5.2}$$

Because of it, the components of the scalar flux vector now depend on the scalar variance. Again, local isotropy is usually invoked to set $\varepsilon_{ic} = 0$.

Under the same conditions that (8.5.1) is valid, the turbulent stress transport equation becomes

$$\mathrm{d}_t\overline{u_i u_j} = \mathcal{P}_{ij} + \wp_{ij} - \tfrac{2}{3}\varepsilon\delta_{ij} + G_{ij}. \tag{8.5.3}$$

This is identical to (7.1.4) except for the last term on the right-hand side,

$$G_{ij} = -\beta(g_i\overline{u_j c} + g_j\overline{u_i c}). \tag{8.5.4}$$

This is often referred to as the gravitational production term. Since $G_{ij}$ contains $\overline{u_j c}$, the second-moment equations for Reynolds stress and for scalar flux are now fully coupled.

Half the trace of (8.5.4) defines the turbulent kinetic energy equation

$$\mathrm{d}_t k = \mathcal{P} + G - \varepsilon, \tag{8.5.5}$$

where $G = -\beta g_i\overline{u_i c}$. The buoyancy terms in (8.5.4) and (8.5.5) are exact once the scalar flux field is known. Hence, the direct influence of the gravitational field on the turbulent field is exactly represented.

Minus the ratio of buoyant to stress production of turbulent kinetic energy in parallel shear flow defines the *flux Richardson number*

$$\mathrm{Ri_f} \equiv \frac{-G}{\mathcal{P}} = \frac{\beta\,\overline{cu_i}\,g_i}{-\overline{u_i u_j}\,\partial_j U_i}. \tag{8.5.6}$$

In the numerator, $-G$ is used so that $\mathrm{Ri_f}$ parameterizes the stabilizing effect of stratification. When $\mathrm{Ri_f} > 0$, the density gradient suppresses turbulence. The kinetic energy equation can be re-expressed as

$$\mathrm{d}_t k = \mathcal{P}(1 - \mathrm{Ri_f}) - \varepsilon.$$

Mixing length ideas have been used previously to argue that, if $\partial_2 U_1 > 0$, the shear stress $\overline{u_1 u_2} < 0$. By that same rationale, if $\partial_2 C > 0$, one expects $\overline{cu_2} < 0$. If $c$ is temperature, this corresponds to downward heat transfer in a parallel shear flow. Downward transfer works against gravity, so in this case the buoyant production becomes negative and $\mathrm{Ri_f} > 0$ (recall that $\boldsymbol{g}$ points downward, in the direction of gravity); turbulent kinetic energy is lost. Similarly, if the heat transfer is upwards, $\overline{cu_2} > 0$, the buoyant production is positive and $\mathrm{Ri_f} < 0$; turbulent kinetic energy is augmented. Observations have shown that turbulence cannot be sustained under the stabilizing effect of buoyancy if $\mathrm{Ri_f} \gtrsim 0.2$ (Turner, 1980).

It is natural also to sensitize the modeled dissipation rate equation to buoyancy. This is usually done by replacing the mean shear production $\mathcal{P}$ with $\mathcal{P} + G$. That gives the revised equation

$$\mathrm{d}_t \varepsilon = \frac{\varepsilon}{k}[C_{\varepsilon 1}(\mathcal{P} + G) - C_{\varepsilon 2}\varepsilon]. \tag{8.5.7}$$

Let us explore the condition of structural equilibrium. Consider a uniform shear flow with mean shear $\mathcal{S}$. The equation for the ratio of mean to turbulent time-scale can be written

$$\frac{1}{\mathcal{S}}\frac{\mathrm{d}}{\mathrm{d}t}\left(\frac{\varepsilon}{\mathcal{S}k}\right) = \left[(1 - \mathrm{Ri_f})(C_{\varepsilon 1} - 1)\frac{\mathcal{P}}{\varepsilon} - (C_{\varepsilon 2} - 1)\right]\left(\frac{\varepsilon}{\mathcal{S}k}\right)^2 \tag{8.5.8}$$

by using (8.5.5)–(8.5.7). Equilibrium requires $\mathrm{d}_t(\varepsilon/\mathcal{S}k) = 0$, giving

$$\frac{\mathcal{P}}{\varepsilon} = \frac{C_{\varepsilon 2} - 1}{(1 - \mathrm{Ri_f})(C_{\varepsilon 1} - 1)}. \tag{8.5.9}$$

The model cannot sustain equilibrium if $\mathrm{Ri_f} \geq 1$, or $G \leq -\mathcal{P}$. This only says that shear production must exceed buoyant suppression if the turbulence is to survive. In practice, stable stratification overcomes shear if $G \lesssim -0.2\mathcal{P}$. The reason that stabilization is stronger than simple energetics would suggest is because it also alters the Reynolds shear stress. To understand this, the Reynolds stress budgets must be examined.

### 8.5.1 Second-moment transport models

Indirect influences of buoyancy are hidden in the redistribution terms, $\wp_{ic}$ in (8.5.1) and $\wp_{ij}$ in (8.5.3). The fluctuating body force caused by density irregularities induces a fluctuating pressure field. Again, consider the free-space solution (7.1.17), page 163, of the Poisson equation for pressure, this time including the gravitational contribution. Now the pressure scalar correlation vector splits into *three* parts,

$$\wp_{ic} = \wp_{ic1} + \wp_{ic2} + \wp_{ic3}.$$

The extra term is associated with the gravitational field. Similarly to the analysis of $\wp_{ic2}$ on page 239, $\wp_{ic3}$ can be expressed formally as

$$\wp_{ic3} = -\frac{1}{4\pi} \iiint_{-\infty}^{\infty} \beta g_k \frac{\partial_i \partial_k [\overline{cc}(\boldsymbol{\xi})]}{|\boldsymbol{\xi}|} \, \mathrm{d}^3 \boldsymbol{\xi}, \tag{8.5.10}$$

where $\overline{cc}(\boldsymbol{\xi})$ stands for a two-point correlation function $\overline{c(\boldsymbol{x})c(\boldsymbol{x}+\boldsymbol{\xi})}$. Comparing this expression for $\wp_{ic3}$ to Eq. (8.4.8) for $\wp_{ic2}$, one sees that the former is of the form

$$\wp_{ic3} = \beta g_k M_{ki},$$

where $M_{ki}$ represents the integral in (8.5.10). Assuming that $M_{ki}$ only depends on the scalar variance, the most general form is then

$$M_{ki} = C_{5c}\, \overline{c^2} \delta_{ki}. \tag{8.5.11}$$

The contraction $M_{kk}$, obtained from the integral (8.5.10), is $M_{kk} = \overline{c^2}$. This follows from the rationale below Eq. (7.1.21)* that $\nabla^{-2}\nabla^2 = 1$. Applying this constraint to (8.5.11) gives $C_{5c} = \frac{1}{3}$. The pressure scalar term can thus be modeled as

$$\wp_{ic3} = C_{5c} \beta g_i \overline{c^2}, \tag{8.5.12}$$

with the theoretical value $C_{5c} = \frac{1}{3}$.

The Reynolds stress is also indirectly affected by the gravitational field through pressure fluctuations. As was done above for the scalar flux equation, the pressure–strain correlation is decomposed into three parts: $\wp_{ij} = \wp_{ij1} + \wp_{ij2} + \wp_{ij3}$. An isotropization of production (IP) type of model can be adopted. The IP formulation is described in Section 7.1.4. Quite simply, redistribution is assumed proportional to the anisotropic part of buoyant production as in

$$\wp_{ij3} = -C_4(G_{ij} - \tfrac{2}{3} G \delta_{ij}). \tag{8.5.13}$$

A value of $C_4 = \frac{3}{5}$ is quoted in Launder (1989), whereas Gibson and Launder (1978) proposed the value $C_4 = 0.5$.

* Alternatively, $f(\boldsymbol{x}) = -\frac{1}{4\pi} \iiint_{-\infty}^{\infty} \frac{\nabla_\xi^2 \overline{cc}(\boldsymbol{\xi})}{|\boldsymbol{\xi} - \boldsymbol{x}|} \mathrm{d}^3 \boldsymbol{\xi}$ is the solution to $\nabla^2 f(\boldsymbol{x}) = \nabla^2 \overline{cc}(\boldsymbol{x})$ and satisfies $f(0) = M_{kk}$. Clearly $f(\boldsymbol{x}) = \overline{cc}(\boldsymbol{x})$, which is the desired result.

As is often the case, model constants are not uniquely specified by experimental measurements. They are interdependent, and must be selected in aggregate to reproduce certain well-founded data; for instance, the constants appearing in Eq. (8.4.14) are related to turbulent Prandtl number. An important quantity in stratified flow is the critical Richardson number. Model predictions of it are discussed in the following.

## 8.5.2 Stratified shear flow

The complete set of equations that generalize the IP model to buoyant flow is obtained by assembling formulas from the previous section. In short,

$$
\begin{aligned}
\mathrm{d}_t\overline{u_i c} &= -C_{1c}\frac{\varepsilon}{k}\,\overline{u_i c} - (1 - C_{4c})\,\overline{u_i u_j}\,\partial_j C - \overline{c u_j}\,\partial_j U_i - (1 - C_{5c})\beta g_i\overline{c^2},\\
\mathrm{d}_t\overline{c^2} &= -2\,\overline{c u_j}\,\partial_j C - \mathcal{R}\frac{\varepsilon}{k}\overline{c^2},\\
\mathrm{d}_t\varepsilon &= [C_{\varepsilon 1}(\mathcal{P} + G) - C_{\varepsilon 2}\varepsilon]\frac{\varepsilon}{k},\\
\mathrm{d}_t\overline{u_i u_j} &= -C_1\frac{\varepsilon}{k}(\overline{u_i u_j} - \tfrac{2}{3}k\delta_{ij}) - (1 - C_2)\left(\overline{u_i u_k}\,\partial_k U_j + \overline{u_j u_k}\,\partial_k U_i\right)\\
&\quad + \tfrac{2}{3}C_2\mathcal{P}\delta_{ij} - (1 - C_4)\beta\left(\overline{c u_j}g_i + \overline{c u_i}g_j\right) + \tfrac{2}{3}C_4 G\delta_{ij} - \tfrac{2}{3}\varepsilon\delta_{ij}.
\end{aligned}
\tag{8.5.14}
$$

These will be solved for uniformly stratified, homogeneous shear flow in order to illustrate the stabilizing effect of stratification. The mean flow and concentration consist of $U$ and $C$ varying linearly in the $y$ direction, which is also the direction of gravity. Under this circumstance, the variables in Eq. (8.5.14) can be non-dimensionalized so that they are functions only of the gradient Richardson number,

$$
\mathrm{Ri_g} \equiv g\beta\,\partial_y C/(\partial_y U)^2,
$$

and of time. Here $g$ is the magnitude of $g_i$ (which is in the $-y$ direction.) If an eddy viscosity and conductivity model were used, then the flux and gradient Richardson numbers would be directly related by

$$
\mathrm{Ri_f} = \frac{g\beta\alpha_\mathrm{T}\,\partial_y C}{\nu_\mathrm{T}(\partial_y U)^2} = \frac{\mathrm{Ri_g}}{\mathrm{Pr_T}}.
$$

The ratio $\mathrm{Ri_g}/\mathrm{Ri_f}$ is a sort of turbulent Prandtl number; hence, one expects it to be of order unity.

After an initial transient, the numerical solution to the system of equations (8.5.14) achieves a state of moving equilibrium, in which $k$ might be growing or decaying, but $\mathcal{S}k/\varepsilon$ becomes constant. The situation is analogous to that described in Section 8.3.3, except now $\overline{c^2}$ and the scalar fluxes also evolve with time, but attain an equilibrium when normalized by $k$. The gradient Richardson number plays the role of a bifurcation parameter in this problem.

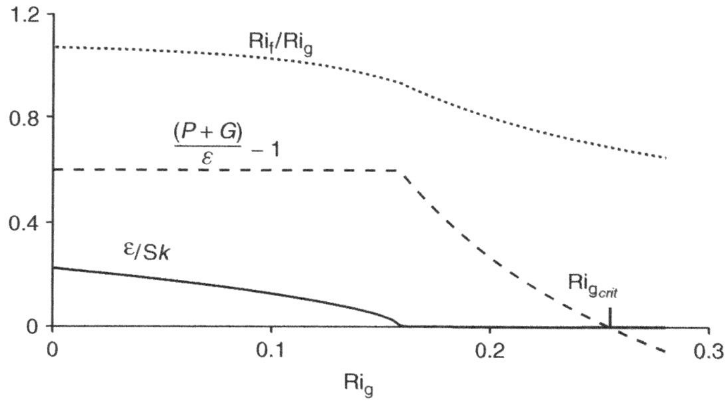


**Figure 8.3** Ratio of production to dissipation versus gradient Richardson number in stratified shear flow. The solid line is $\varepsilon/\mathcal{S}k$; the dotted line is the ratio of the flux to gradient Richardson numbers.

The model was solved for the set of constants

$$C_{1c} = 2.5, C_{4c} = 0.45, C_{5c} = 1/3, C_1 = 1.8, C_2 = 0.6, C_4 = 0.1, \mathcal{R} = 1.5$$

to generate Figure 8.3. As in the homogeneous shear flow computations of Section 7.2.1, the values $C_{\varepsilon 1} = 1.5$ and $C_{\varepsilon 2} = 1.8$ were used to produce the equilibrium value $\mathcal{P}/\varepsilon = 1.6$ in neutrally stratified, homogeneous shear flow. The value of $C_4$ is lower than in Gibson and Launder (1978) in order to obtain a critical gradient Richardson number of about 0.25. Linear stability theory gives $\text{Ri}_\text{g} = 0.25$ for stabilization by negative buoyancy.

Figure 8.3 is analogous to Figure 8.1; it shows how stratification causes the solution to bifurcate from a constant $(\mathcal{P} + G)/\varepsilon$ branch to one on which the turbulence is suppressed. The solid curve shows how $\varepsilon/\mathcal{S}k$ varies on the first branch, and vanishes after bifurcation. The critical Richardson number is defined as the value at which $(\mathcal{P} + G)/\varepsilon = 1$; at larger values of Ri, the turbulence decays. Figure 8.3 shows that the theoretical value of 0.25 is captured reasonably well by the set of constants chosen for the computation. Note that the bifurcation point, where $\varepsilon/\mathcal{S}k$ vanishes, is at the smaller value of $\text{Ri}_\text{g} \approx 0.16$.

As a practical matter, the equilibrium solution can be used to modify an eddy viscosity model such that it mimics suppression of turbulence by stable stratification; this is entirely analogous to the method of representing effects of rotation by a variable $C_\mu$ that was described in Section 8.3.3. The formulation of Mellor and Yamada (1982) is widely used in geophysical fluid dynamics.

# Exercises

**Exercise 8.1.** *Objective tensors.* Show that the absolute vorticity tensor $\Omega^\text{A}_{ij} = \frac{1}{2}(\partial V_i/\partial x_j - \partial V_j/\partial x_i) + \varepsilon_{jik}\Omega^\text{F}_k$ transforms properly under the arbitrary change of reference frame in Eq. (8.1.5).

**Exercise 8.2.** *Realizability by stochastic equations.* If a model can be derived from a well-defined random process, then by definition it is realizable. Consider a random velocity vector that satisfies

$$u_i(t + \mathrm{d}t) = \left(1 - \frac{a\,\mathrm{d}t}{T}\right) u_i(t) + \sqrt{\frac{bk\,\mathrm{d}t}{T}}\,\xi_i(t).$$

Here $\boldsymbol{\xi}$ is a vector version of the random variable used in Eq. (2.2.10), which has the additional property that

$$\overline{\xi_i \xi_j} = \delta_{ij},$$

$a$ and $b$ are coefficients that will be determined, and $k$ is the turbulent kinetic energy, as usual.

(i) Derive the evolution equation for the Reynolds stress tensor $\overline{u_i u_j}$.

(ii) Assume the high Reynolds number formula $T = k/\varepsilon$ and determine $a$ and $b$ such that the Reynolds stress evolution equation is identical to the Rotta model for decaying anisotropic turbulence.

(iii) What realizability constraint is implied by the square root in the first equation above?

**Exercise 8.3.** *Bifurcation.* Verify (8.3.14). Substitute constants and plot the bifurcation curve for the LRR model.

**Exercise 8.4.** *Nonlinear constitutive equations.* In constitutive modeling, the coefficients in Eq. (8.3.8) are regarded as unknown functions of $|\boldsymbol{S}|$ and $|\boldsymbol{W}|$ that can be formulated with complete freedom, provided they are dimensionally consistent. However, useful constraints can be obtained by invoking realizability. Consider a general two-dimensional mean flow field where the mean rate of strain and mean vorticity tensors can be written according to (8.3.3) and let

$$\boldsymbol{b} = 2C_\mu \boldsymbol{S} + B(\boldsymbol{\Omega} \cdot \boldsymbol{S} - \boldsymbol{S} \cdot \boldsymbol{\Omega}) + C(\boldsymbol{S}^2 - \tfrac{1}{3}|\boldsymbol{S}|^2 \boldsymbol{\delta}).$$

Show that, for the flow (8.3.3), the constraints

$$B^2 \leq [(\tfrac{2}{3} + \tfrac{1}{3} c S^2)^2 - (2C_\mu S)^2]/(2\Omega S)^2$$

and

$$C \leq \min[1/S^2, 3(\tfrac{4}{3} - 2C_\mu |S|)/S^2]$$

ensure that a (quadratically) nonlinear model is realizable. Here $S$ and $\Omega$ are understood to be non-dimensionalized by $k/\varepsilon$.

# Part III

# THEORY OF HOMOGENEOUS TURBULENCE

# 9

# Mathematical representations

> Then I come along and try differentiating under the integral sign, and often it worked. So I got a great reputation for doing integrals, only because my box of tools was different ...
>
> –Richard Feynman

The emphasis of this book is on single-point statistical theory and modeling for turbulence. Those subjects sometimes benefit from ideas evolved out of two-point and spectral theory; the latter are introduced in the present chapter to round out the scope of the book. The focus will be on the basic analytical concepts that underlie spectral theory. The intent is to provide exposure to some important areas of turbulence theory, without delving into their greater intricacies. A thorough treatment, including advanced mathematical techniques, can be found in the monograph by McComb (1990).

The theory of statistically homogeneous turbulence forms a backbone for a vast amount of research into fluid dynamical turbulence. The theory gels around spectral representation of two-point correlations. After covering formalisms and definitions in the present chapter, matters of turbulence physics and modeling will be discussed in Chapter 10. The final chapter of Part III is on rapid distortion theory.

The Fourier transform is the central mathematical tool of homogeneous turbulence theory. The classical conception of the spectrum of scales in turbulent flow was explained in Section 2.1. Fourier modes were introduced in that section, equating wavenumber components with "eddies" of differing size. In loosely used terminology, a "large eddy" was synonymous with a low wavenumber; a "small eddy" alluded to a high wavenumber. The Kolmogoroff $\kappa^{-5/3}$ law arose in an intermediate range of scales. Such was a prelude to the more elaborate theory developed below.

Recall that homogeneity means translational invariance of statistics in space (Section 2.2). Single-point statistics are independent of position; two-point statistics are a function of the relative position of the two points, but not of the absolute position

*Statistical Theory and Modeling for Turbulent Flows, Second Edition* P. A. Durbin and B. A. Pettersson Reif

of the individual points. For example, $\overline{u(x)u(x')}$ is a function of $x - x'$, but not of $x$ and $x'$ individually. This property follows from invariance with respect to translation of the origin of the coordinate system. Translational invariance is easily embedded into Fourier representations. Consider an oscillation, $u(x) = \mathrm{e}^{\mathrm{i}kx}$. Its complex conjugate, at $x'$, is $u^*(x') = \mathrm{e}^{-\mathrm{i}kx'}$. Their product is

$$\overline{u(x)u^*(x')} = \mathrm{e}^{\mathrm{i}k(x-x')}.$$

The functional dependence on $x - x'$ is just that required for homogeneous two-point statistics. This is one reason why Fourier analysis is so useful for homogeneous turbulence.

The corollary, that Fourier *spectra* are not useful in non-homogeneous turbulence, should not be confused with the use of Fourier transforms in analysis. In the latter case, Fourier transforms provide a complete basis for solving a certain equation, irrespective of whether or not the equation describes properties of a random process. In the present application, Fourier spectra provide a statistical description of random fields. This is why the restriction is to homogeneity. It will be seen that each random Fourier mode is uncorrelated with all others in homogeneous turbulence: that is another of the properties of Fourier transforms that makes them especially suited to present purposes. A third property is that they convert differential equations into algebraic equations. We begin with a brief discussion of the generalized Fourier transform.

## 9.1 Fourier transforms

The notion of a "generalized" Fourier transform is required for many applications (Lightill, 1958). Why? The conventional Fourier transform is defined for absolutely integrable functions,

$$\int_{-\infty}^{\infty} |f(x)|\,\mathrm{d}x < \infty.$$

However, integrals like

$$\iiint_{-\infty}^{\infty} |u(\boldsymbol{x})|\,\mathrm{d}^3\boldsymbol{x}$$

are not convergent in homogeneous turbulence; a more flexible definition of the Fourier transform is needed.

Alternatively, consider a function like $f(x) = x$. What is its Fourier transform? Transforms of powers of $x$ are not defined in the classical sense. The generalized Fourier transform covers such cases. More significantly for present developments, it encompasses unconventional functions like the $\delta$ function. This function is defined by

$$\int_{-\infty}^{\infty} f(x')\delta(x - x')\,\mathrm{d}x' = f(x), \qquad \text{for any } f(x). \tag{9.1.1}$$

Loosely, $\delta(x) = 0$ when $x \neq 0$ and $\delta(0) = \infty$, such that

$$\int_{-\infty}^{\infty} \delta(x)\,\mathrm{d}x = 1.$$

This is extended in three dimensions to $\delta(\boldsymbol{x}) = 0$ when $x \neq 0$, $y \neq 0$, $z \neq 0$, and

$$\iiint_{-\infty}^{\infty} \delta(\boldsymbol{x})\,\mathrm{d}^3\boldsymbol{x} = 1.$$

One way to think of the $\delta$ function is as the limit $\epsilon \to 0$ of the Gaussian, bell-shaped function

$$\delta(x) = \lim_{\varepsilon \to 0} \frac{\mathrm{e}^{-x^2/\epsilon^2}}{\sqrt{\pi}\,\epsilon} \tag{9.1.2}$$

in one dimension or

$$\delta(\boldsymbol{x}) = \delta(x)\delta(y)\delta(z) = \lim_{\epsilon \to 0} \frac{\mathrm{e}^{-(x^2+y^2+z^2)/\epsilon^2}}{\pi^{3/2}\epsilon^3}$$

in three dimensions. These functions vanish in the limit $\epsilon \to 0$ if $x^2 \neq 0$, or if $x^2 + y^2 + z^2 \neq 0$, respectively. They both integrate to unity.

The generalized Fourier transform is simply the usual Fourier transform with rules for interpreting integrals that do not converge in the classical sense. In some cases, the definition of a generalized Fourier transform can be formalized via a convergence factor: sometimes

$$\hat{f}(k) = \lim_{\epsilon \to 0} \frac{1}{2\pi} \int_{-\infty}^{\infty} \mathrm{e}^{-|x|\epsilon}\,\mathrm{e}^{\mathrm{i}kx} f(x)\,\mathrm{d}x \equiv \mathcal{F}_k(f(x)) \tag{9.1.3}$$

works. If $\epsilon \equiv 0$, this is the classical definition. In practice, the formal limiting process can usually be avoided. A rather useful example of the generalized transform is

$$\mathcal{F}_k(\delta) = \frac{1}{2\pi} \int_{-\infty}^{\infty} \mathrm{e}^{\mathrm{i}kx'}\,\delta(x')\,\mathrm{d}x' = \frac{1}{2\pi}, \tag{9.1.4}$$

which follows from (9.1.1) with $x = 0$ and $f(x) = \mathrm{e}^{\mathrm{i}kx}$.

The inverse Fourier transform is defined by

$$f(x) = \int_{-\infty}^{\infty} \hat{f}(k)\,\mathrm{e}^{-\mathrm{i}kx}\,\mathrm{d}k \equiv \mathcal{F}_x^{-1}\left[\hat{f}(k)\right]. \tag{9.1.5}$$

Note the convention of using a hat (ˆ) for the transformed variable and no hat for the variable in physical space. An important example for present purposes is

$$2\pi\,\delta(x) = \int_{-\infty}^{\infty} \mathrm{e}^{-\mathrm{i}kx}\,\mathrm{d}k. \tag{9.1.6}$$

To see this, invert (9.1.4): $\mathcal{F}_x^{-1}[1/2\pi] = \mathcal{F}^{-1}[\mathcal{F}(\delta(x))] = \delta(x)$. The evaluation (9.1.6) is proved directly in Exercise 9.1–incidentally, that inversion also proves that $\mathcal{F}^{-1}$ is truly the inverse operation to $\mathcal{F}$.

## 9.2 Three-dimensional energy spectrum of homogeneous turbulence

A curious result is that, in homogeneous turbulence, the wavenumber components of the Fourier transform of velocity are uncorrelated: specifically, $\overline{\hat{u}_i(\boldsymbol{k})\hat{u}_i^*(\boldsymbol{k}')} = 0$ if $\boldsymbol{k} \neq \boldsymbol{k}'$. This is an application of (9.1.6). It was mentioned earlier that the lack of correlation between different wavenumber components is one of the properties of Fourier modes that makes them especially suited to statistical description of homogeneous turbulence.

The three-dimensional Fourier transform of the velocity is

$$\hat{u}_i(\boldsymbol{k}) = \frac{1}{(2\pi)^3} \iiint_{-\infty}^{\infty} \mathrm{e}^{\mathrm{i}\boldsymbol{k}\cdot\boldsymbol{x}} u_i(\boldsymbol{x})\,\mathrm{d}^3\boldsymbol{x}. \tag{9.2.1}$$

Since $u(\boldsymbol{x})$ is a real-valued function, the complex conjugate of the integral is the same as the integral with $\boldsymbol{k}$ replaced by $-\boldsymbol{k}$. Therefore, the curious result can also be stated as $\overline{\hat{u}_i(\boldsymbol{k})\hat{u}_i(-\boldsymbol{k}')} = 0$ if $\boldsymbol{k}' \neq \boldsymbol{k}$.

The result is less mysterious if it is derived. The two-wavenumber correlation function can be related to the two-point spatial correlation function. Multiplying Eq. (9.2.1) by its complex conjugate, with $\boldsymbol{k}$ replaced by $\boldsymbol{k}'$, gives

$$\begin{aligned}\overline{\hat{u}_i(\boldsymbol{k})\hat{u}_i^*(\boldsymbol{k}')} &= \frac{1}{(2\pi)^6}\overline{\iiint_{-\infty}^{\infty} u_i(\boldsymbol{x})\,\mathrm{e}^{\mathrm{i}\boldsymbol{k}\cdot\boldsymbol{x}}\,\mathrm{d}^3\boldsymbol{x} \iiint_{-\infty}^{\infty} u_i(\boldsymbol{x}')\,\mathrm{e}^{-\mathrm{i}\boldsymbol{k}'\cdot\boldsymbol{x}'}\,\mathrm{d}^3\boldsymbol{x}'} \qquad (9.2.2)\\ &= \frac{1}{(2\pi)^6}\iiint_{-\infty}^{\infty}\iiint_{-\infty}^{\infty} \overline{u_i(\boldsymbol{x})u_i(\boldsymbol{x}')}\,\mathrm{e}^{\mathrm{i}(\boldsymbol{k}\cdot\boldsymbol{x}-\boldsymbol{k}'\cdot\boldsymbol{x}')}\,\mathrm{d}^3\boldsymbol{x}\,\mathrm{d}^3\boldsymbol{x}'.\end{aligned}$$

By the very definition of homogeneity, the correlation function $\overline{u_i(\boldsymbol{x})u_i(\boldsymbol{x}')}$ depends only on the difference between $\boldsymbol{x}$ and $\boldsymbol{x}'$. Let $\boldsymbol{r} = \boldsymbol{x} - \boldsymbol{x}'$. Then

$$\overline{u_i(\boldsymbol{x})u_i(\boldsymbol{x}')} = q^2 R(\boldsymbol{r}),$$

where $q^2 \equiv \overline{u_i u_i}$ is the turbulence intensity and $R$ is a correlation function.

Continuing with Eq. (9.2.2)

$$\begin{aligned}\overline{\hat{u}_i(\boldsymbol{k})\hat{u}_i^*(\boldsymbol{k}')} &= \frac{q^2}{(2\pi)^6}\iiint_{-\infty}^{\infty}\iiint_{-\infty}^{\infty} \mathrm{e}^{\mathrm{i}\boldsymbol{k}\cdot\boldsymbol{r}}\,\mathrm{e}^{\mathrm{i}(\boldsymbol{k}-\boldsymbol{k}')\cdot\boldsymbol{x}'} R(\boldsymbol{r})\,\mathrm{d}^3\boldsymbol{x}'\,\mathrm{d}^3\boldsymbol{r} \qquad (9.2.3)\\ &= \frac{q^2}{(2\pi)^6}\iiint_{-\infty}^{\infty} \mathrm{e}^{\mathrm{i}(\boldsymbol{k}-\boldsymbol{k}')\cdot\boldsymbol{x}'}\,\mathrm{d}^3\boldsymbol{x}' \iiint_{-\infty}^{\infty} \mathrm{e}^{\mathrm{i}\boldsymbol{k}\cdot\boldsymbol{r}} R(\boldsymbol{r})\,\mathrm{d}^3\boldsymbol{r}\\ &= \delta(\boldsymbol{k}-\boldsymbol{k}')\Phi(\boldsymbol{k}).\end{aligned}$$

The $x'$ integral provides the $\delta$ function. Because the $\delta$ function is zero when $\boldsymbol{k} \neq \boldsymbol{k}'$, the desired result has been proved: the complex conjugate Fourier coefficients of disparate wavenumbers are uncorrelated.

In (9.2.3) the spectral density is defined by

$$\Phi(\boldsymbol{k}) \equiv \frac{q^2}{(2\pi)^3} \iiint_{-\infty}^{\infty} \mathrm{e}^{\mathrm{i}\boldsymbol{k}\cdot\boldsymbol{r}} R(\boldsymbol{r})\, \mathrm{d}^3\boldsymbol{r}. \tag{9.2.4}$$

It is the Fourier transform of the correlation function. Homogeneity implies that $R_{ii}(-\boldsymbol{r}) = R_{ii}(\boldsymbol{r})$; it follows from this and (9.2.4) that $\Phi(\boldsymbol{k}) = \Phi^*(\boldsymbol{k}) = \Phi(-\boldsymbol{k})$.

## 9.2.1 Spectrum tensor and velocity covariances

The spectral density (9.2.4) is a scalar. A Reynolds stress spectrum tensor, $\Phi_{ij}(\boldsymbol{k})$, can be defined by analogy. To this end, the relation (9.2.3) is invoked. The spectrum tensor is related to the correlation of the Fourier components of the velocity by

$$\Re\mathrm{e}\left[\overline{\hat{u}_i(\boldsymbol{k})\hat{u}_j^*(\boldsymbol{k}')}\right] = \delta(\boldsymbol{k}-\boldsymbol{k}')\Phi_{ij}(\boldsymbol{k}). \tag{9.2.5}$$

(Recall that $\hat{u}^*$ denotes the complex conjugate.) This spectrum tensor is the Fourier transform partner of the two-point velocity covariance, as shown by the following argument. The Fourier representation of $\boldsymbol{u}$ is

$$u_i(\boldsymbol{x}) = \iiint_{-\infty}^{\infty} \hat{u}_i(\boldsymbol{k})\, \mathrm{e}^{-\mathrm{i}\boldsymbol{k}\cdot\boldsymbol{x}}\, \mathrm{d}^3\boldsymbol{k} = \iiint_{-\infty}^{\infty} \hat{u}_i^*(\boldsymbol{k})\, \mathrm{e}^{\mathrm{i}\boldsymbol{k}\cdot\boldsymbol{x}}\, \mathrm{d}^3\boldsymbol{k}. \tag{9.2.6}$$

Multiplying this by itself with $i$ replaced by $j$ and $\boldsymbol{x}$ replaced by $\boldsymbol{x}'$ gives

$$\begin{aligned}\overline{u_i(\boldsymbol{x})u_j(\boldsymbol{x}')} &= \frac{1}{2}\iiint_{-\infty}^{\infty}\iiint_{-\infty}^{\infty}\left[\overline{\hat{u}_i(\boldsymbol{k})\hat{u}_j^*(\boldsymbol{k}')} + \overline{\hat{u}_i^*(\boldsymbol{k}')\hat{u}_j(\boldsymbol{k})}\right] \mathrm{e}^{\mathrm{i}(\boldsymbol{k}'\cdot\boldsymbol{x}'-\boldsymbol{k}\cdot\boldsymbol{x})}\, \mathrm{d}^3\boldsymbol{k}\, \mathrm{d}^3\boldsymbol{k}' \\ &= \iiint_{-\infty}^{\infty} \Phi_{ij}(\boldsymbol{k})\, \mathrm{e}^{\mathrm{i}\boldsymbol{k}\cdot(\boldsymbol{x}'-\boldsymbol{x})} \mathrm{d}^3\boldsymbol{k}\end{aligned} \tag{9.2.7}$$

for the two-point velocity covariance. Equation (9.2.5) was used. In homogeneous turbulence, $\overline{u_i(\boldsymbol{x})u_j(\boldsymbol{x}')}$ is a function only of the difference $\boldsymbol{x} - \boldsymbol{x}'$ of measurement points; Eq. (9.2.7) is consistent with this. The $\delta$ function in (9.2.5) ensures homogeneity in (9.2.7). Hence it has been shown that $\overline{u_i u_j}(\boldsymbol{x}-\boldsymbol{x}') = \mathcal{F}^{-1}(\Phi_{ij})$. The statement that the spectrum tensor and two-point velocity covariance are Fourier transform partners means that the inversion of Eq. (9.2.7) gives $\Phi_{ij}$ if $\overline{u_i u_j}(\boldsymbol{r})$ is known: thus, the spectrum tensor is computed as

$$\Phi_{ij}(\boldsymbol{k}) = \frac{1}{(2\pi)^3}\iiint_{-\infty}^{\infty} \overline{u_i u_j}(\boldsymbol{r})\, \mathrm{e}^{\mathrm{i}\boldsymbol{k}\cdot\boldsymbol{r}}\, \mathrm{d}^3\boldsymbol{r}. \tag{9.2.8}$$

The two-point correlation is constructed in (9.2.7) by a superposition of Fourier modes. Each one of these modes extends in an oscillatory manner to $x = \pm\infty$ without decrease in amplitude; it does not have the spatial localization associated with a physical eddy. One might wonder how this is consistent with the fact that the correlation function $R_{ij}(\boldsymbol{r}) \equiv \overline{u_i(\boldsymbol{x})u_j(\boldsymbol{x}')}/q^2$ usually decays as $\boldsymbol{r} \to \infty$. The connection essentially is by the phenomenon of statistical scrambling: a sum of random oscillations, as on the right of Eq. (9.2.6), can produce a localized correlation on the left of (9.2.7). A corollary is that the association of Fourier modes with instantaneous eddies should be treated with some skepticism. Eddies are localized structures; Fourier modes are not. Nevertheless, the sum of random Fourier modes produces a localized correlation function that represents eddies on a statistical level.

Evaluating the trace of (9.2.7) at $\boldsymbol{x} = \boldsymbol{x}'$ gives

$$q^2 = \iiint_{-\infty}^{\infty} \Phi_{ii}(\boldsymbol{k})\,\mathrm{d}^3k. \tag{9.2.9}$$

If $\Phi_{ii}(\boldsymbol{k})$ is only a function of the magnitude $|\boldsymbol{k}|^2 = k_1^2 + k_2^2 + k_3^2$, and the notation $|\boldsymbol{k}| \equiv \kappa$ is used, then this integral simplifies in spherical coordinates to

$$q^2 = \int_0^{\infty} \Phi_{ii}(\kappa) 4\pi\kappa^2\,\mathrm{d}\kappa. \tag{9.2.10}$$

If $\Phi_{ii}(\boldsymbol{k})$ is a function of the direction as well as of the magnitude of $\boldsymbol{k}$, then $\Phi_{ii}(\kappa)$ can be defined as the spectrum averaged over spheres of $|\boldsymbol{k}| = \text{constant}$:

$$\Phi_{ii}(\kappa) = \frac{1}{4\pi}\int_0^{2\pi}\int_0^{\pi} \Phi_{ii}(\kappa\cos\theta,\ \kappa\sin\theta\cos\phi,\ \kappa\sin\theta\sin\phi)\sin\theta\,\mathrm{d}\theta\,\mathrm{d}\phi.$$

The spherically symmetric energy spectral density is defined by

$$E(\kappa) = 2\pi\,\Phi_{ii}(\kappa)\kappa^2. \tag{9.2.11}$$

A factor of $2\pi$ appears here, instead of $4\pi$, so that the integral of $E$ equals the turbulent kinetic energy,

$$\int_0^{\infty} E(\kappa)\,\mathrm{d}\kappa = \tfrac{1}{2}q^2 = k.$$

The energy spectral density $E(\kappa)$ is the same as that which appears in the Kolmogoroff $-5/3$ law (2.1.3), page 18. Its form is sketched in Figure 9.1. Here $E(\kappa)$ is the energy density within spherical shells in spectral space. The term *energy density* is used because $E(\kappa)\,\mathrm{d}\kappa$ is the energy (per unit mass) in between spheres of radius $\kappa$ and $\kappa + \mathrm{d}\kappa$. Summing all these infinitesimal energies gives the total kinetic energy $\frac{1}{2}q^2$.

Note that $E(\kappa)$ is not the quantity plotted in Figure 2.1 on page 17; the latter is a one-dimensional spectrum, $\Theta_{11}(k_1)$. As in Exercise 9.3, a one-dimensional spectrum can be defined by integrating over the other two wavenumber directions,

$$\Theta_{ij}(k_1) = \iint_{-\infty}^{\infty} \Phi_{ij}(k_1, k_2, k_3)\,\mathrm{d}k_2\,\mathrm{d}k_3. \tag{9.2.12}$$

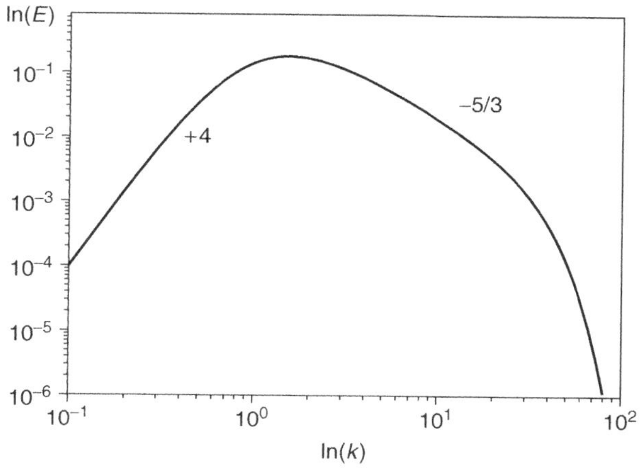


**Figure 9.1** Schematic of the turbulent energy spectrum.

The letter $\Theta$ distinguishes the 1D spectrum from the letter $\Phi$ used for the 3D spectrum. As explained in connection with Figure 2.1, $\Theta_{ij}(k_1)$ is the spectrum that is usually measured in a wind tunnel. From Eqs. (9.2.8) and (9.1.6) it follows that the one-dimensional spectrum is the Fourier transform of the corresponding unidirectional correlation function,

$$\Theta_{ij}(k_1) = \frac{1}{2\pi}\int_{\infty}^{\infty} \overline{u_i u_j}(r_1, 0, 0)\, \mathrm{e}^{\mathrm{i}k_1 r_1}\, \mathrm{d}r_1. \tag{9.2.13}$$

## 9.2.2 Modeling the energy spectrum

The exposition so far has consisted only of definitions. Connections to the properties of turbulent fluid dynamics are met initially by considering constraints imposed by kinematics and by scaling. The role of dynamics is discussed in Chapter 10.

Incompressible flow satisfies the constraint $\partial_i u_i = 0 = \nabla \cdot u$. In Fourier space, the divergence is replaced by a dot product, $\nabla \cdot u \leftrightarrow -\mathrm{i}\boldsymbol{k} \cdot \hat{u}$. The rationale follows from (9.2.6),

$$\partial_i u_i = \partial_i \iiint\limits_{-\infty}^{\infty} \hat{u}_i\, \mathrm{e}^{-\mathrm{i}\boldsymbol{k}\cdot\boldsymbol{x}}\, \mathrm{d}^3k = \iiint\limits_{-\infty}^{\infty} -\mathrm{i}k_i \hat{u}_i\, \mathrm{e}^{-\mathrm{i}\boldsymbol{k}\cdot\boldsymbol{x}}\, \mathrm{d}^3k \tag{9.2.14}$$

upon differentiating under the integral (as recommended by Feynman in the quote that heads this chapter). Incompressibility in physical space $\partial_i u_i = 0$ implies $k_i \hat{u}_i = 0$ in wavenumber space: the velocity is orthogonal to the wavevector – a nice concept. The same reasoning applied to (9.2.7) shows that

$$k_i \Phi_{ij} = 0 = \Phi_{ij} k_j. \tag{9.2.15}$$

This follows because $\partial_{x_i} \overline{u_i(\boldsymbol{x}) u_j(\boldsymbol{x}')} = \overline{[\partial_{x_i} u_i(\boldsymbol{x})] u_j(\boldsymbol{x}')} = 0$ and, similarly, $\partial_{x'_j} \overline{u_i(\boldsymbol{x}) u_j(\boldsymbol{x}')} = 0$. The left-most term in Eq. (9.2.15) is the Fourier transform of $\partial_{x_i} \overline{u_i(\boldsymbol{x}) u_j(\boldsymbol{x}')}$; the right-most is the Fourier transform of $\partial_{x'_i} \overline{u_i(\boldsymbol{x}) u_j(\boldsymbol{x}')}$.

### 9.2.2.1 Isotropic turbulence

The form of $\Phi_{ij}$ is constrained by incompressibility (9.2.15). Symmetry constraints can also be imposed. The most common of these is full isotropic symmetry.

Isotropy implies no directional preference. Hence there can be no external forcing of the flow, and no direction of mean shearing or straining. This limits the immediate applicability of isotropic turbulence, but provides an enormous benefit to theoretical developments. The theory of isotropic turbulence has contributed to the conceptual framework that engineers and scientists use as a foothold on turbulence phenomena.

The basic ideas of isotropy are presented in Section 2.3.1. The same type of intuitive reasoning described there can be applied to two-point correlations. Consider the correlation $\overline{u(x, y)v(x, y')}$. Reflection in the $y$–$z$ plane transforms $u$ to $-u$, $v$ to $v$, and $x$ to $-x$, as illustrated by Figure 9.2. Homogeneity requires that statistics at points $x$ and $-x$ be equivalent. Therefore, reflectional invariance in the $y$–$z$ plane asserts that

$$\overline{u(x, y)v(x, y')} = \overline{-u(x, y)v(x, y')} = 0.$$

Similar considerations about reflection in the $x$–$y$ plane show that the $u$–$v$ correlation is zero unless the velocities are evaluated at two distinct sets of coordinates, $(x, y)$ and $(x', y')$. Invariance under rotation and reflection leads to the requirement that

$$\overline{u(x, y)v(x', y')} \propto (x - x')(y - y').$$

For instance, reflectional invariance in the $x$–$z$ plane gives

$$\overline{u(x, y)v(x', y')} = -\overline{u(x, y')v(x', y)},$$

which is effected in the formula by $(y - y') \to -(y - y')$ under reflection.

Although the sort of reasoning illustrated in Figure 9.2 gives a conceptual basis for isotropic formulas, an algebraic approach is generally more manageable. The algebraic approach will be developed in wavenumber space. If the spectrum tensor $\Phi_{ij}$ is isotropic, it can only be a function of $k_i$ and $\delta_{ij}$; there are no other independent variables available.

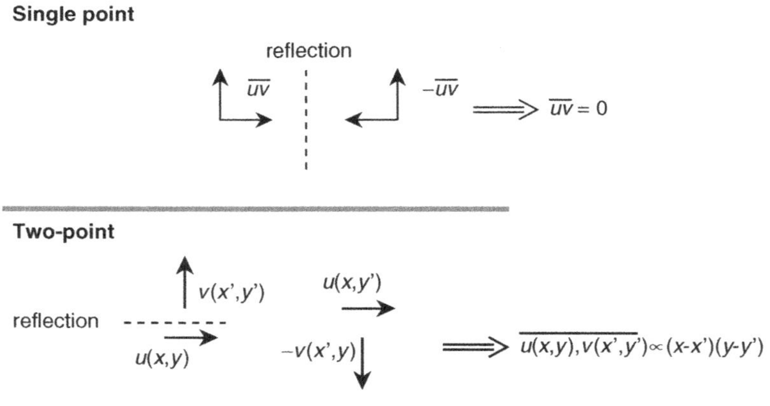


**Figure 9.2** Illustrations for isotropic invariance.

The absence of any distinctive direction immediately leads to isotropic statistics. The only tensor form consistent with the given functional dependence is

$$\Phi_{ij} = A(|\boldsymbol{k}|)\delta_{ij} + B(|\boldsymbol{k}|)k_i k_j. \tag{9.2.16}$$

This is the most general form that has the same free indices, $i$, $j$, on both sides. The coefficients $A$ and $B$ are permitted to be functions of the magnitude of $\boldsymbol{k}$. The continuity constraint (9.2.15) imposes

$$[A(|\boldsymbol{k}|)\delta_{ij} + B(|\boldsymbol{k}|)k_i k_j]k_j = A(|\boldsymbol{k}|)k_i + B(|\boldsymbol{k}|)k_i|\boldsymbol{k}|^2 = 0$$

or $B = -A/|\boldsymbol{k}|^2$. The trace of (9.2.16) must reproduce (9.2.11) with $\kappa = |\boldsymbol{k}|$. It follows that

$$\Phi_{ij} = \frac{E(|\boldsymbol{k}|)}{4\pi|\boldsymbol{k}|^2}\left(\delta_{ij} - \frac{k_i k_j}{|\boldsymbol{k}|^2}\right) \tag{9.2.17}$$

is the most general form for the isotropic, incompressible energy spectrum tensor. In this expression, $E$ is an arbitrary scalar function, which has to be either measured or modeled.

#### 9.2.2.2 How to measure $E(k)$

It would be rather difficult to measure three-dimensional spectra experimentally. For instance, one might have to use two probes, one stationary, and the other traversed along the three coordinate axes. Fortunately, the only unknown in the representation (9.2.17) is the energy spectral density, $E$, and a means exists to measure it with a single fixed probe. The method invokes a relation between $E$ and the one-dimensional spectrum,

$$\Theta_{11}(k_1) = \iint_{-\infty}^{\infty} \Phi_{11}(\boldsymbol{k})\, \mathrm{d}k_2\, \mathrm{d}k_3 = \iint_{-\infty}^{\infty} \frac{E(|\boldsymbol{k}|)}{4\pi|\boldsymbol{k}|^4}(k_2^2 + k_3^2)\, \mathrm{d}k_2\, \mathrm{d}k_3,$$

where (9.2.17) was substituted. Introduce cylindrical coordinates, $k_2^2 + k_3^2 = r^2$, and let $G(k^2) = E(k)/k^4$. Then

$$\Theta_{11}(k_1) = \int_0^{\infty} \frac{G(r^2 + k_1^2)}{4\pi} r^2 2\pi r\, \mathrm{d}r = \int_0^{\infty} \frac{G(y + k_1^2)}{4} y\, \mathrm{d}y.$$

where $y = r^2$. Now

$$\frac{\mathrm{d}\Theta_{11}}{\mathrm{d}k_1^2} = \frac{1}{2k_1}\frac{\mathrm{d}\Theta_{11}}{\mathrm{d}k_1} = \int_0^{\infty} \frac{G'(y + k_1^2)}{4} y\, \mathrm{d}y = -\int_0^{\infty} \frac{G(y + k_1^2)}{4}\, \mathrm{d}y,$$

after integration by parts with $G(\infty) = 0$ assumed. Differentiating and integrating by parts again gives

$$\frac{1}{k_1}\frac{\mathrm{d}}{\mathrm{d}k_1}\left[\frac{1}{k_1}\frac{\mathrm{d}\Theta_{11}}{\mathrm{d}k_1}\right] = -\int_0^{\infty} G'(y + k_1^2)\, \mathrm{d}y = G(k_1^2).$$

Finally, substituting the definition of $G$ gives the desired formula:

$$E(k) = k^3 \frac{\mathrm{d}}{\mathrm{d}k}\left[\frac{1}{k}\frac{\mathrm{d}\Theta_{11}(k)}{\mathrm{d}k}\right]. \tag{9.2.18}$$

To illustrate this, suppose a measurement of $\Theta_{11}(k_1)$ is fitted by $\Theta_{11} = A\,\mathrm{e}^{-L^2 k_1^2}$. Plugging into the above gives $E(k) = 4AL^4k^4\,\mathrm{e}^{-L^2k^2}$. Note that $\Theta_{11}(0) = O(1)$, while $E(k) = O(k^4)$ as $k \to 0$. This explains the difference between Figures 2.1 (page 17) and 9.1, which illustrate one-dimensional and spherically averaged spectra, respectively: the former has its maximum value at $k_1 = 0$; the latter tends to zero as $k_1 \to 0$.

It was explained in connection with Figure 2.1 that the 1D spectrum can be measured as a frequency spectrum using a fixed probe, then converted to a spatial spectrum by invoking Taylor's hypothesis. Thence the three-dimensional energy spectral density is obtained by relation (9.2.18).

#### 9.2.2.3 The Von Karman spectrum

The essential methods to represent homogeneous turbulence by spatial spectra have now been developed. They are general formulas and relationships, phrased in terms of a few unknown functions. To be useful, the functions need to be specified. Sometimes, instead of relying on experimental data, the shape of the spectrum is represented by a formula. Formulas for the energy spectrum are commonly used in rapid distortion theory (Chapter 11), to initialize direct numerical simulations of turbulence, and in kinematic simulations of turbulent mixing. This and the next section describe ways in which spectra are constructed for such purposes.

The Von Karman spectrum is an *ad hoc* formula that interpolates between the energetic and inertial ranges. Its functional form is

$$E(\kappa) = q^2 L C_{\mathrm{vK}} \frac{(\kappa L)^4}{[1 + (\kappa L)^2]^p}, \tag{9.2.19}$$

where $L$ is a length scale that is required for dimensional consistency and $p$ is an exponent to be selected. The limits of this formula are that, as $\kappa \to 0$, $E(\kappa) \propto \kappa^4$, while, as $\kappa \to \infty$, $E(\kappa) \propto \kappa^{4-2p}$. The value $p = 17/6$ reproduces the $-5/3$ law.

The constant $C_{\mathrm{vK}}$ is found from the normalization

$$\int_0^\infty E(\kappa)\,\mathrm{d}\kappa = \tfrac{1}{2}q^2.$$

Substituting expression (9.2.19) gives

$$C_{\mathrm{vK}} = \frac{\Gamma(p)}{\Gamma(5/2)\Gamma(p - 5/2)}, \tag{9.2.20}$$

where $\Gamma$ is the factorial function.

A multiplicative factor of $\mathrm{e}^{-c_\eta(\kappa\eta)^2}$ can be used to produce a viscous cut-off in the dissipative region; $\eta = (\nu^3/\varepsilon)^{1/4}$ is the Kolmogoroff, dissipative length scale described on page 19. The spectrum (9.2.19) with $p = 17/6$, multiplied by $\mathrm{e}^{-0.001(\kappa L)^2}$, is plotted in Figure 9.1. This formula produces the $+4$ spectrum at low $\kappa$, the $-5/3$ spectrum in the inertial subrange, and an exponential fall-off in the viscous, dissipative range. Although the precise form of the viscous cut-off is a matter of speculation, it is generally accepted to be exponential. A $\kappa^2$ in the exponential is suggested by the final period of decay analysis, described in the next chapter. The "Pao spectrum" invokes inertial-range scaling to give an exponent proportional to $\kappa^{4/3}$. The rationale stems from the argument that in the inertial range an eddy viscosity would have the form $\nu_T \propto (\varepsilon\kappa^4)^{1/3}$ because kinematic viscosity has dimensions $\ell^2/t$ and the available dimensional parameters are $\varepsilon \sim \ell^2/t^3$ and $\kappa \sim 1/\ell$. The spectral turbulent Reynolds number, $\nu_T(\kappa)/\nu$, is assumed to be the controlling factor and it suggests the form $e^{-c_\eta(\kappa\eta)^{4/3}}$ for the viscous cut-off. For most purposes, $\kappa^2$ has been found satisfactory and not in disagreement with available data.

Combining (9.2.19) with (9.2.17) gives the 3D spectrum tensor

$$\Phi_{ij} = \frac{q^2 C_{vK} L^5}{4\pi} \frac{(|\boldsymbol{k}|^2\delta_{ij} - k_i k_j)}{[1 + (|\boldsymbol{k}|L)^2]^p} \tag{9.2.21}$$

in the inertial and energy-containing ranges. Corresponding one-dimensional spectra can be computed as in (9.2.12). For instance, integrating in cylindrical coordinates gives

$$\begin{aligned}\Theta_{11}(k_1) &= \frac{q^2 C_{vK} L}{4\pi} \int_0^\infty \frac{x^2}{[1 + x^2 + (k_1 L)^2]^p} 2\pi x \, dx \\ &= \frac{q^2 C_{vK} L}{4(p-1)(p-2)[1 + (k_1 L)^2]^{p-2}}.\end{aligned} \tag{9.2.22}$$

A comparison between the form of 1D and 3D spectra is made in Figure 9.3. The former asymptotes to a non-zero value of $\Theta_{11}(0) \propto q^2 L$ at $k_1 = 0$. This shows how the spectrum at zero wavelength defines a turbulence scale; in fact, it defines the *integral length scale*.

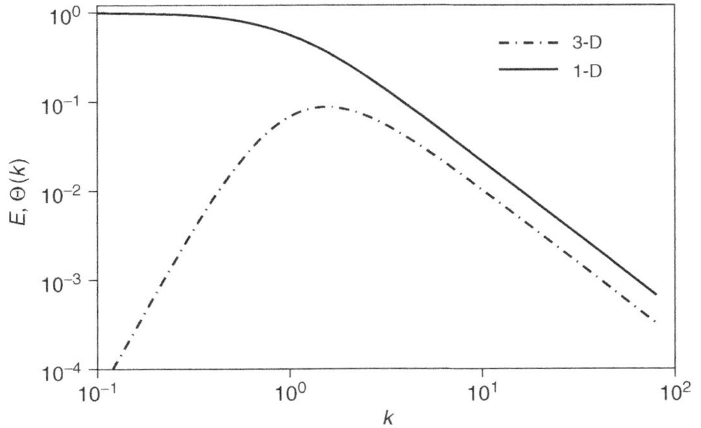


**Figure 9.3** One- and three-dimensional Von Karman energy spectra.

#### 9.2.2.4 Synthesizing spectra from random modes

For some purposes, like kinematic simulation, a random process with a given spectrum is desired (Fung *et al.*, 1992). The term *kinematic simulation* simply means that a random field is generated from a formula, rather than by a dynamical equation. One method is to sum Fourier modes with random amplitudes and phases. In some cases, it suffices to choose phases randomly, while specifying the amplitude to be proportional to $\sqrt{E(k)}$. That is the method described here.

The objective is to synthesize an incompressible, isotropic, random field. To start, let $\boldsymbol{\xi}(\boldsymbol{k})$ be a random Fourier amplitude that satisfies the condition

$$\overline{\xi_i(\boldsymbol{k})\xi_j^*(\boldsymbol{k}')} = \delta_{ij}\delta(\boldsymbol{k} - \boldsymbol{k}'), \tag{9.2.23}$$

where $\boldsymbol{\xi}$ is an isotropic random function in wavenumber space. It can be constructed by analogy to the random processes described in Section 2.2.2.1. One procedure is to set

$$\xi_i(\boldsymbol{k}) = \sqrt{3}\,\mathrm{e}^{\mathrm{i}\phi_\kappa}\frac{\eta_i(\boldsymbol{k})}{\sqrt{\mathrm{d}\kappa}}, \tag{9.2.24}$$

where $\boldsymbol{\eta}$ is a unit vector with random orientation, chosen independently for each $\boldsymbol{k}$. It could be realized as

$$\boldsymbol{\eta} = \left(p,\ \pm\sqrt{1-p^2}\cos\theta,\ \pm\sqrt{1-p^2}\sin\theta\right),$$

where $p$ is a random number distributed uniformly between $-1$ and $+1$, and $\theta$ is uniformly distributed between 0 and $2\pi$. The random orientation of $\boldsymbol{\eta}$ gives it the property $\overline{\eta_i\eta_j} = \frac{1}{3}\delta_{ij}$, and the independence of each $\boldsymbol{k}$ gives $\overline{\eta_i(\boldsymbol{k})\eta_j(\boldsymbol{k}')} = 0$, for $\boldsymbol{k} \neq \boldsymbol{k}'$. These and the factor of $1/\sqrt{\mathrm{d}\kappa}$ produce the behavior desired by Eq. (9.2.23). The phase angle $\phi_\kappa$ in construction (9.2.24) is random and also distributed uniformly between 0 and $2\pi$. Different random numbers are selected for each $\boldsymbol{k}$.

The Fourier modes of a random velocity field are synthesized from the function $\boldsymbol{\xi}$. The condition of incompressibility, $\boldsymbol{k}\cdot\hat{\boldsymbol{u}}$, states that the velocity is orthogonal to the wavenumber vector. This condition is met by

$$\hat{\boldsymbol{u}} = \boldsymbol{k}\wedge\boldsymbol{\xi}\sqrt{\frac{E(\kappa)}{4\pi\kappa^4}} \quad \text{or, in index form,} \quad \hat{u}_i = \varepsilon_{ijl}k_j\xi_l\sqrt{\frac{E(\kappa)}{4\pi\kappa^4}}, \tag{9.2.25}$$

where $\kappa = |\boldsymbol{k}|$. It can be verified that this construction has an isotropic spectrum. First evaluate the velocity correlation

$$\overline{\hat{u}_i\hat{u}_j^*} = \varepsilon_{ilm}\varepsilon_{jnp}k_lk_n\overline{\xi_m\xi_p}\,\frac{E(\kappa)}{4\pi\kappa^4}, \tag{9.2.26}$$

then use relation (9.2.23) in the form $\overline{\xi_m(\boldsymbol{k})\xi_p^*(\boldsymbol{k}')} = \delta_{mp}\delta(\boldsymbol{k}-\boldsymbol{k}')$, along with Exercise 5.1, to obtain

$$\begin{aligned}\overline{\hat{u}_i\hat{u}_j^*} &= (\delta_{ij}\delta_{ln} - \delta_{in}\delta_{lj})k_lk_n\delta(\boldsymbol{k}-\boldsymbol{k}')\frac{E(\kappa)}{4\pi\kappa^4}\\ &= \delta(\boldsymbol{k}-\boldsymbol{k}')\,\frac{E(\kappa)}{4\pi\kappa^4}[\kappa^2\delta_{ij} - k_ik_j].\end{aligned}$$

Comparing to the definition (9.2.5), the spectral tensor is seen to be

$$\Phi_{ij} = \frac{E(\kappa)}{4\pi\kappa^4}[\kappa^2\delta_{ij} - k_i k_j],$$

as desired. So it has been shown that the representation (9.2.25) is a method to synthesize an incompressible, isotropic, random field.

The synthesized random process is not unique. For instance, the factor of $\sqrt{E(\kappa)}$ in (9.2.25) could be replaced by a random variable with variance equal to $E(\kappa)$. If the only constraint is that the random process have a given spectrum, then only its second moment is being prescribed; this does not uniquely define the process.

### 9.2.2.5 Physical space

The discussion on isotropy has so far been in Fourier space. In physical space, one works with covariances and correlation functions; these are simply the inverse Fourier transform of $\Phi_{ij}$. For example, with $\boldsymbol{x} = \boldsymbol{x}'$, (9.2.7) is

$$\overline{u_i u_j} = \iiint_{-\infty}^{\infty} \Phi_{ij}\, \mathrm{d}^3 k. \tag{9.2.27}$$

To see how this works, consider the case of isotropic turbulence. The integral is best evaluated in spherical coordinates. Let

$$\boldsymbol{k} = \kappa(\cos\theta,\ \cos\phi\sin\theta,\ \sin\phi\sin\theta).$$

In terms of the unit vector, $\boldsymbol{e} = (\cos\theta,\ \cos\phi\sin\theta,\ \sin\phi\sin\theta)$, the isotropic spectrum (9.2.17) is

$$\Phi_{ij} = \frac{E(\kappa)}{4\pi\kappa^2}(\delta_{ij} - e_i e_j).$$

The integral (9.2.27) becomes

$$\begin{aligned}
\overline{u_i u_j} &= \int_0^{2\pi}\int_0^{\pi}\int_0^{\infty}(\delta_{ij} - e_i e_j)\frac{E(\kappa)}{4\pi}\,\mathrm{d}\kappa\ \sin\theta\,\mathrm{d}\theta\,\mathrm{d}\phi \\
&= \int_0^{2\pi}\int_0^{\pi}(\delta_{ij} - e_i e_j)\sin\theta\,\mathrm{d}\theta\,\mathrm{d}\phi\int_0^{\infty}\frac{E(\kappa)}{4\pi}\,\mathrm{d}\kappa \\
&= \frac{1}{4\pi}\int_0^{2\pi}\int_0^{\pi}(\delta_{ij} - e_i e_j)\sin\theta\,\mathrm{d}\theta\,\mathrm{d}\phi\,(\tfrac{1}{2}q^2).
\end{aligned}$$

The angular integral can be determined,

$$\int_0^{2\pi}\int_0^{\pi} e_i e_j \sin\theta\,\mathrm{d}\theta\,\mathrm{d}\phi = \tfrac{4}{3}\pi\delta_{ij},$$

so the above integration gives

$$\overline{u_i u_j} = \tfrac{1}{3}q^2\delta_{ij}. \tag{9.2.28}$$

This is the correct result for isotropic turbulence.

The formula for the two-point covariance in physical space follows from (9.2.7) and (9.2.17):

$$\overline{u_i(\boldsymbol{x})u_j(\boldsymbol{x}')} = \iiint_{-\infty}^{\infty} \frac{E(\kappa)}{4\pi\kappa^2}\left(\delta_{ij} - \frac{k_i k_j}{|\boldsymbol{k}|^2}\right) \mathrm{e}^{-\mathrm{i}\boldsymbol{k}\cdot\boldsymbol{r}}\, \mathrm{d}^3k \tag{9.2.29}$$

$$= f(|\boldsymbol{r}|)\delta_{ij} + \frac{|\boldsymbol{r}|}{2} f'(|\boldsymbol{r}|)\left(\delta_{ij} - \frac{r_i r_j}{|\boldsymbol{r}|^2}\right),$$

where $\boldsymbol{r} = \boldsymbol{x} - \boldsymbol{x}'$. The function $f(\boldsymbol{r})$ is related to the energy spectrum by

$$f(\boldsymbol{r}) = \frac{2\mathrm{i}}{|\boldsymbol{r}|^2} \iiint_{-\infty}^{\infty} \boldsymbol{k}\cdot\boldsymbol{r}\, \mathrm{e}^{-\mathrm{i}\boldsymbol{k}\cdot\boldsymbol{r}} \frac{E(\kappa)}{4\pi\kappa^4}\, \mathrm{d}^3k.$$

From this it follows that $f(0) = q^2/3$ and $f'(0) = 0$, so (9.2.29) reproduces (9.2.28) when $r = 0$. For non-zero $r$ it represents the general form of an isotropic correlation function for incompressible turbulence; for instance, it recovers the inference drawn from Figure 9.2 that $\overline{u(x, y)v(x', y')} \propto (x - x')(y - y')$.

Relation (9.2.29) between the covariance and the energy spectrum is obtained simply by rearranging the integrals. To this end, note that the first line of Eq. (9.2.29) can be written

$$\overline{u_i u_j}(\boldsymbol{r}) = \partial_i\partial_j H - \nabla^2 H\delta_{ij} = \partial_i\partial_j H - \frac{1}{r^2}\partial_r[r^2\,\partial_r H]\delta_{ij} \tag{9.2.30}$$

with

$$H(|\boldsymbol{r}|) \equiv \iiint_{-\infty}^{\infty} \frac{E(|\boldsymbol{k}|)}{4\pi|\boldsymbol{k}|^4} \mathrm{e}^{-\mathrm{i}\boldsymbol{k}\cdot\boldsymbol{r}}\, \mathrm{d}^3k.$$

Seeing this is just a matter of differentiating $H$ under the integral. Carrying out the differentiations in expression (9.2.30), using the property that $H$ is a function of $|\boldsymbol{r}|$ alone, gives (9.2.29), in which $f$ is defined to be $-(2/|\boldsymbol{r}|)\,\mathrm{d}_r H$.

The isotropic correlation function (9.2.29) is often stated in terms of *longitudinal* and *transverse* correlations. The longitudinal autocovariance is between a single velocity component and itself, at two points separated in the same direction as the velocity component; for instance

$$\overline{u_1 u_1}(r_1, 0, 0) = f(|\boldsymbol{r}|).$$

If the turbulence is isotropic, it does not matter which velocity component is chosen. The transverse autocovariance is between two points separated in a direction transverse to the velocity component; for instance

$$\overline{u_1 u_1}(0, r_2, 0) = f(|\boldsymbol{r}|) + \frac{|\boldsymbol{r}|}{2} f'(|\boldsymbol{r}|) \equiv g(\boldsymbol{r})$$

according to Eq. (9.2.29). If we define $g(\boldsymbol{r})$ to be the transverse covariance stated above, then (9.2.29) can also be written

$$\overline{u_i(\boldsymbol{x})u_j(\boldsymbol{x}')} = f(|\boldsymbol{r}|)\delta_{ij} + [g(|\boldsymbol{r}|) - f(|\boldsymbol{r}|)]\left(\delta_{ij} - \frac{r_i r_j}{|\boldsymbol{r}|^2}\right). \tag{9.2.31}$$

This shows how a measurement of just the longitudinal and transverse covariance functions provides the entire covariance tensor, under the assumption of isotropy.

#### 9.2.2.6 Integral length scale

The integral length scale is analogous to the integral time-scale (2.2.17) defined on page 29; for instance

$$L_{11}^{1} = \int_0^\infty R_{11}(x_1)\,\mathrm{d}x_1, \tag{9.2.32}$$

where the superscript denotes the direction of integration and the subscripts denote the velocity components. Recall that $R_{ij}$ denotes the components of the correlation tensor. A particularly simple example is provided by the Von Karman form (9.2.22) in the case $p = 3$. The Fourier inversion

$$\overline{u_1^2}R_{11} = \int_{-\infty}^\infty \Theta_{11}(k_1)\,\mathrm{e}^{-\mathrm{i}k_1 x_1}\,\mathrm{d}k_1$$

and (9.2.20) show that

$$\overline{u_1^2}R_{11} = \tfrac{1}{3}q^2\,\mathrm{e}^{-x_1/L}.$$

Under conditions of isotropy, $\overline{u_1^2} = \frac{1}{3}q^2$; then (9.2.32) gives $L_{11}^1 = L$; the length that appears in the Von Karman spectrum is exactly the integral scale for this case of $p = 3$.

Measurement of $L_{11}^1$ is facilitated by a relation between it and the one-dimensional spectrum. It is shown in Exercise 9.3 that

$$L_{11}^1 = \pi\,\Theta_{11}(k_1{=}0)/\overline{u_1^2}.$$

The spectrum on the right is usually measured as a frequency spectrum, invoking Taylor's hypothesis (see page 19).

If the turbulence is isotropic, then $L_{11}^1 = L_{22}^2 = L_{33}^3$, any of which can be called the longitudinal integral scale. The latitudinal scale is $L_{22}^1$, or any similarly non-equal set of super- and subscripts. It can be shown (Exercise 9.7) that $L_{22}^1 = \frac{1}{2}L_{11}^1$. This disparity between longitudinal and latitudinal integral lengths gives rise to a phenomenon known in the literature on heavy-particle dispersion as the "trajectory crossing effect" (Wang and Stock, 1994).

Consider a small solid particle falling through isotropic turbulence under the influence of gravity. Let it settle at the mean velocity $\mathrm{d}y/\mathrm{d}t = -V_\mathrm{s}$, with $-y$ being the direction of gravity. Homogeneous dispersion theory gives rise to the formulas

$$\alpha_{T_{11}} = \int_0^\infty \overline{u_{\mathrm{p}_1}(t)u_{\mathrm{p}_1}(t+\tau)}\,\mathrm{d}\tau,$$

$$\alpha_{T_{22}} = \int_0^\infty \overline{u_{\mathrm{p}_2}(t)u_{\mathrm{p}_2}(t+\tau)}\,\mathrm{d}\tau$$

for the eddy diffusivities, by analogy to Eq. (2.2.21), page 31. The subscript "p" indicates the velocity of the solid particle. For simplicity, assume that the particle fluctuating velocity equals the fluid fluctuating velocity. If both are homogeneous and isotropic, and if furthermore $\overline{u^2} \ll V_s^2$, then $d\tau$ can be substituted by $dy/V_s$ and the integrals evaluated as follows:

$$\alpha_{T_{11}} = \int_0^\infty \overline{u_1 u_1}(y) \frac{dy}{V_s} = \frac{\overline{u_1^2} L_{11}^2}{V_s},$$

$$\alpha_{T_{22}} = \int_0^\infty \overline{u_2 u_2}(y) \frac{dy}{V_s} = \frac{\overline{u_2^2} L_{22}^2}{V_s}.$$

But $\overline{u_1^2} = \overline{u_2^2}$ and $L_{22}^2 = 2L_{11}^2$ in isotropic turbulence. Hence $\alpha_{T_{22}} = 2\alpha_{T_{11}}$; a cloud of particles will disperse twice as fast in the direction of gravity as transverse to it.

This behavior has its origin in two causes, one physical and one mathematical. Dispersion is effected by random convection. A particle that does not settle under gravity moves coherently within an eddy; but a settling particle moves for only a short time within an eddy before it crosses into the next. The duration of coherent convection is reduced thereby. This decreases the correlation time-scale and hence reduces the extent to which a cloud of particles disperses: the dispersion coefficient decreases inversely with the settling velocity. This is the effect of trajectory crossing.

Longitudinal, isotropic correlations are more persistent in space than are latitudinal correlations; thus the velocity that a falling particle experiences remains coherent in the direction of settling for a longer duration than the velocity in perpendicular directions. This mathematical property causes greater longitudinal dispersion.

# Exercises

**Exercise 9.1.** *Familiarization with $\delta$ function.* Show that

$$\lim_{\epsilon \to 0} \int_{-\infty}^{\infty} e^{-|x|\epsilon} e^{-ikx} \, dx = 2\pi \delta(k)$$

by explicitly evaluating the integral, then applying the definition of the $\delta$ function.

**Exercise 9.2.** *Direction of independence.* Let $\boldsymbol{r} = (r_x, r_y, r_z)$ be the components of the separation between two hot-wire anemometers. Suppose that $R_{ii}(\boldsymbol{r}) = e^{-(r_x^2 + r_y^2)/L^2}$, independently of $r_z$. This means that the eddying motion is independent of the $z$ axis. What is $\Phi(k_x, k_y, k_z)$? What is the $k_z$-space analog to independence of $z$ in physical space?

**Exercise 9.3.** *Integral length scale.* The integral length scale provides a measure of the size of energetic eddies. This length scale can be defined as a vector with $x_1$ component

$$L_{11}^1 = \int_0^\infty R_{11}(x_1; x_2 = 0, x_3 = 0) \, dx_1,$$

and similarly for $L^2_{11}$ and $L^3_{11}$. This notation means "the integral scale in the $x_2$ ($x_3$) direction of the 11 correlation." For instance,

$$L^2_{11} = \tfrac{1}{2}\int_{-\infty}^{\infty} R_{11}(0, x_2, 0)\, \mathrm{d}x_2.$$

Show that

$$\overline{u_1^2} L^2_{11} = \pi \iint_{-\infty}^{\infty} \Phi_{11}(k_x, 0, k_z)\, \mathrm{d}k_x\, \mathrm{d}k_z,$$

and hence that $\Theta_{11}(k_2{=}0) = \overline{u_1^2} L^2_{11}/\pi$. Often it is easier to measure or compute $\Theta(0)$ than to obtain $L$ by integrating the correlation function.

**Exercise 9.4.** *Practice with isotropy.* Derive (9.2.29) in physical space by assuming that $\overline{u_i u_j}(\boldsymbol{r})$ is a function of $r_k$ and $\delta_{kl}$. To this end:

(i) Write the most general tensoral form for this function.

(ii) Apply incompressibility: why is the incompressibility constraint now $\partial_{r_i}\overline{u_i u_j}(\boldsymbol{r}) = 0 = \partial_{r_j}\overline{u_i u_j}(\boldsymbol{r})$?

In your derivation, simply obtain the form (9.2.29) with $f$ an unspecified function; you need not relate it to the energy spectrum.

**Exercise 9.5.** *Experimental test of isotropy.* The 1D spectra $\Theta_{11}(k_1)$ and $\Theta_{22}(k_1)$ can be measured with a fixed probe. The method is to measure frequency spectra and then use Taylor's hypothesis to replace $\omega$ by $k_1 U_c$, where $U_c$ is the mean convection velocity measured by the anemometer. Show that, if the turbulence is isotropic, then

$$\Theta_{22}(k_1) = \tfrac{1}{2}[\Theta_{11}(k_1) - k_1\, \partial_{k_1}\Theta_{11}(k_1)].$$

The 1D spectrum is defined by Eq. (9.2.12).

The connection between spectral components has been used to test experimentally for isotropy. Both $\Theta_{11}$ and $\Theta_{22}$ are measured and it is checked whether they agree with the relation above.

**Exercise 9.6.** *The Von Karman spectrum.* What is the proportionality between the length scale $L$ that appears in the Von Karman spectrum and the 1D integral length scale $L^1_{11}$? Recall the definition

$$\overline{u_1^2} L^1_{11} = \int_0^{\infty} \overline{u_1(x) u_1(x + r_x)}\, \mathrm{d}r_x.$$

**Exercise 9.7.** *One-dimensional spectra.* Use the Von Karman form to evaluate the 1D spectrum, $\Theta_{22}(k_1)$. From this, show that the integral scale in the $x_1$ direction of the 22 correlation ($L^1_{22}$) is one half the integral scale in the $x_1$ direction of the 11 correlation ($L^1_{11}$). In general, the transverse integral scale in isotropic turbulence is one-half the longitudinal integral scale.

**Exercise 9.8.** *Synthetic spectra.* Another way to synthesize an isotropic, incompressible random field is

$$\hat{\boldsymbol{u}} = \left[ \boldsymbol{\xi} - \frac{(\boldsymbol{k} \cdot \boldsymbol{\xi})\boldsymbol{k}}{|\boldsymbol{k}|^2} \right] \sqrt{\frac{E(\kappa)}{4\pi\kappa^2}}.$$

Verify that this equation has an isotropic spectrum.

# 10

# Navier–Stokes equations in spectral space

... for it is harder to discover the elements than to develop the science.
– Alfred North Whitehead

The first two parts of this book focused on single-point statistics. Evolution equations for the Reynolds stresses in space and time were derived in Section 3. The dynamics of interest were those that governed the transport of mean momentum and those that governed development of the turbulent stresses. The manner in which turbulent energy was distributed among the scales of random motion was at best of passing interest. The energy cascade was introduced in Section 2.1 as an example of an insight into turbulence dynamics that derived from inspired dimensional analysis. Nothing was said about its causes. Most research into homogeneous isotropic turbulence is concerned precisely with that – with how the Navier–Stokes equations describe the transfer of energy between the infinite degrees of freedom embodied in the Fourier spectrum. Energy transfer has its origin in the quadratic nonlinearity of the Navier–Stokes equations.

## 10.1 Convolution integrals as triad interaction

The energy cascade can be described qualitatively as a flow of kinetic energy from large-scale eddies to small-scale eddies. In physical space, it could be associated with vortex stretching, with instabilities, with folding of material surfaces, or with chaotic convection, depending on one's viewpoint. In Fourier space, the energy cascade has a more concrete analytical form: nonlinear coupling transfers energy between Fourier modes. However, the fact that it cascades from large to small scale is to some extent an empirical and intuitive observation.

*Statistical Theory and Modeling for Turbulent Flows, Second Edition* P. A. Durbin and B. A. Pettersson Reif

Recall the convolution theorem for the Fourier transform of the product of two functions: if $\hat{u}_i(\boldsymbol{k})$ and $\hat{u}_j(\boldsymbol{k})$ are the transforms of $u_i(\boldsymbol{x})$ and $u_j(\boldsymbol{x})$, then the transform of their product is

$$\mathcal{F}_k(u_i(\boldsymbol{x})u_j(\boldsymbol{x})) = \iiint_{-\infty}^{\infty} \hat{u}_i(\boldsymbol{k}-\boldsymbol{k}')\,\hat{u}_j(\boldsymbol{k}')\,\mathrm{d}^3\boldsymbol{k}' \tag{10.1.1}$$
$$= \iiint_{\boldsymbol{p}+\boldsymbol{q}=\boldsymbol{k}} \hat{u}_i(\boldsymbol{p})\hat{u}_j(\boldsymbol{q})\,\mathrm{d}\boldsymbol{p}\,\mathrm{d}\boldsymbol{q} \equiv \hat{u}_i \star \hat{u}_j|_{\boldsymbol{p}+\boldsymbol{q}=\boldsymbol{k}}.$$

The notation $\star$ is defined to be shorthand for the integral. This type of quadratic nonlinearity appears in the Navier–Stokes equations. It causes the energy cascade because a Fourier mode with wavevector $\boldsymbol{p}$ times a mode with wavevector $\boldsymbol{q}$ produces a mode with wavevector $\boldsymbol{k}$ if $\boldsymbol{k} = \boldsymbol{p} + \boldsymbol{q}$. This is readily seen by multiplying complex exponentials:

$$\mathrm{e}^{\mathrm{i}\boldsymbol{p}\cdot\boldsymbol{x}}\,\mathrm{e}^{\mathrm{i}\boldsymbol{q}\cdot\boldsymbol{x}} = \mathrm{e}^{\mathrm{i}(\boldsymbol{p}+\boldsymbol{q})\cdot\boldsymbol{x}} = \mathrm{e}^{\mathrm{i}\boldsymbol{k}\cdot\boldsymbol{x}}.$$

Thus nonlinearity transfers energy between Fourier modes. Expectations, and observations, are that energy is transferred from energetic, large-scale modes to less energetic, smaller-scale modes, where energy is removed by viscous dissipation. It is quite natural to posit this process. It is quite another matter to prove it. The objective here will be solely to make it plausible.

The second form of the integral in (10.1.1) is over wavenumber vector triangles, $\boldsymbol{p} + \boldsymbol{q} = \boldsymbol{k}$. This is just a reinterpretation of the first form. Such integrals are interpreted as interactions within triads of eddies of differing length scales: small $\boldsymbol{k}$ corresponds to "large eddies," and large $\boldsymbol{k}$ corresponds to "small eddies." Possible triads are illustrated in Figure 10.1. Types A and C are formally the same; what is meant here is that the horizontal leg represents the scale of interest. It can interact with comparable (B), larger (C) or smaller (A) eddies.

Kolmogoroff's inertial-range theory assumes that the dominant interactions in the inertial subrange are local. This corresponds to case B in Figure 10.1. Eddies in this range are not significantly influenced by large, energy-containing, or small, viscous, scales. Numerical simulations suggest that this type of interaction is indeed dominant in the inertial subrange. There is evidence that type A might be dominant in the small-scale, dissipation range. However, that interaction probably amounts to convection of small

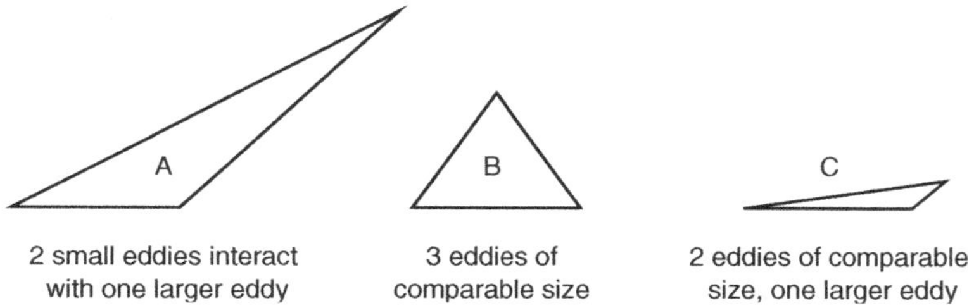


**Figure 10.1** Triad interactions in homogeneous turbulence.

eddies by large ones – referred to as "sweeping" – rather than to the energy cascade (Kraichnan 1968). In particular, it is not clear that these long-range interactions violate Kolmogoroff's universality hypothesis (Section 2.1). His hypothesis is that the smallest scales of motion are insensitive to the flow geometry; hence their structure is not influenced by the large-scale eddies. Long-range interactions that impose anisotropy on the small eddies would invalidate the hypothesis.

In general, all the triad interactions of Figure 10.1 take place and can transfer energy in any direction between different scales. The averaged direction of transfer depends on the relative amplitudes of the Fourier modes. Evidence is that triad interactions like A transfer energy from large scales to small scales on average (Domaradski and Rogallo, 1991).

## 10.2 Evolution of spectra

In hopes of a more definitive understanding of nonlinear processes, we will develop the Navier–Stokes equations in Fourier space and, from them, the spectrum evolution equations.

The Navier–Stokes equations for fluctuations in homogeneous turbulence are

$$\partial_t u_i + \partial_j u_j u_i = -\partial_i p/\rho + \nu\nabla^2 u_i \tag{10.2.1}$$

in physical space. These are Fourier-transformed using the rule $\partial_i \leftrightarrow -\mathrm{i}k_i$ stated on page 257 above Eq. (9.2.14), and the convolution theorem (10.1.1), to find

$$\mathrm{d}_t\hat{u}_i - \mathrm{i}k_j\hat{u}_j \star \hat{u}_i = \mathrm{i}k_i\hat{p}/\rho - \nu\kappa^2\hat{u}_i, \tag{10.2.2}$$

where $\kappa = |\boldsymbol{k}|$. Incompressibility requires that $k_i\hat{u}_i = 0$. Taking the dot product of Eq. (10.2.2) with $\boldsymbol{k}$ therefore gives

$$\hat{p} = -\rho\frac{k_l k_j}{\kappa^2}\hat{u}_j \star \hat{u}_l \tag{10.2.3}$$

for the pressure. This permits the pressure to be eliminated from (10.2.2), giving the evolution equation

$$\mathrm{d}_t\hat{u}_i + \nu\kappa^2\hat{u}_i = \mathrm{i}k_j\left(\delta_{il} - \frac{k_i k_l}{\kappa^2}\right)\hat{u}_j \star \hat{u}_l \tag{10.2.4}$$

for Fourier velocity components. Equation (10.2.4) is the incompressible, Navier–Stokes momentum equation in spectral space.

### 10.2.1 Small-$k$ behavior and energy decay

The behavior of the Fourier coefficient $\hat{u}$ as $k$ tends to zero has been a matter of discussion among turbulence researchers because it might have a controlling influence on the decay rate of isotropic turbulence. Consider the magnitude, in powers of $k$, of $\hat{u}_i$ at small $k$. Viscous terms in (10.2.4) are of order $k^2$, and hence smaller than those retained. For short times, an integration of (10.2.4) from some initial state to time $t$ is

$$\hat{u}_i = \hat{u}(0)_i + \mathrm{i}k_j t\left(\delta_{il} - \frac{k_i k_l}{\kappa^2}\right)\hat{u}_j \star \hat{u}_l = \hat{u}(0)_i + k_j t \times (\text{something})_{ij}, \tag{10.2.5}$$

where "something" is $O(1)$ as $k \to 0$. If $\hat{u}_i(0)$ is smaller than $O(k)$, then the second term above dominates. Squaring and averaging shows that

$$\overline{\hat{u}_i^2} \propto k^2 \qquad \text{and} \qquad E(k) \propto k^4.$$

If the extra factor of $k^2$ in $E$ is unclear, see Eq. (9.2.11) on page 256. Under these assumptions, $E(k) \to k^4$ as $k \to 0$. Thus the nonlinear terms in the Navier–Stokes equations will generate a $k^4$ power law. If $E(k)$ goes initially like a higher power of $k$, it will promptly revert to $k^4$.

However, if $E(k) \propto k^s$ with $s < 4$ at $t = 0$, then the first term in (10.2.5) dominates and $\hat{u}_i$ remains equal to its initial value, $\hat{u}_i(0)$ as $k \to 0$. The Navier–Stokes equations are unable to change the initial power law because they can only generate $k^4$ and higher powers: the low-wavenumber energy spectrum remains defined by its initial condition. When $s < 4$, $E(k) \to C_s k^s$, as $k \to 0$, for all $t$. The coefficient $C_s$ is independent of time; in other words, $C_s$ is an invariant under these assumptions.

If $s < 4$ were true at $t = 0$, then the exponent $s$ would determine the decay law for isotropic turbulence as follows: $E(k)$ has dimensions of velocity$^2$ × length, so by dimensional analysis $C_s \propto q^2 L^{s+1}$. The observation that $C_s$ is constant if $s < 4$ means that $q^2 L^{s+1}$ is invariant with time, or

$$L \propto q^{-2/(s+1)}.$$

In self-similar, decaying isotropic turbulence, kinetic energy is simply dissipated. If the scaling $\varepsilon \propto q^3/L$ is invoked (as explained on page 103), then

$$\tfrac{1}{2}\, \mathrm{d}_t q^2 = -\varepsilon \propto q^3/L \propto q^{(3s+5)/(s+1)}.$$

This is readily integrated to obtain

$$q^2 \propto (t + t_0)^{-(2s+2)/(s+3)}. \tag{10.2.6}$$

In principle, a measurement of the decay exponent in grid turbulence determines $s$.

Experimental data are fit quite well by the form $q^2 \propto (t + t_0)^{-n}$. The exponent $n$ depends somewhat on Reynolds number, but a typical value is $n = 1.2$. This $n$ is obtained if $s = 2$. It is tempting to conclude that $s = 2$ and hence that $q^2 L^3$ is an invariant of self-similar turbulence. However, it is hard to see how $s = 2$ could be produced as an initial state.

If turbulence is generated by processes like shear flow instability, that are governed by the Navier–Stokes equations, then only $s = 4$ can be produced; thus it seems more likely that $s = 4$. In that case, the decay exponent predicted by (10.2.6) is $n = 10/7$, which is larger than found in most experiments. However, when $s = 4$, Eq. (10.2.6) would be valid only if the combination $q^2 L^5$ were independent of time; but we have already discussed the fact that, if $s = 4$, the Navier–Stokes equations do not permit this invariant. Hence, formula (10.2.6) is not justified when $s = 4$ and the inference of decay exponent equal to $10/7$ is not consistent.

One approach to maintaining the small-$k$ theory with $s = 4$ but to bring it into agreement with experiment is to allow $C_4$ to increase with time (Lesieur 1990). If $C_4 = t^\gamma$, then a modification of the rationale leading to expression (10.2.6) gives

$q^2 \propto (t + t_0)^{-(10-2\gamma)/7}$. The value $\gamma = 0.8$ provides a decay exponent of 1.2. It is likely that the small-$k$ theory is only suggestive of the true processes that determine the decay exponent. Also, it may well be that $n$ is not exactly a constant.

### 10.2.2 Energy cascade

The energy cascade is a statistical concept, alluding to the dynamics of the kinetic energy spectrum. Equation (10.2.4) is for the instantaneous, fluctuating velocity. The dynamical equation of the spectrum is derived by multiplying (10.2.4) by $\hat{u}_i^*$, adding the result to its complex conjugate, and averaging. This gives

$$\begin{aligned} \mathrm{d}_t\overline{\hat{u}_i\hat{u}_i^*} &+ 2\nu\kappa^2\overline{\hat{u}_i\hat{u}_i^*} \\ &= \mathrm{i}k_l\left\{\overline{\hat{u}_i^*(\boldsymbol{k})[\hat{u}_l(\boldsymbol{p})\star\hat{u}_i(\boldsymbol{q})]}\Big|_{p+q=k} - \overline{\hat{u}_i(\boldsymbol{k})[\hat{u}_l^*(\boldsymbol{p})\star\hat{u}_i^*(\boldsymbol{q})]}\Big|_{p+q=k}\right\}. \end{aligned} \tag{10.2.7}$$

Strictly speaking, all these terms are infinite due to the $\delta$ function in (9.2.5), page 255. But since

$$\iiint \overline{u_i(\boldsymbol{k})u_i^*(\boldsymbol{k}')}\,\mathrm{d}^3k' = \frac{E(\kappa)}{2\pi\kappa^2}$$

(recall that $|\boldsymbol{k}| = \kappa$) in isotropic turbulence, the $\delta$ function can be integrated out and (10.2.7) can be rewritten as

$$\underbrace{\mathrm{d}_t E}_{\mathrm{I}} + \underbrace{2\nu\kappa^2 E}_{\mathrm{II}} = \underbrace{T(\boldsymbol{k})}_{\mathrm{III}}, \tag{10.2.8}$$

where $T$ is defined formally as

$$\begin{aligned} T(\boldsymbol{k}) = 2\pi\mathrm{i}\kappa^2 k_l \iiint_{-\infty}^{\infty} & \overline{\hat{u}_i^*(\boldsymbol{k}')[\hat{u}_l(\boldsymbol{p})\star\hat{u}_i(\boldsymbol{q})]}\Big|_{p+q=k} \\ & - \overline{\hat{u}_i(\boldsymbol{k}')[\hat{u}_l^*(\boldsymbol{p})\star\hat{u}_i^*(\boldsymbol{q})]}\Big|_{p+q=k} \quad \mathrm{d}^3\boldsymbol{k}'. \end{aligned} \tag{10.2.9}$$

The integration simply indicates that the $\delta$ function has been fired at $\boldsymbol{k}' = \boldsymbol{k}$.

We now consider the physical meaning of the terms in Eq. (10.2.8), one by one:

(I) The first term of (10.2.8) is the evolution of $E$. This is the variation of the energy density in each spherical shell of wavenumber space with time. The term "spherical shell" simply means the volume between $|\boldsymbol{k}| = \text{constant} = \kappa$ and $|\boldsymbol{k}| = \text{constant} = \kappa + \mathrm{d}\kappa$. The surface $|\boldsymbol{k}| = \text{constant}$ defines a sphere in three-dimensional $k$ space; and $E(\kappa)\,\mathrm{d}\kappa$ is the energy in that shell. The energy density, $E$, varies with time due to dissipation within the shell and due to transfer to, or from, other shells.

Integrating the energy spectrum over all wavenumbers gives the total energy, and this first term becomes

$$\int_0^\infty \mathrm{d}_t E(\kappa)\,\mathrm{d}\kappa = \tfrac{1}{2}\,\mathrm{d}_t q^2.$$

(II) The second term is the viscous dissipation within the spherical shell $\kappa \le |\boldsymbol{k}| \le \kappa + \mathrm{d}\kappa$. In this case, integration over all shells gives the total rate of energy dissipation, $2\nu \int_0^\infty \kappa^2 E(\kappa)\,\mathrm{d}\kappa = \varepsilon$. This relation can be derived from (9.2.7) as follows. Differentiating under the integrals gives

$$\overline{\partial_l u_i(\boldsymbol{x})\,\partial'_m u_j(\boldsymbol{x}')} = \iiint_{-\infty}^{\infty} k_l k_m \Phi_{ij}(\boldsymbol{k})\,\mathrm{e}^{\mathrm{i}\boldsymbol{k}\cdot(\boldsymbol{x}'-\boldsymbol{x})}\,\mathrm{d}^3k.$$

Setting $x = x'$, contracting on $i - j$ and $l - m$, and substituting (9.2.17) gives

$$\overline{\partial_l u_i\,\partial_l u_i} = \iiint_{-\infty}^{\infty} k_l k_l \frac{E(k)}{4\pi k^2}\left[\delta_{ii} - \frac{k_i k_i}{k^2}\right]\mathrm{d}^3k = \iiint_{-\infty}^{\infty} \frac{E(k)}{2\pi}\,\mathrm{d}^3k. \qquad (10.2.10)$$

In spherical coordinates, $\mathrm{d}^3k = 4\pi\kappa^2\,\mathrm{d}\kappa$. The rate of turbulent dissipation is $\varepsilon = \nu\,\overline{\partial_j u_i\,\partial_j u_i}$, which is now seen to be

$$\varepsilon = 2\nu \int_0^\infty \kappa^2 E(\kappa)\,\mathrm{d}\kappa,$$

as was to be shown.

In the inertial subrange, where $E \propto \varepsilon^{2/3}\kappa^{-5/3}$, the dissipation spectrum $2\nu\kappa^2 E$ varies as $\varepsilon^{2/3}\kappa^{1/3}$. The dissipation rate *increases* with $k$ through the inertial subrange, peaking in the viscous range. This behavior is illustrated by Figure 10.2. To stop this increase, introduce an exponential cut-off at high wavenumber, represented by

$$\kappa^2 E(\kappa) \propto \kappa^{1/3}\,\mathrm{e}^{-c_\eta(\kappa\eta)^2}.$$

Now the dissipation spectrum has a maximum at $\kappa = 1/(\sqrt{6c_\eta}\,\eta)$. Hence, the maximum rate of energy dissipation is affected by eddies of order $\eta$ in size. These are the smallest scales of turbulent motion.

(III) The right-hand side of (10.2.8) causes energy transfer among eddies of wavenumbers $\boldsymbol{p}$, $\boldsymbol{q}$, and $\boldsymbol{k}$ that form triads. This redistributes energy between spherical shells, but causes no net production or destruction. Energy removed from one shell by $T$ must be deposited into another. In isotropic turbulence, the total energy satisfies $\tfrac{1}{2}\,\mathrm{d}_t q^2 = -\varepsilon$. Integrating (10.2.8) proves that

$$\int_0^\infty T(\kappa)\,\mathrm{d}\kappa = 0.$$

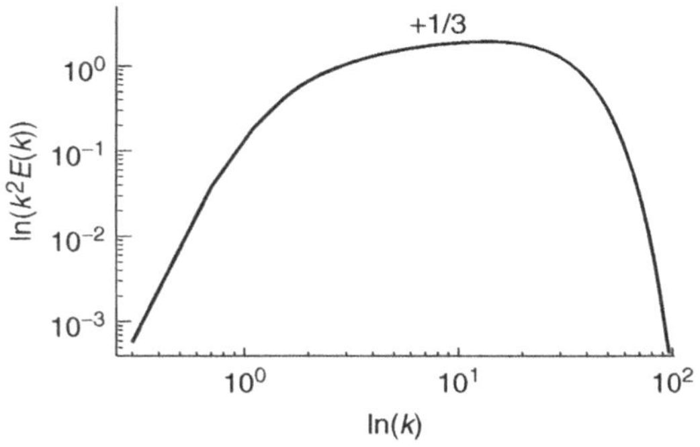


**Figure 10.2** Schematic of energy dissipation spectrum.

This is why $T(\kappa)$ must be understood to transfer energy in $k$ space, neither creating nor destroying it. Our intuitive understanding of the form of this redistribution is that $T$ removes energy from large scales (low $k$) and transfers it to small scales (large $k$), where it is dissipated. That intuition is supported by the following line of reasoning.

In the inertial subrange, viscous and unsteady effects are negligible, so the left-hand side of (10.2.8) vanishes and $T = 0$, for $\kappa$ in this range. In the viscous dissipation range, where $\kappa \sim 1/\eta$, unsteadiness is negligible and $T = 2\nu\kappa^2 E(\kappa)$ from (10.2.8). Integrating over the dissipation range gives

$$\int_{k_\mathrm{I}}^{\infty} T(\kappa)\,\mathrm{d}\kappa = \int_{k_\mathrm{I}}^{\infty} 2\nu\kappa^2 E(\kappa)\,\mathrm{d}\kappa \approx \varepsilon, \tag{10.2.11}$$

where $k_\mathrm{I}$ is chosen to be in the inertial range, and hence $k_\mathrm{I} \ll 1/\eta$. The integral on the right covers the range where the bulk of the dissipation occurs; hence at high Re it asymptotes to $\varepsilon$. Then, integrating over the large scales gives

$$\int_{0}^{k_\mathrm{I}} T(\kappa)\,\mathrm{d}\kappa = \int_{0}^{\infty} T(\kappa)\,\mathrm{d}\kappa - \int_{k_\mathrm{I}}^{\infty} T(\kappa)\,\mathrm{d}\kappa = 0 - \varepsilon = -\varepsilon. \tag{10.2.12}$$

It follows from (10.2.12) and (10.2.11) that energy is transferred from small $k$, where $T$ is negative, to large $k$, where it is positive (Figure 10.3). This provides analytical support for the idea of an energy cascade.

Further examination shows that $T$ redistributes energy at a “fine-grained” level as well. It is common to define a third-order spectrum tensor that gets rid of the $\delta(\boldsymbol{k} - \boldsymbol{k}')$, analogously to $\Phi_{ij}$ in (9.2.5), page 255. For present purposes, it suffices to define $q_{ili}$ by

$$q_{ili}(\boldsymbol{k}, \boldsymbol{q})\delta(\boldsymbol{k} - \boldsymbol{k}') = \overline{\hat{u}_i^*(\boldsymbol{k}')[\hat{u}_l(\boldsymbol{k} - \boldsymbol{q})\hat{u}_i(\boldsymbol{q})]}. \tag{10.2.13}$$

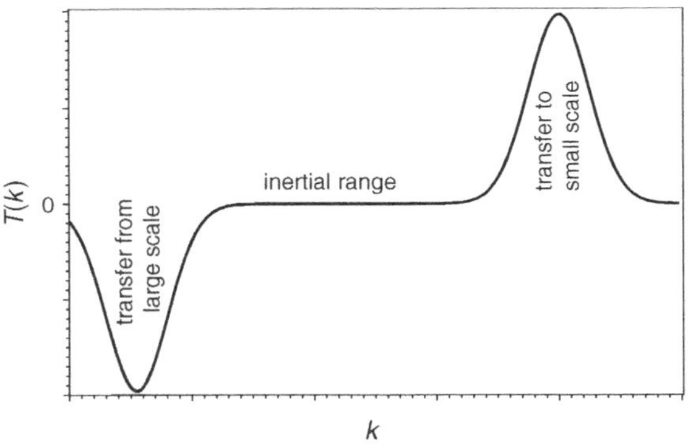


**Figure 10.3** Schematic of spectral energy transfer.

With this, after the convolution in (10.2.9) is made explicit using (10.1.1), the definition of $T$ becomes

$$T(\boldsymbol{k}) = 2\pi i\kappa^2 \iiint_{-\infty}^{\infty} k_l\, q_{ili}(\boldsymbol{k},\boldsymbol{q}) - k_l\, q^*_{ili}(\boldsymbol{k},\boldsymbol{q})\, d^3q$$
$$= 2\pi i\kappa^2 \iiint_{-\infty}^{\infty} k_l\, q_{ili}(\boldsymbol{k},\boldsymbol{q}) - k_l\, q_{ili}(\boldsymbol{q},\boldsymbol{k})\, d^3q, \tag{10.2.14}$$

where $q^*_{ili}(\boldsymbol{k},\boldsymbol{q}) = q_{ili}(\boldsymbol{q},\boldsymbol{k})$ has been used. [To show that $q^*_{ili}(\boldsymbol{k},\boldsymbol{q}) = q_{ili}(\boldsymbol{q},\boldsymbol{k})$, reverse $\boldsymbol{q}$ and $\boldsymbol{k}$ on the right-hand side of (10.2.13) to get

$$\overline{\hat{u}^*_i(\boldsymbol{q})\hat{u}_l(\boldsymbol{q}-\boldsymbol{k})\hat{u}_i(\boldsymbol{k})} = \overline{\hat{u}_i(\boldsymbol{k})\hat{u}^*_l(\boldsymbol{k}-\boldsymbol{q})\hat{u}^*_i(\boldsymbol{q})},$$

given that $\hat{u}(-\boldsymbol{k}) = \hat{u}^*(\boldsymbol{k})$. This is the complex conjugate of the original term.]

A further insight into $T(\boldsymbol{k})$ is now possible: the integrand of (10.2.14) represents spectral energy transfer *from* wavenumber $\boldsymbol{q}$ *to* wavenumber $\boldsymbol{k}$. This transfer is equal and opposite to that *from* $\boldsymbol{k}$ *to* $\boldsymbol{q}$ because reversing $\boldsymbol{k}$ and $\boldsymbol{q}$ reverses the sign of transfer:

$$k_l\, q_{ili}(\boldsymbol{k},\boldsymbol{q}) - k_l\, q_{ili}(\boldsymbol{q},\boldsymbol{k}) = -[k_l\, q_{ili}(\boldsymbol{q},\boldsymbol{k}) - k_l\, q_{ili}(\boldsymbol{k},\boldsymbol{q})].$$

At this fine-grained level $T(\boldsymbol{k})$ accounts for a conservative transfer of energy between wavenumber components.

### 10.2.3 Final period of decay

The dynamical equation (10.2.8) for the spectrum is unclosed because the right-hand side involves third moments of $\hat{u}$. Closing this equation has proved formidable. Early attempts consisted of spectral eddy viscosities; these were unsuccessful. Approaches based on a quasi-normal approximation for fourth moments fared better, but are fraught with difficulties. Successful instances of this approach are the "direct interaction approximation" and the "eddy damped quasi-normal model." These approaches are beyond the scope of this book; a reference is McComb (1990).

A grossly simplistic closure consists in just ignoring the right-hand side. This can only be justified by assuming a very low Reynolds number. Although that approximation has limited bearing on fully turbulent flow, it has appeared in the literature under the present heading of *final period of decay*. In a sense, the analysis in this section shows that the velocity decay exponent must depend on turbulent Reynolds number and gives the limiting low Reynolds number value.

Setting its right-hand side to zero, (10.2.8) becomes $\mathrm{d}_t E + 2\nu\kappa^2 E = 0$. This has the solution

$$E(\kappa, t) = E(\kappa, 0)\,\mathrm{e}^{-2\nu\kappa^2 t}.$$

The turbulent kinetic energy equals the integral of this energy spectrum, given by

$$\tfrac{1}{2} q^2 = \int_0^\infty E(\kappa, 0)\,\mathrm{e}^{-2\nu\kappa^2 t}\,\mathrm{d}\kappa.$$

As $t \to \infty$ the dominant contribution to the integral is from $\kappa \to 0$. To see this, let $\zeta = \kappa\sqrt{2\nu t}$ so that

$$\tfrac{1}{2} q^2 = (2\nu t)^{-1/2} \int_0^\infty E(\zeta/\sqrt{2\nu t}, 0)\,\mathrm{e}^{-\zeta^2}\,\mathrm{d}\zeta.$$

If $E$ has the limiting form $E \to C_s \kappa^s$ as $\kappa \to 0$, then

$$\begin{aligned}\tfrac{1}{2} q^2 &\to (2\nu t)^{-(s+1)/2} \int_0^\infty C_s \zeta^s\,\mathrm{e}^{-\zeta^2}\,\mathrm{d}\zeta \\ &= \tfrac{1}{2} C_s (2\nu t)^{-(s+1)/2}\Gamma[(s+1)/2] \propto t^{-(s+1)/2}.\end{aligned}$$

In the case $s = 4$ this gives $q^2 \propto t^{-2.5}$. The value $n = 2.5$ is the decay exponent for the final period of decay. It is generally considered to be the upper bound to $n$. The lower bound is $n = 1.0$, corresponding to strict scale similarity at infinite Reynolds number (Exercise 10.6). Experimental data all fall between these two extremes.

# Exercises

**Exercise 10.1.** *Triad interactions.* The Navier–Stokes equations contain the term $u_j\,\partial_j u_i$. Write the Fourier transform of this term as a convolution integral. Also write it as an integral over wavenumber triangles.

**Exercise 10.2.** *Energy decay.* Considered a crude model for the spectrum. Let

$$\begin{aligned} E(\kappa) &= C_s \kappa^s, && \kappa < 1/L(t), \\ E(\kappa) &= C_K \varepsilon^{2/3} \kappa^{-5/3}, && 1/\eta > \kappa > 1/L(t), \\ E(\kappa) &= 0, && \kappa > 1/\eta. \end{aligned}$$

Here $C_K$ is supposed to be a universal constant. Consider the case $\eta \approx 0$. How is $C_s$ related to $C_K$, $\varepsilon$, and $L(t)$? How is $q^2$ related to $\varepsilon$ and $L(t)$? Assume that $\varepsilon$ and $q^2$ follow power-law decays. Derive a formula relating the exponent in $L \propto t^m$ to $s$.

**Exercise 10.3.** *Two-point correlations.* Consider the two-point correlation

$$q^2 R_{ii} = \overline{u_i(\boldsymbol{x}) u_i(\boldsymbol{x}')}$$

in homogeneous turbulence. If $\boldsymbol{r} = \boldsymbol{x} - \boldsymbol{x}'$ then

$$\partial_{x_j} R_{ii}(\boldsymbol{r}) = -\partial_{x'_j} R_{ii}(\boldsymbol{r})$$

and $u_i(x')\,\partial_x u_i(x) = \partial_x[u_i(x)u_i(x')]$ because $x$ and $x'$ can be regarded as independent variables.

Let $\boldsymbol{u}$ be statistically homogeneous in all three directions. A "microscale," $\lambda$, was defined in Exercise 2.5. It could also be defined by

$$\lim_{|\boldsymbol{\xi}| \to 0} \partial_{\xi_i} \partial_{\xi_i} R(\boldsymbol{\xi}) = -\frac{1}{\lambda^2}.$$

Derive a formula to relate $\lambda$ to the rate of energy dissipation $\varepsilon$, $q^2$, and $\nu$.

**Exercise 10.4.** *Physical space.* The spectral evolution equations have a physical space corollary. Two-point correlations and spectra form Fourier transform pairs. Because of this, spectral closure models are called "two-point" closures. An equation for the two-point autocorrelation in homogeneous, isotropic turbulence, analogous to (10.2.8) is

$$\partial_t \overline{u_i u_i}(\boldsymbol{r}) + 2\nu \nabla_r^2 \overline{u_i u_i}(\boldsymbol{r}) = -2\,\partial_{r_j} \overline{u_j u_i u'_i}(\boldsymbol{r}),$$

where

$$\overline{u_j u_i u'_i}(\boldsymbol{r}) = \overline{u_j(\boldsymbol{x}) u_i(\boldsymbol{x}) u_i(\boldsymbol{x}')}.$$

Derive the above equation.

**Exercise 10.5.** *A dissipation formula for isotropic turbulence.* Streamwise derivatives of streamwise velocity $(\partial_1 u_1)$ are the easiest to measure. Show that this derivative can be used to infer the rate of dissipation by deriving the formula

$$\varepsilon = 15\nu\,\overline{(\partial_1 u_1)^2}$$

for isotropic turbulence. Your derivation *must* start from the energy spectrum tensor, Eq. (9.2.17). The "velocity derivative skewness" is defined by

$$S_d = \overline{(\partial_1 u_1)^3} / (\overline{(\partial_1 u_1)^2})^{3/2}.$$

High Reynolds number experimental values of $S_d$ are about 0.4. In isotropic turbulence, the exact equation for $\mathrm{d}_t \varepsilon$ contains a term proportional to $\nu\overline{(\partial_1 u_1)^3}$ (the self-stretching term). Show that this term non-dimensionalized by $\varepsilon^2/k$ is of order $R_T^{1/2}$, where $R_T = k^2/\varepsilon\nu$.

The velocity derivative skewness relation above can also be derived by invoking the definition of isotropy in physical space, rather than via Fourier space. The general form for an isotropic fourth-order tensor is (2.3.4)

$$\overline{\partial_i u_j\, \partial_k u_l} = A\delta_{ik}\delta_{jl} + B\delta_{il}\delta_{jk} + C\delta_{ij}\delta_{kl}.$$

Determine $A$ and $B$ in terms of $C$ by invoking incompressibility and symmetry – symmetry demands that the functional form be unaltered by the exchange $i \leftrightarrow k$.

**Exercise 10.6.** *Scale similarity.* Strict scale similarity would require that $E(\boldsymbol{k})$ be of the form $q^2(t)L(t)\tilde{E}(\eta)$ with $\eta = \boldsymbol{k}L(t)$ and where $\tilde{E}$ is a non-dimensional, similarity solution. Assume that $q^2 \propto t^{-n}$ and find $n$ by requiring strict self-similarity in Eq. (10.2.8).

# 11

# Rapid distortion theory

It is science that has taught us the way to substitute tentative truth for cocksure error.
–Bertrand Russell

Rapid distortion theory (RDT) developed out of ideas about how grid turbulence behaves as it passes through a wind tunnel contraction. The basic idea is quite simple. To conserve mass, the flow accelerates continuously along the length of the contraction. The increase of velocity with distance along the tunnel creates a rate of strain along the axis of the tunnel. Turbulence within the flow is subjected to this strain. Thereby, turbulent eddies are distorted as they flow downstream. The nature of that distortion can be reasoned out. If turbulence is thought of as a random tangle of vortex filaments, then the contraction consists of a straining flow that will compress or elongate the filaments (Figure 1.6, page 11): they are stretched along the axis of the tunnel and compressed perpendicular to it. The distorted vorticity induces corresponding distortions of the velocity field. Basic fluid dynamic considerations provide an understanding of the evolution of turbulence in this and similar situations. The subject of this chapter is a theory that describes such processes.

On a statistical level, the Reynolds stress tensor also will be altered by the distortion. Intuitive reasoning about vorticity, and the Biot–Savart relation between vorticity and velocity (Batchelor 1967) lead to an understanding of how Reynolds stresses evolve in a contracting nozzle. A similar rationale provides insight into turbulence in flow round bluff bodies and in many other geometries in which turbulence undergoes rapid distortion. RDT provides a mathematical framework to analyze the evolution of turbulence when it is deformed by flow gradients or by boundaries.

The mathematical approximation made in RDT consists of linearizing the equations about a given mean flow. In a wind tunnel contraction, the turbulence is considered to be strained by the mean flow, but its self-induced distortion is omitted. This is the exact opposite of the phenomenology in Section 10.1, which addresses the energy cascade. The energy cascade is dominated by nonlinearity, whereas RDT analysis is linear.

*Statistical Theory and Modeling for Turbulent Flows, Second Edition* P. A. Durbin and B. A. Pettersson Reif

Rapid distortions have been studied experimentally by sending grid turbulence through a variable cross-section duct (Gence and Mathieu, 1979), or by introducing turbulence into the flow upstream of an impingement plate (Britter *et al.*, 1979). In both cases, the turbulence scale and rate of distortion are established so as to approximate the requirements of the theory. The basic requirement is that the distortion occurs on a time-scale short compared to the eddy lifetime, $T = k/\varepsilon$.

The analytical approach makes use of the spectral representations developed in Chapter 9. Straining flow RDT is the most tractable and illustrates the concepts: it will be described first.

## 11.1 Irrotational mean flow

Consider an irrotational straining flow with a mean rate of strain of magnitude $S$. The essential RDT assumption is that $ST \gg 1$, where $T$ is the turbulence time-scale. Then the mean distortion is large compared to the self-distortion of the turbulence. The turbulent Reynolds number is also assumed to be large, so that effects of viscous dissipation can be ignored.

The inviscid, barotropic vorticity equation is (Batchelor 1967)

$$\mathrm{D}_t\left(\frac{\boldsymbol{\omega}}{\rho}\right) = \left(\frac{\boldsymbol{\omega}}{\rho}\right)\cdot\nabla\boldsymbol{u}. \tag{11.1.1}$$

The right-hand side of equation (11.1.1) represents stretching and rotation of vorticity by velocity gradients. An equivalent way to write the barotropic vorticity equation is

$$\mathrm{D}_t\boldsymbol{\omega} = \boldsymbol{\omega}\cdot\nabla\boldsymbol{u} - \boldsymbol{\omega}\nabla\cdot\boldsymbol{u}.$$

The continuity equation, $\mathrm{D}_t\rho = -\rho\nabla\cdot\boldsymbol{u}$, shows this to be the same as (11.1.1). The last term vanishes in incompressible flow.

If (11.1.1) is linearized about an irrotational flow, then $U$ and $\rho$ are mean quantities and $\omega$ is turbulent:

$$\overline{D}_t\left(\frac{\boldsymbol{\omega}}{\overline{\rho}}\right) = \left(\frac{\boldsymbol{\omega}}{\overline{\rho}}\right)\cdot\nabla\boldsymbol{U} \tag{11.1.2}$$

or, equivalently,

$$\partial_t\boldsymbol{\omega} + \boldsymbol{U}\cdot\nabla\boldsymbol{\omega} = \boldsymbol{\omega}\cdot\nabla\boldsymbol{U} - \boldsymbol{\omega}\nabla\cdot\boldsymbol{U}.$$

Terms like $\boldsymbol{\omega}\cdot\nabla\boldsymbol{u}$, which are quadratic in the fluctuating quantities, have been dropped. Because of the linearization, this equation describes the stretching and rotation of *turbulent* vortex tubes by an irrotational *mean* flow. This is illustrated by Figure 11.1. The solid line represents a vortex tube that moves along the streamlines represented as dashed curves, being stretched in length and rotated in direction.

### 11.1.1 Cauchy form of vorticity equation

The right-hand side of (11.1.2) describes the *rate* of vortex stretching and rotation. Another form of the vorticity equation is particularly effective for RDT problems in which the mean flow is irrotational; it is called the "Cauchy" form. The Cauchy form of

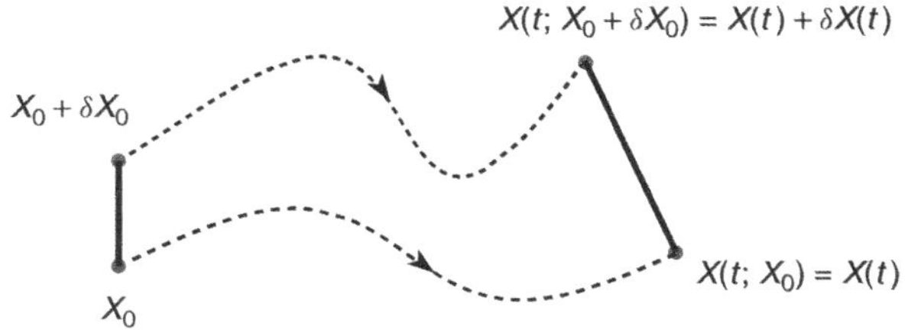


**Figure 11.1** Schematic of vortex line distortion. The solid line represents an infinitesimal segment of a material line; the dashed lines represent trajectories of its endpoints.

the vorticity equation describes the *net* stretching and rotation of vortex filaments: it can be thought of as an integrated version of (11.1.2).

The approach is to relate the evolution of the vorticity to the deformation of infinitesimal material line elements. Denote the trajectory of a fluid particle by $\boldsymbol{X}(t;\ \boldsymbol{X}_0)$. This is a convected particle, initially located at $\boldsymbol{X}_0$. Following the particle, $\mathrm{d}\boldsymbol{X}/\mathrm{d}t = \boldsymbol{U}(\boldsymbol{X})$, subject to the initial position $\boldsymbol{X}(t = 0) = \boldsymbol{X}_0$. By considering two neighboring fluid particles, an equation relating the trajectory to the distortion of an infinitesimal line segment, $\delta\boldsymbol{X}$, will be obtained.

Figure 11.1 shows how the separation of the two particles at any time is a function of their initial separation. Consider particles with a small initial separation $\delta\boldsymbol{X}_0$. They will follow slightly different trajectories, which determine their subsequent separation. Mathematically

$$\delta X_i = X_i(t;\ \boldsymbol{X}_0 + \delta\boldsymbol{X}_0) - X_i(t;\ \boldsymbol{X}_0) \rightarrow \frac{\partial X_i}{\partial X_{0j}}\delta X_{0j} \tag{11.1.3}$$

as $\delta\boldsymbol{X}_0 \rightarrow 0$. The matrix $\partial X_i/\partial X_{0j}$ describes the net stretching and rotation of the material line shown in the figure. The meaning of this matrix is found by deriving the evolution equation for $\delta\boldsymbol{X}$. This also draws a connection between (11.1.3) and the vorticity equation.

The motion of particles located at $\boldsymbol{X} + \delta\boldsymbol{X}$ and at $\boldsymbol{X}$ is determined by

$$\begin{aligned} \mathrm{d}(X_i + \delta X_i)/\mathrm{d}t &= U_i(\boldsymbol{X} + \delta\boldsymbol{X}), \\ \mathrm{d}X_i/\mathrm{d}t &= U_i(\boldsymbol{X}). \end{aligned} \tag{11.1.4}$$

Subtracting the second from the first gives

$$\mathrm{d}\delta X_i/\mathrm{d}t = U_i(\boldsymbol{X} + \delta\boldsymbol{X}) - U_i(\boldsymbol{X}).$$

Then letting $\delta\boldsymbol{X} \rightarrow 0$ results in

$$\mathrm{D}_t\delta X_i = \delta X_j\, \partial_j U_i$$

or

$$\mathrm{D}_t\delta\boldsymbol{X} = (\delta\boldsymbol{X}\cdot\nabla)\boldsymbol{U}. \tag{11.1.5}$$

The Lagrangian time derivative $\mathrm{d}/\mathrm{d}t$ has been restated as the Eulerian material derivative $\mathrm{D}/\mathrm{D}t$. By comparing (11.1.5) to (11.1.2), it can be seen that $\boldsymbol{\omega}/\rho$ evolves in the same way as an infinitesimal line element. If the two equations have proportionate initial conditions, $(\boldsymbol{\omega}/\rho)(0) \propto \delta \boldsymbol{X}(0)$, then they have proportionate solutions at later times: $\boldsymbol{\omega}/\rho \propto \delta \boldsymbol{X}$. Consequently (11.1.3) is a formal solution to the vorticity equation if $\delta \boldsymbol{X}$ is replaced by $\boldsymbol{\omega}/\rho$. Hence the vorticity equation can be written

$$\omega_i(\boldsymbol{X}, t) = \frac{\rho}{\rho_0} \frac{\partial X_i}{\partial X_{0j}} \omega_{0j}(\boldsymbol{X}_0); \tag{11.1.6}$$

this is Cauchy's form. Equation (11.1.6) is a general relation between Lagrangian coordinates and vortex line stretching and rotation. If deformation of material lines $\partial X_i/\partial X_{0j}$ is known, the evolution of the vorticity vector is obtained by (11.1.6).

Generally, the particle paths required by Cauchy's formula would depend on the vorticity, so that (11.1.6) would be an integral equation for the vorticity, not a solution *per se*. However, in rapid distortion theory Cauchy's formula is linearized about an irrotational mean flow. Then the Lagrangian coordinates, $\boldsymbol{X}(t; \boldsymbol{X}_0)$, are particle trajectories following the mean flow and $\omega$ is the turbulent vorticity. The Jacobian derivative, $\partial \boldsymbol{X}/\partial \boldsymbol{X}_0$, and the density, $\rho$, are computed solely from the mean flow. This provides an elegant analysis of the present case, in which turbulence is distorted by an irrotational mean flow. If the mean flow is rotational, the analysis is more involved, and Cauchy's formula is not helpful.

#### 11.1.1.1 Example: Lagrangian coordinates for linear distortions

The meaning of Eq. (11.1.6) can be clarified by computing the Lagrangian coordinate and its Jacobian derivative for a constant-direction, irrotational straining flow. In this case, the velocity components in the principal axes of the rate of strain are $U_i = \alpha_i x_i$, with no summation on $i$. Then the particle trajectories satisfy

$$\begin{aligned} \mathrm{d}X_1/\mathrm{d}t &= \alpha_1 X_1, \\ \mathrm{d}X_2/\mathrm{d}t &= \alpha_2 X_2, \\ \mathrm{d}X_3/\mathrm{d}t &= \alpha_3 X_3. \end{aligned} \tag{11.1.7}$$

The mean dilatation is $\nabla \cdot U \equiv D = \alpha_1 + \alpha_2 + \alpha_3$. The solutions to (11.1.7) for the individual $X_i$ are

$$X_1 = X_{0_1} \mathrm{e}^{\alpha_1 t}, \qquad X_2 = X_{0_2} \mathrm{e}^{\alpha_2 t}, \qquad X_3 = X_{0_3} \mathrm{e}^{\alpha_3 t}. \tag{11.1.8}$$

The deformation matrix occurring in (11.1.6) is then found to be

$$\left( \frac{\partial X_i}{\partial X_{0j}} \right) = \begin{pmatrix} \mathrm{e}^{\alpha_1 t} & 0 & 0 \\ 0 & \mathrm{e}^{\alpha_2 t} & 0 \\ 0 & 0 & \mathrm{e}^{\alpha_3 t} \end{pmatrix}. \tag{11.1.9}$$

Mass conservation can also be expressed in terms of this matrix: density times volume is the mass of a fluid element. Equating its initial and subsequent density,

$$\rho\, \delta X_1\, \delta X_2\, \delta X_3 = \rho_0\, \delta X_{0_1}\, \delta X_{0_2}\, \delta X_{0_3},$$

which can also be stated as

$$\frac{\rho}{\rho_0} = \frac{1}{\mathrm{Det}(\partial X_i/\partial X_{0j})}.$$

This form of mass conservation permits (11.1.6) to be written in terms of the deformation matrix alone:

$$\omega_i(x,t) = \frac{(\partial X_i/\partial X_{0j})}{\mathrm{Det}(\partial X_k/\partial X_{0_l})}\,\omega_{0j}(X_0). \tag{11.1.10}$$

In the case of (11.1.9)

$$\mathrm{Det}(\partial X_i/\partial X_{0j}) = \mathrm{e}^{(\alpha_1+\alpha_2+\alpha_3)t} = \mathrm{e}^{Dt};$$

hence $\rho = \rho_0\,\mathrm{e}^{-Dt}$. For compression, $D < 0$; for expansion, $D > 0$; and for incompressible flow, $D = 0$.

As an example of an irrotational distortion, consider a compression in the $x_1$ direction alone: $\alpha_2 = 0 = \alpha_3$ and $D = \alpha_1 < 0$. Then

$$\frac{\rho}{\rho_0}\left(\frac{\partial X_i}{\partial X_{0j}}\right) = \begin{pmatrix} 1 & 0 & 0 \\ 0 & \mathrm{e}^{-\alpha_1 t} & 0 \\ 0 & 0 & \mathrm{e}^{-\alpha_1 t} \end{pmatrix} \tag{11.1.11}$$

is used on the right-hand side of (11.1.6). It is then found that this distortion gives $\omega_1 = \omega_{0_1}$, $\omega_2 = \mathrm{e}^{-\alpha_1 t}\omega_{0_2}$, and $\omega_3 = \mathrm{e}^{-\alpha_1 t}\omega_{0_3}$. Even though the compression is along the $x_1$ axis, it intensifies $\omega_2$ and $\omega_3$ but does not alter $\omega_1$. It is the increase in density that intensifies the vorticity. This might play a role in piston engines (see Figure 11.9).

## 11.1.2 Distortion of a Fourier mode

Considerations on the stretching and rotation of vorticity are only the first step. The next step is to obtain the corresponding velocity field and, from that, the Reynolds stress tensor.

To proceed with this agenda, the randomness of the flow must be represented. This can be done by the techniques of Chapter 9: the instantaneous turbulence is represented by a sum of Fourier modes. The velocity field of a single mode can be obtained from the Cauchy solution for the vorticity. Second moments of the velocity can then be formed and averaged, using spectral representations. Finally, the solution is integrated over the turbulence spectrum to find the Reynolds stress, as in Eq. (9.2.27) on page 263. The net result will relate the Reynolds stress after the distortion to that prior to the distortion.

Pursuing this program, we next obtain a solution for the velocity of an individual component Fourier component. Owing to linearity, the full solution is a sum of these individual solutions.

The initial vorticity field of a single Fourier component is

$$\boldsymbol{\omega}_0 = \hat{\boldsymbol{\omega}}_0(\boldsymbol{k})\,\mathrm{e}^{-\mathrm{i}\boldsymbol{k}\cdot\boldsymbol{X}_0}.$$

Consider a particle located at the field point $\boldsymbol{x}$; that is, $\boldsymbol{X}(t) = \boldsymbol{x}$. Substituting the initial vorticity and (11.1.9) into the Cauchy formula (11.1.6) gives

$$\omega_\gamma = \mathrm{e}^{-Dt}\,\mathrm{e}^{\alpha_\gamma t}\,\mathrm{e}^{-\mathrm{i}\boldsymbol{k}\cdot\boldsymbol{X}_0}\hat{\omega}_{0\gamma}, \qquad \gamma = 1, 2, 3. \tag{11.1.12}$$

For a homogeneous mean straining flow, from (11.1.8) the initial coordinates of the particle currently at $\boldsymbol{x}$ are $X_{0_\gamma} = x_\gamma \, \mathrm{e}^{-\alpha_\gamma t}$. Using this in the result (11.1.12) gives

$$\omega_\gamma(\boldsymbol{x}, t) = \mathrm{e}^{-Dt} \, \mathrm{e}^{\alpha_\gamma t} \, \mathrm{e}^{-\mathrm{i}\boldsymbol{\kappa}\cdot\boldsymbol{x}} \hat{\omega}_{0\gamma}, \qquad \gamma = 1, 2, 3 \tag{11.1.13}$$

for the vorticity at position $\boldsymbol{x}$, at time $t$. In the exponent of (11.1.13), $k_j X_{0j} = k_j \, \mathrm{e}^{-\alpha_j t} x_j$ has been substituted and then

$$\kappa_\gamma(t) \equiv k_\gamma \, \mathrm{e}^{-\alpha_\gamma t}, \qquad \gamma = 1, 2, 3 \tag{11.1.14}$$

was defined to simplify the algebra.

Note that $\boldsymbol{\kappa}(t)$ is a time-dependent wavevector, while $\boldsymbol{k}$ is constant in time. If $\alpha_i > 0$ then $\kappa_i$ decreases with time. A physical explanation follows from noting that $\kappa = 2\pi/\lambda$, where $\lambda$ is the wavelength. When $\alpha_i > 0$, the wavelength in the $i$ direction is stretched; hence the wavenumber, $\kappa$, decreases. If $\alpha_i < 0$, the wavelength is reduced and $\kappa$ increases. An example of wavelength reduction occurs when turbulence flows toward a stagnation point; the flow slows down as it approaches the surface, corresponding to $\alpha_1 < 0$. In this case, wavelength shortening can also can be interpreted as vortices piling up round the surface, as sketched later by Figure 11.3 (Hunt, 1973).

We define $\hat{\omega}$ by $\omega_i = \hat{\omega}_i \, \mathrm{e}^{\mathrm{i}\kappa\cdot x}$, that is, just drop the exponential to get $\hat{\omega}$. Then solution (11.1.13) becomes

$$\hat{\omega}_\gamma = \hat{\omega}_{0_\gamma} \, \mathrm{e}^{-Dt} \, \mathrm{e}^{\alpha_\gamma t}, \qquad \gamma = 1, 2, 3.$$

In Fourier space, $-\mathrm{i}\boldsymbol{\kappa}$ is the "gradient operator" because $\nabla\boldsymbol{\omega} = -\mathrm{i}\boldsymbol{\kappa}\boldsymbol{\omega}$ in the case of (11.1.13). Thus

$$\nabla \wedge \boldsymbol{\omega} \Rightarrow -\mathrm{i}\boldsymbol{\kappa} \wedge \hat{\boldsymbol{\omega}} \qquad \text{and} \qquad \nabla^2 \boldsymbol{u} \Rightarrow -|\boldsymbol{\kappa}|^2 \hat{\boldsymbol{u}}. \tag{11.1.15}$$

Note that both $\boldsymbol{\kappa}$ and $\hat{\boldsymbol{u}}$ are functions of $t$ and of the initial wavevector, $\boldsymbol{k}$.

In physical space, the kinematic relation between velocity and vorticity is $\nabla^2 \boldsymbol{u} = -\nabla \wedge \boldsymbol{\omega}$ if the turbulence is incompressible. The corresponding equation in Fourier space follows from the correspondences (11.1.15),

$$\hat{\boldsymbol{u}} = \frac{-\mathrm{i}\boldsymbol{\kappa} \wedge \hat{\boldsymbol{\omega}}}{|\boldsymbol{\kappa}|^2}$$

or, in index form,

$$\hat{u}_j = -\varepsilon_{jkl} \frac{\mathrm{i}\kappa_k \hat{\omega}_l}{|\boldsymbol{\kappa}|^2}. \tag{11.1.16}$$

For instance, in the incompressible case, where $D = 0$,

$$\begin{aligned} \hat{u}_1 &= \mathrm{i}\frac{\kappa_3 \hat{\omega}_2 - \kappa_2 \hat{\omega}_3}{|\boldsymbol{\kappa}|^2} = \mathrm{i}\frac{\kappa_3 \, \mathrm{e}^{\alpha_2 t} \hat{\omega}_{0_2} - \kappa_2 \, \mathrm{e}^{\alpha_3 t} \hat{\omega}_{0_3}}{|\boldsymbol{\kappa}|^2} \\ &= \mathrm{i}\frac{k_3 \, \mathrm{e}^{(\alpha_2 - \alpha_3)t} \hat{\omega}_{0_2} - k_2 \, \mathrm{e}^{(\alpha_3 - \alpha_2)t} \hat{\omega}_{0_3}}{|\boldsymbol{\kappa}|^2} \end{aligned}$$

after substituting $k_\gamma\, \mathrm{e}^{-\alpha_\gamma t}$ for $\kappa_\gamma$. Multiplying by the complex conjugate and averaging give

$$\overline{\hat{u}_1\hat{u}_1^*} = \frac{k_3^2\, \mathrm{e}^{2(\alpha_2-\alpha_3)t}\,\overline{\hat{\omega}_{0_2}\hat{\omega}_{0_2}^*} + k_2^2\, \mathrm{e}^{2(\alpha_3-\alpha_2)t}\,\overline{\hat{\omega}_{0_3}\hat{\omega}_{0_3}^*} - k_2k_3(\overline{\hat{\omega}_{0_2}\hat{\omega}_{0_3}^*} + \overline{\hat{\omega}_{0_3}\hat{\omega}_{0_2}^*})}{|\kappa|^4}. \tag{11.1.17}$$

### 11.1.3 Calculation of covariances

The randomness of the turbulence can now be introduced via a model for the vorticity correlations, $\overline{\hat{\omega}_{0_i}(\boldsymbol{k})\hat{\omega}_{0_j}^*(\boldsymbol{k}')}$, that appear in Eq. (11.1.7). In homogeneous turbulence, these are of the form

$$\Re\mathrm{e}\left[\overline{\hat{\omega}_{0_i}(\boldsymbol{k})\hat{\omega}_{0_j}^*(\boldsymbol{k}')}\right] = \delta(\boldsymbol{k}-\boldsymbol{k}')\Phi_{ij}^{\omega}(\boldsymbol{k})$$

as in (9.2.5), page 255. From $\hat{\omega}_j = -\mathrm{i}\varepsilon_{jkl}k_k\hat{u}_l$, it is seen that the vorticity spectrum can be obtained from the velocity spectrum as

$$\Phi_{ij}^{\omega} = k_k k_m \varepsilon_{ikl}\varepsilon_{jmn}\Phi_{lm}.$$

Substituting the isotropic velocity spectrum (9.2.17), page 259, it follows that the isotropic vorticity spectrum is

$$\Phi_{ij}^{\omega} = |\boldsymbol{k}|^2\Phi_{ij} = \frac{E(|\boldsymbol{k}|)}{4\pi}\left(\delta_{ij} - \frac{k_ik_j}{|\boldsymbol{k}|^2}\right). \tag{11.1.18}$$

Though the turbulence is initially isotropic, it will not remain so: formula (11.1.17) gives the evolution of the 11 component of the velocity spectrum,

$$\Phi_{11}(t) = \left\{\frac{k_2^2(k_1^2+k_2^2)\,\mathrm{e}^{2(\alpha_3-\alpha_2)t} + k_3^2(k_1^2+k_3^2)\,\mathrm{e}^{2(\alpha_2-\alpha_3)t} + 2k_2^2k_3^2}{(k_1^2\,\mathrm{e}^{-2\alpha_1 t} + k_2^2\,\mathrm{e}^{-2\alpha_2 t} + k_3^2\,\mathrm{e}^{-2\alpha_3 t})^2}\right\}\frac{E(|\boldsymbol{k}|)}{4\pi|\boldsymbol{k}|^2}, \tag{11.1.19}$$

with similar solutions for other components.

The factor in curly brackets determines the time evolution. Note that this factor is independent of the shape of the energy spectrum, $E(|\boldsymbol{k}|)$. It is a general result of inviscid RDT that the turbulence evolution is not a function of the initial spectrum shape. An equivalent observation is that the factor in curly brackets is independent of the magnitude of $\boldsymbol{k}$. Thus, if $e_i \equiv k_i/|\boldsymbol{k}|$, then

$$\Phi_{11}(t) = \left\{\frac{e_2^2(e_1^2+e_2^2)\,\mathrm{e}^{2(\alpha_3-\alpha_2)t} + e_3^2(e_1^2+e_3^2)\,\mathrm{e}^{2(\alpha_2-\alpha_3)t} + 2e_2^2e_3^2}{(e_1^2\,\mathrm{e}^{-2\alpha_1 t} + e_2^2\,\mathrm{e}^{-2\alpha_2 t} + e_3^2\,\mathrm{e}^{-2\alpha_3 t})^2}\right\}\frac{E(|\boldsymbol{k}|)}{4\pi|\boldsymbol{k}|^2}, \tag{11.1.20}$$

where $\boldsymbol{e}$ is a unit vector in the direction of the initial wavevector. Only this direction, not the magnitude of $\boldsymbol{k}$, affects the time evolution.

The integration of Eq. (11.1.20) can be performed conveniently in spherical polar coordinates: $e_1 = \cos\theta$, $e_2 = \sin\theta\cos\phi$, $e_3 = \sin\theta\sin\phi$. With this substitution, the relation (9.2.27) between the Reynolds stress and the spectrum tensor becomes

$$\overline{u_1^2} = \int_0^{2\pi}\int_0^{\pi}\int_0^{\infty} \Phi_{11}(\boldsymbol{e}, k; t)k^2\,\mathrm{d}k\,\sin\theta\,\mathrm{d}\theta\,\mathrm{d}\phi.$$

After substituting (11.1.20), it is seen that the $k$ integral can be done independently of the angular integrals. It gives a factor of $\frac{1}{2}q_0^2$ irrespective of the form of $E(k)$. The subscript "0" indicates the initial value.

Thus the velocity variance obtained from (11.1.20) is given by the very messy integral (Batchelor and Proudman 1954):

$$\overline{u_1^2}(t) = \frac{q_0^2}{8\pi}\int_0^{2\pi}\int_0^{\pi} \frac{\cos^2\phi\,(\cos^2\theta + \sin^2\theta\cos^2\phi)\,\mathrm{e}^{2(\alpha_3-\alpha_2)t} + \sin^2\phi\,(\cos^2\theta + \sin^2\theta\sin^2\phi)\,\mathrm{e}^{2(\alpha_2-\alpha_3)t} + 2\sin^2\theta\cos^2\phi\sin^2\phi}{(\mathrm{e}^{-2\alpha_1 t}\cos^2\theta + \mathrm{e}^{-2\alpha_2 t}\sin^2\theta\cos^2\phi + \mathrm{e}^{-2\alpha_3 t}\sin^2\theta\sin^2\phi)^2}\sin^3\theta\,\mathrm{d}\theta\,\mathrm{d}\phi \tag{11.1.21}$$

Analogous results can be derived for $\overline{u_2^2}$ and $\overline{u_3^2}$. The angular integrations can be performed numerically.

Figures 11.2 and 11.4 plot the rapid distortion solution, $\overline{u_i^2}(t)/\overline{u_0^2}$, for turbulence subjected to incompressible, irrotational distortions. The initial intensity is defined by $\overline{u_0^2} = \frac{1}{3}q_0^2$. Incompressibility, $\alpha_1 + \alpha_2 + \alpha_3 = 0$, shows that the total strains, $s_i \equiv \mathrm{e}^{\alpha_i t}$, satisfy $s_1 s_2 s_3 = 1$.

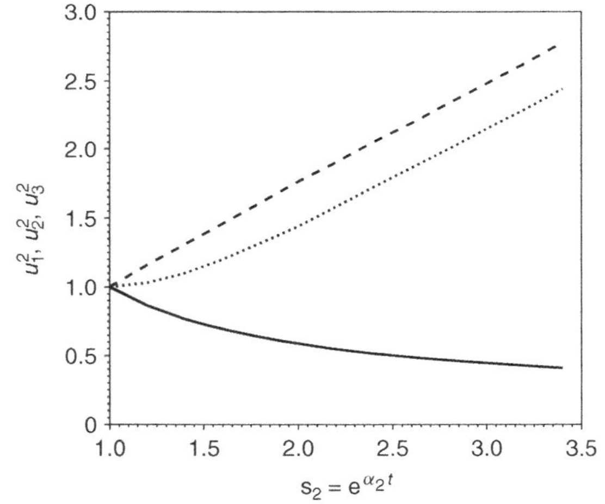


**Figure 11.2** Normalized normal stresses provided by the RDT solution for a plane strain: $\overline{u_1^2}$ (- - - -); $\overline{u_2^2}$ (——); and $\overline{u_3^2}$ (········).

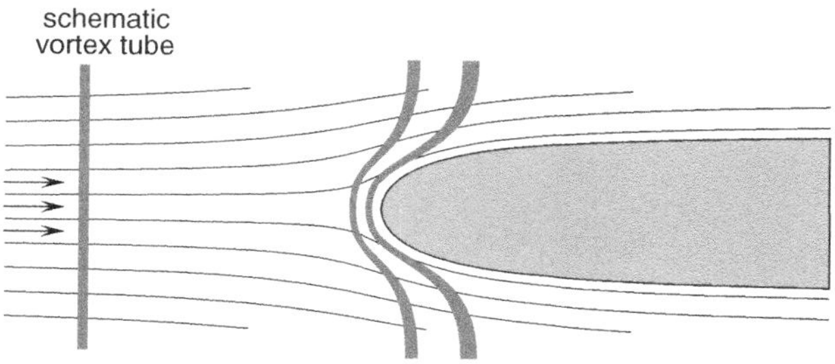


**Figure 11.3** Sketch of vortex tube distortion in flow round a bluff body. Vortices piling up at the stagnation point are synonymous with wavelength reduction.

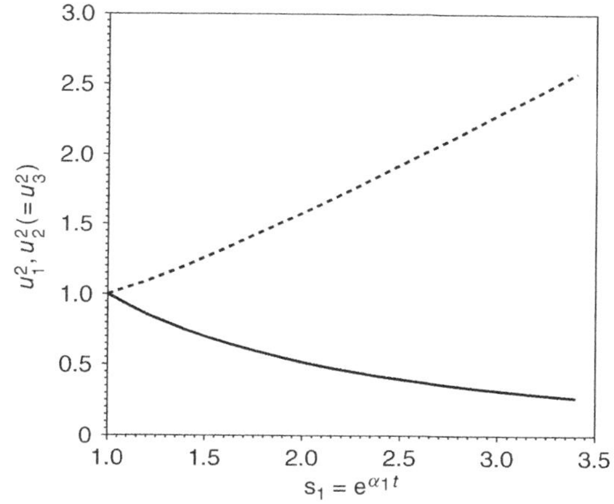


**Figure 11.4** Normalized normal stresses provided by the RDT solution for axisymmetric strain: $\overline{u_1^2}$ (———); and $\overline{u_2^2}$, $\overline{u_3^2}$ (- - - -).

Figure 11.2 is a computation for the plane strain, $\alpha_1 = -\alpha_2$, $\alpha_3 = 0$ with $\alpha_2 > 0$. A qualitative understanding of this figure comes from considering the distortion of vorticity. Vortex lines are being stretched in the $x_2$ direction and compressed in $x_1$. Correspondingly $\omega_2$ is increasing with $s_2$ and $\omega_1$ is decreasing.

The associated effect on velocity can be inferred by the "right-hand rule:" the velocity associated with the vortex line lies in a plane perpendicular to it. Analytically, since

$$\nabla^2 u_1 = \partial_3 \omega_2 - \partial_2 \omega_3$$

and $\omega_2$ is increased by the plane strain, one might anticipate that $\overline{u_1^2}$ would increase with $s_1$, as indeed it does. Similarly, because

$$\nabla^2 u_2 = \partial_1 \omega_3 - \partial_3 \omega_1$$

and $\omega_1$ is decreased, $\overline{u_2^2}$ decreases. Finally,

$$\nabla^2 u_3 = \partial_2 \omega_1 - \partial_1 \omega_2.$$

It might be less clear what this component should do: $\omega_2$ increases, but $\omega_1$ decreases. The amplified component wins out and $\overline{u_3^2}$ increases, but not so much as $\overline{u_1^2}$.

Asymptotic approximation to angular integrals like (11.1.21) gives the limiting large-strain behaviors

$$\overline{u_1^2}/\overline{u_0^2} \to \frac{3}{4}s_2 + \frac{3}{8s_2}\left(\log 4s_2 - \frac{1}{2}\right),$$

$$\overline{u_2^2}/\overline{u_0^2} \to \frac{3}{4s_2}(\log 4s_2 - 1), \tag{11.1.22}$$

$$\overline{u_3^2}/\overline{u_0^2} \to \frac{3}{4}s_2 - \frac{3}{8s_2}\left(\log 4s_2 - \frac{3}{2}\right),$$

as $s_2 \to \infty$ (Townsend, 1976). These are concrete formulas for the relative magnitudes of the normal stresses.

The plane-strain result can be interpreted as the solution for small-scale turbulence on the stagnation line of a two-dimensional, bluff body as in Figure 11.3. A "quasi-homogeneous" approximation is invoked (Goldstein and Durbin 1980) when applying the strictly homogeneous solution to such a case. The rate of strain along the axis is $\mathrm{d}U/\mathrm{d}x$. The time a particle takes to travel along the stagnation streamline from $x$ to $x + \mathrm{d}x$ is $\mathrm{d}x/U(x)$. Hence, a vortex tube is subjected to a total strain

$$\int \alpha_1 \mathrm{d}t = \int_{-\infty}^{x} \frac{\partial U}{\partial x}\frac{\mathrm{d}x}{U} = \log[U(x)/U_\infty].$$

This is a nice result because it relates the straining to the local potential flow velocity as

$$s_1 = \mathrm{e}^{\int \alpha_1 \mathrm{d}t} = U(x)/U_\infty. \tag{11.1.23}$$

By incompressibility $s_2 = 1/s_1 = U_\infty/U(x)$.

In the two-dimensional stagnation point flow of Figure 11.3, the streamwise component of intensity, $\overline{u_1^2}$, is amplified as vorticity stretches around the nose of the body. At the stagnation point, $U \to 0$, and $s_2 \to \infty$. Invoking the large-strain result (11.1.22), one finds $\overline{u_1^2}/\overline{u_0^2} \to 3U_\infty/4U$. The turbulence intensity increases inversely to the velocity as stagnation point is approached.

The asymptotic approximation (11.1.22) shows that the cross-stream component, $\overline{u_2^2}$, is suppressed near the stagnation point. Further around the surface, the stretched vorticity component rotates into the $x_1$ direction and $\overline{u_2^2}$ is amplified. Of course, the singularity predicted by (11.1.22) at the surface is unphysical, but the amplification of $\overline{u_1^2}$ and suppression of $\overline{u_2^2}$ on the stagnation streamline is consistent with experiments (Britter *et al.* 1979).

Figure 11.4 shows the solution for axisymmetric strain, $\alpha_2 = \alpha_3 = -\frac{1}{2}\alpha_1$ with $\alpha_1 > 0$. Again, $\omega_1$ is increased by the strain so that $\overline{u_2^2}$ $(= \overline{u_3^2})$ increases with $s_1$. Both $\omega_3$ and $\omega_2$

are reduced by negative strain, so $\overline{u_1^2}$ is decreased by the strain. The asymptotic behaviors for large $s_1$ are

$$\overline{u_1^2}/\overline{u_0^2} \to \frac{3}{4s_1^2}(\log 4s_1^3 - 1),$$

$$\overline{u_2^2}/\overline{u_0^2},\ \overline{u_3^2}/\overline{u_0^2} \to \frac{3}{4}s_1.$$

These provide reasonable approximations to the exact solution when $s_1 > 2$.

The solution for an axisymmetric rate of strain could be interpreted as the solution for small-scale turbulence on the centerline of a contracting pipe. Again, the straining is equal to the ratio of upstream to downstream velocities: $s_1 = U_{\text{down}}/U_{\text{up}}$. By mass conservation, $UA$ is constant, where $A$ is the cross-section of the pipe. If the cross-sectional area is $A_{\text{up}}$ upstream of the contraction and $A_{\text{down}}$ downstream, then $s_1 = A_{\text{up}}/A_{\text{down}}$. The turbulence intensity is directly related to the duct area. In fact, this was the original RDT problem discussed by Taylor (1935). Figure 11.4 indicates the form of anisotropy that can be expected for this type of configuration if the upstream turbulence is isotropic grid turbulence. The component $\overline{u_1^2}$, along the axis of the pipe is decreased relative to the other two components.

When the flow is an axisymmetric expansion, $s_1 < 1$ and $s_2 = s_3 = 1/s_1^{1/2}$, two components of vorticity are stretched. Consequently, all components of turbulent intensity amplify. The situation is like Figure 11.3, except the body is axisymmetric, so that a vortex perpendicular to the page will also wrap around the nose. The relation between strain and velocity corresponding to (11.1.23) is seen to be (Durbin 1981)

$$s_1 = U(x)/U_\infty, \qquad s_2 = s_3 = s_1^{-1/2} = \sqrt{U_\infty/U(x)}.$$

As the stagnation point is approached, $U \to 0$ and the solution asymptotes to that for large strain. The result, analogous to (11.1.22), is found to be

$$\overline{u_1^2} \to \frac{3\pi s_2}{8} = \frac{3\pi}{8}\sqrt{\frac{U_\infty}{U(x)}},$$

$$\overline{u_2^2} = \overline{u_3^2} \to \frac{3\pi s_2}{16} = \frac{3\pi}{16}\sqrt{\frac{U_\infty}{U(x)}}, \tag{11.1.24}$$

as $s_2 \to \infty$. For instance, all components of intensity amplify on the stagnation streamline of a sphere.

## 11.2 General homogeneous distortions

In the general case, the mean flow vorticity is not zero. The intuitive, analytically elegant, approach of the previous section does not work. Up to this point, only distortion of

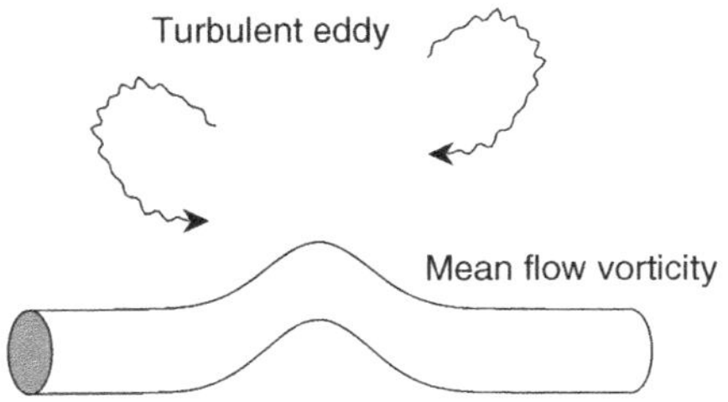


**Figure 11.5** Schematic of mean vorticity distortion by turbulent eddies.

turbulent vorticity by mean velocity gradients arose; but when mean vorticity is present, it will be distorted by turbulent velocity gradients. When mean vorticity is distorted, it becomes fluctuating vorticity, as suggested by Figure 11.5. The tube represents mean vorticity that has been kinked by an eddy. The kink represents the creation of fluctuating vorticity from the mean. This can be considered a mechanism of turbulence production.

The mean velocity is generally of the form

$$U_i = A_{ij} x_j \tag{11.2.1}$$

in homogeneous turbulence (Section 3.3). Consider a Fourier mode

$$\hat{u}_i(t; \boldsymbol{\kappa})\, \mathrm{e}^{-\mathrm{i}\boldsymbol{\kappa}(t)\cdot\boldsymbol{x}}. \tag{11.2.2}$$

Substituting this and the mean flow (11.2.1) into the linearized, inviscid momentum equation

$$\partial_t u_i + U_j\, \partial_j u_i + u_j\, \partial_j U_i = -\partial_i p.$$

(for constant density with $\rho \equiv 1$) gives

$$\mathrm{d}_t \hat{u}_i - \mathrm{i}\, \mathrm{d}_t \kappa_j\, x_j \hat{u}_i - \mathrm{i} A_{kj} x_j \kappa_k \hat{u}_i + \hat{u}_j A_{ij} = \mathrm{i}\kappa_i \hat{p}. \tag{11.2.3}$$

The evolution equation for $\boldsymbol{\kappa}$ is found by setting the sum of the two terms containing $x_j$ to zero. This eliminates the secular term from (11.2.3):

$$\mathrm{d}_t \kappa_j = -\kappa_k A_{kj}, \qquad \mathrm{d}_t |\boldsymbol{\kappa}|^2 = -2\kappa_k A_{kj} \kappa_j. \tag{11.2.4}$$

The pressure is eliminated by invoking continuity, $\kappa_i \hat{u}_i = 0$. This gives

$$\hat{p} = -\frac{2\mathrm{i}\kappa_i \hat{u}_j A_{ij}}{|\boldsymbol{\kappa}|^2},$$

which was derived from (11.2.3) using (11.2.4). Eliminating the pressure from (11.2.3) gives the evolution equation of the Fourier amplitude,

$$\mathrm{d}_t \hat{u}_i = \hat{u}_j A_{kj} \left\{ \frac{2\kappa_i \kappa_k}{|\boldsymbol{\kappa}|^2} - \delta_{ik} \right\} = \hat{u}_j A_{kj} \left\{ 2 e_i e_k - \delta_{ik} \right\}, \tag{11.2.5}$$

where $e_i = \kappa_i / |\boldsymbol{\kappa}|$. In fact, it is readily shown that, again, the evolution of $\hat{\boldsymbol{u}}$ is a function only of the direction, $\boldsymbol{e}$, of the wavevector and not of its magnitude $|\boldsymbol{\kappa}|$. The general homogeneous RDT problem is to solve (11.2.5) and then to compute turbulent statistics from a given initial spectrum.

Since (11.2.5) is linear, its solution can be written symbolically as

$$\hat{u}_i = \mathcal{M}_{ij}(t; \boldsymbol{e})\hat{u}_j^0.$$

This shows the general approach. The distortion matrix $\mathcal{M}_{ij}$ must first be found for a particular velocity gradient matrix $A_{ij}$. Then the Reynolds stresses at any time are related to the initial energy spectrum tensor by

$$\overline{u_i u_j} = \iiint_{-\infty}^{\infty} \mathcal{M}_{ik}\mathcal{M}_{jl}(t; \boldsymbol{e})\Phi_{kl}^0(\boldsymbol{k})\,\mathrm{d}^3\boldsymbol{k}. \tag{11.2.6}$$

Hunt and Carruthers (1990) develop the theory via this formalism. Homogeneous shear flow provides a case in point.

### 11.2.1 Homogeneous shear

The case of homogeneously sheared turbulence is of some practical interest (see Chapter 7). In this case, $U_1 = \mathcal{S}x_2$, so $A_{12} = \mathcal{S}$ and all other components of $A_{ij}$ are zero. Then (11.2.5) becomes

$$\begin{aligned} \mathrm{d}_t\hat{u}_1 &= \mathcal{S}\hat{u}_2(2e_1^2 - 1), \\ \mathrm{d}_t\hat{u}_2 &= 2\mathcal{S}\hat{u}_2 e_1 e_2, \\ \mathrm{d}_t\hat{u}_3 &= 2\mathcal{S}\hat{u}_2 e_1 e_3. \end{aligned} \tag{11.2.7}$$

The solution to (11.2.4) with the initial values $\kappa_i(0) = k_i$, is

$$\kappa_2 = k_2 - \mathcal{S}tk_1, \qquad \kappa_1 = k_1, \qquad \kappa_3 = k_3.$$

Only the $\kappa_2$ component evolves in time, due to the mean shearing. Figure 11.6 illustrates how shearing distorts wavecrests. The crests rotate, changing their spacing in the $x_2$ direction, but not altering the $x_1$ spacing. Hence $\kappa_1$ is not altered, while $\kappa_2$ increases or decreases to maintain that the wavevector is perpendicular to the crests.

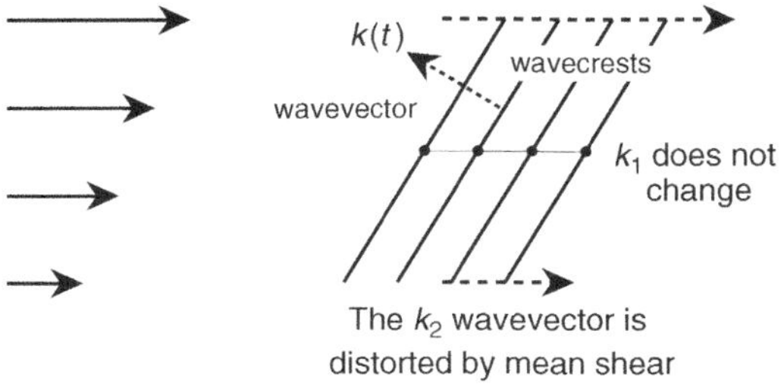


**Figure 11.6** Shearing of the $\kappa_2$ wavevector component. The other components of $\kappa_i$ are unaffected.

It follows from the second of (11.2.7) and (11.2.4) that $d_t(|\boldsymbol{\kappa}|^2\hat{u}_2) = 0$ and hence

$$\hat{u}_2 = \hat{u}_2^0 \frac{|\boldsymbol{k}|^2}{|\boldsymbol{\kappa}|^2} = \hat{u}_2^0 \frac{k_1^2 + k_2^2 + k_3^2}{k_1^2 + (k_2 - \mathcal{S}tk_1)^2 + k_3^2}. \tag{11.2.8}$$

The denominator increases with time for most values of $k_1$ ($k_1 > k_2/\mathcal{S}t$ or $k_1/k_2 < 0$), hence one expects that $\overline{u_2^2}$ will be decreased by shear, as is the case. However, it can be seen that, if $k_1 = k_2/\mathcal{S}t$, then $\hat{u}_2$ is independent of time. Values of $k_1$ that satisfy this will become increasingly dominant as time progresses. This leads to a conclusion that turbulent intensity is increasingly associated with small $k_1$ as time evolves. But small $k_1$ corresponds to long wavelength in the streamwise direction. The very important conclusion is that eddies that are *elongated in the streamwise direction* are preferred by a strongly sheared flow.

A suggestive way to see why streamwise elongation develops in time is to analyze how $\overline{u_2^2}$ is calculated from (11.2.8). If that equation is squared and averaged,

$$\overline{\hat{u}_2(\boldsymbol{k})\hat{u}_2^*(\boldsymbol{k}')} = \left[\frac{k_1^2 + k_2^2 + k_3^2}{k_1^2 + (k_2 - \mathcal{S}tk_1)^2 + k_3^2}\right]^2 \Phi_{22}^0(\boldsymbol{k})\delta(\boldsymbol{k} - \boldsymbol{k}')$$

is obtained. The velocity variance is the integral over $\boldsymbol{k}$ and $\boldsymbol{k}'$. If the change of variables $\eta = \mathcal{S}tk_1$ is made in the $\boldsymbol{k}$ integral,

$$\overline{u_2^2} = \frac{1}{\mathcal{S}t} \iiint_{-\infty}^{\infty} \Phi_{22}(\eta/\mathcal{S}t, k_2, k_3) \left[\frac{(\eta/\mathcal{S}t)^2 + k_2^2 + k_3^2}{(\eta/\mathcal{S}t)^2 + (k_2 - \eta)^2 + k_3^2}\right]^2 d\eta\, dk_2\, dk_3$$

results. As $\mathcal{S}t \to \infty$, the portion of the spectrum that contributes to the integral becomes approximately $\Phi_{22}(0, k_2, k_3)$, showing the dominance of $k_1 = 0$. This change of variables also shows that $\overline{u_2^2}$ decreases at long time because of the factor in front of the integral. [The simplistic reasoning of setting $\eta/\mathcal{S}t$ to zero gives the correct understanding that $\overline{u_2^2}$ is decreased by shear, although it is flawed mathematically. Letting $\eta/\mathcal{S}t \to 0$ inside the integral suggests the long-time behavior $\overline{u_2^2} \sim 1/\mathcal{S}t$. However, the resulting integral is logarithmically divergent if $\Phi_{22}$ has the isotropic proportionality to $(\eta/\mathcal{S}t)^2 + k_3^2$. Consequently, the actual behavior is $\overline{u_2^2} \sim \log(\mathcal{S}t)/\mathcal{S}t$ as $\mathcal{S}t \to \infty$ (Rogers 1991).]

The other components of Eqs. (11.2.7) have closed-form, although more complicated, solutions (Townsend, 1976). After forming moments, as in (11.2.6), Reynolds stresses can be computed by numerical integration. Solutions are plotted in Figure 11.7. The short-time behaviors that can be derived by Taylor series solution to Eqs. (11.2.7) are

$$\begin{aligned}
\overline{u_1u_2}/\overline{u_0^2} &= -\tfrac{2}{5}\mathcal{S}t + O(t)^3,\\
\overline{u_1^2}/\overline{u_0^2} &= 1 + \tfrac{2}{7}(\mathcal{S}t)^2 + O(t)^4,\\
\overline{u_2^2}/\overline{u_0^2} &= 1 - \tfrac{4}{35}(\mathcal{S}t)^2 + O(t)^4,\\
\overline{u_3^2}/\overline{u_0^2} &= 1 + \tfrac{8}{35}(\mathcal{S}t)^2 + O(t)^4,
\end{aligned} \tag{11.2.9}$$

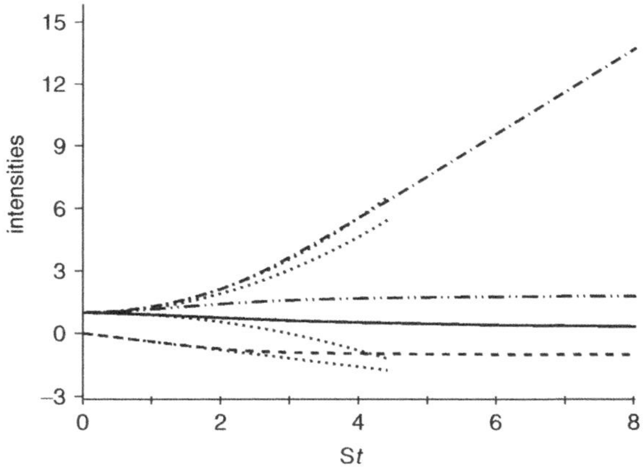


**Figure 11.7** RDT solution for homogeneously sheared isotropic turbulence: $\overline{u^2}$ (— · —); $\overline{v^2}$ (———); $\overline{w^2}$ (—·· —); and $\overline{uv}$ (- - - -). The dotted lines are the short-time asymptote stated in Eq. (11.2.9) (the dotted line corresponding to $\overline{u^2}$ is almost on top of the full RDT solution).

as $St \to 0$. The first of these is analogous to the result (7.1.29) on page 166, and as such is widely used in turbulence modeling – in the literature this connection has led to the factor of 2/5 in (7.1.29) being called a "rapid distortion constraint." The second and third of Eqs. (11.2.9) illustrate the general result that $\overline{u_1^2}$ increases and $\overline{u_2^2}$ decreases under the action of mean shearing in the $x_1$ direction; $\overline{u_3^2}$ increases only slightly. Approximations (11.2.9) are included as dotted lines in Figure 11.7. They are reasonably accurate for $St < 2.5$.

This solution and Figure 11.7 show that $\overline{u_1^2}$ is substantially increased by shear. As time progresses, $\overline{u_1^2}$ becomes the largest Reynolds stress, at long times growing in proportion to $t$. We have previously concluded that Fourier components with $k_1 \approx 0$ become increasingly dominant; now it is also seen that $u_1$ is the dominant velocity component. In conjunction, these observations are that the theory predicts persistent growth of streamwise velocity fluctuations that have a long length in the $x_1$ direction, that is, it predicts streamwise streaks (see Figure 5.10, page 101).

The mechanism of streak formation is quite simple. For components with $k_1 \approx 0$,

$$\hat{u}_2 = \hat{u}_2^0 \qquad \text{and} \qquad \mathrm{d}_t\hat{u}_1 = S\hat{u}_2^0 = \hat{u}_2^0\,\mathrm{d}U/\mathrm{d}x_2 \tag{11.2.10}$$

according to (11.2.7). This equation for $\hat{u}_1$ simply equates its evolution to vertical displacement of mean momentum by the $u_2$ component of velocity. For these low-wavenumber components, growth of $\overline{u_1^2}$ is simply due to random displacement of mean momentum by cross-stream velocity fluctuations; the pressure gradient is negligible.

Equation (11.2.10) shows that $\mathrm{d}_t\overline{u_1^2} \to -2\overline{u_1u_2}S$ in this large-$St$ limit. It can be shown that $\overline{u_1u_2}/u_0^2 \to -\frac{1}{2}\log 8$ as $St \to \infty$ (Rogers, 1991). Thus the turbulent kinetic energy grows linearly with time, asymptoting to

$$k \sim k_0\, St \log 8,$$

and this growth is due to amplification of the streamwise component, $\overline{u_1^2}$. The qualitative behavior that $\overline{u_1^2}$ grows and $\overline{u_2^2}$ diminishes is more important than the precise mathematical asymptotes.

This overview of shear flow RDT provides an understanding of the structure of sheared turbulence. It tends to develop jet-like eddies with fluctuating velocity of the functional form $u_1(y, z)$. This is a physical interpretation of the dominance of the streamwise Reynolds stress component, $u_1$, and of the independence from the streamwise wavenumber, $k_1 \approx 0$. Visualizations of turbulent shear flow consistently show streamwise-elongated, jet-like eddies. The prevalence of such structure in shear flow turbulence is indeed remarkable. RDT provides a convincing explanation of this phenomenon.

### 11.2.2 Turbulence near a wall

This final section describes a linear analysis of kinematic blocking. The analysis of blocking is subsumed under the heading of *non-homogeneous* rapid distortion theory even though it does not involve turbulence distortion *per se*. It serves as a good example of the non-homogeneous theory.

The classical rapid distortion theory was developed for strictly homogeneous turbulence. Section 11.1.3 referred implicitly to a quasi-homogeneous approximation, when replacing the total strain by the local mean velocity (see Eqs. (11.1.23) and (11.1.24)). That serves as an introduction to non-homogeneous RDT, which body of work is reviewed in Hunt and Carruthers (1990). We conclude this chapter on rapid distortion by discussing a prototypical example of full non-homogeneity.

The kinematic wall effect is that the normal component of velocity must vanish on an impenetrable surface. This was mentioned in Section 7.3.4. For a plane, infinite boundary, the no-normal-velocity condition is accommodated by an image system of vorticity as in Figure 7.7, page 185. Turbulence is a three-dimensional, random field of vorticity. The idea of image vorticity gives some insight into wall effects.

A preliminary understanding can be had by considering a two-dimensional problem. The method of images solution for the complex conjugate velocity field of a point vortex at $(0, y_v)$ above a wall along $y = 0$ is

$$u - \mathrm{i}v = \frac{\mathrm{i}\gamma}{2\pi}\left[\frac{1}{x + \mathrm{i}(y + y_v)} - \frac{1}{x + \mathrm{i}(y - y_v)}\right]. \tag{11.2.11}$$

The second term is the vortex in the flow; the first is its image in the wall. Evaluating expression (11.2.11) on $y = 0$, we find

$$v = 0 \qquad \text{and} \qquad u = \frac{\gamma y_v}{\pi(x^2 + y_v^2)}.$$

Right beneath the vortex, at $x = 0$, the velocity tangential to the boundary is $u = \gamma/\pi y_v$. Without a wall, $u$ would equal $\gamma/2\pi y_v$ at this same point. The horizontal velocity is doubled and the normal velocity is canceled by the wall effect. In a boundary layer, shear production probably overwhelms enhancement of $\overline{u^2}$ by the image effect, but suppression of $\overline{v^2}$ by this mechanism very likely plays a role.

The full RDT analysis is of unsheared turbulence near a boundary (Hunt and Graham, 1978). Viscous effects are assumed to be confined to a very thin layer next to the wall, which is ignored.

The point vortex plus image, Eq. (11.2.11), is a solution to $\nabla^2\phi = 0$ with $\boldsymbol{u} = \nabla\phi$. The corresponding rapid distortion analysis is as follows. Let

$$\boldsymbol{u}_v = \iiint_{-\infty}^{\infty} \hat{\boldsymbol{u}}_v(\boldsymbol{k})\, \mathrm{e}^{-\mathrm{i}(kx+ly+mz)}\, \mathrm{d}k\, \mathrm{d}l\, \mathrm{d}m \tag{11.2.12}$$

be a field of homogeneous, vortical turbulence. A wall is instantaneously inserted at $y = 0$. Far from the wall, the turbulence remains homogeneous and given by $\boldsymbol{u}_v$. To satisfy the no-penetration condition, an irrotational turbulent field, $\boldsymbol{u}_p = \nabla\phi$, is added. The total velocity is $\boldsymbol{u} = \boldsymbol{u}_v + \boldsymbol{u}_p$.

The total velocity is divergence-free, as is the turbulence far from the wall, $\boldsymbol{u}_v$. Hence

$$0 = \nabla\cdot\boldsymbol{u} = \nabla\cdot\boldsymbol{u}_v + \nabla\cdot\boldsymbol{u}_p = \nabla^2\phi.$$

That is, the irrotational velocity potential, $\phi$, must solve

$$\nabla^2\phi = 0. \tag{11.2.13}$$

Consider a Fourier representation in directions of homogeneity:

$$\phi(x, y, z) = \iint_{-\infty}^{\infty} \tilde{\phi}(y)\, \mathrm{e}^{-\mathrm{i}(kx+mz)}\, \mathrm{d}k\, \mathrm{d}m. \tag{11.2.14}$$

Note that a tilde is used to represent a Fourier mode in $k$ and $m$, while a hat is used for the full 3D Fourier transform in (11.2.12). Using the representation (11.2.14) in the Laplace equation and differentiating under the integral give

$$\iint_{-\infty}^{\infty} [\partial_y^2\tilde{\phi} - (k^2 + m^2)\tilde{\phi}]\, \mathrm{e}^{-\mathrm{i}(kx+mz)}\, \mathrm{d}k\, \mathrm{d}m = 0.$$

As the Fourier modes are independent, the bracketed term itself must vanish:

$$\partial_y^2\tilde{\phi} - (k^2 + m^2)\tilde{\phi} = 0.$$

The solution to this equation, which is bounded as $y \to \infty$, is

$$\tilde{\phi} = \tilde{\phi}_0\, \mathrm{e}^{-\lambda y}, \tag{11.2.15}$$

where $\lambda^2 = k^2 + m^2$.

The wall condition is that the *total* normal velocity vanishes, $v = v_v + v_p = 0$; or, taking the 2D Fourier transform,

$$\tilde{v}_p = -\tilde{v}_v$$

on $y = 0$. From the representation (11.2.12), the vortical velocity is

$$\tilde{v}_v = \int_{-\infty}^{\infty} \hat{v}_v \, \mathrm{d}l$$

on $y = 0$. Then, from Eq. (11.2.15), the potential velocity is

$$\tilde{v}_p = \partial_y \tilde{\phi} = -\lambda \tilde{\phi}_0$$

on $y = 0$. Substituting these wall values into the above boundary condition gives

$$\lambda \tilde{\phi}_0 = \int_{-\infty}^{\infty} \hat{v}_v \, \mathrm{d}l. \tag{11.2.16}$$

The solution for $\phi$ is completed by substituting (11.2.15) with (11.2.16) into (11.2.14). It is now

$$\phi = \iiint_{-\infty}^{\infty} \frac{1}{\lambda} \hat{v}_v(\boldsymbol{k}) \, \mathrm{e}^{-\lambda y} \, \mathrm{e}^{-\mathrm{i}(kx+mz)} \, \mathrm{d}k \, \mathrm{d}l \, \mathrm{d}m.$$

With this, the Fourier representation of the full turbulent fluctuating velocity, $\boldsymbol{u}_v + \nabla\phi$, is

$$(u, v, w) = \iiint_{-\infty}^{\infty} \left\{ (\hat{u}_v, \hat{v}_v, \hat{w}_v) \, \mathrm{e}^{-\mathrm{i}ly} - \frac{(\mathrm{i}k, \sqrt{k^2+m^2}, \mathrm{i}m)}{\sqrt{k^2+m^2}} \hat{v}_v \, \mathrm{e}^{-\sqrt{k^2+m^2}\,y} \right\} \times \mathrm{e}^{-\mathrm{i}(kx+mz)} \, \mathrm{d}k \, \mathrm{d}l \, \mathrm{d}m. \tag{11.2.17}$$

This is the essence of the kinematic blocking solution. The rest of the RDT analysis consists in computing statistics. The methodology for the further analysis is that used previously for homogeneous turbulence. It will be illustrated by showing that blocking increases the tangential velocity by 50%, in the sense that $\overline{u^2}(y = 0) = \frac{3}{2}\overline{u^2}(y = \infty)$.

On $y = 0$ the solution (11.2.17) gives $v = 0$ and

$$u(0) = \iiint_{-\infty}^{\infty} \left( \hat{u}_v - \frac{\mathrm{i}k}{\sqrt{k^2+m^2}} \hat{v}_v \right) \mathrm{e}^{-\mathrm{i}(kx+mz)} \, \mathrm{d}k \, \mathrm{d}l \, \mathrm{d}m. \tag{11.2.18}$$

By "squaring" this and assuming that the turbulence far from the wall (i.e., $\hat{u}_v$) is isotropic, it can be shown that

$$\overline{u(0)^2} = \iiint_{-\infty}^{\infty} \Phi_{11} + \frac{k^2}{k^2+m^2} \Phi_{22} \, \mathrm{d}^3\boldsymbol{k}. \tag{11.2.19}$$

The contribution from $\Phi_{12}$ that would seem to arise from the bracketed term in the integrand of (11.2.18) vanishes upon integration. For the isotropic spectrum (9.2.17), page 259, the integration can be done in spherical coordinates,

$$\overline{u(0)^2} = \iiint_{-\infty}^{\infty} \frac{E(|\boldsymbol{k}|)}{4\pi|\boldsymbol{k}|^4}\left[(l^2+m^2) + \frac{k^2}{k^2+m^2}(k^2+m^2)\right] \mathrm{d}^3\boldsymbol{k}$$

$$= \int_0^{\infty} \frac{E(\kappa)}{4\pi\kappa^2} 4\pi\kappa^2 \,\mathrm{d}\kappa = \tfrac{1}{2}q^2,$$

where $\kappa^2 = k^2 + l^2 + m^2$.

The turbulence at infinity is isotropic, which means that $\overline{u_\infty^2} = \frac{1}{3}q^2$. Thus,

$$\overline{u^2}(0) = \tfrac{3}{2}\overline{u_\infty^2},$$

as was to be shown. The intensity at the wall is 50% higher than the intensity far above the wall. This result is consistent with the point vortex reasoning below Eq. (11.2.11), but the 50% amplification has to be derived by statistical averaging over the random velocity field (Hunt and Graham, 1978).

When $y$ is not zero, the full solution (11.2.17) is used. To compute the variances, each component of the integrand is multiplied by its complex conjugate and averaged. This gives

$$\overline{u^2} = \overline{w^2} = \iiint_{-\infty}^{\infty} \left\{ \Phi_{11} + \Phi_{22}\,\mathrm{e}^{-2\sqrt{k^2+m^2}\,y} \frac{k^2}{k^2+m^2} + \frac{2k\sin(ly)}{\sqrt{k^2+m^2}}\Phi_{12}\,\mathrm{e}^{-\sqrt{k^2+m^2}\,y} \right\} \mathrm{d}k\,\mathrm{d}l\,\mathrm{d}m, \qquad (11.2.20)$$

$$\overline{v^2} = \iiint_{-\infty}^{\infty} \Phi_{22} \left\{ 1 + \mathrm{e}^{-2\sqrt{k^2+m^2}\,y} - 2\cos(ly)\,\mathrm{e}^{-\sqrt{k^2+m^2}\,y} \right\} \mathrm{d}k\,\mathrm{d}l\,\mathrm{d}m,$$

and $\overline{uv}$ is zero because there is no mean shear. The turbulence is axisymmetric about the $y$ axis. The integrals (11.2.20) have been computed using the isotropic form (9.2.17) for $\Phi_{ij}$, with the Von Karman spectrum (9.2.19), defined on page 260, for $E(\boldsymbol{k})$. The result is plotted in Figure 11.8. Note that the non-homogeneous solution for wall blocking does depend on the shape of the energy spectrum, unlike the case of homogeneous RDT.

If $L_\infty$ is the integral scale of the turbulence far above the wall, then $\overline{v^2}$ starts to be suppressed at a height of order $(L_\infty)$ above the surface, and tends to 0 at the wall. The horizontal intensities, $\overline{u^2} = \overline{w^2}$, are amplified in the same region. The behavior is qualitatively consistent with a simple understanding of the influence of image vorticity. As in other RDT analyses, the conceptual understanding of wall blocking is as important as the detailed solution.

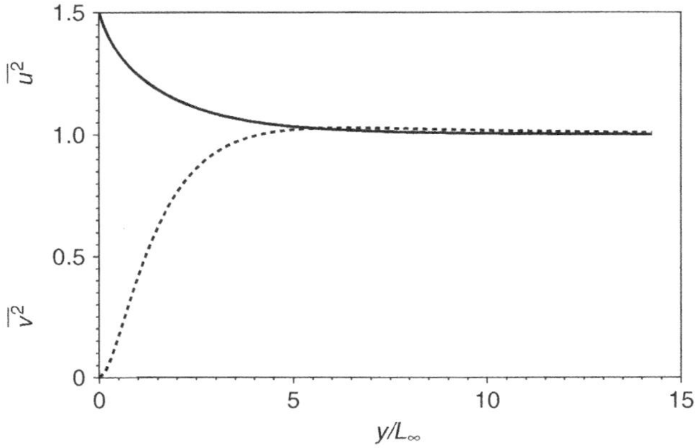


**Figure 11.8** Kinematic solutions for $\overline{v^2}$ (- - - -) and $\overline{u^2} = \overline{w^2}$ (——) near a plane boundary.

# Exercises

**Exercise 11.1.** *Turbulence in a piston; 1D compression.* RDT analysis can be applied to a unidirectional compression (Hunt, 1977). This models the effect of compression on turbulence in a piston (away from the walls, because the analysis is of homogeneous turbulence).

Consider the mean flow

$$U = \frac{x\,\mathrm{d}H/\mathrm{d}t}{H(t)}$$

corresponding to Figure 11.9. Show that $s_1 = H/H_0$, that is, that the total strain is just the expansion ratio. This permits the solution to be written in terms of the physical dimension $H$ instead of time.

Show that the distorted wavevector is

$$\kappa_1 = \frac{H_0}{H}k_1, \qquad \kappa_2 = k_2, \qquad \kappa_3 = k_3,$$

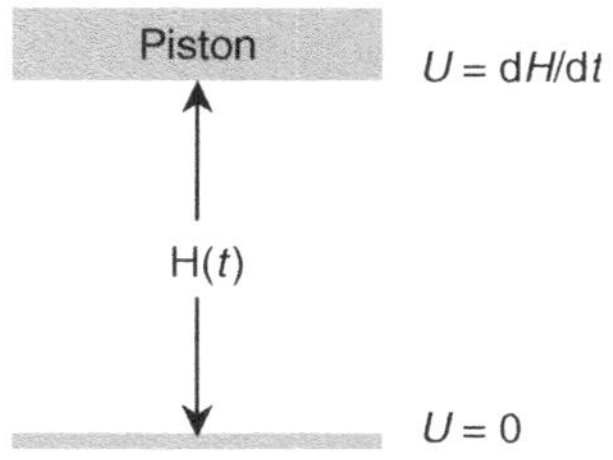


**Figure 11.9** Schematic for compression of turbulence by a piston.

and, from Cauchy's formula, that

$$\hat{\omega}_1 = \hat{\omega}_{0_1}, \qquad \hat{\omega}_2 = \frac{H_0}{H}\hat{\omega}_{0_2}, \qquad \hat{\omega}_3 = \frac{H_0}{H}\hat{\omega}_{0_3}.$$

By invoking (11.1.16) obtain a complete solution for the component $\hat{u}_1$ and obtain the integrand in

$$\overline{u_1^2} = \iiint_{-\infty}^{\infty} \overline{\hat{u}_1\hat{u}_1^*}\, \mathrm{d}^3k$$

explicitly. Evaluate the integral for initially isotropic turbulence.

**Exercise 11.2.** *Homogeneous shear RDT.* Use a two-term series solution to (11.2.7), starting with isotropic turbulence, to obtain the short-time solutions (11.2.9) for $\overline{v^2}$ and $\overline{uv}$.

# Part IV

# TURBULENCE SIMULATION

# 12

# Eddy-resolving simulation

> I am an old man now, and when I die and go to heaven there are two matters on which I hope for enlightenment. One is quantum electrodynamics, and the other is the turbulent motion of fluids. And about the former I am rather optimistic
>
> – Sir Horace Lamb

In 1932, Sir Horace Lamb lamented the incomprehensibility of the turbulent motion of fluids. Even at that time, it was supposed that turbulence was governed by the same laws as laminar flow; but how equations of such apparent simplicity could invoke such complex behavior was a puzzle. In the course of time, the evidence that the Navier–Stokes equations are the appropriate laws of physics has become indisputable. Now, with sufficient computing power, turbulence can be simulated via numerical solution to these equations. Highly resolved, time-accurate computations are able to simulate the chaotic eddying motions of turbulent flow. This is not the enlightenment sought by Horace Lamb; but it does mean that we know the governing laws, and have to accept that they have very complex, deterministic solutions that can be found numerically.

Computer simulation provides a laboratory in which to study turbulence. Computations with grid spacings that are small enough to resolve Kolmogoroff scale eddies are called direct numerical simulations (DNS). That terminology is synonymous with "fully resolved simulation:" the spacing and number of grid points suffice to capture the entire spectrum of scales.

Fully resolved simulation is quite demanding of computer resources. A good part of the expense is incurred in capturing the smallest scales. The role of those scales is to dissipate fluctuation energy. Expense can be reduced by capturing the largest scales and using a dissipative model in place of the smaller eddies. This is called large eddy simulation (LES). For instance, if the grid spacing is five times that of direct simulation in each of three directions, the number of grid points is reduced by a factor of 125. To the extent that the dissipative model does not contaminate the large scales, LES can provide Navier–Stokes simulations of satisfactory accuracy for many purposes. However, the

*Statistical Theory and Modeling for Turbulent Flows, Second Edition* P. A. Durbin and B. A. Pettersson Reif

results depend inherently on the grid. Selecting a grid that is coarse enough for efficiency and fine enough for accuracy is an art.

Near to walls, a separation between large- and small-scale eddies becomes impossible. Accurate large eddy simulation requires the same grid resolution as direct numerical simulation within this region. To avoid the expense of large grids, the notion of hybrid RANS–LES has been developed. The near-wall region is represented by a formulation suited to Reynolds averaged description; away from the wall, large eddies are simulated. Detached eddy simulation (DES) is a version of this hybrid approach.

The two chapters in Part IV describe aspects of these various versions of eddy-resolving simulation.

## 12.1 Direct numerical simulation

### 12.1.1 Grid requirements

The endeavor to simulate turbulence by direct numerical solution of the Navier–Stokes equations is predicated on the availability of suitable algorithms and a fine enough discretization of space and time to faithfully reproduce turbulence dynamics. It also demands sufficiently powerful computers.

The overall computational cost depends strongly on the number of discrete points in space and the number of discrete steps in time. These numbers grow rapidly with Reynolds number. Conventional estimates of grid requirements have been developed from ideas that were presented in earlier chapters of this text.

Our discussion of direct simulation starts with estimates of grid requirements. First we consider homogeneous turbulence, then turbulent boundary layers.

The subject of Exercise 2.2 is Reynolds number scaling of grid points and time steps in homogeneous turbulence. Let $N_x$, $N_y$, and $N_z$ denote the number of grid points in the $x$, $y$, and $z$ directions; then $N = N_x N_y N_z$ is the total number of points. The dashed grid in Figure 12.1 schematically represents the resolution required to capture both large- and small-scale eddies. The grid spacing must be on the order of the Kolmogoroff scale (2.1.6). The number of points must suffice to capture several of the largest-scale eddies.

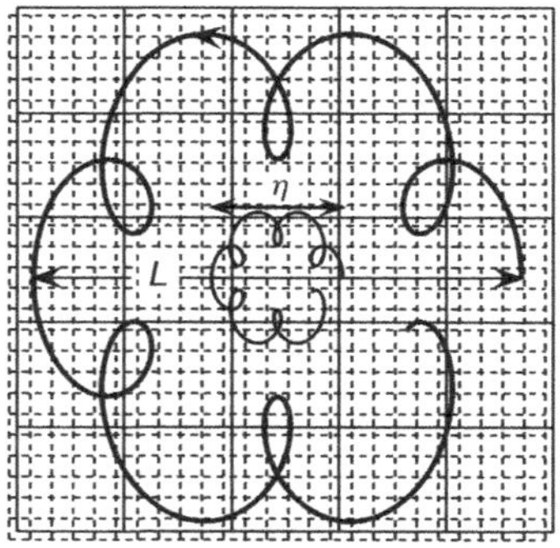


**Figure 12.1** Schematic of DNS and LES grid resolution. Dashed lines are a DNS grid; solid lines are an LES grid.

Thus

$$N_x \gg \frac{L}{\eta} = \frac{L}{(\nu^3/\varepsilon)^{1/4}} \sim R_{\mathrm{T}}^{3/4}.$$

This is true for all three directions. Thus, $N \sim N_x^3$ grows with Reynolds number as

$$N \sim R_{\mathrm{T}}^{9/4}.$$

Similarly, the number of time steps grows as $R_{\mathrm{T}}^{1/2}$. The overall expense scales as $R_{\mathrm{T}}^{11/4}$, or almost the third power of Reynolds number.

Historically, the rapid increase of expense restricted direct simulation to low Reynolds numbers. However, at the time of writing, simulation of isotropic turbulence with $4028^3 \sim 6.5 \times 10^{10}$ grid points has become feasible. Reynolds numbers of order $R_{\mathrm{T}} \sim 10^5$–$10^6$ can be computed. This brings computer simulation of homogeneous turbulence up to typical laboratory Reynolds numbers. The large number of grid points is feasible because highly efficient, pseudo-spectral methods can be used to simulate spatially homogeneous turbulence (Canuto *et al.*, 2006).

Such is not the case for inhomogeneous flows. Consider a channel flow, or flat-plate boundary layer. The grid requirements are governed by statistical correlation lengths; however, reference to streak widths and wall units provides a more intuitive perspective and leads to the same conclusions. Streaks have a spacing of about 100 plus units in the direction across the span – the $z$ direction (Section 5.1.2). If 20 grid points are needed to resolve a streak, $\Delta z_+ \sim 5$. If 10 steaks are needed to represent statistical homogeneity, $N_z \sim 200$. This number is typical of boundary-layer and channel flow DNS.

Streaks are long in the streamwise, or $x$, direction. However, to capture their irregularities requires about the same spacing as that in the $z$ direction: $\Delta x_+ \approx \Delta z_+ \sim 5$. The streak length is about 1000 plus units. To capture four streaks requires $N_x \sim 800$. The long streamwise correlation length creates a demand for large $N_x$, that often is not met (Del Alamo and Jiménez, 2003). However, tests show that local statistics are not highly sensitive to streamwise domain length.

If $N_y \approx N_z$, the overall requirement is of order $3 \times 10^7$; this is typical of channel flow simulations to date (Moser *et al.*, 1999). But the estimate should scale with Reynolds number. Consider how such an estimate might proceed.

If $\Lambda_+^z$ is the spanwise correlation length and the domain width is a fixed multiple of this, say $10\Lambda_+^z$, then $N_z = 10\Lambda_+^z/\Delta z_+$ is independent of Reynolds number. However, if the domain width is fixed, say $W$, then $N_z = W/\Delta z = W u_*/\nu \Delta z_+$. In boundary layers, Eq. (4.2.8) on page 68 states the empirical relation $C_{\mathrm{f}} \propto R_\theta^{-1/4}$. Because $u_* = U_\infty \sqrt{2C_{\mathrm{f}}}$,

$$N_z \sim U_\infty W \sqrt{2C_{\mathrm{f}}}/\nu \Delta z_+ \sim (W/\Theta) R_\theta^{7/8}$$

for a fixed $\Delta z_+$ (Chapman, 1979). The same argument applies in the $x$ direction: $N_x \sim R_\theta^{7/8}$.

The wall normal direction, $y$, is more ambiguous. Adjacent to the wall, $\Delta y_+ < 1$ is needed to capture the viscous sublayer. Grid spacing expands slowly until it reaches $\Delta y_+ \sim 5$ in the law-of-the-wake region. If it expands to some fixed $y_+$ in the log region, then $N_y \sim y_+/\Delta y_+$ is independent of Reynolds number. If most grid points are below this level, then $N = N_x N_y N_z \sim R_\theta^{7/4}$. However, the number of grid points in the outer portion of the boundary layer increases with Reynolds number in the same manner as

$N_z$. So the $y$ grid splits into two parts that scale as $R_\theta^0$ and as $R_\theta^{7/8}$. The overall Reynolds number scaling is in the range

$$N \sim R_\theta^{7/4}\text{–}R_\theta^{21/8}. \tag{12.1.1}$$

Turbulent boundary-layer structure is fully established for $R_\theta \gtrsim 3000$; hence, direct simulations of boundary-layer and channel flow to date have required $10^6$–$10^9$ grid points. The larger grids create quite stringent demands for computer time.

Computational expense is a function of both the number of grid points and the efficiency of the solution algorithm. Spectral methods are accurate and are efficient when fast Fourier transforms (FFTs) are used. FFTs are applicable if the grid spacing is uniform along a Cartesian direction. Homogeneous turbulence can be simulated on a grid that is uniformly spaced in all three directions. Highly efficient pseudo-spectral methods permit large grids.

To understand the origin of computational expense, consider solving Poisson's equation $\nabla^2 p = S$: for instance, a Poisson equation is solved for the pressure in incompressible flow. It can be solved exactly by Fourier transforms (see page 271):

$$p = -\mathcal{F}_k^{-1}\left(\frac{\mathcal{F}_k(S)}{|\boldsymbol{k}|^2}\right).$$

The equation is Fourier-transformed and solved by dividing by $-|\boldsymbol{k}^2|$, then inverse-transformed.

If the grid is uniform in two directions but not in the third, the Poisson equation can be Fourier-transformed in those two directions, providing an ordinary differential equation in the third:

$$\mathrm{d}_y^2 \hat{p} - (k_x^2 + k_y^2)\hat{p} = \hat{S}.$$

This can be solved directly, but pressure cannot be eliminated from the momentum equation. The efficiency is less than in homogeneous turbulence, but still relatively high. Grids that are uniform in two spatial directions arise in channel flow simulations.

When the flow is inhomogeneous in two or three directions, non-uniform grids are required. The Poisson equation cannot be solved directly; iterative methods are required, and the computational time increases in proportion to the number of iterations. Thus, the solution algorithms for non-homogeneous turbulence are greatly more costly per time step than those for homogeneous turbulence. The expense usually scales with the total number of grid points (times $\log N$ for FFTs) *times* the number of iterations, or substeps per time step. Iteration adds expense. The increased expense per time step must be compensated by a smaller grid. That is why the number of grid points for inhomogeneous turbulence simulation must be one or two orders of magnitude fewer than for homogeneous turbulence for numerical simulation to be feasible.

### 12.1.2 Numerical dissipation

Direct numerical simulation is the art of accurately solving the full, time-dependent Navier–Stokes equations. To capture the smallest scales of motion, the computational algorithm should have low numerical dissipation. Indeed, too much numerical dissipation can cause turbulence to decay spuriously, confounding the effort to simulate it.

One way to characterize numerical dissipation is to apply the discretization to the equation

$$\partial_t \phi + c\,\partial_x \phi = 0, \tag{12.1.2}$$

with $\phi = \mathrm{e}^{ikx}$ at $t = 0$. The exact solution is $\phi = \mathrm{e}^{ik(x-ct)}$. Consider a discrete approximation of the $x$ derivative:

$$\partial_t \phi + c\,\delta_x \phi = 0, \tag{12.1.3}$$

with the notation that $\delta_x$ is a discrete approximation of $\partial_x$. For instance, it might be the central difference formula $\delta_x \phi = [\phi(j+1) - \phi(j-1)]/2\Delta x$. If the solution to this discrete equation has the form

$$\phi = \mathrm{e}^{ik(x-c_{\mathrm{eff}}t)-\sigma t}, \tag{12.1.4}$$

the difference between $c_{\mathrm{eff}}$ and $c$ in (12.1.3) is called a dispersive error and $\sigma$ is called the dissipative error.

Centered difference discretization only has a dispersive error. For instance, consider a set of uniformly spaced grid points, $x = [0, \Delta x, 2\Delta x, \ldots, j\Delta x]$ and let

$$\delta_x \phi = \frac{\phi(j+1) - \phi(j-1)}{2\Delta x}. \tag{12.1.5}$$

It is found by substituting (12.1.4) into (12.1.3) that

$$ikc_{\mathrm{eff}} + \sigma = c\,\frac{\mathrm{e}^{ik\Delta x} - \mathrm{e}^{-ik\Delta x}}{2\Delta x}.$$

Then the solution to (12.1.3) has $c_{\mathrm{eff}} = c\sin(k\Delta x)/(k\Delta x)$ and $\sigma = 0$. The dispersive error is characterized alternatively by the effective wavenumber

$$k_{\mathrm{eff}} = k\frac{c_{\mathrm{eff}}}{c} = \frac{\sin(k\Delta x)}{\Delta x}.$$

For small $\Delta x$, $k_{\mathrm{eff}} \approx k$; for larger $\Delta x$, $k_{\mathrm{eff}} < k$. The exact derivative of $\phi = \mathrm{e}^{ikx}$ is $ik\phi$; the numerical derivative is $ik_{\mathrm{eff}}\phi$. It is in substantial error unless $\Delta x$ is sufficiently small. An objective of high-accuracy discretization methods is to make $k_{\mathrm{eff}}$ close to $k$ for large $\Delta x$.

If, instead of (12.1.5), the derivative is discretized as

$$\delta_x \phi = \frac{\phi(j+1) - \phi(j)}{\Delta x}, \tag{12.1.6}$$

then $k_{\mathrm{eff}} = \sin(k\Delta x)/\Delta x$, again, but $\sigma = c[1 - \cos(k\Delta x)]/\Delta x$. Now there is a dissipative error. The approximation (12.1.6) is called a first-order finite-difference approximation because $\sigma \to ck^2\Delta x/2$ as $\Delta x \to 0$: the error diminishes as $\Delta x$ to the first power.

The discretization

$$\delta_x \phi = \frac{2\phi(j+1) + 3\phi(j) - 6\phi(j-1) + \phi(j-2)}{6\Delta x} \tag{12.1.7}$$

is a third-order, upwind approximation, for which

$$\sigma = c[3 - 4\cos(k\Delta x) + \cos(2k\Delta x)]/(6\Delta x) \to O(k\Delta x)^3 \qquad \text{as } \Delta x \to 0.$$

The common element of (12.1.6) and (12.1.7) is that they are asymmetric about the grid point $x_j$; for example, the points $[j-2, j-1, j, j+1]$ are biased to the left of $j$. On a uniformly spaced grid, *asymmetric* schemes have a dissipative error, whereas *symmetric* schemes have only dispersive error.

Centered, symmetric discretization is more suited to direct numerical simulation because it is not dissipative on a uniform grid. However, it can be unstable, or can lead to large grid-point to grid-point oscillations (called $2\Delta$ waves). Often, some degree of numerical dissipation is needed; the demands of DNS require it to be minimized, subject to constraints of stability and fidelity.

### 12.1.3 Energy-conserving schemes

Conservation of kinetic energy provides another perspective on discretization schemes for turbulence simulation. If kinetic energy is conserved, then the scheme will not dissipate turbulence spuriously. Also, the method will be stable because the total kinetic energy is bounded by its initial value. Even though the exact, inviscid equations conserve kinetic energy, such may not be true of a numerical approximation.

The exact, inviscid momentum and continuity equations are (page 46)

$$\begin{aligned} \partial_t \tilde{u}_i + \tilde{u}_j\, \partial_j \tilde{u}_i &= -\frac{1}{\rho}\, \partial_i \tilde{p}, \\ \partial_i \tilde{u}_i &= 0. \end{aligned} \tag{12.1.8}$$

Energy conservation is derived from these. The product of the momentum equation with $u_i$, summed over $i$, gives

$$\partial_t \frac{1}{2}|\tilde{\boldsymbol{u}}|^2 + \partial_j\left(\tilde{u}_j \frac{1}{2}|\tilde{\boldsymbol{u}}|^2\right) = -\frac{1}{\rho}\, \partial_i \tilde{u}_i \tilde{p}$$

upon invoking the second equation of (12.1.8). Letting $K \equiv \frac{1}{2}|\tilde{\boldsymbol{u}}|^2$ and writing this in vector form gives

$$\partial_t \rho K + \nabla\cdot(\tilde{\boldsymbol{u}}\rho K) + \nabla\cdot(\tilde{\boldsymbol{u}}\tilde{p}) = 0. \tag{12.1.9}$$

The two terms that are divergences transport kinetic energy without creating or destroying it. Integrated over the flow domain, they become equal to the flux of energy into the domain, minus the flux out. Energy is conserved within the domain. Let us see whether this carries over to numerical schemes.

To isolate convection, consider the quantity

$$\boldsymbol{u}\cdot\nabla\phi. \tag{12.1.10}$$

Then, for a divergence-free velocity field,

$$\phi\boldsymbol{u}\cdot\nabla\phi = \tfrac{1}{2}\boldsymbol{u}\cdot\nabla\phi^2 = \tfrac{1}{2}\nabla\cdot\boldsymbol{u}\phi^2.$$

The last is a conservative form because it is written as the divergence of a flux.

The question of whether numerical schemes conserve energy arises because the steps of this derivation are not generally respected by discretized equations. For instance, central differencing

$$\tilde{u}\,\delta_x\phi = \tilde{u}(j)\frac{\phi(j+1)-\phi(j-1)}{2\Delta x}$$

implies that $\phi \cdot (\tilde{u}\,\delta_x\phi) \neq \frac{1}{2}\tilde{u}\,\delta_x\phi^2$. Explicitly,

$$\begin{aligned}\phi \cdot (\tilde{u}\,\delta_x\phi) &= \phi(j)\tilde{u}(j)\frac{\phi(j+1)-\phi(j-1)}{2\Delta x}\\ &\neq \frac{1}{2}\tilde{u}\,\delta_x\phi^2 = \tilde{u}(j)\frac{\phi^2(j+1)-\phi^2(j-1)}{4\Delta x}.\end{aligned}$$

By analogy, the convective term in (12.1.9) would not be in conservation form in the discrete approximation.

A resolution, originally introduced by Arakawa (see Morinishi *et al*., 1998), is to discretize as

$$\tilde{u}_i\,\partial_i\phi = \tfrac{1}{2}\tilde{u}_i\,\delta_i\phi + \tfrac{1}{2}\delta_i(\tilde{u}_i\phi),$$

which goes under the name of the "skew form." Continuity is invoked to show that $\delta_i\tilde{u}_i\phi = \tilde{u}_i\,\delta_i\phi$, so the two sides of this equation are equivalent.

For the skew form, $\phi \cdot \tilde{u}_i\,\delta_i\phi$ becomes

$$\begin{aligned}\phi \cdot \tilde{u}\,\delta_x\phi = \frac{\phi(j)}{2}\Bigg(&\tilde{u}(j)\frac{\phi(j+1)-\phi(j-1)}{2\Delta x}\\ &+\frac{\tilde{u}(j+1)\phi(j+1)-\tilde{u}(j-1)\phi(j-1)}{2\Delta x}\Bigg),\end{aligned} \tag{12.1.11}$$

plus similar terms in the $y$ and $z$ directions. If

$$F(j+\tfrac{1}{2}) \equiv \tfrac{1}{4}[u(j)\phi(j)\phi(j+1) + u(j+1)\phi(j)\phi(j+1)],$$

then the right-hand side of (12.1.11) has the conservation form

$$\frac{F(j+\frac{1}{2})-F(j-\frac{1}{2})}{\Delta x}.$$

Hence, the discretized convection equation

$$\partial_t\phi + u_i\,\partial_i\phi = 0$$

conserves variance because it gives the equation

$$\tfrac{1}{2}\,\partial_t\phi^2 = -\phi u_i\,\delta_i\phi = -\frac{F(j+\frac{1}{2})-F(j-\frac{1}{2})}{\Delta x}.$$

for the variance. Global conservation is confirmed by summing: in one dimension

$$\Delta x\sum_{j=1}^{J}\partial_t\phi_j^2 = -2\sum_{j=1}^{J}F(j+\tfrac{1}{2}) - F\left(j-\tfrac{1}{2}\right).$$

But

$$\sum_{j=1}^{J} F(j+\tfrac{1}{2}) - F(j-\tfrac{1}{2}) = F(J+\tfrac{1}{2}) - F(\tfrac{1}{2}). \qquad (12.1.12)$$

Thus, the net flux is that crossing the boundaries. In directions of statistical homogeneity, periodic boundary conditions are applied, so the right-hand side vanishes. Then the integrated variance is constant in time.

It follows by similar reasoning that

$$\tilde{u}_j \partial_j \tilde{u}_i = \tfrac{1}{2}\big(\tilde{u}_j\, \delta_j \tilde{u}_i + \delta_j(\tilde{u}_j \tilde{u}_i)\big) \qquad (12.1.13)$$

is an energy-conserving form of the convection term of the momentum equation.

Another approach to energy conservation for (12.1.8) is to rewrite convection in rotational form:

$$\tilde{u}_j\, \partial_j \tilde{u}_i = \tilde{u}_j(\partial_j \tilde{u}_i - \partial_i \tilde{u}_j) + \tfrac{1}{2}\, \partial_i \tilde{u}_j^2. \qquad (12.1.14)$$

This ensures that kinetic energy is conserved in incompressible flow, as follows. The first term on the right is the cross-product of velocity and vorticity, $\tilde{\boldsymbol{u}} \wedge \boldsymbol{\omega}$. From this, or from antisymmetry in $i$ and $j$ of the rotation tensor, it follows by taking the product of (12.1.14) with $\tilde{u}_i$ that

$$\tilde{u}_i \tilde{u}_j\, \partial_j \tilde{u}_i = \tfrac{1}{2} \tilde{u}_j\, \partial_j |\tilde{\boldsymbol{u}}|^2.$$

So conservation hinges on putting the last term in divergence form. That is the same as putting the pressure contribution into divergence form, as in (12.1.9); in fact, $|\tilde{\boldsymbol{u}}|^2/2$ is just the dynamic pressure, divided by density.

Consider the finite-difference representation

$$\tilde{u}\, \delta_x \tilde{p} = \tilde{u}(i)\left(\frac{\tilde{p}(i+1) - \tilde{p}(i-1)}{2\Delta x}\right).$$

Adding $p(i)$ times the continuity equation, discretized as

$$\delta_j u_j = \frac{\tilde{u}(i+1) - \tilde{u}(i-1)}{2\Delta x} + \frac{\tilde{v}(j+1) - \tilde{v}(j-1)}{2\Delta y} + \frac{\tilde{w}(k+1) - \tilde{w}(k-1)}{2\Delta z}, \qquad (12.1.15)$$

to this gives

$$\begin{aligned}\tilde{u}\, \delta_x \tilde{p} &= \tilde{u}(i)\left(\frac{\tilde{p}(i+1) - \tilde{p}(i-1)}{2\Delta x}\right) + \tilde{p}(i)\left(\frac{\tilde{u}(i+1) - \tilde{u}(i-1)}{2\Delta x}\right)\\ &= \left(\frac{\tilde{u}(i)\tilde{p}(i+1) + \tilde{u}(i+1)\tilde{p}(i)}{2\Delta x}\right) - \left(\frac{\tilde{u}(i-1)\tilde{p}(i) + \tilde{u}(i)\tilde{p}(i-1)}{2\Delta x}\right),\end{aligned}$$

and similarly in the $y$ and $z$ directions. If

$$G_x(i+\tfrac{1}{2}) \equiv \tfrac{1}{2}[\tilde{u}(i)p(i+1) + \tilde{u}(i+1)\tilde{p}(i)],$$

then the pressure term assumes the divergence form

$$\begin{aligned}\tilde{u}_i\,\delta_i\,\tilde{p} &= \frac{G_x(i+\frac{1}{2}) - G_x(i-\frac{1}{2})}{\Delta x} + \frac{G_y(j+\frac{1}{2}) - G_y(j-\frac{1}{2})}{\Delta y}\\ &\quad + \frac{G_z(k+\frac{1}{2}) - G_z(k-\frac{1}{2})}{\Delta z}\\ &\equiv \delta_i G_i.\end{aligned}$$

So energy conservation is obeyed if the discrete continuity equation (12.1.15) is satisfied.

## 12.2 Illustrations

Some numerical concepts will be illustrated via an equation that is much simpler than Navier–Stokes. The Burgers equation is

$$\partial_t u + u\,\partial_x u = \partial_x(\nu\,\partial_x u). \tag{12.2.1}$$

Its convective nonlinearity has some semblance to the Navier–Stokes momentum equation, but there is no pressure gradient. The domain is $-1/2 \le x \le 1/2$. The initial condition is $u = 1 + \sin(2\pi x)$ and periodic conditions are applied:

$$u(-\tfrac{1}{2} - \xi) = u(\tfrac{1}{2} - \xi), \qquad u(\tfrac{1}{2} + \xi) = u(-\tfrac{1}{2} + \xi).$$

Thus a point outside the range $-1/2 \le x \le 1/2$ that appears in a finite-difference formula can be replaced by one inside the range. Variables are non-dimensional, so the viscosity is equivalent to $1/\mathrm{Re}$.

First derive the equation for kinetic energy. In differential form,

$$\tfrac{1}{2}\,\partial_t u^2 + \tfrac{1}{3}\,\partial_x u^3 = \partial_x(\nu\,\partial_x \tfrac{1}{2}u^2) - \nu(\partial_x u)^2$$

follows from (12.2.1) upon multiplication by $u$. If $\nu = 0$, energy is conserved. If it is not zero, the last term on the right dissipates energy.

Next consider the discrete form. We will use Arakawa's skew treatment of convection:

$$\begin{aligned}u\,\partial_x u &\approx a u\,\delta_x u + (1-a)\,\delta_x \tfrac{1}{2}u^2\\ &= a\frac{u(j)}{2\Delta x}[u(j+1) - u(j-1)] + (1-a)\frac{1}{4\Delta x}[u(j+1)^2 - u(j-1)^2].\end{aligned}$$

If $a = 1/3$ this becomes

$$\frac{u(j)}{6\Delta x}[u(j+1) - u(j-1)] + \frac{1}{6\Delta x}[u(j+1)^2 - u(j-1)^2]. \tag{12.2.2}$$

Multiplying by $u(j)$ shows this to be energy-conserving, with the flux function

$$F(j+\tfrac{1}{2}) = \tfrac{1}{6}[u^2(j)u(j+1) + u(j)u(j+1)^2].$$

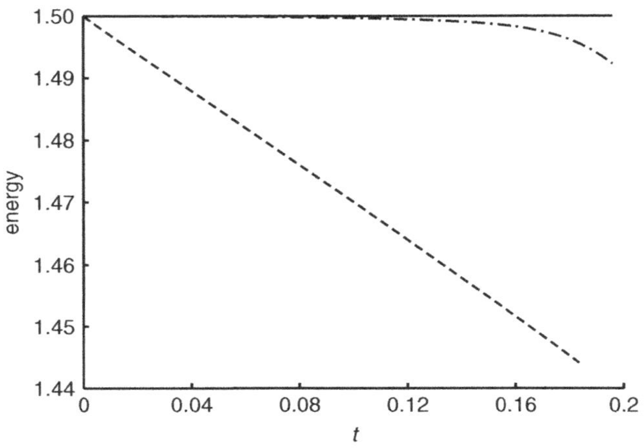


**Figure 12.2** Energy versus time for the Burgers equation with skew (——), central (— · —), and upwind (- - - -) differencing.

Figure 12.2 compares the treatment (12.2.2) to the central and upwind forms,

$$\frac{1}{4\Delta x}[u(j+1)^2 - u(j-1)^2] \qquad \text{and} \qquad \frac{1}{\Delta x}u(j)[u(j) - u(j-1)].$$

The total number of grid points is $J = 128$, with periodicity: $u(128) = u(1)$. The viscosity is set to zero in order to illustrate energy conservation. The discrete equations were integrated by fourth-order Runge–Kutta (see Exercise 12.2). The integrated energy

$$\Delta x \sum_{j=1}^{J-1} u(j)^2 \tag{12.2.3}$$

is plotted versus time. It is constant for the skew form, but decreases for the other two. The first-order, upwind scheme has a dissipative error and energy decays rapidly. The central form initially conserves energy; but, as time progresses, the Burgers equation develops steep gradients, which cause energy to decay.

Next, the Burgers equation is solved with the energy-conserving scheme and with the random initial condition

$$u = 1 + \tfrac{1}{2}[1 + \xi(x)]\sin(2\pi x),$$

where $\xi$ is a random number between 0 and 1. This is an analogy to DNS. Profiles at two times are illustrated by the dashed curves in Figure 12.3.

For future reference, in addition to the raw data, data averaged according to (13.1.1) with $N = 2$ are shown by the solid curves in Figure 12.3. The running average removes short wavelength wiggles, leaving smoother velocity profiles. The smoother field is the subject of the next section, on large eddy simulation. Exercise 12.2 asks the reader to explore DNS of the Burgers equation.

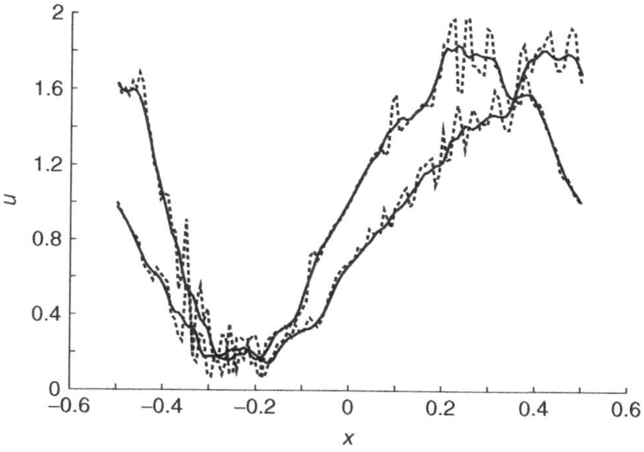


**Figure 12.3** Velocity profiles at two times for the Burgers equation. Instantaneous (- - - -) and filtered (———) fields.

The numerical techniques and resolution requirements described in this chapter are the starting point of direct simulation. To go further is beyond our present scope. The discussion of direct simulation concludes by showing a few examples of instantaneous fields of eddies. Other illustrations of DNS fields can be found in Chapters 1 and 5.

Figure 12.4 is from a spectral simulation of isotropic turbulence: the pseudo-spectral method is the topic of the next section. The grid is $128^3$. The Reynolds averaged description is rather less intriguing than this instantaneous flow field: it is $k \propto t^{-1.22}$ (see discussion of the decay exponent on pages 124 and 273). This simulation is at the

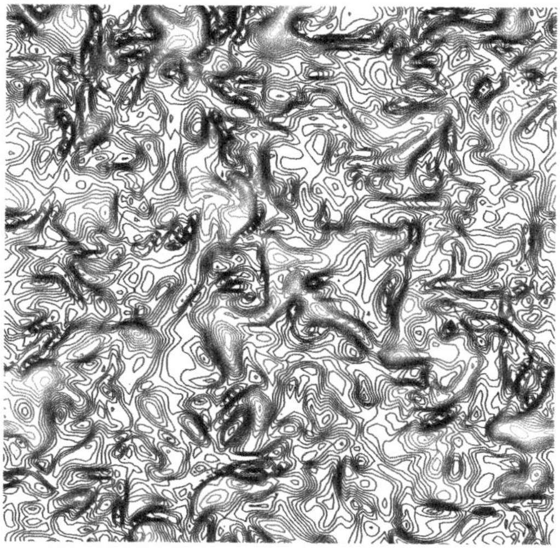

**Figure 12.4** DNS of homogeneous, isotropic turbulence. Contours of velocity magnitude. Courtesy of Dr. Yang Liu.

microscale Reynolds number $R_\lambda = 27.13$, which is low compared to typical laboratory data. Nevertheless, one can detect large and small features in the contours. DNS captures the full range of eddy scales.

Vorticity contours in a DNS of plane channel flow are illustrated by Figure 12.5. A variation in the scales of turbulence can be seen. The eddies are large in the center of the channel and become smaller near the walls. Figure 12.6 displays elongated velocity contours just above the wall, at $y_+ = 10$. There, $x$ is the streamwise direction and $z$ is the spanwise direction. Contours of the $u$ component emphasize elongated streaks. The $v$ contours contain substantial irregularities of a size much less than the streak length. The streamwise grid resolution must be adequate to capture these small scales; hence it cannot be significantly coarser than the resolution in the spanwise direction. Moser *et al.* (1999) provide other data and the grid requirements for simulations at various Reynolds numbers. Statistics from a channel flow DNS are displayed in Figure 4.3.

A wide range of eddy sizes is captured in Figure 12.7. This is from a simulation of flow in a channel that has ribs along the lower wall. The ribs lift the near-wall streaks away from the surface, producing small scale features higher in the flow field. The upper wall is plane. Large eddies are seen in the center of the channel. In this example, the geometrical complexity is minimal; however, the turbulence structure is complex and computationally demanding. The grid for this simulation was $1024 \times 352 \times 192$ in the streamwise, cross-stream, and spanwise directions.

Direct simulation has proved to be especially effective for studying transition from laminar to turbulent flow. By its nature, transition occurs at relatively low Reynolds number, which reduces the demand for large grids. Early studies did not progress beyond nonlinear growth of instability waves and were even less demanding. More recently, it

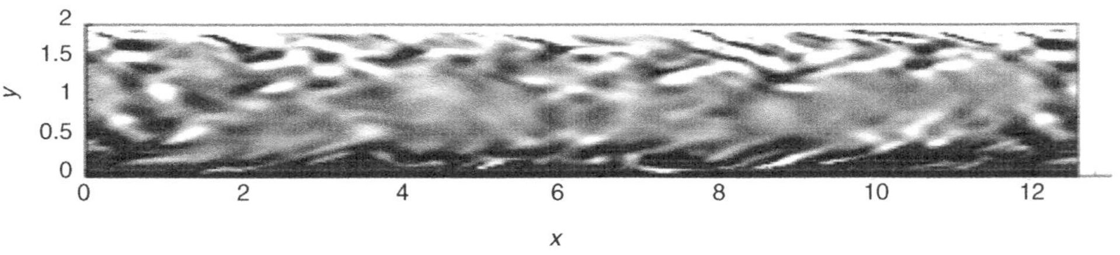


**Figure 12.5** Contours of horizontal vorticity in DNS of a plane channel. Courtesy of Dr. Greg Laskowski. Flow from left to right.

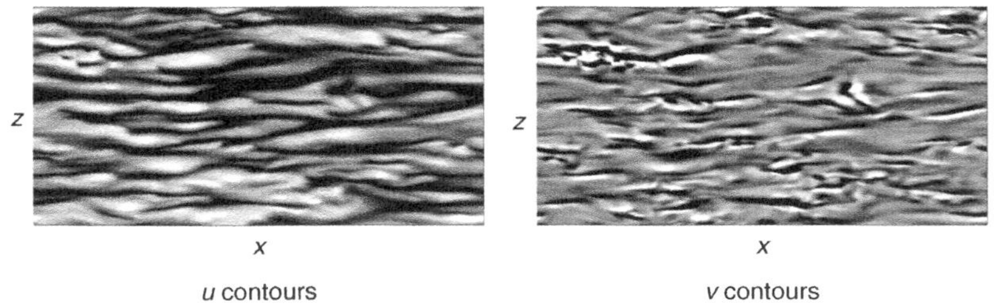


**Figure 12.6** Contours of streamwise and normal velocities near the wall of a plane channel at $y^+ = 10$. Courtesy of Dr. Tomoaki Ikeda.

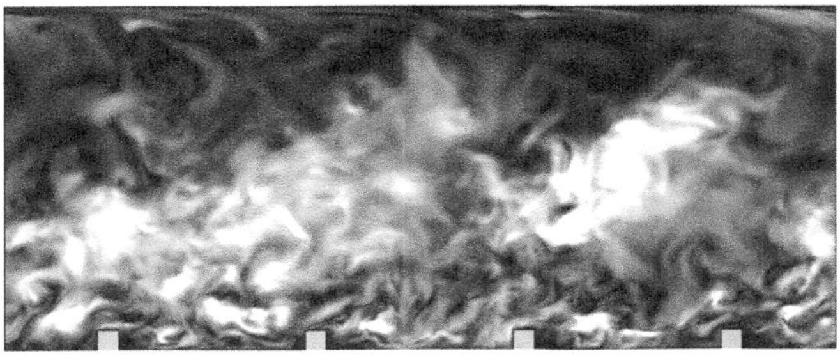

**Figure 12.7** Fluctuating velocity field in DNS of a plane channel with ribs on the lower wall. Flow from left to right. *Source*: Tomoaki Ikeda and Paul A. Durbin, 'Direct simulations of a rough wall channel flow', *The Journal of Fluid Mechanics*, **571**, 235–263. 2007 © Cambridge Journals, published by Cambridge University Press.

has become possible to carry simulations all the way through to a turbulent state. Grids become larger, but simulations are done at realistic Reynolds numbers. DNS has been performed for both orderly and bypass transition (Section 6.5), for separated and attached flow, and for incompressible and compressible flow (Criminale *et al.*, 2003).

If the computation is to be carried right through from instability to turbulence, then grid resolution must be adequate to capture eddies that appear after transition. The wavelength of instabilities is on the the order of the shear-layer thickness, $\delta$. Eddies of order the Kolmogoroff scale, $\eta$, or in a wall-bounded flow of order $\nu/u_*$, must be resolved. Thus the number of grid points scales as $(\delta/\eta)^3 \sim R_\delta^{9/4}$ in a free-shear layer and as $(u_*\delta/\nu)^3 \sim R_\delta^{21/8}$ in a boundary layer, per Eq. (12.1.1). The saving grace is that Reynolds numbers are low.

Figure 12.8 illustrates transition in the boundary layer on a compressor blade (Zaki *et al*., 2009). Instability is initiated by free-stream turbulence. Perturbations to the region next to the leading edge are large-scale and relatively smooth. But a high degree of irregularity develops on the latter part of the blade. In the lower image, contours of $Q \equiv |\boldsymbol{S}|^2 - |\boldsymbol{\Omega}|^2 =$ constant are superimposed on contours of wall normal velocity; $Q$ is an indicator of vortices and thus of small scale features. In transitional flow, vortices lift from the surface and breakdown into turbulence.

In this geometry, transition occurs for chord Reynolds numbers in the range $10^5$–$10^6$ and momentum thickness Reynolds numbers of order $10^2$–$10^3$. In the case of Figure 12.8, the Reynolds number based on axial chord and free-stream velocity is $\mathrm{Re} = 1.385 \times 10^5$. A grid of dimensions $1025 \times 641 \times 129$ in the streamwise, cross-stream, and spanwise directions was employed. A rather large number of points is needed in the streamwise direction. Streamwise resolution is especially important in the neighborhood of the location of transition; the abrupt rise in skin friction (Figures 6.11 and 6.13) can be difficult to capture.

Free-shear layers are subject to an inviscid instability. It is more powerful than viscous instability in boundary layers. The critical Reynolds number may be less than 100. A mixing layer is unstable at any Reynolds number. Primary instabilities grow rapidly and develop secondary instabilities that break down to turbulence.

The primary instability of a mixing layer is a spanwise roll. The upper panel of Figure 12.9 is a side view of a mixing layer; it highlights the formation of rolls. The

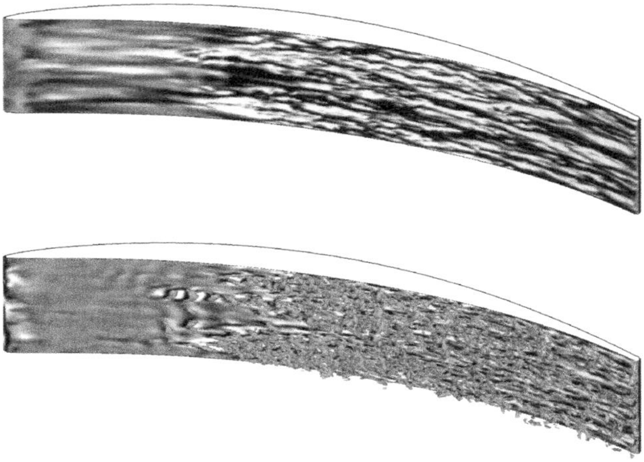

**Figure 12.8** Transition on the pressure side of a compressor blade. Flow from left to right. Top: contours of tangential velocity. Bottom: contours of normal velocity with vorticity surfaces superimposed. Figure courtesy of T. Zaki and J. Wissink.

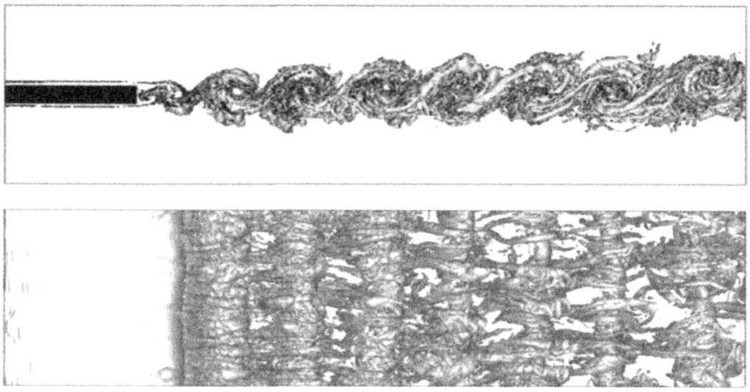

**Figure 12.9** Transition in the mixing layer downstream of a splitter plate. Reprinted with permission from 'Direct numerical simulation of a mixing layer downstream a thick splitter plate' by Sylvain Laizet, Sylvain Lardeau, Eric Lamballais, in *Physics of Fluids 22, 015104*. Copyright 2010, American Institute of Physics.

lower panel is a view looking down on the layer; it illustrates the development of small-scale, three-dimensional motion. These views are plots of vorticity magnitude. The Reynolds number based on velocity difference and splitter plate thickness is 400. Shear-layer thickness grows by a factor of 10 in the computational domain. So on the order of $10^8$ grid points are needed if the grid is uniform. This grid for this simulation was $961 \times 257 \times 120$ (Laizet *et al.*, 2010).

## 12.3 Pseudo-spectral method

Figure 12.4 shows the simplest, and earliest application of direct numerical simulation. The idea that homogeneous turbulence might be simulated numerically occurred to the mathematician John Von Neuman in 1949, shortly after the first digital electronic computer was built. He made this proposal at the end of a review of then-recent literature on the theory of homogeneous turbulence. The idea is seemingly easy: the set of ordinary differential equations (10.2.4) for the evolution of the Fourier coefficients of the velocity field are integrated in time. In order to make use of fast Fourier transforms (FFTs), coefficients are evaluated only at a discrete set of wavenumbers,

$$k_n = 2\pi n/(N\Delta x), \qquad -N/2 \le n < N/2, \tag{12.3.1}$$

in each of the three directions. Thus, the problem reduces to that of solving a set of $N = N_x \times N_y \times N_z$ coupled ordinary differential equations. The computation starts with a random set of Fourier coefficients, selected so that their mean-square amplitude follows a model spectrum, such as the Von Karman spectrum (9.2.19) with a viscous cut-off in the dissipation range. Whenever a velocity field is desired in physical space, it is found by inverse transformation of the Fourier coefficients.

Despite being conceptually attractive, computing power was not up to the task for over two decades, until, in 1972, Orszag and Patterson performed a spectral simulation with $N = 64^3$ grid points. Their approach was to replace the Fourier integrals of Chapter III with a truncated Fourier series. Then the evolution is solved numerically. Efficiency was gained by combining a collocation method with FFTs, which is now called a "pseudo-spectral method." We will describe the gist of pseudo-spectral simulation of homogeneous turbulence. A comprehensive treatment of spectral methods can be found in Canuto *et al*. (2006).

For the present case, the starting point is the truncated Fourier series representation

$$u(x) = u(j\Delta x) = \sum_{n=-N/2}^{N/2-1} \hat{u}(n)\, e^{i2\pi n(j/N)}. \tag{12.3.2}$$

Here, the $x$ axis is discretized such that grid points are at $x = j\Delta x$ and the domain is an interval of length $N\Delta x$. The $N$ Fourier coefficients,*

$$\hat{u}(n), \qquad -N/2 \le n \le N/2 - 1,$$

are determined by the values of $u(x)$ at $N$ equally spaced points $j = 0, 1, 2, \ldots, N-1$ through the transform

$$\hat{u}(n) = \frac{1}{N}\sum_{j=0}^{N-1} u(j)\, e^{i2\pi n(j/N)}, \tag{12.3.3}$$

where the argument of $u$ is simplified from $j\Delta x$ to $j$. Note that periodicity $u(N) = u(0)$ is assumed. The transformed Navier–Stokes equations are (10.2.4) restricted to the discrete values of the truncated series expansion:

$$d_t\hat{\boldsymbol{u}}(n) + \nu\kappa^2\hat{\boldsymbol{u}}(n) = \boldsymbol{k}\cdot\hat{\boldsymbol{u}}\star\hat{\boldsymbol{u}} - \boldsymbol{k}\frac{\boldsymbol{k}\cdot\hat{\boldsymbol{u}}\star\hat{\boldsymbol{u}}\cdot\boldsymbol{k}}{\kappa^2}. \tag{12.3.4}$$

* For real-valued functions, $u(-N/2)$ should be set to 0 so that the range of summation is symmetric about 0.

As previously, $\kappa^2 = |\boldsymbol{k}|^2$ and values of $k$ are of the form (12.3.1). On a grid with equal spacing in the $x$, $y$, and $z$ directions, $\boldsymbol{k}$ is represented by three integers as $2\pi(n, m, p)/N\Delta x$. The primary question is how to evaluate the convolution product, $\hat{\boldsymbol{u}} \star \hat{\boldsymbol{u}}$.

Consider a product of Fourier sums

$$
\begin{aligned}
u(j)v(j) &= \sum_{n=-N/2}^{N/2-1} \hat{u}(n)\, \mathrm{e}^{\mathrm{i}2\pi n(j/N)} \sum_{m=-N/2}^{N/2-1} \hat{v}(m)\, \mathrm{e}^{\mathrm{i}2\pi m(j/N)} \\
&= \sum_n \sum_m \hat{u}(n)\hat{v}(m)\, \mathrm{e}^{\mathrm{i}2\pi(m+n)(j/N)}.
\end{aligned}
\tag{12.3.5}
$$

A convolution sum arises when this is Fourier-transformed. The transform at wavenumber $p$ simply extracts the coefficient of $\mathrm{e}^{\mathrm{i}2\pi p(j/N)}$:

$$
\widehat{uv}(p) = \sum_{n+m=p} \hat{u}(n)\hat{v}(m) = \sum_n \hat{u}(n)\hat{v}(p-n). \tag{12.3.6}
$$

This is the convolution sum $\hat{u} \star \hat{v}$.

Equation (12.3.5) warrants consideration. As $n$ and $m$ vary from $-N/2$ to $N/2 - 1$, the sum in their product ranges from $-N$ to $N - 2$. However, (12.3.4) only applies in the range of $-N/2$ to $N/2$, so the convolution extends to wavenumbers higher than those being computed; in physical space, wavelengths shorter than those resolved on the grid are generated by the product. The shortest resolved wavelength occurs when $n = N/2$. The cosine part of (12.3.2) then becomes $\cos(\pi j)$, which equals $(-i)^j$: it oscillates between $+1$ and $-1$ from one grid point to the next. This is the most rapid oscillation that can be captured on the grid. The condition that $n$ must be less than $N/2$ is called the Nyquist criterion. The convolution product generates $n$ larger than the Nyquist condition.

What should be done about this? One approach is to cut out those wavenumbers created by convolution that are greater in magnitude than $N/2$. Expression (12.3.6) is set to zero if $p > N/2 - 1$ or if $p < -N/2$. For instance, if $n = N/3$, values of $m \geq N/6$ are omitted. This is called a Galerkin approximation in Canuto *et al.* (2006). Under this approximation, it is assumed that the omitted modes have little influence on the turbulence dynamics.

Another approach stems from the need for efficiency. In practice, the sum (12.3.6) is not evaluated. For each $p$, it is a sum of $N$ terms. There are $N$ values of $p$. So the total work would increase like $N^2$. That is too high for large simulations. In efficient computational algorithms, the number of operations increases about in proportion to $N$; the FFT requires of order $N \log N$ operations.

A reduction in work is based on the observation that the convolution sum is the Fourier transform of $u(x)v(x)$ (see Eq. (10.1.1)). Given the Fourier coefficients, $\hat{u}$, the FFT algorithm permits an efficient transformation to physical space via (12.3.2). The convolution can be evaluated by transforming to physical space, forming products, then going back to Fourier space. So, when a term like $u(x)\,\partial_x u(x)$ is needed, two FFTs to physical space are performed, to provide $u$ and $\partial_x u$, then the product is evaluated at each grid point of physical space, after which an FFT back to wavenumber space is performed. In this context, the grid points in physical space are called *collocation* points.

The convolution sum has been obtained with operation count proportional to $N \log N$. This is called an *aliased* pseudo-spectral algorithm. It takes three FFTs and a product at all $x$, but those are far cheaper to compute than the $O(N^2)$ direct evaluation of a convolution sum.

Aliasing refers to the fact that the Nyquist criterion has been ignored. No truncation of the sum (12.3.6) has occurred. If the unresolved wavenumbers have not been truncated, what has become of them? The answer is given by Figure 12.10: they have shown up as an additional contribution to the wavenumbers that can be resolved on the grid. In Figure 12.10, the diamonds are located at the grid points. Two sine waves are sampled, one above the Nyquist wavenumber and one below. At the sample points, they are indistinguishable. The dynamic equation (12.3.4) treats both as having the lower wavenumber. The higher wavenumber is said to have been *aliased* into the lower.

Mathematically, the product $e^{i2\pi nj/N} \times e^{i2\pi mj/N}$ is equal to $e^{i2\pi(n+m-N)j/N}$. If $n+m$ is greater than $N/2$, the product appears at $n+m-N$, rather than at $m+n$. That is illustrated in Figure 12.11 at the bottom. The range between $N/2+1$ and $N$ is aliased into the range between $-N/2+1$ and 0. Looking at the magnitude of wavenumber, this is described as folding across $N/2$: the energy at $N/2+n$ is folded to $N/2-n$. The turbulent energy spectrum has low amplitude at high wavenumbers. Hence the aliasing from high $n$ might be harmless. For instance, if $N/2$ is a wavenumber in the dissipation range, high wavenumbers are aliased by folding dissipation-range energy into the inertial and energetic ranges, which has negligible effect. The primary contamination is near $N/2$, where the aliased amplitude is comparable to that which is resolved. If the energy is already low at $N/2$, aliasing might not corrupt the simulation. Aliased DNS may be viable. However, as a rule, it is preferable to *de-alias*.

Several techniques have been developed to de-alias a spectral simulation. We mention only the simplest, the 3/2 rule. This technique is illustrated at the top of Figure 12.11. In physical space, products are formed on a grid that has a spacing of $\frac{2}{3}\Delta x$. In Fourier space, the wavenumber range is extended to $\pm\frac{3}{4}N$ solely for the purpose of evaluating products. Then aliasing folds wavenumbers between $3N/4$ and $N$ into the range below $-N/2$, so that the computational interval is uncontaminated. The simulation itself continues on the grid with spacing $\Delta x$. The method is implemented by padding the Fourier coefficients with zeros between $N/2$ and $3N/4$ before transforming to physical space and evaluating products at the $3N/2$ collocation points. After transforming the product back to Fourier space, modes beyond $\pm N/2$ are deleted.

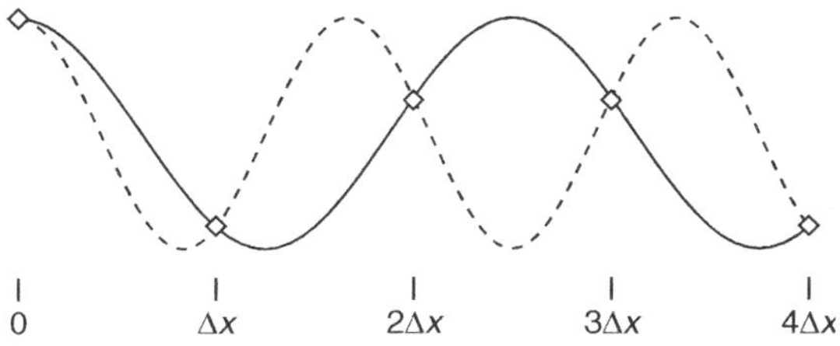


**Figure 12.10** At the computational points, the cosine waves, $\cos(2n\pi x/L)$ with $n = 0.4$ (——) and 0.6 (- - - -) are indistinguishable.

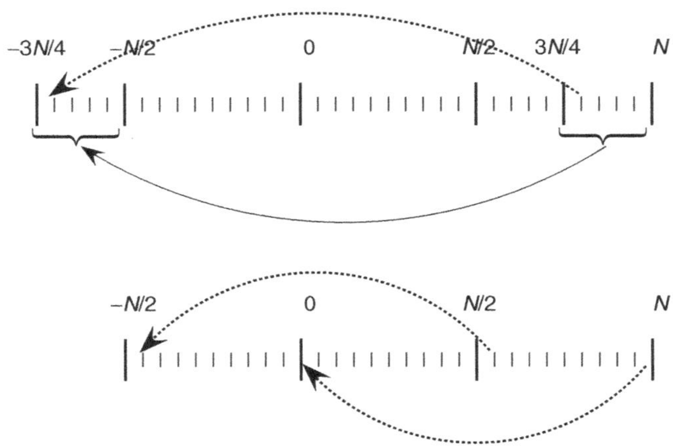


**Figure 12.11** In the lower part, energy in the range $N/2 \leq n \leq N$ is aliased into the range $-N/2 \leq n \leq 0$. The upper part illustrates the 3/2 method for de-aliasing. No energy is aliased into the range $-N/2 \leq n \leq N/2$.

The primary motive for spectral simulation is efficiency. Pressure can be eliminated exactly and the problem reduced to integrating a system of ordinary differential equations (12.3.4). Usually that is done with a Runge–Kutta method (Exercise 12.2). Not only is this more efficient than finite-difference methods in physical space, but it is more accurate. The Riemann–Lebesgue lemma says that the coefficients of a Fourier series converge exponentially toward zero as $n \to \infty$ if the function is infinitely differentiable. Hence, spectral methods are said to have exponential order of accuracy. The high accuracy is because derivatives are evaluated exactly, rather than by finite differences. For instance, differentiating (12.3.2) under the sum,

$$\frac{\mathrm{d}u(x)}{\mathrm{d}x} = \frac{\mathrm{i}2\pi}{N\Delta x} \sum_{n=-N/2}^{N/2-1} n\hat{u}(n)\, \mathrm{e}^{\mathrm{i}2\pi n(j/N)}. \tag{12.3.7}$$

In the pseudo-spectral method, when $u\,\partial_x u$ is computed in physical space, this sum is evaluated by FFT to obtain the derivative.

# Exercises

**Exercise 12.1.** *Energy-conserving schemes.* Consider the equation

$$\partial_t \phi + u\,\partial_x \phi + \partial_x(u\phi) = 0.$$

Show that $\phi^2$ is conserved.

The velocity is stored at the half grid points, $x = (j + \frac{1}{2})\Delta x$, $j = 1, 2, 3, \ldots, J$, and $\phi$ is stored at the grid points, $x = j$. Consider the discretizations

$$u\,\delta_x\phi\big|_j = \frac{1}{2}\left(u(j+\tfrac{1}{2})\frac{\phi(j+1)-\phi(j)}{\Delta x} + u(j-\tfrac{1}{2})\frac{\phi(j)-\phi(j-1)}{\Delta x}\right)$$

and

$$\delta_x(u\phi)\big|_j = \frac{1}{\Delta x}\left(u(j+\tfrac{1}{2})\frac{\phi(j+1)+\phi(j)}{2} - u(j-\tfrac{1}{2})\frac{\phi(j)+\phi(j-1)}{2}\right).$$

Show that $\phi^2$ is conserved for this discretization.

**Exercise 12.2.** *DNS via Burgers equation.* Solve the Burgers equation (12.2.1) numerically with $\nu = 0.01$ and $0.001$, imposing the initial condition

$$\tfrac{1}{2} + \left\{e^{-20x^2} + e^{-20(x+1)^2} + e^{-20(x-1)^2}\right\}(0.5 + 0.5\xi), \qquad (12.3.8)$$

where $\xi$ is a random number between 0 and 1. The flow domain is $-1/2 \le x \le 1/2$. Apply periodic boundary conditions: $u(1/2 + j\Delta x) = u(-1/2 + j\Delta x)$. Use about 160 grid points. Integrate to a non-dimensional time of 0.12, plotting the solution at $t = 0, 0.04, 0.08$, and 0.12.

Looking ahead to Section 13.1.1: compare the DNS velocity field to its filtered field at four equally spaced times, $t = 0, 0.4, 0.8$, and 0.12. Try the running-average filter (13.1.2) with $N = 2$ and 3, and the Padé filter (13.1.3) with $\alpha = 0.45$.

[Note: DNS commonly employs the Runge–Kutta method for time integration. The discrete equations

$$\partial_t u_j = -\frac{u_{j+1}^2 - u_{j-1}^2}{4\Delta x} + \frac{1}{\Delta x}\left(\nu_{j+1/2}\frac{u_{j+1}-u_j}{\Delta x} - \nu_{j-1/2}\frac{u_j - u_{j-1}}{\Delta x}\right)$$

are a system of ordinary differential equations of the form

$$\partial_t u_j = \mathrm{RHS}(u_i).$$

The second-order Runge–Kutta method is as follows.

```
up(:)  = RHS(u(:),t)
u1(:)  = u(:)+up(:)Δt/2
up(:)  = RHS(u1(:),t)
u(:)   = u(:)+up(:)Δt   ;    t = t+Δt
```

At the end, $u$ has been advanced one time step. This is a simple integration method, which facilitates tests of various convection schemes.]

**Exercise 12.3.** *Spectral methods.* Use fast Fourier transforms (FFT) to evaluate $\partial_x u^2$ and $\partial_x^2 u$. In the pseudo-spectral method, $u^2$ is evaluated in physical space to avoid convolution sums; derivatives are evaluated in Fourier space, as in (12.3.7). Specify $u$ by the initial condition (12.3.8) of the previous problem.

Optional: Replace the finite-difference discretization used in the previous problem by pseudo-spectral evaluation and solve the Burgers equation.

# 13

# Simulation of large eddies

We haven't the money, so we've got to think.

– Ernest Rutherford

Fully resolved simulation of the spectrum of turbulent eddies is quite demanding of computer resources, not to say of manpower. Tending the simulation, accumulating statistics, and post-processing the data can take months of effort. For those reasons, less expensive methods of approximate simulation have been developed. Those addressed in this chapter are large eddy and detached eddy simulation. The former has been in use for many years. Indeed, in 1970, before direct numerical simulation of turbulent channel flow was feasible, James Deardorff published a seminal study that initiated the concept of large eddy simulation (LES). From then until now, LES has been viewed as a pragmatic version of eddy simulation. As of any approximate method, one can point to shortcomings. Citing certain limitations to LES, in 1999 Philippe Spalart proposed the method of detached eddy simulation (DES). We will review the motives and methods of LES and DES in this chapter.

The numerical requirements for capturing large eddies are much the same as those discussed in the previous chapter. Low-dissipation or energy-conserving schemes are required. Although the grids are coarser, they still must be designed to capture turbulent eddies.

## 13.1 Large eddy simulation

The expense of direct simulation owes to the need to resolve dissipation-range eddies. The bulk of energy often resides in scales more than an order of magnitude larger. If the small scales are omitted, the grid can be coarsened in each of the three coordinate directions. Increasing grid spacing by a factor of 5 reduces the number of grid points by a factor of 125. In Figure 12.1, the grid shown by solid lines is fives times coarser

*Statistical Theory and Modeling for Turbulent Flows, Second Edition* P. A. Durbin and B. A. Pettersson Reif

than the DNS grid, represented by dashed lines. It captures the larger eddy, but is unable to resolve the smaller eddy. If the objective is to capture energetic eddies completely, but to delete dissipation-range eddies, then it makes sense to choose the grid spacing to lie within the inertial range. In practice, large eddy simulation is effected by performing DNS on the coarse grid, with the addition of a model that represents effects of the small eddies.

In the context of large eddy simulation, the notion of deleting small scales is called *filtering*, and the deleted scales are said to have been *cut off*; those that are retained are called *resolved*.

A conceptual picture emerges by reference to the energy spectrum of homogeneous isotropic turbulence. Consider Figure 2.1, on page 17. When $R_\lambda = 600$, the spectral energy density falls by $10^{-4}$ from the energetic range to the start of the dissipation range. It would seem adequate to resolve scales with $\kappa\eta < 10^{-1}$. Call the cut-off scale $\Delta_c$: that is $\kappa_c\eta = 2\pi\eta/\Delta_c$. Then $\Delta_c \approx 10 \times 2\pi\eta$. It is clear from Figure 2.1 that $\Delta_c \approx 2\pi\eta$ is sufficient resolution for DNS; so LES resolution could be 10 times coarser in this case.

A caveat must be made: this reasoning is based on the *energy* spectrum. The *vorticity* spectrum is $\kappa^2$ times the energy spectrum; it increases as $\kappa^{1/3}$ in the inertial range and peaks in the dissipation range. Cutting off the spectrum in the inertial range omits a large portion of the vorticity. Fortunately, the primary role of small-scale vorticity is to dissipate energy. A premise of LES is that this omission can be overcome in a fairly simple manner. The exact dissipation is to be replaced by an alternative that produces the correct level of dissipation without requiring the smallest scales to be simulated.

The overall rate of dissipation is controlled by the energy cascade. The fundamental assumption of LES is that, if the energy cascade is captured at large scales, the precise nature of the small-scale processes is not critical. They need only dissipate energy as it arrives through the cascade. In particular, an artificial model can replace the exact, Navier–Stokes, dynamics. The LES approach is to use a grid that is under-resolved, then add a *subgrid* model to represent the small-scale dissipative processes. Subgrid models are discussed in Section 13.1.2 after the following discussion of filtering.

### 13.1.1 Filtering

Formally, cutting off the small scales is described as filtering. A low-pass filter removes small scales and leaves large scales unscathed. In physical space, the filter is a running average. For instance, a three-point, running average replaces $u(x)$ by

$$\hat{u}(x) = \tfrac{1}{4}[u(x + \Delta x) + 2u(x) + u(x - \Delta x)],$$

where the hat over $u$ indicates a filtered variable. An $N$-point average replaces $u(x)$ by

$$\hat{u}(x) = \frac{1}{2N}\left[\frac{u(x + N\Delta x) + u(x - N\Delta x)}{2} + \sum_{j=1-N}^{N-1} u(x + j\Delta x)\right]. \tag{13.1.1}$$

The filter in spectral space is evaluated by applying the average to the complex exponential $u = \mathrm{e}^{ikx}$:

$$\widehat{\mathrm{e}^{ikx}} = \frac{1}{2N}\left[1 + \cos(Nk\Delta x) + 2\sum_{j=1}^{N-1}\cos(jk\Delta x)\right]\mathrm{e}^{ikx}. \tag{13.1.2}$$

Equation (13.1.2) has the form $\hat{u} = F_k u_k$, where the filter $F_k$ is the bracketed factor on its right-hand side. In Fourier space, the running average becomes a multiplicative function of $k$. A low-pass filter is an $F_k$ that is nearly unity for small $k$ and becomes very small at large $k$.

When $N = 1$, the filter of the three-point running average,

$$F_k = \frac{1 + \cos(k\Delta x)}{2},$$

is obtained. Note that this vanishes at $k = \pi/\Delta x$. That is the shortest wavelength that can appear on the grid; it corresponds to a disturbance $u = (-1)^j$, which oscillates between $+1$ and $-1$ from one grid point to the next. The filter removes this component entirely, and reduces the amplitude at wavenumbers near to it.

A sharper filter can be constructed by the Padé average:

$$\begin{aligned}\hat{u}(x) &+ \alpha[\hat{u}(x+\Delta x) + \hat{u}(x-\Delta x)] \\ &= au(x) + \tfrac{1}{2}b[u(x+\Delta x) + u(x-\Delta x)] + \tfrac{1}{2}c[u(x+2\Delta x) + u(x-2\Delta x)].\end{aligned} \tag{13.1.3}$$

This is an implicit formula. The filtered variable is found by inverting the left-hand side with a tridiagonal matrix algorithm. Lele (1992) gives $a = (5+6\alpha)/8$, $b = (1+2\alpha)/2$, and $c = (2\alpha - 1)/8$ for fourth-order accuracy. He shows that $\alpha = 0.45$ provides a filter that is flat for small $k$ and cuts off sharply at large $k$. The functional form of the filter is

$$F_k = \frac{a + b\cos(k\Delta x) + c\cos(2k\Delta x)}{1 + 2\alpha\cos(k\Delta x)} \tag{13.1.4}$$

in Fourier space.

Running-average and Padé filters are compared in Figure 13.1. The Padé filter has the desired form: near unity for small wavenumbers, cutting off steeply at high wavenumbers.

More generally, a weighted running average has the form

$$\hat{u}(x) = \sum_{j=-N}^{N} u(x + j\Delta x)w_j,$$

where the weights sum to unity, $\sum w_j = 1$. In (13.1.1) the weights are $1/2N$ for $j \neq \pm N$ and $1/4N$ for $j = \pm N$. The general running average can be represented as an integral

$$\hat{u}(x) = \int_{-\Delta_f}^{\Delta_f} F(x - \xi; x)u(\xi)\,d\xi, \tag{13.1.5}$$

where $\Delta_f$ is the filter width. If $F$ is a function of only $x - \xi$, the filter is said to be homogeneous.

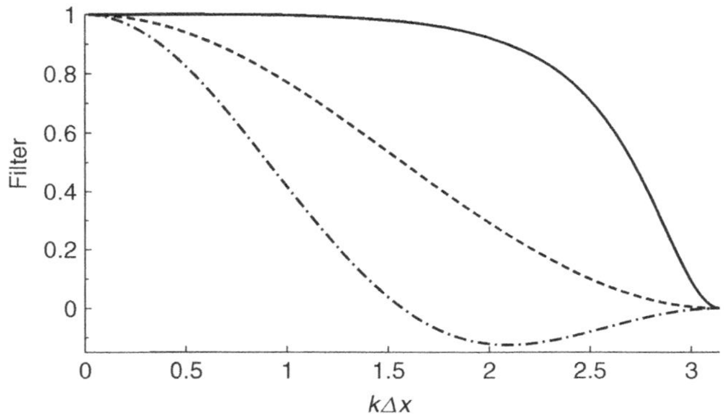


**Figure 13.1** Running-average filter (13.1.2) with $N = 2$ (- - - -) and $N = 3$ (— · —), and Padé filter (13.1.4) with $\alpha = 0.45$ (———).

For a homogeneous filter, the average of $e^{ikx}$ is the Fourier transform of $F$ (see Chapter 9):

$$\widehat{e^{ikx}} = F_k \, e^{ikx},$$

where $F_k$ is the Fourier transform of the filter. The top-hat filter

$$F(x-\xi) = \frac{1}{2\Delta_f}, \qquad |\xi| < \Delta_f,$$
$$F_k = \frac{\sin(k\Delta_f)}{k\Delta_f}$$

is a simple example. The running average (13.1.1) is a discrete version of the top-hat filter. An example with Gaussian weights is provided by Exercise (13.2). That exercise requests the reader to obtain a curve corresponding to Figure 13.1.

An expression is filtered formally by drawing a hat over it. For instance, the filtered Navier–Stokes equations (3.1.1) are written as

$$\partial_t \hat{u}_i + \partial_j \hat{u}_j \hat{u}_i = -\frac{1}{\rho}\,\partial_i \hat{p} + \nu\nabla^2 \hat{u}_i - \partial_j \tau_{ij}^{\mathrm{SGS}},$$
$$\partial_i \hat{u}_i = 0. \qquad (13.1.6)$$

That is, a hat is drawn over each term. Here

$$\tau_{ij}^{\mathrm{SGS}} = \widehat{u_i u_j} - \hat{u}_i \hat{u}_j \qquad (13.1.7)$$

is called the subgrid stress. Strictly, (13.1.6) is exact only for a homogeneous filter because the hat has been moved inside the derivatives. Differentiation does not commute with filtering unless the filter is homogeneous, so extra terms arise from non-commutation. They are usually ignored.

Although the filtered Navier–Stokes equations (13.1.6) have a formal similarity to the Reynolds averaged Navier–Stokes equations, the two are quite distinct. The filter is

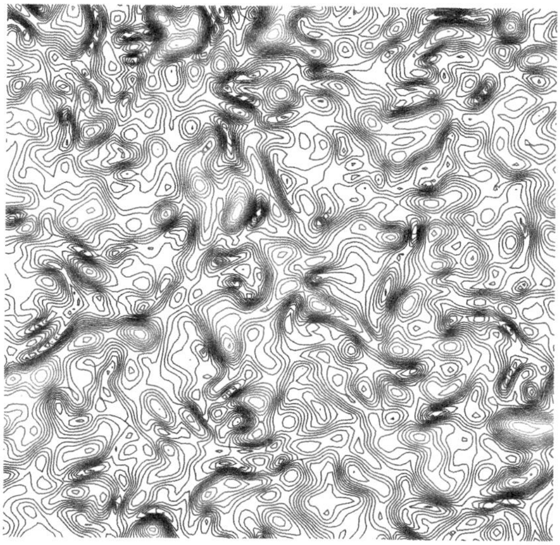

**Figure 13.2** Filtered version of Figure 12.4.

a local smoother that leaves an irregular, turbulent velocity field; the Reynolds average sums over an ensemble, creating a regular, smooth velocity field. Figure 13.2 was created by cutting off three-quarters of the Fourier modes of the data in Figure 12.4 – which is a top-hat filter in Fourier space. The field in Figure 13.2 has a similar appearance to the original DNS; only the smallest-scale irregularity has been removed.

Filtering once again would make the field in Figure 13.2 smoother still; filtering twice produces a different result from filtering once: $\widehat{\hat{u}} \neq \hat{u}$. This contrasts to properties of the Reynolds average, stated in Eq. (2.2.3): averaging an average does not change its value: $\overline{\overline{u}} = \overline{u}$.

By corollary, the filtered value of the fluctuation $u' = u - \hat{u}$ does not vanish: $\widehat{u'} = \hat{u} - \widehat{\hat{u}} \neq 0$. As an example, consider the formula

$$\mathcal{L}_{ij} = \widehat{\hat{u}_i \hat{u}_j} - \hat{u}_i \hat{u}_j,$$

which is called the Leonard stress. It is not Galilean invariant (see Section 8.1): if a constant is added to $u$, it becomes

$$\widehat{(\hat{u}_i + c_i)(\hat{u}_j + c_j)} - (\hat{u}_i + c_i)(\hat{u}_j + c_j) = \mathcal{L}_{ij} + c_i(\widehat{\hat{u}}_j - \hat{u}_j) + c_j(\widehat{\hat{u}}_i - \hat{u}_i).$$

The last two terms do not vanish.

In the filtered Navier–Stokes equations (13.1.6), variables with a hat are simply the dependent variables. No explicit filtering is involved. Once a closure model is provided for $\tau_{ij}^{\mathrm{SGS}}$, the equations can be solved. The result is a turbulent field. Instead of referring to it as the filtered field, it can be called the *resolved field*.

The resolved field could be explicitly filtered. The filter denoted by a hat is a notional filter. A different filter could be applied to the $\hat{u}(x)$ field. Denote the explicit filter by a tilde: $\widetilde{u}$.

A computable stress can be defined by applying this new filter to the LES field:

$$L_{ij} = \widetilde{\hat{u}_i \hat{u}_j} - \tilde{\hat{u}}_i \tilde{\hat{u}}_j. \tag{13.1.8}$$

The known, resolved velocity is processed explicitly by the filter. A rather useful identity was derived by Germano (see Germano *et al.*, 1991; Sagaut, 2001). Let

$$T_{ij} = \widetilde{\widehat{u_i u_j}} - \tilde{\hat{u}}_i \tilde{\hat{u}}_j.$$

Then Germano's identity is

$$\widetilde{\tau_{ij}^{\text{SGS}}} = T_{ij} - L_{ij}. \tag{13.1.9}$$

It can be verified by substituting the definition of each term. Germano's identity will be used in the next section to describe the dynamic procedure for subgrid modeling.

## 13.1.2 Subgrid models

The closure problem for subgrid stress (13.1.7) is far less demanding than the closure problem for Reynolds stress. The primary dictum is: do not let the subgrid model undermine the Navier–Stokes physics; or the Hippocratic, "first, do no harm." Large eddy simulation relies on the Navier–Stokes equations to capture most of the physics. The primary role for the subgrid model is to replace dissipation by the smallest-scale eddies. In fact, one approach, which is called implicit large eddy simulation (ILES), invokes a dissipative numerical algorithm and no subgrid closure model at all (Grinstein *et al.*, 2007).

### 13.1.2.1 Smagorinsky model

Among explicit models, the Smagorinsky model is most popular. This is an eddy viscosity

$$\tau_{ij}^{\text{SGS}} = -2\nu_{\text{SGS}} \widehat{S}_{ij}, \tag{13.1.10}$$

where the resolved rate-of-strain tensor is

$$\widehat{S}_{ij} = \tfrac{1}{2}(\partial_i \hat{u}_j + \partial_j \hat{u}_i).$$

(Note that the traces of both sides of (13.1.10) are not equal because the subgrid kinetic energy is omitted; $\frac{2}{3} k_{\text{SGS}} \delta_{ij}$ could be added on the right, but without a model for $k_{\text{SGS}}$, the subgrid energy must be absorbed into the definition of pressure.)

Smagorinsky's model is the mixing length form (see Section 6.1.2, on page 115)

$$\nu_{\text{SGS}} = (c_s \Delta)^2 \sqrt{2|\widehat{\boldsymbol{S}}|^2}, \tag{13.1.11}$$

where $|\widehat{\boldsymbol{S}}|^2 = \widehat{S}_{ij} \widehat{S}_{ji}$ is the magnitude of the rate-of-strain tensor. For a pure shear, $\sqrt{2|\widehat{\boldsymbol{S}}|^2}$ becomes $|\mathrm{d}U/\mathrm{d}y|$, as in Prandtl's model (6.1.6).

The mixing length $c_s \Delta$ is proportional to the grid spacing, and $c_s$ is an empirical constant. On a finite-volume mesh, the grid spacing can be defined as $\Delta = (\Delta \text{Vol})^{1/3}$, or as $\Delta = (\Delta x \Delta y \Delta z)^{1/3}$ on a Cartesian grid.

Returning to the Burgers equation (12.2.1) for a simple illustration, the Smagorinsky subgrid viscosity is added to the molecular viscosity in (12.2.1):

$$\partial_t u + u\,\partial_x u = \partial_x[(\nu + \nu_{\mathrm{SGS}})\,\partial_x u],$$
$$\nu_{\mathrm{SGS}} = c_{\mathrm{Burgers}}(c_{\mathrm{s}}\Delta x)^2|\partial_x u|. \tag{13.1.12}$$

An additional constant, $c_{\mathrm{Burgers}}$, was incorporated because the empirical value of $c_{\mathrm{s}}$ is not appropriate to the Burgers equation. The constant $c_{\mathrm{s}}$ is set to the typical value 0.2, but $c_{\mathrm{Burgers}}$ is set to 100 in order to approximate the filtered field obtained in Figure 12.3. Figure 13.3 compares the solution of (13.1.12) with 32 grid points to the filtered data from Figure 12.3. The LES grid is four times coarser than the DNS grid.

The rate of subgrid energy production is stress times rate of strain, $-\tau_{ij}S_{ji}$. Invoking an equilibrium assumption, equating this to the rate of subgrid dissipation, and using (13.1.10) gives

$$\varepsilon^{\mathrm{SGS}} = -\tau_{ij}S_{ji} = 2\nu_{\mathrm{SGS}}\widehat{S}_{ji}\widehat{S}_{ji} = (c_{\mathrm{s}}\Delta)^2(2|\widehat{S}|^2)^{3/2}. \tag{13.1.13}$$

Note that this is consistent with inertial-range scaling, $|\widehat{S}|^3 \sim \varepsilon/\Delta^2$. Hence, the Smagorinsky model is consistent with a cut-off in the inertial range. The eddy viscosity formula then introduces subgrid dissipation that is consistent with the energy cascade.

Formula (13.1.13) permits an estimate of $c_{\mathrm{s}} \sim 0.18$ from measurements of inertial-range spectra (Sagaut 2001). However, it has been found that the empirical constant $c_{\mathrm{s}}$ in (13.1.11) depends on the flow, ranging from 0.1 for plane channel flow to 0.2 for isotropic turbulence. For this and other reasons, a dynamic procedure has been developed to compute $c_{\mathrm{s}}$ in the course of the simulation. The method is derived from (13.1.9) and (13.1.11).

Consider two filter scales $\Delta$ and $\widetilde{\Delta}$ and assume that $c_{\mathrm{s}}$ is the same on both. Substitute (13.1.10) and

$$T_{ij} = 2^{3/2}(c_{\mathrm{s}}\widetilde{\Delta})^2|\widetilde{\widehat{S}}|\widetilde{\widehat{S}}_{ij}$$

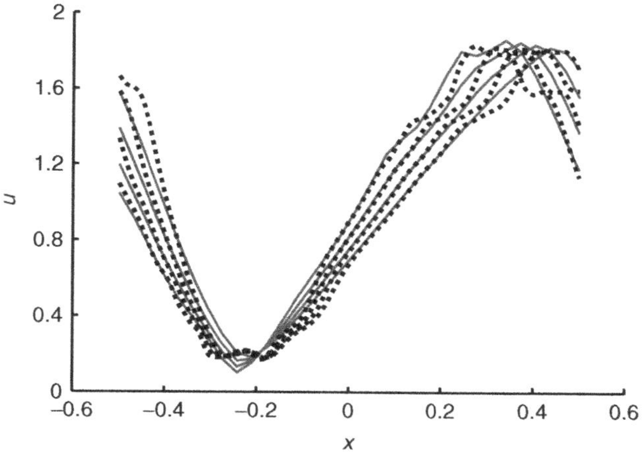


**Figure 13.3** Burgers equation with Smagorinsky model, compared to filtered DNS: LES (———) and filtered DNS (········).

into (13.1.9):

$$2^{3/2}(c_s\tilde{\Delta})^2|\tilde{\hat{S}}|\tilde{\hat{S}}_{ij} - 2^{3/2}(c_s\Delta)^2\widetilde{|\hat{S}|\hat{S}_{ij}} = L_{ij} - \tfrac{1}{3}\delta_{ij}L_{kk}. \quad (13.1.14)$$

Recall that the hat denotes dependent variables of the simulation. The tilde denotes an explicit filter, such as a running average (13.1.1). Typically, $\tilde{\Delta}$ is twice the grid spacing. The stress $L_{ij}$ is computed from the resolved field per (13.1.8). Hence, all quantities in (13.1.14) can be computed. In the dynamic procedure, this is used to obtain $c_s$ (Germano *et al.*, 1991). A different value is obtained at each grid point – which is not entirely consistent with the derivation, as it treated $c_s$ as a constant.

The tensor equation (13.1.14) does not provide a unique value for the scalar $c_s$. Select a unique value by least squares: then

$$c_s^2 = -\frac{L_{ij}m_{ji}}{m_{kl}m_{lk}}, \quad (13.1.15)$$

where

$$m_{ij} = 2^{3/2}\Delta^2\widetilde{|\hat{S}_{ij}|\hat{S}_{ij}} - 2^{3/2}\tilde{\Delta}^2|\tilde{\hat{S}}_{ij}|\tilde{\hat{S}}_{ij}.$$

The right-hand side of (13.1.15) is evaluated at each computational node. However, it can be negative, which is unphysical. To avoid this, the formula is clipped above zero and values of $c_s$ are averaged over some region, such as a direction of homogeneity.

The dynamic procedure has several virtues: $c_s$ is no longer a prescribed constant, it adapts to the flow; in laminar regions, $c_s = 0$, so the subgrid viscosity is switched off; as a wall is approached, $c_s \to 0$, so the viscous sublayer is recovered. All of these are improvements to the original formation by Smagorinsky.

Despite the final virtue, no-slip walls present a challenge to LES. In the vicinity of a wall, the large eddies are not distinct from the small eddies. The notion of cutting off the energy spectrum in the inertial range relies inherently on a high Reynolds number. The relevant Reynolds number is that based in the integral scale of the turbulence. In the log layer, the integral scale decreases linearly with decreasing distance from the wall. Hence, LES loses its rationale as a wall is approached. Two resolutions are available. The first is to decrease the grid spacing near the wall. The grid resolution must become comparable to that needed for DNS. The second solution is to introduce a wall model. In early efforts, either Van Driest damping or wall functions were invoked, by analogy to Reynolds averaged (RANS) modeling (Section 6.2.2.1). More elaborate wall models are discussed in Sagaut (2001). Generally, they invoke ideas from RANS. The notion of hybridizing RANS and LES leads to the method of detached eddy simulation, which is the topic of Section 13.2.

#### 13.1.2.2 Scale-similar models

Although the Smagorinsky model is very popular, many alternatives have been proposed. The similarity between the computable stress $L_{ij}$ in Eq. (13.1.8) and the unclosed subgrid stress (13.1.7) motivates the scale-similar model

$$\tau_{ij}^{\mathrm{SGS}} = c_{\mathrm{ssm}}L_{ij}. \quad (13.1.16)$$

The tensoral form of the subgrid stress is assumed to be that of the explicitly filtered, resolved velocity. The coefficient $c_{ssm}$ is empirical. It is found to be near to unity.

By itself, the scale-similar model is not sufficiently dissipative. It does not prevent energy from accumulating at small scales and can lead to numerical instability. This is corrected by the mixed model,

$$\tau_{ij}^{SGS} = c_{ssm} L_{ij} - 2\nu_{SGS} \widehat{S}_{ij}. \tag{13.1.17}$$

It adds the Smagorinsky eddy viscosity (13.1.11) to the scale-similar stress. The constant in the Smagorinsky model can be prescribed or found dynamically; the latter generally provides improved predictions.

The notion of an *a priori* test is to compare the formula of the subgrid model to subgrid stresses evaluated by filtering DNS data. The model is not solved; DNS data provide $\widehat{S}_{ij}$ and $L_{ij}$ also. Such tests show that the mixed model is an improvement on the Smagorinsky model.

When the model is implemented into an LES code and computed statistics are compared to data, the test is called an *a posteriori* test. For instance, Figure 13.3 has the flavor of an *a posteriori* test. *A posteriori* tests are less flattering to the mixed model; little benefit is found. In complex geometries, the inherent grid dependence of LES overwhelms any perceived benefit to elaborations of the Smagorinsky subgrid model beyond the dynamic procedure.

#### 13.1.2.3 Wall-adapting local eddy viscosity model

Near to a wall, the subgrid shear stress should approach zero as $y^3$; see Section 7.3.3. As the wall is approached,

$$u \to O(y), \qquad v \to O(y^2), \qquad w \to O(y).$$

Thus, the magnitude of the rate of strain is $O(1)$. Consequently, the eddy viscosity should approach zero as $y^3$. The Smagorinsky model is not consistent with this scaling. It makes eddy viscosity proportional to rate of strain, and hence of $O(1)$.

Nicoud and Ducros (1999) introduced a subgrid viscosity with the correct scaling, in their wall-adapting local eddy (WALE) viscosity model. They first noted that the tensor

$$\mathscr{S}_{ij} = \tfrac{1}{2}(\partial_k u_i\, \partial_j u_k + \partial_k u_j\, \partial_i u_k) - \tfrac{1}{3}\delta_{ij}\, \partial_k u_l\, \partial_l u_k, \tag{13.1.18}$$

constructed from the velocity gradient, is $O(y)$ as the wall is approached. The reasoning is of the following form. If all derivatives are in the $x$ and $z$ directions, then the term $\partial_k u_i\, \partial_j u_k$ is $O(y^2)$. If $j$ is the $y$ direction and $k$ is the $x$ direction, then this term is $\partial_x u_i\, \partial_y u = O(y)$. Such considerations lead to $\mathscr{S} = O(y)$.

The tensor (13.1.18) can be rewritten in terms of the rate of rotation and rate of strain,

$$\Omega_{ij} = \tfrac{1}{2}(\partial_i u_j - \partial_j u_i), \qquad S_{ij} = \tfrac{1}{2}(\partial_i u_j + \partial_j u_i),$$

as

$$\mathscr{S}_{ij} = S_{ik}S_{kj} + \Omega_{ik}\Omega_{kj} - \tfrac{1}{3}(S_{ik}S_{ki} + \Omega_{ik}\Omega_{ki})\delta_{ij} \tag{13.1.19}$$

(see Section 2.3). In a parallel shear flow,

$$\boldsymbol{S} = \begin{pmatrix} 0 & \frac{1}{2}\partial_y u & 0 \\ \frac{1}{2}\partial_y u & 0 & 0 \\ 0 & 0 & 0 \end{pmatrix}, \qquad \boldsymbol{\Omega} = \begin{pmatrix} 0 & \frac{1}{2}\partial_y u & 0 \\ -\frac{1}{2}\partial_y u & 0 & 0 \\ 0 & 0 & 0 \end{pmatrix},$$

so $\boldsymbol{S}^2 = -\boldsymbol{\Omega}^2$ and $\mathscr{S}_{ij} = 0$.

Nicoud and Ducros (1999) started the formulation of their WALE viscosity model with the idea that it was more suitable for $\nu_{\mathrm{SGS}}$ to depend on $|\mathscr{S}|$ than on $|\boldsymbol{S}|$. To obtain the scaling $\nu_{\mathrm{SGS}} \sim O(y^3)$, they devised the expression

$$\nu_{\mathrm{SGS}} = (c_{\mathrm{w}}\Delta)^2 \frac{(\mathscr{S}_{ij}\mathscr{S}_{ji})^{3/2}}{(S_{kl}S_{lk})^{5/2} + (\mathscr{S}_{mn}\mathscr{S}_{mn})^{5/4}}. \tag{13.1.20}$$

This is dimensionally correct and has the correct near-wall asymptote. The denominator was devised to prevent a singularity if $|\boldsymbol{S}| = 0$. Near a wall (13.1.20) becomes

$$\nu_{\mathrm{SGS}} = (c_{\mathrm{w}}\Delta)^2 \frac{(\mathscr{S}_{ij}\mathscr{S}_{ji})^{3/2}}{(S_{kl}S_{lk})^{5/2}}.$$

The empirical constant $c_{\mathrm{w}} = 0.5$ was selected.

The WALE subgrid model has been found to provide predictions similar to the dynamic Smagorinsky model (13.1.15). The latter also provides the correct near-wall asymptotic behavior, through its dependence on $L_{ij}$. The WALE formulation does not require explicit filtering and hence is well suited to unstructured grids.

#### 13.1.2.4 What are the grid requirements?

In scientific computing, the term "grid independence" refers to demonstrating accuracy by decreasing the grid spacing until the solution practically does not change. The grid requirements of large eddy simulation are somewhat nebulous. There is no concept of grid independence, *per se*: the subgrid model is an explicit function of the grid. Sometimes DNS is described as the grid-independent limit of LES; but that is not helpful. Nevertheless, ideas about proper LES grid resolution do exist, and grid refinement is sometimes used to provide a sense of accuracy.

The sufficiency of an LES grid is tested by checking that statistics are relatively insensitive to resolution. Spalart (2000) refers to such accuracy studies as physical, rather than numerical, tests. The objective is to make the large eddies accurate, knowing that the subgrid model is grid-dependent. An assessment is made by comparing data from the LES to DNS or to experiment.

Figure 13.4 is from a study in which grid refinement and comparisons to experiment were used for validation. As in many similar studies, the grid was fine near the walls, as shown by the inset, but even then was considered by the computationalists to be marginal. Large scales in the turbulent wake are captured well.

Another test provides a rule of thumb when there are no data with which to compare. As the grid is refined, the subgrid viscosity decreases. When the ratio of the subgrid viscosity to molecular viscosity is not more than the order of 10, LES is found to be very accurate. When it reaches 100, the simulation often is found to be somewhat inaccurate.

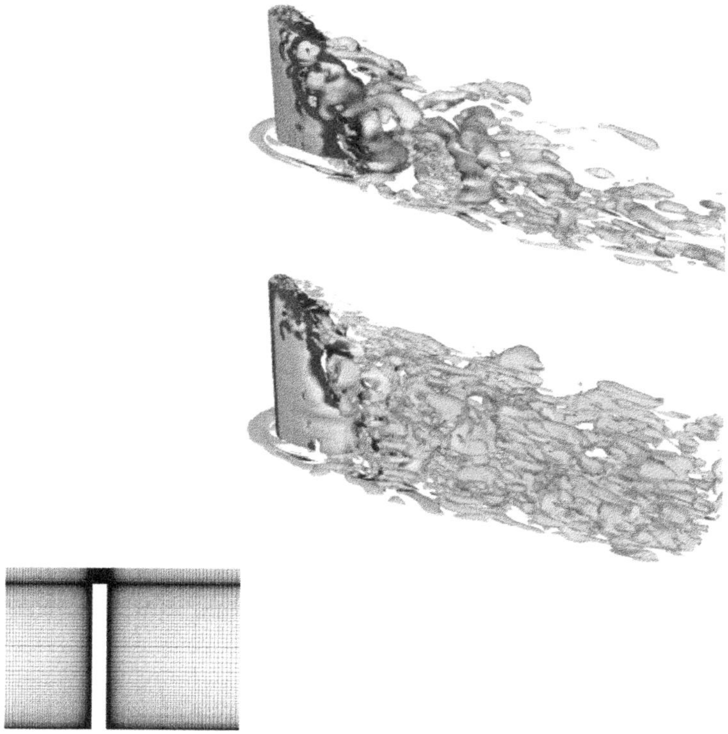

**Figure 13.4** Large eddy simulation of flow over a cylinder attached to a wall. The surface $Q \equiv |S|^2 - |\boldsymbol{\Omega}|^2 = 1$ is shaded by magnitude of vorticity. The height of the cylinder is 6 (top) and 10 (bottom) times its diameter. The grid is shown bottom left. Reprinted from *International Journal of Heat and Fluid Flow*, **28**, Imran Afgan, Charles Moulinec, Robert Prosser and Dominique Laurence, 'Large eddy simulation of turbulent flow for wall mounted cantilever cylinders of aspect ratio 6 and 10', 561–574, Copyright 2007, with permission from Elsevier.

Figure 13.5, from Le Ribault *et al.* (1999), is an example of the ratio of subgrid to molecular viscosity in a jet, using the Smagorinsky and mixed models, with coefficients obtained by the dynamic procedure. The grid is more than adequate, as the maximum ratio is less than 10 in all cases.

The ratio $\mu_{SGS}/\mu$ is a measure of the role of the subgrid model. One can define a subgrid activity parameter

$$s = \frac{\mu_{SGS}}{\mu_{SGS} + \mu},$$

which ranges between zero and unity. Zero corresponds to DNS and unity to very coarse-grid LES. The parameter $s$ can alternatively be defined via the rate of energy dissipation, $\varepsilon = -\tau_{ij} S_{ij}$, as

$$s = \frac{\varepsilon_{SGS}}{\varepsilon_{SGS} + \varepsilon_{\mu}},$$

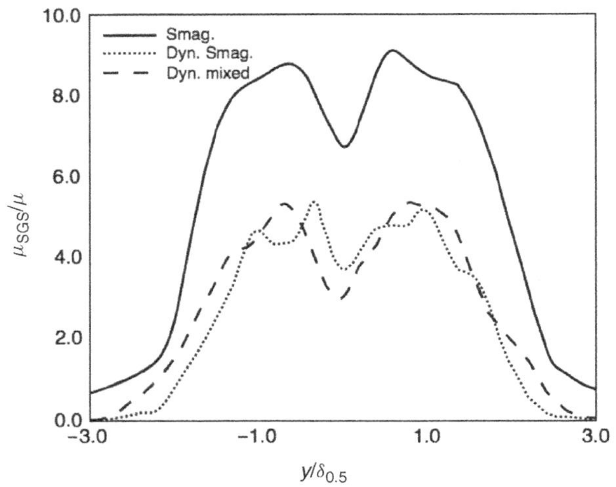


**Figure 13.5** Ratio of subgrid viscosity to molecular viscosity in a turbulent plane jet. Curves are for the Smagorinsky, dynamic Smagorinsky, and mixed models, as indicated. Reprinted with permission from 'Large eddy simulation of a plane jet' by C. Le Ribault, S. Sarkar, S. A. Stanley, in *Physics of Fluids (11)* Copyright 1999, American Institute of Physics.

where $\varepsilon_\mu$ is dissipation by molecular viscosity. Geurts and Fröhlich (2002) provide an example in which the LES error correlates with this definition of $s$. When $s < 1/2$, the error is very low and LES produces the accuracy of DNS. After that, the error rises exponentially with $s$.

The study cited in Figure 13.5 consisted of a DNS of a low Reynolds number jet on a $205 \times 189 \times 60$ grid and an LES on a $61 \times 105 \times 16$ grid. In this case, comparison to DNS provided a direct assessment of the suitability of the LES grid. The subgrid stress was evaluated exactly from formula (13.1.7) by filtering the DNS to the LES resolution. In an *a priori* test, the dynamic mixed model (13.1.17) was found to agree quite well with the DNS data, while the dynamic Smagorinsky model greatly underpredicted the stress.

However, this assessment proved to be unrepresentative of *a posteriori* performance. The dynamic Smagorinsky model provided an accurate prediction when used in a full computation: it is not safe to rely on *a priori* assessments; a proper evaluation demands that the model be used in a full simulation. Figure 13.6 is an example in which the jet thickness predicted by LES is compared to DNS. Both the dynamic Smagorinsky and dynamic mixed models are faithful to the DNS. Other variables, including the mean velocity profiles and profiles of Reynolds stresses, were also found to be accurate when the dynamic procedure was applied either to the Smagorinsky or to the mixed model. As in Figure 13.6, the Smagorinsky model with a fixed constant performed very poorly. The dynamic procedure produced a coefficient $c_s$ that decreased with downstream distance and at the edge of the jet. This variation appears to be needed to obtain an accurate large eddy simulation.

Inaccuracy of a large eddy simulation has two sources: the numerical discretization, and limitations of the subgrid model. By nature, LES is an under-resolved simulation of turbulent eddies, so numerical error is inevitable. The standard definition of numerical

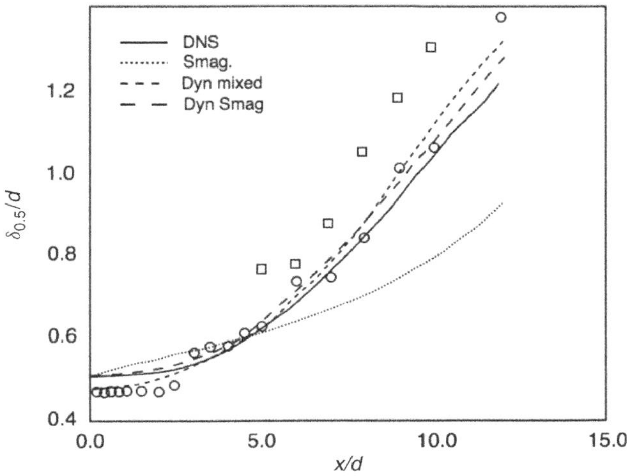


**Figure 13.6** Evolution of jet thickness as predicted by three subgrid models, compared to DNS and experimental data. Reprinted with permission from 'Large eddy simulation of a plane jet' by C. Le Ribault, S. Sarkar, S. A. Stanley, in *Physics of Fluids (11)* Copyright 1999, American Institute of Physics.

error applies within the neighborhood of a well-resolved computation. For instance, if a method is accurate to second order, the error tends to zero in proportion to $\Delta^3$ as $\Delta/L \to 0$, where $L$ is scale of velocity variation. But in large eddy simulation, $\Delta/L \sim 1$ and we are not considering its approach to zero. The definition of order of accuracy is cloudy. Ambiguity could be removed by making the computational grid fine compared to the filter width (Geurts and Fröhlich 2002); but that is not common practice. Grid spacing and filter width are usually the same. If the grid is quite coarse, the numerical error can overwhelm the subgrid model. One should appreciate that the two causes of inaccuracy are not easily separated.

Another cause of inaccuracy is slow statistical convergence. Eddy simulations start from an arbitrary initial condition. In time, through nonlinear scrambling and the energy cascade, the flow field becomes representative of fluid dynamical turbulence. If the computational domain is between inflow and outflow boundaries, the duration of the transient is measured in *through-flow* times. Through-flow time is defined as the domain length divided by the bulk velocity. Typically 5–10 through-flow times are required to reach statistically steady state. After that, statistics are accumulated. It is common to cite the duration of averaging in through-flow times, as well; typically averaging occurs over another 5–100 through-flow times. This is a bit misleading. The rate of statistical convergence is determined by the correlation time, as discussed on page 20. If the domain were 20 eddy correlation lengths, 10 through flows would sample 200 eddies; that is a more relevant characterization. Statistical convergence is slow, improving as the square root of the number of samples. So averaging over 200 eddies reduces error only by a factor of about 14. This makes LES, and eddy simulation in general, costly. For instance, the curves in Figure 13.5 are averaged data. They should be smooth and symmetric; clearly a much longer time would be needed to obtain statistically converged plots.

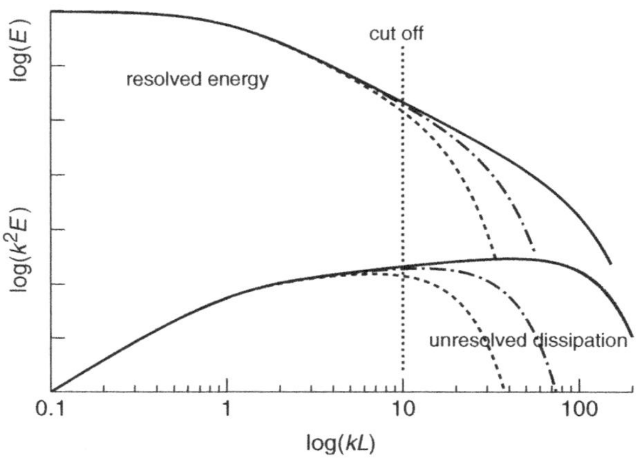


**Figure 13.7** Energy and dissipation spectra at three Reynolds numbers, increasing from dashed to solid. The vertical line marked "cut off" divides resolved from subgrid scales.

The question "What are the grid requirements?" that heads this section could be addressed in the same vein as Section 12.1.1: How do grid requirements scale with Reynolds number? Scaling for LES is not clear-cut. The viscosity entered DNS scaling via the dissipation length, $\eta$. LES does not resolve eddies of order $\eta$ in size. The perspective offered by Figure 12.1 is that dissipation eddies are smaller than the grid spacing. Another perspective is provided by Figure 13.7. It illustrates the variation of the spectrum of homogeneous turbulence with Reynolds number, when the wavenumber is non-dimensionalized by the integral scale. Below the cut-off, the spectrum is insensitive to Reynolds number. Hence, requirements on grid resolution are independent of Reynolds number. Above the cut-off, the spectrum extends to smaller scales because $L/\eta$ increases as $R_T^{3/4}$ (see page 19). The energy is low above the cut-off, but dissipation has its maximum at subgrid scales. The subgrid model must account for an increasing proportion of the energy dissipation as $R_T$ increases.

The curves in Figure 13.7 are plotted with the same level of $E$ at small $kL$. Were they not normalized so, this level would vary with Reynolds number. Hence Figure 13.7 suggests that the grid requirement is independent of $R_T$, not that the simulation is.

The spectral reasoning is valid for homogeneous turbulence, but it also applies to free-shear flow. Free-shear layers tend to be tolerant of low resolution. Consider the schematic in Figure 1.8. It suggests how the small scales become more intense as the Reynolds number increases, but the large scales are unchanged. If only the large scales are resolved, the grid should be nearly independent of Reynolds number. The number of grid points should scale on the shear-layer thickness. The thickness of canonical, fully developed free-shear layers is insensitive to Reynolds number–see Section 4.3.1. Thus the number of grid points within a computational domain of height $H$ scales on the ratio $H/\delta$ of height to shear-layer thickness. For instance, 30 grid points across the jet might resolve the large eddies. Then $\Delta y = \delta/30$, so that $30H/\delta$ points are needed if a uniform grid spans the domain. Similar grid spacing is needed in spanwise and streamwise directions, so $N \approx 30^3\, W \times H \times L/\delta^3$. Usually a coarser grid will be used outside the shear layer to reduce this number.

LES is well suited for free-shear layers because the resolution requirements are nearly independent of Reynolds number. Again, the solution can depend on Reynolds number via the split of dissipation between resolved and subgrid scales. The total, $2(\nu + \nu_{SGS})|S|^2$, includes contributions from both resolved and subgrid scales.

The most challenging issue in gridding for LES is near-wall resolution. The *outer layer* of a turbulent boundary layer is analogous to a free-shear layer: $N \sim (L/\delta)^3$. Boundary-layer thickness grows as $x/R_x^{1/5}$ (see page 68). If we interpret $x$ as $L$, then $N \sim \mathrm{Re}^{3/5}$. Alternatively, if the domain height scales on the boundary-layer thickness, then $L_y/\delta$ is constant and $N \sim \mathrm{Re}^{2/5}$ (Piomelli 2008). The grid requirements are not highly demanding, even at high Reynolds number.

The expense arises from resolving the inner layer. If the outer grid scales on $\delta$, then the expense is due to a requirement for finer spacing, relative to $\delta$, near the wall. In the direction normal to the wall, the eddy size scales as $\nu/u_*$; hence

$$N_y \sim \delta/(\nu/u_*) \sim R_\delta\sqrt{C_f} \sim R_\delta^{7/8}.$$

If the same resolution is applied in the spanwise direction, but the outer scaling is retained in the streamwise direction, then

$$N_x N_y N_z \sim R_\delta^{1/5} R_\delta^{7/8} R_\delta^{7/8} = R_\delta^{1.95}.$$

In practice, the expense is comparable to DNS. LES of wall-bounded flows is expensive because the energetic eddies become small as the wall is approached. Efforts to reduce expense by replacing simulation by a wall model are surveyed by Piomelli (2008). We will not discuss wall models. Rather, we move on to the topic of detached simulation, which has the same motivation. The region next to walls is modeled by RANS, avoiding the stringent demands of eddy-resolving simulation.

## 13.2 Detached eddy simulation

Several versions of detached eddy simulation (DES) have been developed since it was originated in 1999 (see Spalart, 2009). The original DES was based on the Spalart–Allmaras model of Section 6.6. That is an eddy viscosity model for Reynolds averaged flow. Detached eddy simulation consists of limiting the eddy viscosity, so as to permit large-scale, turbulent eddies to occur. In the Spalart–Allmaras model, the distance to the wall, $d$, is used as a length scale. For detached eddy simulation, this same model is solved, but with $d$ replaced by

$$\tilde{d} = \min[d, C_{\mathrm{DES}} \max(\Delta x, \Delta y, \Delta z)],$$

with $C_{\mathrm{DES}} = 0.65$. Near to a wall, this reduces to the original value $d$. Farther from the wall, it becomes $C_{\mathrm{DES}}$ times the maximum grid spacing. For instance, if $\Delta x^+ \sim 100$, then $\tilde{d}$ would be $d$ for $y^+ \lesssim 65$ and $C_{\mathrm{DES}}\Delta x$ for $y^+ \gtrsim 65$. Then a computation that used the DES limiter would transition from RANS to eddy simulation within the log layer.

The bound on length scale has the effect of bounding the eddy viscosity. If the eddy viscosity is kept low, natural instabilities within separated shear layers are able to evolve into turbulence. Also, the computation must be three-dimensional and accurate in time to permit eddying.

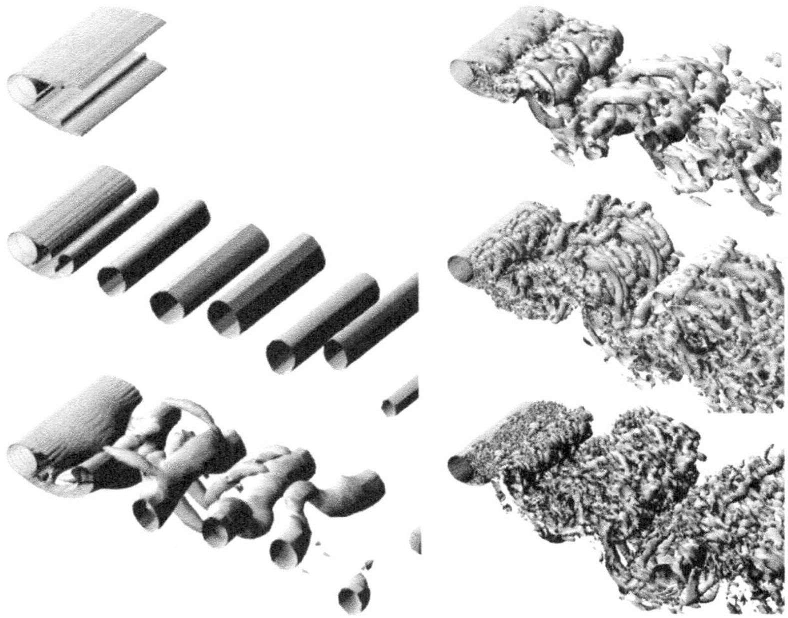

**Figure 13.8** Computations of flow over a cylinder. First column (from top): 2D steady RANS; 2D unsteady RANS; and 3D, unsteady RANS. Second column (from top): DES on a coarse grid; and two DES models on a fine grid. Reproduced with permission of Annual Review of Fluid Mechanics, Vol. 41, 2009.

A zone near to the wall is called the RANS region, and the rest of the flow is the eddy simulation region; this behavior is illustrated by Figure 13.8. Figure 13.8 compares steady and unsteady RANS in the left column to DES in the right column. As the flow passes over the cylinder, shear layers detach from the surface. In the RANS computation, they roll into vortices in the wake. The detached eddy simulation shows highly three-dimensional structure in the wake, with rolls and ribs similar to those discussed in Chapter 5 and shown in Figure 5.2. The shear layer transitions promptly upon separating from the surface. The automatic transition to eddy simulation is what makes this such an attractive method. No boundary between RANS and LES regions need be drawn; the same model serves as both the Reynolds averaged and subgrid eddy viscosity.

Indeed, there cannot be a clear boundary between RANS and eddy simulation regions. Although, the notion of detached eddy simulation is that the computation shifts from a Reynolds averaged formulation in attached boundary layers to eddy simulation in separated shear layers, in practice the eddy simulation region extends into the boundary layer. There is no RANS region, *per se*, because the whole flow is unsteady. With sufficient grid resolution, and sufficiently low Reynolds number, DES behaves quite like LES all the way to the wall.

The transition to eddying is effected by limiting the eddy viscosity. In the original DES formulation, this was done indirectly, through a term in a transport equation. Rather than being indirect, the eddy viscosity can be bounded directly. The eddy viscosity is naturally expressed as $\nu_T = kT$, where $T$ is the correlation time-scale (see page 31). If this is re-expressed in terms of a mixing length, as $\nu_T = \sqrt{k}\ell$, then the DES model is

$$\nu_T = \sqrt{k}\,\tilde{\ell} \qquad \text{with} \qquad \tilde{\ell} = \min[\ell, C_{\text{DES}} \max(\Delta x, \Delta y, \Delta z)]. \tag{13.2.1}$$

For instance, in the $k$–$\omega$ model (6.3.2), the mixing length is $\ell = C_\mu \sqrt{k}/\omega$. Placing a bound on $\ell$ allows eddying in detached shear layers.

A more elaborate variant of the $k$–$\omega$ DES model was developed by Menter and co-workers (Menter *et al.*, 2003), who called it scale adaptive simulation (SAS). It was derived from the SST model described on page 138. There were two primary motives for SAS. One was to replace the grid spacing with a physical length scale. The Von Karman length

$$L_{\text{vK}}^2 = \frac{|\kappa \boldsymbol{S}|^2}{|\nabla^2 \boldsymbol{U}|^2} \tag{13.2.2}$$

plays that role. Here $\kappa = 0.41$ is the Von Karman constant.

The second motive for SAS was to reduce the severe grid sensitivity created by (13.2.1). In a typical boundary-layer grid, $\Delta x$ will be larger than the other grid spacings. If $\Delta x$ is larger than the boundary-layer thickness, then (13.2.1) will ensure that the RANS model is used in the boundary layer. However, if $\Delta x$ is smaller than the boundary thickness, the model will switch to eddy simulation inside the boundary layer. That is contrary to the intent of DES; it is meant to simulate eddies in detached shear layers. The SAS formulation invokes the interpolation functions of the SST model to forestall the switch to eddy simulation. Spalart (2009) refers to this as delayed detached eddy simulation (DDES).

Switching inside the boundary layer is undesirable because it creates a stress depletion layer. The total stress is a sum of that due to the eddy viscosity and that due to the resolved eddies; say

$$\tau_{ij} = -2\nu_T S_{ij} + \overline{\hat{u}_i \hat{u}_j},$$

where the overbar denotes a Reynolds average. Turbulent eddies begin to form when the eddy viscosity is suppressed, but initially they are weak and the second contribution to the stress tensor does not make up for the reduction of the first. Hence, the total stress is underpredicted at the start of the eddying region. That effect is benign if it occurs outside the boundary layer. The general notion of delayed DES is to maintain the RANS eddy viscosity throughout most of the boundary layer, so that the stress depletion region is away from the steep shear near the wall.

A stark version of DDES was described by Spalart *et al.* (2006). They define the parameter

$$r_d = \frac{\nu_T + \nu}{\sqrt{\partial_i u_j\, \partial_i u_j}\; \kappa^2 d^2}.$$

By analogy to the SST interpolation functions, they let

$$f_d = 1 - \tanh(8^3 r_d^3)$$

and interpolate as

$$\tilde{d} = (1 - f_d)d + f_d \min[d, C_{\text{DES}} \max(\Delta x, \Delta y, \Delta z)]. \tag{13.2.3}$$

**Figure 13.9** Contours of spanwise velocity in flow over a backstep using the SAS model. Courtesy of Dr. Jongwook Joo.

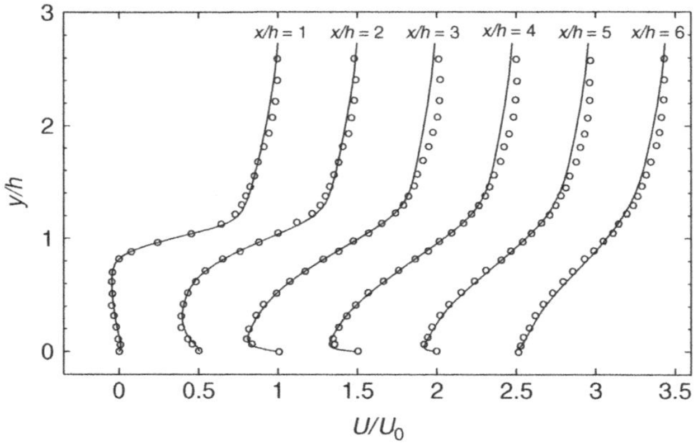


**Figure 13.10** Mean velocity profiles in flow over a backstep using the SAS model (solid lines) compared to data. Courtesy of Dr. Jongwook Joo.

When $d \to 0$, $r_d \to \infty$ and $f_d \to 0$. When $d \to \infty$, $r_d \to 0$ and $f_d \to 1$. In the log layer,

$$r_d \sim \frac{\nu_T}{\partial_y u \kappa^2 y^2} = \frac{\kappa u_* y}{(u_*/\kappa y)\kappa^2 y^2} = 1,$$

and hence $f_d \sim 1 - \tanh 8^3 \approx 0$. The switch to eddy simulation cannot occur in the log layer, irrespective of the grid resolution.

A second example of DES is provided by Figures 13.9 and 13.10. This is flow over a backward-facing step. It was computed with the SAS model. The instantaneous velocity contours show the unsteady eddying that develops behind the step. They can be compared to the steady, Reynolds averaged flow presented in Section 7.4.2.2. Fluctuations were added at the inlet in order to trigger instability in the detached shear layer. Eddy fields are averaged to obtain statistics, just as in all other forms of eddy simulation. Figure 13.10 shows mean velocity profiles that were obtained by averaging the simulation over a long time.

Detached eddy simulation is an extension of Reynolds averaged computation. As such, often it is implemented into a computer code that already solves the underlying model, with some changes to the numerics. For instance, a low-dissipation treatment of convection is needed to permit the eddying to develop. Schemes for convection are similar to those described in Section 12.1.3.

# Exercises

**Exercise 13.1.** *Filtering in physical space.* The top-hat filter is

$$\hat{u} = \frac{1}{\Delta_f}\int_{-\Delta_f/2}^{\Delta_f/2} u(x+\xi)\,d\xi.$$

Let $u = \sin(kx)$. Evaluate $\hat{u}$ and $\hat{\hat{u}}$.

Let

$$u = \begin{cases} |1-x|, & x < 1, \\ 0, & |x| \geq 1. \end{cases}$$

Evaluate $\hat{u}$ and compare it to the unfiltered function.

**Exercise 13.2.** *Filter in Fourier space.* The Gaussian filter is expressed by the running average

$$\hat{u} = \frac{1}{\sqrt{\pi}\,\Delta_f}\int_{-\infty}^{\infty} e^{-\xi^2/\Delta_f^2} u(x+\xi)\,d\xi.$$

What is $F_k$ of this filter in Fourier space?

**Exercise 13.3.** *Filtered spectrum.* Convince yourself that the filtered and unfiltered one-dimensional energy spectra (see page 256) are related by

$$\widehat{\Theta}_{11} = F_k F_k^* \Theta_{11}.$$

On a single graph, plot the filtered and unfiltered spectrum (9.2.22) with $p = 17/6$ for various ratios of $\Delta_f/L$. Use either the top-hat or Gaussian filter.

**Exercise 13.4.** *Filtering and vorticity.* Let

$$u = A\cos(kx)\sin(ky)\cos(kz),$$

$$v = -A\sin(kx)\cos(ky)\cos(kz),$$

$$w = 0.$$

This is called the Taylor–Green vortex. Evaluate the vorticity vector field. Average $|\boldsymbol{u}|^2$ and $|\boldsymbol{\omega}^2|$ over one period of the sine waves in each direction.

Let

$$A_2 = \frac{k^3}{[1+(k\eta)^2]^2}.$$

Filter the velocity and vorticity fields with a filter width equal to $5\eta$. Average $|\hat{\boldsymbol{u}}|^2$ and $|\hat{\boldsymbol{\omega}}^2|$ over a period and compare to their unfiltered values.

**Exercise 13.5.** *WALE subgrid model.* Show that

$$\mathscr{S}_{ij}\mathscr{S}_{ji} = \tfrac{1}{6}(S^2S^2 + \Omega^2\Omega^2) - \tfrac{2}{3}\Omega^2S^2 + 2S_{ij}^2\Omega_{ji}^2,$$

where $\mathscr{S}_{ij}$ is defined in (13.1.19), $\Omega^2_{ij} = \Omega_{ik}\Omega_{kj}$, $\Omega^2 = \Omega^2_{ii}$, and similarly for $S$. The Cayley–Hamilton theorem (2.3.7) with the invariants defined in Eq. (2.3.9) is needed. Why is $\Omega^2 \leq 0$? Show that $\mathscr{S}_{ij}\mathscr{S}_{ji}$ is zero in parallel shear flow, $u(y)$.

**Exercise 13.6.** *LES via Burgers equation.* Repeat Exercise 12.2, adding the Smagorinsky model (13.1.12) and using a four times coarser grid. Select a value of $c_{\text{Burgers}}$ to optimize agreement with the filtered DNS field. Compare the LES solution to the filtered DNS data at several times.

# References

ABU-GANNAM, B. J. and SHAW, R. 1980. Natural transition of boundary layers – the effect of turbulence, pressure gradient and flow history. *J. Mech. Eng. Sci.* **22**, 213–228.

AFGAN, I., MOULINEC, C., PROSSER, R. and LAURENCE, D. 2007. Large eddy simulation of turbulent flow for wall mounted cantilevered cylinders of aspect ratio 6 and 10. *Int. J. Heat Fluid Flow* **28**, 561–574.

ANTONIA, R. A. and KIM, J. 1994. Low-Reynolds-number effects on near-wall turbulence. *J. Fluid Mech.* **276**, 61–80.

BALDWIN, B. S. and BARTH, T. J. 1990. A one-equation turbulence transport model for high Reynolds number wall-bounded flows. NASA TM-102847, National Aeronautics and Space Administration.

BALDWIN, B. S. and LOMAX, H. 1978. Thin layer approximation and algebraic model for separated turbulent flows. AIAA Paper 78-257, American Institute of Aeronautics and Astronautics.

BARDINA, J. E., HUANG, P. G. and COAKLEY, T. J. 1997. Turbulence modeling validation. AIAA Paper 97-2121, American Institute of Aeronautics and Astronautics.

BARLOW, R. S. and JOHNSTON, J. P. 1988. Structure of a turbulent boundary layer on a concave surface. *J. Fluid Mech.* **191**, 137–176.

BATCHELOR, G. K. 1967. *An Introduction to Fluid Dynamics*. Cambridge University Press.

BATCHELOR, G. K. and PROUDMAN, I. 1954. The effect of rapid distortion on a fluid in turbulent motion. *Q. J. Mech. Appl. Math.* **7**, 83–103.

BATCHELOR, G. K. and TOWNSEND, A. A. 1947. Decay of vorticity in isotropic turbulence. *Proc. R. Soc. A.* **190**, 534–550.

BEHNIA, M., PARNEIX, S. and DURBIN, P. A. 1998. Prediction of heat transfer in an axisymmetric turbulent jet impinging on a flat plate. *Int. J. Heat Mass Transfer.* **41**, 1845–1855.

BELL, J. H. and MEHTA, R. D. 1990. Development of a two-stream mixing layer from tripped and untripped boundary layers. *AIAA J.* **28**, 2034–2042.

BORGAS, M. S. and SAWFORD, B. L. 1996. Molecular diffusion and viscous effects on concentration statistics in grid turbulence. *J. Fluid Mech.* **324**, 25–54.

BRADSHAW, P. (ed.) 1976. *Turbulence*, Topics in Applied Physics, vol. 12. Springer.

BRITTER, R. E., HUNT, J. C. R. and MUMFORD, J. C. 1979. The distortion of turbulence by a circular cylinder. *J. Fluid Mech.* **92**, 269–301.

*Statistical Theory and Modeling for Turbulent Flows, Second Edition* P. A. Durbin and B. A. Pettersson Reif

Brown, G. L. and Roshko, A. 1974. On density effects and large structure in turbulent mixing layers. *J. Fluid Mech.* **64**, 775–816.

Buice, C. U. and Eaton, J. K. 1997. Experimental investigation of flow through an asymmetric plane diffuser. Report TSD-107, Department of Mechanical Engineering, Stanford University.

Cantwell, B. J. 1981. Organized motion in turbulent flow. *Annu. Rev. Fluid Mech.* **13**, 457–515.

Canuto, C. G., Husaini, M. Y., Quateroni, A. and Zang, T. A. 2006. *Spectral Methods: Fundamentals in Single Domains*. Springer.

Cazalbou, J. B., Spalart, P. R. and Bradshaw, P. 1994. On the behavior of two-equation models at the edge of a turbulent region. *Phys. Fluids.* **6**, 1797–1953.

Chapman, D. R. 1979. Computational aerodynamics development and outlook. *AIAA J.* **12**, 1293–1313.

Chen, C.-J. and Jaw, S.-Y. 1998. *Fundamentals of Turbulence Modeling*. Taylor & Francis.

Chen, C.-J. and Patel, V. C. 1988. Near-wall turbulence models for complex flows including separation. *AIAA J.* **26**, 641–648.

Choi, K. S. and Lumley, J. L. 1984. On velocity correlations and the solution of equations of turbulent fluctuations. *Turbulence and Chaotic Phenomena in Fluids* (ed. T. Tatsumi), 267. North-Holland.

Chou, P. Y. 1945. On velocity correlations and the solution of equations of turbulent fluctuations. *Q. Appl. Math.* **3**, 38–54.

Clauser, F. H. 1956. The turbulent boundary layer. *Adv. Mech.* **4**, 1–51.

Comte, P., Silvestrini, J. H. and Bégou, P. 1998. Streamwise vortices in large-eddy simulations of mixing layers. *Eur. J. Mech. B.* **17**, 615–637.

Cooper, D., Jackson, D., Launder, B. and Liao, G. 1993. Impinging jet studies for turbulence model assessment – I. Flow-field experiments. *Int. J. Heat Mass Transfer.* **36**, 2675–2684.

Copson, E. T. 1975. *Partial Differential Equations*. Cambridge University Press.

Corrsin, S. and Kisler, A. L. 1954. The free-stream boundaries of turbulent flows. NACA Tech. Note 3133, National Advisory Committee for Aeronautics.

Craft, T., Graham, L. and Launder, B. 1993. Impinging jet studies for turbulence model assessment – II. An examination of the performance of four turbulence models. *Int. J. Heat Mass Transfer.* **36**, 2685–2697.

Craft, T., Launder, B. and Suga, K. 1996. Development and application of a cubic eddy-viscosity model of turbulence, *Int. J. Heat Fluid Flow.* **17**, 108–115.

Craik, A. D. D. 1989. The stability of unbounded two- and three-dimensional flows subject to body forces: some exact solutions. *J. Fluid Mech.* **198**, 275–292.

Criminale, W. O., Jackson, T. L. and Joslin, R. D. 2003. *Theory and Computation in Hydrodynamic Stability*. Cambridge University Press.

Crow, S. C. 1968. Viscoelastic properties of fine grained incompressible turbulence. *J. Fluid Mech.* **33**, 1–20.

Cutler, A. D. and Johnston, J. P. 1989. The relaxation of a turbulent boundary layer in an adverse pressure gradient. *J. Fluid Mech.* **200**, 367–387.

Daly, B. J. and Harlow, F. H. 1970. Transport equations of turbulence. *Phys. Fluids.* **13**, 2634–2649.

Danaila, I. and Boersma, B. J. 1998. Mode interaction in a forced homogeneous jet at low Reynolds numbers. Center for Turbulence Research, Summer Program, 1998, Stanford University.

DeGraaff, D. B. and Eaton, J. K. 2000. Reynolds Number scaling of the flat plate turbulent boundary layer. *J. Fluid Mech.* **422**, 319–386.

Del Alamo, J. C. and Jiménez, J. 2003. Spectra of the very large anisotropic scales in turbulent channels. *Phys. Fluids A.* **15**, L41–L44.

Demuren, A. O. and Wilson, R. V. 1995. On elliptic relaxation near wall models. *Transition, Turbulence and Combustion*, vol. II, *Turbulence and Combustion* (eds M. Y. Hussaini, T. B. Gatski and T. L. Jackson) 61–71. Kluwer Academic.

Domaradski, J. A. and Rogallo, R. S. 1991. Local energy transfer and nonlocal interactions in homogeneous isotropic turbulence. *Phys. Fluids A.* **2**, 413–426.

Driver, D. and Seegmiller, H. L. 1985. Features of a reattaching turbulent shear layer in divergent channel flow. *AIAA J.* **23**, 163–171.

Durbin, P. A. 1980. A stochastic model of two-particle dispersion and concentration fluctuations in homogeneous turbulence. *J. Fluid Mech.* **100**, 279–302.

Durbin, P. A. 1981. Distorted turbulence in axisymmetric flow. *Q. J. Mech. Appl. Math.* **34**, 489–500.

Durbin, P. A. 1991. Near-wall turbulence closure modeling without "damping functions". *Theor. Comput. Fluid Dyn.* **3**, 1–13.

Durbin, P. A. 1993. A Reynolds stress model for near-wall turbulence. *J. Fluid Mech.* **249**, 465–498.

Durbin, P. A. 1995. Separated flow computations with the $k$–$\varepsilon$–$\overline{v^2}$ model. *AIAA J.* **33**, 659–664.

Durbin, P. A. 1996. On the $k$–$\varepsilon$ stagnation point anomaly. *Int. J. Heat Fluid Flow.* **17**, 89–90.

Durbin, P. A. and Pettersson Reif, B. A. 1999. On algebraic second moment models. *Flow, Turbul. Combust.* **63**, 23–37.

Durbin, P. A. and Speziale, C. G. 1994. Realizability of second moment closures by stochastic analysis. *J. Fluid Mech.* **280**, 395–407.

Eringen, A. C. 1980. *Mechanics of Continua*, 2nd edn, vol. 1. R. E. Krieger.

Fletcher, C. A. J. 1991. *Computational Techniques for Fluid Dynamics*. Springer.

Fox, R. O. 2003. *Computational Models for Turbulent Reacting Flows*. Cambridge University Press.

Fung, J. C. H., Hunt, J. C. R., Malik, N. A. and Perkins, R. J. 1992. Kinematic simulation of homogeneous turbulence by unsteady random Fourier modes. *J. Fluid Mech.* **236**, 281–318.

Gatski, T. B. and Speziale, C. G. 1993. On explicit algebraic stress models for complex turbulent flows. *J. Fluid Mech.* **254**, 59–78.

Gence, J. N. and Mathieu, J. 1979. On the application of successive plane strains to grid-generated turbulence. *J. Fluid Mech.* **93**, 501–513.

Germano, M., Piomelli, U., Moin, P. and Cabot, W. H. 1991. A dynamics subgrid scale eddy viscosity model. *Phys. Fluids A.* **3**, 1760–1765.

Geurts, B. J. and Fröhlich, J. 2002. A framework for predicting accuracy limitations in large-eddy simulation. *Phys. Fluids.* **14**, L41–L44.

Gibson, M. M. and Launder, B. E. 1978. Ground effects on pressure fluctuations in the atmospheric boundary layer. *J. Fluid Mech.* **86**, 491–511.

Giel, P. W., Thurman, D. R., Van Fossen, G. J., Hippensteele, A. A. and Boyle, R. J. 1998. Endwall heat transfer measurements in a transonic turbine cascade. *Trans. ASME, J. Turbomach.* **120**, 305–313.

Gillis, J. C., Johnston, J. P., Kays, W. M. and Moffat, R. J. 1980. Turbulent boundary layer on a convex curved surface. Report HMT-31, Department of Mechanical Engineering, Stanford University.

Goldstein, M. E. and Durbin, P. A. 1980. The effect of finite turbulence spatial scale on the amplification of turbulence by a contracting stream. *J. Fluid Mech.* **98**, 473–508.

Goodbody, A. M. 1982. *Cartesian Tensors*. Ellis Horwood.

GRINSTEIN, F. F., MARGOLIN, L. G. and RIDER, W. J. (eds) 2007. *Implicit Large Eddy Simulation*. Cambridge University Press.

HANJALIC, K. 1994. Advanced turbulence closure models: a view of current status and future prospects. *Int. J. Heat Fluid Flow*. **15**, 178–203.

HANJALIC, K. and LAUNDER, B. E. 1980. Sensitizing the dissipation to irrotational strains. *Trans. ASME, J. Fluids Eng*. **102**, 34–40.

HANJALIC, K., POPOVAC, M. and HADZIABDIC, M. 2005. A robust near-wall elliptic relaxation eddy-viscosity turbulence model for CFD. *Int. J. Heat Fluid Flow*. **25**, 1047–1051.

HEAD, M. R. and BANDYOPADHYAY, P. 1981. New aspects of turbulent boundary-layer structure. *J. Fluid Mech*. **107**, 497–532.

HUNT, J. C. R. 1973. A theory of turbulent flow round two-dimensional bluff bodies. *J. Fluid Mech*. **61**, 625–706.

HUNT, J. C. R. 1977. A review of rapidly distorted turbulent flows and its applications. *Fluid Dyn. Trans*. **9**, 121–152.

HUNT, J. C. R. and CARRUTHERS, D. J. 1990. Rapid distortion theory and the "problems" of turbulence. *J. Fluid Mech*. **212**, 497–532.

HUNT, J. C. R. and GRAHAM, J. M. R. 1978. Free-stream turbulence near plane boundaries. *J. Fluid Mech*. **84**, 209–235.

HUSER, A. and BIRINGEN, S. 1993. Direct numerical simulation of turbulent flow in a square duct. *J. Fluid Mech*. **257**, 65–95.

HUSSAIN, A. K. M. F. 1978. Coherent structures and turbulence. *J. Fluid Mech*. **173**, 303–356.

HUSSEIN, J. H., CAPP, S. P. and GEORGE, W. K. 1994. Velocity measurements in a high-Reynolds number momentum-conserving axisymmetric, turbulent jet. *J. Fluid Mech*. **258**, 31–75.

IACCARINO, G., OOI, A., DURBIN, P. A. and BEHNIA, M. 2003. Reynolds averaged simulation of unsteady separated flow. *Int. J. Heat Fluid Flow*. **24**, 147–156.

IKEDA, T. and DURBIN, P. A. 2007. Direct simulations of a rough wall channel flow. *J. Fluid Mech*. **571**, 235–263.

JACOBS, R. G. and DURBIN, P. A. 2000. Bypass transition phenomena studied by computer simulation. Report TF-77, Department of Mechanical Engineering, Stanford University.

JIMÉNEZ, J. 1999. Self-similarity and coherence in the turbulent cascade. *Proc. IUTAM Symp. on Geometry and Statistics of Turbulence*, Hayama, Japan (ed. T. Kambe) 58–66. Kluwer.

JOHNSON, D. A. and KING, L. S. 1985. A mathematically simple turbulence closure model for attached and separated turbulent boundary layers. *AIAA J*. **23**, 1684–1692.

JOHNSON, P. L. and JOHNSTON, J. P. 1989. The effects of grid-generated turbulence on flat and concave turbulent boundary layers. Report MD-53, Department of Mechanical Engineering, Stanford University.

JOHNSON, T. A. and PATEL, V. C. 1999. Flow past a sphere up to a Reynolds number of 300. *J. Fluid Mech*. **378**, 19–70.

JONES, W. P. and LAUNDER, B. E. 1972. The prediction of laminarization with a two-equation model. *Int. J. Heat Mass Transfer* **15**, 301–314.

KADER, B. A. 1981. Temperature and concentration profiles in fully turbulent boundary layer. *Int. J. Heat Mass Transfer* **24**, 1541–1544.

KAYS, W. M. 1994. Turbulent Prandtl number – Where are we? *J. Heat Transfer* **116**, 284–295.

KAYS, W. M. and CRAWFORD, M. E. 1993. *Convective Heat and Mass Transfer*, 3rd edn. McGraw-Hill.

KIM, J. and MOIN, P. 1989. Transport of passive scalars in a turbulent channel flow. *Turbul. Shear Flows*. **6**, 85–95.

KLINE, S. J., MORKOVIN, M. V., SOVRAN, G. and COCKRELL, D. J. (eds) 1968. *Computation of Turbulent Boundary Layers, Proc. AFOSR–IFP–Stanford Conf.*, vol. I, *Methods, Predictions, Evaluation and Flow Structure*.

KOLMOGOROFF, A. N. 1941. The local structure of turbulence in incompressible viscous fluid for very large Reynolds number. *Dokl. Akad. Nauk SSSR*. **30**, 301–305.

KRAICHNAN, R. H. 1968. Small-scale structure convected by turbulence. *Phys. Fluids*. **11**, 945–953.

KRISTOFFERSEN, R. and ANDERSSON, H. I. 1993. Direct simulations of low-Reynolds-number turbulent flow in a rotating channel. *J. Fluid Mech*. **256**, 163–197.

LAI, Y. G. and SO, R. M. C. 1990. On near-wall turbulent flow modelling. *AIAA J*. **34**, 2291–2298.

LAIZET, S., LARDEAU, S. and LAMBALLAIS, E. 2010. Direct numerical simulation of a mixing layer downstream a thick splitter plate. *Phys. Fluids*. **22**, 015104.

LAUNDER, B. E. 1989. Second-moment closure: present... and future. *Int. J. Heat Fluid Flow*. **10**, 282–300.

LAUNDER, B. E. and KATO, M. 1993. Modelling flow-induced oscillations in turbulent flow around a square cylinder. *Proc. Forum on Unsteady Flows*, ASME FED. **157**, 189–199.

LAUNDER, B. E. and LI, B. L. 1994. On the elimination of wall topography parameters from second-moment closure. *Phys. Fluids*. **6**, 999–1006.

LAUNDER, B. E. and RODI, W. 1983. The turbulent wall-jet: measurements and modelling. *Annu. Rev. Fluid Mech*. **15**, 429–459.

LAUNDER, B. E. and SHARMA, B. I. 1974. Application of the energy-dissipation model of turbulence to the calculation of flow near a spinning disk. *Lett. Heat Mass Transfer*. **1**, 131–138.

LAUNDER, B. E. and SHIMA, N. 1989. Second-moment closure for the near-wall sublayer: development and application. *AIAA J*. **27**, 1319–1325.

LAUNDER, B. E., REECE, G. J. and RODI, W. 1975. Progress in the development of a Reynolds stress turbulence closure. *J. Fluid Mech*. **68**, 537–566.

LELE, S. K. 1992. Compact difference schemes with spectral like resolution. *J. Comput. Phys*. **103**, 16–42.

LESIEUR, M. 1990. *Turbulence in Fluids*, 2nd edn. Kluwer Academic.

LE RIBAULT, C., SARKAR, S. and STANLEY, S. A. 1999. Large eddy simulation of a plane jet. *Phys. Fluids*. **11**, 3069–3083.

LIEN, F. S. and LESCHZINER, M. A. 1994. A general non-orthogonal finite volume algorithm for turbulent flow at all speeds incorporating second-moment closure, Part I: Numerical implementation. *Comput. Meth. Mech. Eng*. **114**, 123–148.

LIEN, F. S. and LESCHZINER, M. A. 1996a. Second-moment closure for three-dimensional turbulent flow around and within complex geometries. *Comput. Fluids*. **25**, 237–262.

LIEN, F. S. and LESCHZINER, M. A. 1996b. Low-Reynolds-number eddy-viscosity modeling based on non-linear stress–strain/vorticity relations. *Proc. 3rd Symp. on Engineering Turbulence Modeling and Measurements*. 91–101.

LIEN, F. S., DURBIN, P. A. and PARNEIX, S. 1997. Non-linear $v^2\!-\!f$ modeling with application to high-lift. *Proc. 11th Int. Conf. on Turbulent Shear Flows*.

LIGHTHILL, M. J. 1958. *Introduction to Fourier Analysis and Generalised Functions*. Cambridge Unversity Press.

LIGRANI, P. M. and MOFFAT, R. J. 1986. Structure of transitionally rough and fully rough turbulent boundary layers. *J. Fluid Mech*. **162**, 69–98.

LUMLEY, J. L. 1978. Computation modeling of turbulent flows. *Adv. Appl. Mech*. **18**, 126–176.

LUNDGREN, T. S. 2007. Asymptotic analysis of the constant pressure turbulent boundary layer. *Phys. Fluids A*. **19**, 055105.

MANCEAU, R., WANG, M. and LAURENCE, D. 1999. Assessment of inhomogeneity effects on the pressure term using DNS database: implication for RANS models. *Proc. 1st Int. Symp. on Turbulence and Shear Flow Phenomena*. 239–244.

MANSOUR, N. N., KIM, J. and MOIN, P. 1988. Reynolds-stress and dissipation budgets in a turbulent channel flow. *J. Fluid Mech*. **194**, 15–44.

MANTEL, T. and BILGER, R. W. 1994. Conditional statistics in a turbulent premixed flame derived from direct numerical simulation. Center for Turbulence Research, Annual Briefs, 1994, Stanford University 3–28.

MCCOMB W. D. 1990. *The Physics of Fluid Turbulence*. Oxford University Press.

MELLOR, G. L. and HERRING, H. J. 1973. A survey of mean turbulent field closure models. *AIAA J*. **11**, 590–599.

MELLOR, G. L. and YAMADA, T. 1982. Development of a turbulence closure model for geophysical fluid problems. *Rev. Geophys. Space Phys*. **20**, 851–875.

MENTER, F. R. 1994. Two-equation eddy-viscosity turbulence models for engineering applications. *AIAA J*. **32**, 1598–1605.

MENTER, F. R., KUNTZ, M. and BENDER, R. 2003. A scale-adaptive simulation model for turbulent flow predictions. AIAA Paper 2003-0767, American Institute of Aeronautics and Astronautics.

MENTER, F. R., LANGTRY, R. B., LIKKI, S. R., SUZEN, Y. B., HUANG, P. G. and VÖLKER, S. 2004. A correlation based transition model using local variables. *ASME Turbo Expo 2004*, Vienna, Paper GT2004-53452.

MONIN, A. S. and YAGLOM, A. M. 1975. *Statistical Fluid Dynamics*. MIT Press.

MORINISHI, Y., LUND, T. S., VASILYEV, O. V. and MOIN, P. 1998. Fully conservative higher order finite difference schemes for incompressible flow. *J. Comput. Phys*. **143**, 90–124.

MOSER, M. M. and ROGERS, R. D. 1991. Mixing transition in the cascade to small scales in a plane mixing layer. *Phys. Fluids A*. **3**, 1128–1134.

MOSER, R. D., KIM, J. and MANSOUR, N. N. 1999. Direct numerical simulation of turbulent channel flow up to $Re_\tau = 590$. *Phys. Fluids*. **11**, 943–945.

MUMFORD, J. C. 1982. The structure of the large eddies in fully developed turbulent shear flows. Part I. The plane jet. *J. Fluid Mech*. **118**, 241–268.

MYDLARSKI, L. and WARHAFT, Z. 1996. On the onset of high-Reynolds-number grid-generated wind tunnel turbulence. *J. Fluid Mech*. **320**, 331–368.

NICOUD, F. and DUCROS, F. 1999. Subgrid-scale stress modelling based on the square of the velocity gradient tensor. *Flow, Turbul. Combust*. **62**, 183–200.

OBI, S., OHIMUZI, H., AOKI, K. and MASUDA, S. 1993. Experimental and computational study of turbulent separating flow in an asymmetric plane diffuser. *Proc. 9th Symp. on Turbulent Shear Flows*, Kyoto, Japan, P305-1–4.

OFFEN, G. R. and KLINE, S. J. 1975. A proposed model of the bursting process in turbulent boundary layers. *J. Fluid Mech*. **70**, 209–228.

PARNEIX, S., DURBIN, P. A. and BEHNIA, M. 1998. Computation of 3-D turbulent boundary layers using the $v^2$–$f$ model. *Flow, Turbul. Combust*. **10**, 19–46.

PATEL, V. C., RODI, W. and SCHEURER, G. 1984. Turbulence modeling for near-wall and low Reynolds number flows: a review. *AIAA J*. **23**, 1308–1319.

PETTERSSON, B. A. and ANDERSSON, H. I. 1997. Near-wall Reynolds-stress modelling in noninertial frames of reference. *Fluid Dyn. Res*. **19**, 251–276.

PETTERSSON REIF, B. A. and ANDERSSON, H. I. 1999. Second-moment closure predictions of turbulence-induced secondary flow in a straight square duct. *Eng. Turbul. Model. Meas*. **4**, 349–358.

PETTERSSON REIF, B. A., DURBIN, P. A. and OOI, A. 1999. Modeling rotational effects in eddy-viscosity closures. *Int. J. Heat Fluid Flow*. **20**, 563–573.

PIOMELLI, U. 2008. Wall-layer models for large-eddy simulations. *Prog. Aerospace Sci.* **44**, 437–446.

POPE, S. B. 1975. A more general effective-viscosity hypothesis. *J. Fluid Mech.* **72**, 331–340.

POPE, S. B. 1985. PDF methods for turbulent reactive flows. *Prog. Energy Combust. Sci.* **11**, 119–192.

POPE, S. B. 1994. On the relationship between stochastic Lagrangian models of turbulence and second-moment closures. *Phys. Fluids.* **6**, 973–985.

PURTELL, L. P., KLEBANOFF, S. and BUCKLEY, F. T. 1981. Turbulent boundary layer at low Reynolds number. *Phys. Fluids.* **24**, 802–811.

REYNOLDS, O. 1883. On the experimental investigation of the circumstances which determine whether the motion of water shall be direct or sinuous, and the law of resistance in parallel channels. *Phil. Trans. R. Soc. Lond. A.* **174**, 935–982.

RICHARDSON, L. F. 1922. *Weather Prediction by Numerical Process*. Cambridge University Press.

ROACH, P. E. and BRIERLEY, D. H. 1990. The influence of a turbulent freestream on zero pressure gradient transitional boundary layer development, Part I: Test cases T3A and T3b. *ERCOFTAC Workshop: Numerical Simulation of Unsteady Flows and Transition to Turbulence*, Lausanne, Switzerland 319–347. Cambridge University Press.

RODI, W. 1976. A new algebraic relation for calculating the Reynolds stresses. *Z. Angew. Math. Mech.* **56**, T219–221.

RODI, W. 1991. Experience using two-layer models combining the $k-\varepsilon$ model with a one-equation model near the wall. AIAA Paper 91-0609, American Institute of Aeronautics and Astronautics.

ROGERS, M. M. 1991. The structure of a passive scalar field with a uniform mean gradient in rapidly sheared homogeneous turbulent homogeneous. *Phys. Fluids A.* **3**, 144–154.

ROGERS, M. M. and MOSER, R. D. 1994. Direct simulation of a self-similar turbulent mixing layer. *Phys. Fluids.* **6**, 903–923.

SADDOUGHI, S. G. and VEERAVALLI, V. S. 1994. Local isotropy in turbulent boundary layers at high Reynolds number. *J. Fluid Mech.* **268**, 333–272.

SAGAUT, P. 2001. *Large Eddy Simulation for Incompressible Flows*. Springer.

SAMUEL, A. E. and JOUBERT, P. N. 1974. A boundary layer developing in an increasingly adverse pressure gradient. *J. Fluid Mech.* **66**, 481–505.

SCHLICHTING, H. 1968. *Boundary Layer Theory*. McGraw-Hill.

SCHUMANN, U. 1977. Realizability of Reynolds stress turbulence models. *Phys. Fluids.* **20**, 721–725.

SHABANY, Y. and DURBIN, P. A. 1997. Explicit algebraic scalar flux approximation. *AIAA J.* **35**, 985–989.

SHIH, T. H., ZHU, J. and LUMLEY, J. L. 1995. A new Reynolds stress algebraic equation model. *Comput. Meth. Appl. Mech. Eng.* **125**, 287–302.

SPALART, P. R. 1988. Direct simulation of a turbulent boundary layer up to $R_\theta = 1400$. *J. Fluid Mech.* **187**, 61–98.

SPALART, P. R. 2000. Strategies for turbulence modelling and simulations. *Int. J. Heat Fluid Flow.* **21**, 252–263.

SPALART, P. R. 2009. Detached eddy simulation. *Annu. Rev. Fluid Mech.* **41**, 181–202.

SPALART, P. R. and ALLMARAS, S. R. 1992. A one-equation turbulence model for aerodynamic flows. AIAA Paper 92-0439, American Institute of Aeronautics and Astronautics.

SPALART, P. R., DECK, S., SHUR, M. L., SQUIRES, K. D., STRELETS, M. KH. and TRAVIN, A. 2006. A new version of detached-eddy simulation, resistant to ambiguous grid densities. *Theor. Comput. Fluid Dyn.* **20**, 181–195.

SPENCER, A. J. M. and RIVLIN, R. S. 1959. The theory of matrix polynomials and its application to the mechanics of isotropic continua. *J. Ration. Mech. Anal.* **2**, 309–336.

SPEZIALE, C. G. and MAC GIOLLA MHUIRIS, N. 1989. On the prediction of equilibrium states in homogeneous turbulence. *J. Fluid Mech.* **209**, 591–615.

SPEZIALE, C. G., SARKAR, S. and GATSKI, T. B. 1991. Modeling the pressure–strain correlation of turbulence: an invariant dynamical systems approach. *J. Fluid Mech.* **227**, 245–272.

STEINER, H. and BUSCHE, W. K. 1998. LES of non-premixed turbulent reacting flows with conditional source term estimation. Center for Turbulence Research, Annual Briefs, 1998, Stanford University.

SU, L. K. and MUNGAL, M. G. 1999. Simultaneous measurements of velocity and scalar fields: application in crossflowing jets and lifted jet diffusion flames. Center for Turbulence Research, Annual Briefs, 1999, Stanford University.

SULUKSNA, K., DECHAUMPHAI, P. and JUNTASARO, E. 2009. Correlations for modeling transitional boundary layers under influences of free stream turbulence and pressure gradient. *Int. J. Heat Fluid Flow.* **30**, 66–72.

TANNEHILL, J. C., ANDERSON, D. A. and PLETCHER, R. H. 1997. *Computational Fluid Mechanics and Heat Transfer*. Taylor & Francis.

TAVOULARIS, S. and KARNIK, U. 1989. Further experiments on the evolution of turbulent stresses and scales in uniformly sheared turbulence. *J. Fluid Mech.* **204**, 457–478.

TAYLOR, G. I. 1921. Diffusion by continuous movements. *Proc. Lond. Math. Soc.* **20**, 196–212.

TAYLOR, G. I. 1935. Turbulence in a contracting stream. *Z. Angew. Math. Mech.* **15**, 91–96.

TAYLOR, G. I. 1954. The dispersion of matter in turbulent flow through a pipe. *Proc. R. Soc. Lond. A* **223**, 446–468.

TAYLOR, G. I. and GREEN, A. E. 1937. Mechanism of the production of small eddies from large ones. *Proc. R. Soc. Lond. A.* **158**, 499–521.

TOWNSEND, A. A. 1976. *The Structure of Turbulent Shear Flow*. Cambridge University Press.

TURNER, J. S. 1980. *Buoyancy Effects in Fluids*. Cambridge University Press.

TURNER, M. G. and JENIONS, I. K. 1993. An investigation of turbulence modeling in transsonic fans including a novel implementation of an implicit $k$–$\varepsilon$ turbulence model. *ASME J. Turbomach.* **115**, 248–260.

WALTERS, D. K. and COKALJAT, D. 2008. A three-equation eddy-viscosity model for Reynolds-averaged Navier–Stokes simulations of transitional flow. *J. Fluids Eng.* **130**, 121401.

WANG, L. and STOCK, D. E. 1994. Dispersion of heavy particles by turbulent motion. *J. Atmos. Sci.* **50**, 1897–1913.

WARHAFT, Z. 2000. Passive scalars in turbulent flows. *Annu. Rev. Fluid Mech.* **32**, 203–240.

WEYGANDT, J. H. and MEHTA, R. D. 1995. Three-dimensional structure of straight and curved plane wakes. *J. Fluid Mech.* **282**, 279–311.

WHITE, F. M. 1991. *Viscous Fluid Flow*, 2nd edn. McGraw-Hill.

WILCOX, D. C. 1993. *Turbulence Modeling for CFD*. DCW Industries.

WIZMAN, V., LAURENCE, D., DURBIN, P. A., DEMUREN, A. and KANNICHE, M. 1996. Modeling near wall effects in second moment closures by elliptic relaxation. *J. Heat Fluid Flow.* **17**, 255–266.

WYGNANSKI, I., CHAMPAGNE, F. and MARASLI, B. 1986. On the large scale structures in two-dimensional, small-deficit, turbulent wakes. *J. Fluid Mech.* **168**, 31–71.

ZAKI, T. A., WISSINK, J. G., DURBIN, P. A. and RODI, W. 2009. Direct computations of boundary layers distorted by migrating wakes in a linear compressor cascade. *Flow, Turbul. Combust.* **83**, 307–332.

ZHOU, J., ADRIAN, R. J., BALACHANDAR, S. and KENDALL, T. 1999. Hairpin vortices in near-wall turbulence and their regeneration mechanisms. *J. Fluid Mech.* **387**, 353–396.

# Index

Printed and bound by CPI Group (UK) Ltd, Croydon, CR0 4YY

12/01/2025

14624501-0004